# American Society of Civil Engineers

# Seismic Rehabilitation of Existing Buildings

This document uses both the International System of Units (SI) and customary units.

Published by the American Society of Civil Engineers

**Library of Congress Cataloging-in-Publication Data**

Seismic rehabilitation of existing buildings.
    p. cm.
  "ASCE standard ASCE/SEI 41-06."
  Includes bibliographical references and index.
  ISBN-13: 978-0-7844-0884-1
  ISBN-10: 0-7844-0884-X
  1. Buildings—Repair and reconstruction—Standards. 2. Earthquake resistant design—Standards. 3. Buildings—Earthquake effects. I. American Society of Civil Engineers.

TH420.S43 2007
693.8'52—dc22

2007009080

Published by American Society of Civil Engineers
1801 Alexander Bell Drive
Reston, Virginia 20191
www.pubs.asce.org

# STANDARDS

In 2003, the Board of Direction approved the revision to the ASCE Rules for Standards Committees to govern the writing and maintenance of standards developed by the Society. All such standards are developed by a consensus standards process managed by the Society's Codes and Standards Committee (CSC). The consensus process includes balloting by a balanced standards committee made up of Society members and nonmembers, balloting by the membership of the Society as a whole, and balloting by the public. All standards are updated or reaffirmed by the same process at intervals not exceeding five years.

The following standards have been issued:

ANSI/ASCE 1-82 N-725 Guideline for Design and Analysis of Nuclear Safety Related Earth Structures

ASCE/EWRI 2-06 Measurement of Oxygen Transfer in Clean Water

ANSI/ASCE 3-91 Standard for the Structural Design of Composite Slabs and ANSI/ASCE 9-91 Standard Practice for the Construction and Inspection of Composite Slabs

ASCE 4-98 Seismic Analysis of Safety-Related Nuclear Structures

Building Code Requirements for Masonry Structures (ACI 530-02/ASCE 5-02/TMS 402-02) and Specifications for Masonry Structures (ACI 530.1-02/ASCE 6-02/TMS 602-02)

ASCE/SEI 7-05 Minimum Design Loads for Buildings and Other Structures

SEI/ASCE 8-02 Standard Specification for the Design of Cold-Formed Stainless Steel Structural Members

ANSI/ASCE 9-91 listed with ASCE 3-91

ASCE 10-97 Design of Latticed Steel Transmission Structures

SEI/ASCE 11-99 Guideline for Structural Condition Assessment of Existing Buildings

ASCE/EWRI 12-05 Guideline for the Design of Urban Subsurface Drainage

ASCE/EWRI 13-05 Standard Guidelines for Installation of Urban Subsurface Drainage

ASCE/EWRI 14-05 Standard Guidelines for Operation and Maintenance of Urban Subsurface Drainage

ASCE 15-98 Standard Practice for Direct Design of Buried Precast Concrete Pipe Using Standard Installations (SIDD)

ASCE 16-95 Standard for Load Resistance Factor Design (LRFD) of Engineered Wood Construction

ASCE 17-96 Air-Supported Structures

ASCE 18-96 Standard Guidelines for In-Process Oxygen Transfer Testing

ASCE 19-96 Structural Applications of Steel Cables for Buildings

ASCE 20-96 Standard Guidelines for the Design and Installation of Pile Foundations

ANSI/ASCE/T&DI 21-05 Automated People Mover Standards—Part 1

ASCE 21-98 Automated People Mover Standards—Part 2

ASCE 21-00 Automated People Mover Standards—Part 3

SEI/ASCE 23-97 Specification for Structural Steel Beams with Web Openings

ASCE/SEI 24-05 Flood Resistant Design and Construction

ASCE/SEI 25-06 Earthquake-Actuated Automatic Gas Shutoff Devices

ASCE 26-97 Standard Practice for Design of Buried Precast Concrete Box Sections

ASCE 27-00 Standard Practice for Direct Design of Precast Concrete Pipe for Jacking in Trenchless Construction

ASCE 28-00 Standard Practice for Direct Design of Precast Concrete Box Sections for Jacking in Trenchless Construction

ASCE/SEI/SFPE 29-05 Standard Calculation Methods for Structural Fire Protection

SEI/ASCE 30-00 Guideline for Condition Assessment of the Building Envelope

SEI/ASCE 31-03 Seismic Evaluation of Existing Buildings

SEI/ASCE 32-01 Design and Construction of Frost-Protected Shallow Foundations

EWRI/ASCE 33-01 Comprehensive Transboundary International Water Quality Management Agreement

EWRI/ASCE 34-01 Standard Guidelines for Artificial Recharge of Ground Water

EWRI/ASCE 35-01 Guidelines for Quality Assurance of Installed Fine-Pore Aeration Equipment

CI/ASCE 36-01 Standard Construction Guidelines for Microtunneling

SEI/ASCE 37-02 Design Loads on Structures During Construction

CI/ASCE 38-02 Standard Guideline for the Collection and Depiction of Existing Subsurface Utility Data

EWRI/ASCE 39-03 Standard Practice for the Design and Operation of Hail Suppression Projects

ASCE/EWRI 40-03 Regulated Riparian Model Water Code

ASCE/SEI 41-06 Seismic Rehabilitation of Existing Buildings

ASCE/EWRI 42-04 Standard Practice for the Design and Operation of Precipitation Enhancement Projects

ASCE/SEI 43-05 Seismic Design Criteria for Structures, Systems, and Components in Nuclear Facilities

ASCE/EWRI 44-05 Standard Practice for the Design and Operation of Supercooled Fog Dispersal Projects

ASCE/EWRI 45-05 Standard Guidelines for the Design of Urban Stormwater Systems

ASCE/EWRI 46-05 Standard Guidelines for the Installation of Urban Stormwater Systems

ASCE/EWRI 47-05 Standard Guidelines for the Operation and Maintenance of Urban Stormwater Systems

ASCE/SEI 48-05 Design of Steel Transmission Pole Structures

# FOREWORD

In 2003, the Board of Direction approved the revision to the ASCE Rules for Standards Committees to govern the writing and maintenance of standards developed by the Society. All such standards are developed by a consensus standards process managed by the Society's Codes and Standards Committee (CSC). The consensus process includes balloting by a balanced standards committee made up of Society members and nonmembers, balloting by the membership of the Society as a whole, and balloting by the public. All standards are updated or reaffirmed by the same process at intervals not exceeding five years.

The material presented in this Standard has been prepared in accordance with recognized engineering principles. This Standard should not be used without first securing competent advice with respect to its suitability for any given application. The publication of the material contained herein is not intended as a representation or warranty on the part of the American Society of Civil Engineers, or of any other person named herein, that this information is suitable for any general or particular use or promises freedom from infringement of any patent or patents. Anyone making use of this information assumes all liability from such use.

# ACKNOWLEDGMENTS

The Structural Engineering Institute (SEI) of the American Society of Civil Engineers (ASCE) acknowledges the devoted efforts of the membership of the Seismic Rehabilitation of Existing Buildings Standards Committee of the Codes and Activities Division of SEI. This group comprises individuals from many backgrounds, including consulting engineering, research, construction industry, education, government, design, and private practice.

Balloting for this standard began with FEMA 356, *Prestandard and Commentary for the Seismic Rehabilitation of Buildings*, prepared by ASCE for the Federal Emergency Management Agency (FEMA). FEMA 356 was developed from FEMA 273, *NEHRP Guidelines for the Seismic Rehabilitation of Buildings*, developed for FEMA by the Applied Technology Council (ATC). ASCE acknowledges and is grateful for the over ten years of support provided by FEMA to the development of a new generation rehabilitation standard, and particularly for their support during this final step, the development of this consensus standard.

This standard was prepared through the consensus standards process in compliance with the procedures established by the ASCE Codes and Standards Committee and accredited by the American National Standards Institute (ANSI). Those individuals who served on the standards committee are:

Bechara Elias Abboud, Ph.D., P.E., M.ASCE
Michael D. Blakely, P.E., M.ASCE
David Clare Breiholz, P.E., F.ASCE
James Brown, P.E., L.S., F.ASCE
Thomas Marvin Bykonen, P.E., M.ASCE
Hashu H. Chandwaney, P.E., F.ASCE
Chang Chen, Ph.D., P.E., M.ASCE
Kevin C. K. Cheung V, Ph.D., P.E., M.ASCE
James Hamilton Collins, M.ASCE
W. Gene Corley, Ph.D., P.E., Hon.M.ASCE
Mark W. Fantozzi, P.E., M.ASCE
Hans Gesund, Ph.D., P.E., F.ASCE
Satyendra K. Ghosh, M.ASCE
Sergio Gonzalez-Karg, P.E., F.ASCE
Phillip Gould, P.E., F.ASCE
Melvyn Green, P.E., F.ASCE
D. Kirk Harman, P.E., S.E., M.ASCE
John R. Hayes, Jr., PhD, PE, M.ASCE
Jon A. Heintz, P.E., M.ASCE
Richard L. Hess, P.E., F.ASCE
Darrick Bryan Hom, P.E., M.ASCE
Jen-Kan Hsiao, Ph.D., P.E., S.E., M.ASCE
Tom Chi-Tong Hui, P.E., M.ASCE
Roy J. Hunt, P.E., M.ASCE
Mohammad Iqbal, Ph.D., P.E., F.ASCE
Robert C. Jackson, P.E., M.ASCE
Wen-Chen Jau, Ph.D., A.M.ASCE
Martin W. Johnson, P.E., M.ASCE
John C. Kariotis, P.E., M.ASCE
Brian Edward Kehoe, S.E., F.ASCE
Patrick J. Lama, P.E., M.ASCE
Jim Eugene Lapping, P.E., M.ASCE
Feng-Bao Lin, M.ASCE
Philip Line, M.ASCE
Roy F. Lobo, P.E., M.ASCE

Charles R. Magadini, P.E., L.S., F.ASCE
Ayaz H. Malik, P.E., M.ASCE
Rusk Masih, Ph.D., P.E., Aff.M.ASCE
Vicki Vance May, P.E., A.M.ASCE
Bruce Herman McCracken, P.E., M.ASCE
Richard McConnell, Ph.D., P.E., M.ASCE
Thomas Harold Miller, P.E., M.ASCE
Andy Hess Milligan, P.E., M.ASCE
Andrew Douglass Mitchell, P.E., M.ASCE
Myles A. Murray, P.E., M.ASCE
Joseph F. Neussendorfer, Aff.M.ASCE
Glen John Pappas, Ph.D., M.ASCE
James C. Parker, P.E., M.ASCE
Celina Ugarte Penalba, M.ASCE
Mark Allan Pickett, Ph.D., P.E., M.ASCE
Chris Donald Poland, M.ASCE
Daniel E. Pradel, Ph.D., P.E., F.ASCE
Timothy Edward Roecker, M.ASCE
Charles W. Roeder, Ph.D., P.E., M.ASCE
Abdulreza A. Sadjadi, P.E., M.ASCE
Ashvin A. Shah, P.E., F.ASCE
Richard Lee Silva, P.E., M.ASCE
Thomas David Skaggs, P.E., M.ASCE
Glenn R. Smith, Jr., Ph.D., P.E., M.ASCE
Eric Christian Stovner, P.E., M.ASCE
Donald R. Strand, P.E., F.ASCE
Peter Tian, P.E., A.M.ASCE
Frederick Michael Turner, S.E., M.ASCE
Michael T. Valley, P.E., M.ASCE
Thomas George Williamson, P.E., F.ASCE
Lyle L. Wilson, F.ASCE
Lisa A. Wipplinger, P.E., M.ASCE
Tom Chuan Xia, P.E., M.ASCE
Wade Wesley Younie, P.E., M.ASCE

# CONTENTS

The Structural Engineering Institute (SEI) of the American Society of Civil Engineers is committed to providing accurate, up-to-date information to its readers. To that end, SEI maintains a listing of errata at http://www.seinstitute.org/publications/errata.cfm.

# Seismic Rehabilitation of Existing Buildings

## 1.0 REHABILITATION REQUIREMENTS

### 1.1 SCOPE

This standard for the *Seismic Rehabilitation of Existing Buildings*, referred to herein as "this standard," specifies nationally applicable provisions for the seismic rehabilitation of buildings. Seismic rehabilitation is defined as improving the seismic performance of structural and/or nonstructural components of a building by correcting deficiencies identified in a seismic evaluation. Seismic evaluation is defined as an approved process or methodology of evaluating deficiencies in a building, which prevent the building from achieving a selected Rehabilitation Objective. Seismic evaluation using ASCE 31 (ASCE 2002), the procedures and criteria of this standard, or other procedures and criteria approved by the authority having jurisdiction is permitted.

Seismic rehabilitation of existing buildings shall comply with requirements of this standard for selecting a Rehabilitation Objective and conducting the seismic rehabilitation process to achieve the selected Rehabilitation Objective. This standard does not preclude a building from being rehabilitated by other procedures approved by the authority having jurisdiction.

Symbols, acronyms, definitions, and references used throughout this standard are cited separately in sections located at the end of this standard.

### C1.1 SCOPE

This standard is intended to serve as a nationally applicable tool for design professionals, code officials, and building owners undertaking the seismic rehabilitation of existing buildings. In jurisdictionally mandated seismic rehabilitation programs, the code official serves as the authority having jurisdiction. In voluntary seismic rehabilitation programs, the building owner, or the owner's designated agent, serves as the authority having jurisdiction.

This standard consists of two parts: Provisions, which contain the technical requirements, and Commentary, intended to explain the provisions. Commentary for a given section is located immediately following the section and is identified by the same section number preceded by the letter C.

It is expected that most buildings rehabilitated in accordance with this standard would perform within the desired levels when subjected to the design earthquakes. However, compliance with this standard does not guarantee such performance; rather, it represents the current standard of practice in designing to attain this performance. The practice of earthquake engineering is rapidly evolving, and both our understanding of the behavior of buildings subjected to strong earthquakes and our ability to predict this behavior are advancing. In the future, new knowledge and technology will improve the reliability of accomplishing these goals.

The procedures contained in this standard are specifically applicable to the rehabilitation of existing buildings and, in general, are more appropriate for that purpose than are new building codes. New building codes are primarily intended to regulate the design and construction of new buildings; as such, they include many provisions that encourage or require the development of designs with features important for good seismic performance, including regular configuration, structural continuity, ductile detailing, and materials of appropriate quality. Many existing buildings were designed and constructed without these features and contain characteristics such as unfavorable configuration and poor detailing that preclude application of building code provisions for their seismic rehabilitation.

Although it is intended to be used as a follow-up to a previous seismic evaluation, this standard can also be used as an evaluation tool to ascertain compliance with a selected rehabilitation objective. An ASCE 31, Tier 3 evaluation is an example of this use. It should be noted, however, that an evaluation using this standard may be more stringent than other evaluation methodologies because the provisions have been calibrated for use in design. Historically, criteria for evaluation have been set lower than those for design to minimize the need to strengthen buildings that would otherwise have only modest deficiencies.

The expertise of the design professional in earthquake engineering is an important prerequisite for the appropriate use of this standard in assisting a building owner to select voluntary seismic criteria or to design and analyze seismic rehabilitation projects, whether voluntary or required. The analytical work required by this standard must be performed under the responsible charge of a licensed professional engineer; however, that does not preclude a design professional without a professional engineering license, but with responsible charge, from leading a seismic rehabilitation project. For example, an architect with responsible charge can lead a seismic rehabilitation project conducted in accordance with the simplified rehabilitation described in Chapter 10.

This standard is intended to be generally applicable to seismic rehabilitation of all buildings—regardless of importance, occupancy, historic status, or other classifications of use. However, application of these provisions should be coordinated with other requirements that may be in effect, such as ordinances governing historic structures or hospital construction. In addition to the direct effects of ground shaking, this standard also addresses the effects of local geologic site hazards such as liquefaction.

This standard is arranged such that there are four analysis procedures that can be used, including the Linear Static Procedure, Linear Dynamic Procedure, Nonlinear Static Procedure, and Nonlinear Dynamic Procedure. The linear analysis procedures are intended to provide a conservative estimate of building response and performance in an earthquake, though they are not always accurate. Since the actual response of buildings to earthquakes is not typically linear, the nonlinear analysis procedures should provide a more accurate representation of building response and performance. In recognition of the improved representation of building behavior when nonlinear analysis is conducted, the nonlinear procedures have less-conservative limits on permissible building response than do linear procedures. Buildings that are found to be seismically deficient based on linear analysis may comply with this standard if a nonlinear analysis is performed. Therefore, performing a nonlinear analysis can minimize or eliminate unnecessary seismic rehabilitation and potentially lower construction costs.

This standard applies to the seismic rehabilitation of both the overall structural system of a building and its nonstructural components, including ceilings, partitions, mechanical, electrical, and plumbing systems.

With careful extrapolation, the procedures of this standard may also be applied to many nonbuilding structures such as pipe racks, steel storage racks, structural towers for tanks and vessels, piers, wharves, and electrical power generating facilities. However, the applicability of these procedures has not been fully examined for every type of structure—particularly those that have generally been covered by specialized codes or standards, such as bridges and nuclear power plants.

Jurisdictions will adopt this standard as an ordinance that only applies to the seismic rehabilitation of existing buildings or adopt this standard by reference as part of a comprehensive code addressing all aspects of rehabilitating existing buildings. In adopting this standard, the jurisdiction will select one or more rehabilitation objectives which must be met by buildings that have either been targeted by the jurisdiction for mandated seismic rehabilitation or—by reason of owner-initiated activities, such as major structural modifications—have come under the jurisdiction's rehabilitation ordinance. Since codes for new buildings have chapters that briefly address existing buildings, care must be taken in coordinating and referencing the adoption of this standard to avoid ambiguity and confusion with other ordinances and codes.

Since almost all structural seismic rehabilitation work requires a building permit, the code official will become an important part of the process. For voluntary rehabilitation efforts, the building owner and the code official need to come to agreement about the intended rehabilitation objective. The code official will verify that the owner's stated objective is met in the design and construction phases of the work. For jurisdictionally required rehabilitation efforts, whether caused by passive or active programs (see Appendix A), the code official will verify that the required objective is met. Because the approaches and technology of this standard are not yet in the mainstream of design and construction practices of the United States, it is imperative that the code official either develop the expertise in this methodology or utilize a peer review type of process to verify the appropriate application of this standard. A jurisdiction must also remain flexible and open to other analyses and evaluations, which provide a reasonable assurance of meeting the appropriate rehabilitation objective.

In addition to techniques for increasing the strength and ductility of systems, this standard provides techniques for reducing seismic demand, such as the introduction of isolation or damping devices. Design of new buildings and evaluation of components for gravity and wind forces in the absence of earthquake demands are beyond the scope of this standard.

This standard does not explicitly address the determination of whether or not a rehabilitation project should be undertaken for a particular building. Guidance on the use of this standard in voluntary or directed risk-mitigation programs is provided in Appendix A. Determining where these provisions should be required is beyond the scope of this standard. Once the decision to rehabilitate a building has been made, this standard can be referenced for detailed engineering guidance on how to conduct a seismic rehabilitation analysis and design.

Featured in this standard are descriptions of damage states in relation to specific performance levels. These descriptions are intended to aid the authority having jurisdiction, design professionals, and owners in selecting appropriate performance levels for rehabilitation design. They are not intended to be used for condition assessment of earthquake-damaged buildings. Although there may be similarities between these damage descriptions and those used for postearthquake damage assessment, many factors enter into the processes of assessing seismic performance. No single parameter in this standard should be cited as defining either a performance level or the safety or usefulness of an earthquake-damaged building.

Techniques for repair of earthquake-damaged buildings are not included in this standard, but are referenced in the commentary pertaining to Chapters 5 through 8 where such guidelines exist. Any combination of repaired components, undamaged existing components, and new components can be modeled using this standard, and each checked against performance level acceptance criteria. If the mechanical properties of repaired components are known, acceptance criteria for use with this standard can be either deduced by comparison with other similar components or derived.

## 1.2 DESIGN BASIS

The selection of a seismic Rehabilitation Objective and the performance-based design of rehabilitation measures to achieve the selected Rehabilitation Objective shall be in accordance with the rehabilitation process specified in Section 1.3. The use of alternative performance-based criteria and procedures approved by the authority having jurisdiction shall be permitted.

## C1.2 DESIGN BASIS

Provisions of this standard for seismic rehabilitation are based on a performance-based design methodology that differs from seismic design procedures for the design of new buildings currently specified in national building codes and standards.

The framework in which these requirements are specified is purposefully broad so that Rehabilitation Objectives can accommodate buildings of different types, address a variety of performance levels, and reflect the variation of seismic hazards across the United States and U.S. territories.

The provisions and commentary of this standard are based primarily on the FEMA 356 *Prestandard* (FEMA 2000) with limited material taken from the FEMA 274 (FEMA 1997) Commentary. This standard is intended to supersede FEMA 356, but FEMA 274 remains a valid explanation for the provisions in this standard unless indicated otherwise in the relevant

commentary of this standard. For this reason, section numbers in this standard remain essentially the same as in FEMA 356.

FEMA 356 was based on FEMA 273 (FEMA 1997), which was developed by a large team of specialists in earthquake engineering and seismic rehabilitation. The most advanced analytical techniques considered practical for production use have been incorporated. The acceptance criteria have been specified using actual laboratory test results, where available, supplemented by the engineering judgment of various development teams. Certain buildings damaged in the 1994 Northridge earthquake and a limited number of designs using codes for new buildings have been checked using the procedures of FEMA 273. A comprehensive program of case studies was undertaken by FEMA in 1998 to test more thoroughly the various analysis techniques and acceptability criteria. The results of this study are reported in FEMA 343, *Case Studies: An Assessment of the NEHRP Guidelines for the Seismic Rehabilitation of Buildings*. The results of the FEMA 343 case studies have been incorporated in the provisions of this standard, where possible. Similarly, information from FEMA 350 (FEMA 2000), FEMA 351 (FEMA 2000), and other reports published by the SAC Joint Venture project, formed as a result of the Northridge steel moment frame damage, has been incorporated where applicable. Engineering judgment should be exercised in determining the applicability of various analysis techniques and material acceptance criteria in each situation.

The commentary to this standard contains specific references to many other documents. In addition, this standard is related generically to the following publications.

1. FEMA 450, *2003 NEHRP Recommended Provisions for Seismic Regulations for New Buildings and Other Structures,* also referred to herein as the *2003 NEHRP Recommended Provisions* (FEMA 2004).
2. FEMA 237, *Development of Guidelines for Seismic Rehabilitation of Buildings, Phase I: Issues Identification and Resolution* (FEMA 1992), which underwent an American Society of Civil Engineers (ASCE) consensus approval process and provided policy direction for this standard.
3. Applied Technology Council (ATC), ATC-28-2, *Proceedings of the Workshop to Resolve Seismic Rehabilitation Sub-Issues* (ATC 1993) provided recommendations to the writers of this standard on more detailed sub-issues.
4. FEMA 172, *NEHRP Handbook of Techniques for the Seismic Rehabilitation of Existing Buildings*

(FEMA 1992), originally produced by URS/Blume and Associates and reviewed by the Building Seismic Safety Council (BSSC), contains construction techniques for implementing engineering solutions to the seismic deficiencies of existing buildings.

5. FEMA 178, *NEHRP Handbook for the Seismic Evaluation of Existing Buildings* (FEMA 1992), which was originally developed by ATC and underwent the consensus approval process of the BSSC, covered the subject of evaluating existing buildings to determine if they are seismically deficient in terms of life safety. This document has been updated by FEMA and ASCE, and is now ASCE 31, *Seismic Evaluation of Existing Buildings* (ASCE 2002), which underwent an ASCE consensus approval process. (The model building types and other information from ASCE 31 are used or referred to extensively in this standard in Chapter 10.)

6. FEMA 156 and 157, *Typical Costs for Seismic Rehabilitation of Existing Buildings*, Second Edition (FEMA 1995), reports statistical analysis of the costs of rehabilitation of more than 2,000 buildings based on construction costs or detailed studies. Several different seismic zones and performance levels are included in the data. Since the data were developed in 1994, none of the data is based on buildings rehabilitated specifically in accordance with the FEMA 273 *Guidelines* document. Performance levels defined in this standard are not intended to be significantly different from parallel levels used previously, and costs still should be reasonably representative.

7. FEMA 275, *Planning for Seismic Rehabilitation: Societal Issues* (FEMA 1998), discusses societal and implementation issues associated with rehabilitation and describes several case histories.

8. FEMA 276, *Guidelines for the Seismic Rehabilitation of Buildings: Example Applications* (FEMA 1999), intended as a companion document to FEMA 273 and FEMA 274, describes examples of buildings that have been seismically rehabilitated in various seismic regions and for different Rehabilitation Objectives. Costs of the work are given and references made to FEMA 156 and 157. Because this document is based on previous case histories, none of the examples was rehabilitated specifically in accordance with this standard. However, performance levels defined in this standard are not intended to be significantly different from parallel levels used previously, and the case studies are therefore considered representative.

9. ATC 40, *Seismic Evaluation and Retrofit of Concrete Buildings* (ATC 1996), incorporates performance levels almost identical to those shown in Table C1-8 and employs "pushover" nonlinear analysis techniques. The capacity spectrum method for determining the displacement demand is treated in detail. This document covers only concrete buildings.

## 1.3 SEISMIC REHABILITATION PROCESS

Seismic rehabilitation of an existing building shall be conducted in accordance with the process outlined in Sections 1.3.1 through 1.3.6.

## C1.3 SEISMIC REHABILITATION PROCESS

The steps are presented in this section in the order in which they would typically be followed in the rehabilitation process. However, the criteria for performing these steps are presented in a somewhat different order to facilitate presentation of the concepts.

Figure C1-1 depicts the rehabilitation process specified in this standard and shows specific chapter references in parentheses at points where input from this standard is to be obtained. Although Fig. C1-1 is written for voluntary rehabilitations, it can also be used as a guide for mandatory rehabilitations.

This standard requires the selection of a Rehabilitation Objective for a building that has been previously identified as needing seismic rehabilitation.

Prior to embarking on a rehabilitation program, an evaluation should be performed to determine whether the building, in its existing condition, has the desired seismic performance capability. ASCE 31 contains an evaluation methodology that may be used for this purpose. It should be noted, however, that a building may meet certain performance objectives using the methodology of ASCE 31, but may not meet those same performance objectives when an evaluation is performed using the procedures of this standard. This is largely because ASCE 31 is specifically intended to accept somewhat greater levels of damage within each performance level than permitted by this standard, which is consistent with the historic practice of evaluating existing buildings for slightly lower criteria than those used for design of new buildings. ASCE 31 quantifies this difference with the use of a 0.75 factor on demands when using this standard in a Tier 3 evaluation. This essentially lowers the reliability of achieving the selected performance level from about 90% to about 60%. This practice minimizes the need to rehabilitate structures with relatively modest deficiencies relative to the desired performance level.

### 1.3.1 Initial Considerations

The design professional shall review initial considerations with the authority having jurisdiction to determine any restrictions that exist on the design of

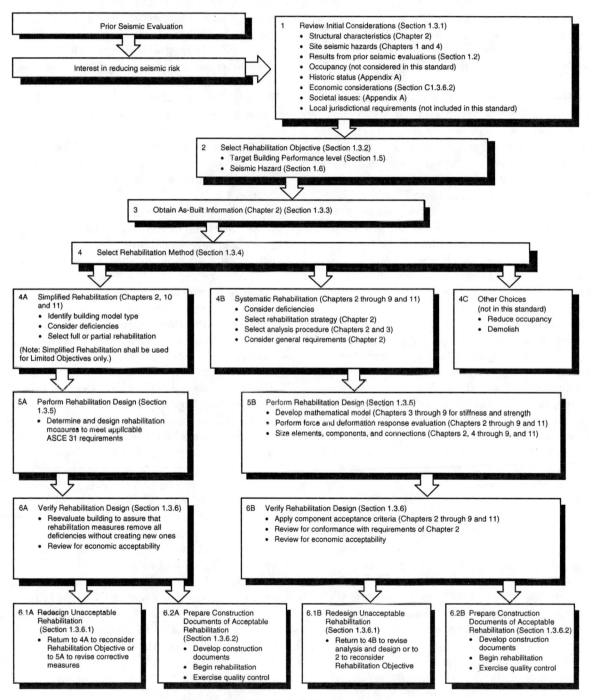

**FIGURE C1-1. Rehabilitation Process.**

rehabilitation measures. Initial considerations shall include structural characteristics of the building, seismic hazards including geologic site hazards known to be present at the site, results of prior seismic evaluations, building use and occupancy requirements, historic status, economic considerations, societal issues, and local jurisdictional requirements.

**C1.3.1 Initial Considerations**

The process of building rehabilitation will be simplified and made more efficient if information that significantly affects the rehabilitation design is obtained and considered prior to beginning the process. Rehabilitation requirements mandated by local jurisdictions would be particularly important to determine in the initial stages of a project.

The building owner should be aware of the range of costs and impacts of rehabilitation, including both the variation associated with different Rehabilitation Objectives and the potential additional costs often associated with seismic rehabilitation, such as other life safety upgrades, hazardous material removal, work associated with the Americans with Disabilities Act, and nonseismic building remodeling. Also to be considered are potential federal tax incentives for the rehabilitation of historic buildings and for some other older nonresidential buildings.

Seismic hazards other than ground shaking may exist at the building site. The risk and possible extent of damage from geologic site hazards identified in Section 4.2.2 should be considered before undertaking a rehabilitation aimed solely at reducing damage due to shaking. In some cases it may be feasible to mitigate the site hazard or rehabilitate the building and still meet the selected performance level. In other cases, the risk due to site hazards may be so extreme and difficult to control that rehabilitation is neither cost-effective nor feasible.

The use of the building must be considered in weighing the significance of potential temporary or permanent disruptions associated with various risk-mitigation schemes. Other limitations on modifications to the building due to historic or aesthetic features must also be understood. The historic status of every building at least 50 years old should be determined (see Appendix A, Section A.6, Considerations for Historic Buildings). This determination should be made early as it could influence the choices of rehabilitation approaches and techniques.

There are many ways to reduce seismic risk, whether the risk is to property, life safety, or post-earthquake use of the building. The occupancy of vulnerable buildings can be reduced, redundant facilities can be provided, and nonhistoric buildings can be demolished and replaced. The risks posed by nonstructural components and contents can be reduced. Seismic site hazards other than shaking can be mitigated.

Most often, however, when all alternatives are considered, the options of modifying the building to reduce the risk of damage should be studied. Such corrective measures include stiffening or strengthening the structure, adding local components to eliminate irregularities or tie the structure together, reducing the demand on the structure through the use of seismic isolation or energy dissipation devices, and reducing the height or mass of the structure. Rehabilitation strategies are discussed in Chapter 2.

**1.3.2 Selection of Rehabilitation Objective**

A seismic Rehabilitation Objective shall be selected for the building in accordance with Section 1.4.

**C1.3.2 Selection of Rehabilitation Objective**

The concepts and terminology of performance-based design are new and should be carefully studied and discussed with building owners before use. The terminology used for target Building Performance Levels is intended to represent goals of design. The actual ground motion will seldom be comparable to that specified in the Rehabilitation Objective, so in most events, designs targeted at various damage states may only determine relative performance. Even given a ground motion similar to that specified in the Rehabilitation Objective and used in design, variations from stated performance objectives should be expected and compliance with this standard should not be considered a guarantee of performance. Variations in actual performance could be associated with unknown geometry and member sizes in existing buildings, deterioration of materials, incomplete site data, variation of ground motion that can occur within a small area, and incomplete knowledge and simplifications related to modeling and analysis. Information on the expected reliability of achieving various target Building Performance Levels when the requirements are followed can be found in Chapter 2 of FEMA 274 (FEMA 1997).

The determination of the Rehabilitation Objective differs depending on whether the rehabilitation is mandated or voluntary. For a voluntary building rehabilitation, the building owner shall select a seismic rehabilitation for the building as specified in Section 1.4. In a mandated rehabilitation project, the rehabilitation objective is either stipulated directly by local code or ordinance or the code official is provided with guidelines for negotiating the rehabilitation objective.

### 1.3.3 As-Built Information

Available as-built information for the building shall be obtained and a site visit shall be conducted as specified in Section 2.2.

### 1.3.4 Rehabilitation Method

An applicable rehabilitation method shall be determined in accordance with Section 2.3.

### C1.3.4 Rehabilitation Method

Rehabilitation can consist of the Simplified Rehabilitation Method or the Systematic Rehabilitation Method. These methods are defined in Section 2.3 and further explained in the associated commentary of that section.

### 1.3.5 Rehabilitation Measures

Rehabilitation measures shall be designed using the applicable rehabilitation method.

### 1.3.6 Verification of Rehabilitation Design

The design of rehabilitation measures shall be verified to meet the requirements of this standard through an analysis of the building, including the rehabilitation measures. The analysis shall be consistent with the procedures for the applicable rehabilitation method specified in Section 2.3. A separate analytical evaluation shall be performed for each combination of building performance and seismic hazard specified in the selected Rehabilitation Objective.

### C1.3.6 Verification of Rehabilitation Design

At this stage, a cost estimate can be made using a conceptual or schematic design to verify economic acceptability.

### 1.3.6.1 Unacceptable Rehabilitation

If the design of rehabilitation measures fails to comply with the acceptance criteria for the selected Rehabilitation Objective, the rehabilitation measures shall be redesigned or an alternative rehabilitation strategy with a different Rehabilitation Objective shall be implemented. This process shall be repeated until the design is in compliance with the acceptance criteria for the selected Rehabilitation Objective.

### 1.3.6.2 Construction Documents

If the design of rehabilitation measures meets the acceptance criteria for the selected Rehabilitation Objective, and the decision is made to proceed with the rehabilitation, construction documents shall be prepared and shall include requirements for construction quality assurance in accordance with Section 2.7.

### C1.3.6.2 Construction Documents

At this stage, a cost estimate can be made to review the economic acceptability of the design. Cost estimating or reviewing economic acceptability of the rehabilitation design is not included in this standard, but is an essential part of the rehabilitation process shown in Fig. C1-1.

Construction costs are discussed in FEMA 276, *Example Applications* (FEMA 1999), and FEMA 156 and 157, *Typical Costs for Seismic Rehabilitation of Buildings* (FEMA 1995).

If the design proves uneconomical or otherwise not feasible, further refinement may be considered in analysis, a different rehabilitation scheme may be designed or a different Rehabilitation Objective may be considered.

A successful rehabilitation project requires a good set of construction documents with a quality assurance program to ensure that the design is implemented properly. Section 2.7 specifies provisions for a quality assurance program during the construction or implementation of the rehabilitation design. Other aspects of the implementation process, including details of the preparation of construction documents, obtaining a building permit, selection of a contractor, details of historic preservation techniques for particular kinds of materials, and financing are not covered in this standard.

## 1.4 REHABILITATION OBJECTIVES

A seismic Rehabilitation Objective shall consist of one or more rehabilitation goals. Each goal shall consist of a target Building Performance Level defined in Section 1.5 and an Earthquake Hazard Level defined in Section 1.6. Goals shall be selected considering basic, enhanced, or limited objectives as defined in Sections 1.4.1 through 1.4.3.

## C1.4 REHABILITATION OBJECTIVES

Recommendations regarding the selection of a Rehabilitation Objective for any building are beyond the scope of this standard. FEMA 274 (FEMA 1997) discusses issues to consider when combining various performance and seismic hazard levels. It should be noted that not all combinations constitute reasonable or cost-effective Rehabilitation Objectives. This standard is written under the premise that greater flexibility is required in seismic rehabilitation than in the design of new buildings. However, given that flexibility, once a Rehabilitation Objective is selected, this standard provides internally consistent procedures with the necessary specificity to perform a rehabilitation analysis and design.

Building performance can be described qualitatively in terms of the safety afforded building occupants during and after the event; the cost and feasibility of restoring the building to its pre-earthquake condition; the length of time the building is removed from service to effect repairs; and economic, architectural, or historic impacts on the larger community. These performance characteristics are directly related to the extent of damage that would be sustained by the building.

In this standard, the extent of damage to a building is categorized as a Building Performance Level. A broad range of target Building Performance Levels may be selected when determining Rehabilitation Objectives.

Probabilistic Earthquake Hazard Levels frequently used in this standard and their corresponding mean return periods (the average number of years between events of similar severity) are as follows:

| Earthquake Having Probability of Exceedance | Mean Return Period (years) |
| --- | --- |
| 50%/50 year | 72 |
| 20%/50 year | 225 |
| 10%/50 year | 474 |
| 2%/50 year | 2,475 |

These mean return periods are typically rounded to 75, 225, 500, and 2,500 years, respectively.

The Rehabilitation Objective selected as a basis for design will determine, to a great extent, the cost and feasibility of any rehabilitation project, as well as the benefit to be obtained in terms of improved safety, reduction in property damage, and interruption of use in the event of future earthquakes. Table C1-1 indicates the range of Rehabilitation Objectives that may be used in this standard.

### 1.4.1 Basic Safety Objective

The Basic Safety Objective (BSO) is a Rehabilitation Objective that achieves the dual rehabilitation goals of Life Safety Building Performance Level (3-C) for the BSE-1 Earthquake Hazard Level and Collapse Prevention Building Performance Level (5-E) for the BSE-2 Earthquake Hazard Level.

### Table C1-1. Rehabilitation Objectives

| | | Target Building Performance Levels | | | |
| --- | --- | --- | --- | --- | --- |
| | | Operational Performance Level (1-A) | Immediate Occupancy Performance Level (1-B) | Life Safety Performance Level (3-C) | Collapse Prevention Performance Level (5-E) |
| Earthquake Hazard Level | 50%/50 year | a | b | c | d |
| | 20%/50 year | e | f | g | h |
| | BSE-1 (~ 10%/50 year) | i | j | k | l |
| | BSE-2 (~ 2%/50 year) | m | n | o | p |

[1]Each cell in the above matrix represents a discrete Rehabilitation Objective.

[2]The Rehabilitation Objectives in the matrix above may be used to represent the three specific Rehabilitation Objectives defined in Sections 1.4.1, 1.4.2, and 1.4.3, as follows:

| | |
| --- | --- |
| Basic Safety Objective (BSO) | k and p |
| Enhanced Objectives | k and m, n, or o |
| | p and i or j |
| | k and p and a, b, e, or f |
| | m, n, or o alone |
| Limited Objectives | k alone |
| | p alone |
| | c, d, g, h, or l alone |

### C1.4.1 Basic Safety Objective (BSO)

The BSO is intended to approximate the earthquake risk to life safety traditionally considered acceptable in the United States. Buildings meeting the BSO are expected to experience little damage from relatively frequent, moderate earthquakes, but significantly more damage and potential economic loss from the most severe and infrequent earthquakes that could affect them. The level of damage and potential economic loss experienced by buildings rehabilitated to the BSO may be greater than that expected in properly designed and constructed new buildings.

### 1.4.2 Enhanced Rehabilitation Objectives

Rehabilitation that provides building performance exceeding that of the BSO is termed an Enhanced Objective. Enhanced Rehabilitation Objectives shall be achieved using one or both of the following two methods:

1. By designing for target Building Performance Levels that exceed those of the BSO at the BSE-1 hazard level, the BSE-2 hazard level, or both.
2. By designing for the target Building Performance Levels of the BSO using an Earthquake Hazard Level that exceeds either the BSE-1 or BSE-2 hazard levels, or both.

### C1.4.2 Enhanced Rehabilitation Objectives

Enhanced Rehabilitation Objectives can be obtained by designing for higher target Building Performance Levels (method 1) or by designing using higher Earthquake Hazard Levels (method 2), or a combination of these methods.

### 1.4.3 Limited Rehabilitation Objectives

Rehabilitation that provides building performance less than that of the BSO is termed a Limited Objective. Limited Rehabilitation Objectives shall be achieved using Reduced Rehabilitation specified in Section 1.4.3.1 or Partial Rehabilitation specified in Section 1.4.3.2, and shall comply with the following conditions:

1. The rehabilitation measures shall not result in a reduction in the performance level of the existing building;
2. The rehabilitation measures shall not create a new structural irregularity or make an existing structural irregularity more severe;
3. The rehabilitation measures shall not result in an increase in the seismic forces to any component that is deficient in capacity to resist such forces; and
4. All new or rehabilitated structural components shall be detailed and connected to the existing structure in compliance with the requirements of this standard.

### C1.4.3 Limited Rehabilitation Objectives

Reduction in performance should not necessarily be measured based strictly on a single component but, rather, on the overall building performance. A partial or limited rehabilitation could increase forces on some noncritical components without a reduction in the overall performance of the building.

### 1.4.3.1 Reduced Rehabilitation Objective

Rehabilitation that addresses the entire building structural and nonstructural systems, but uses a lower seismic hazard or lower target Building Performance Level than the BSO, is termed Reduced Rehabilitation Objective. Reduced Rehabilitation shall be designed for one or more of the following objectives:

1. Life Safety Building Performance Level (3-C) for earthquake demands that are equal to the BSE-1, or Collapse Prevention Building Performance Level (5-E) for earthquake demands that are equal to the BSE-2, but not both;
2. Life Safety Building Performance Level (3-C) for earthquake demands that are less severe (more probable) than the BSE-1;
3. Collapse Prevention Building Performance Level (5-E) for earthquake demands that are less severe (more probable) than the BSE-2; or
4. Building Performance Levels 4-C, 4-D, 4-E, 5-C, 5-D, 5-E, 6-C, or 6-D for BSE-1 or less severe (more probable) earthquake demands.

### C1.4.3.1 Reduced Rehabilitation Objective

Rehabilitation for the Life Safety Building Performance Level at the BSE-1 is a commonly used reduced rehabilitation objective.

### 1.4.3.2 Partial Rehabilitation Objective

Rehabilitation that addresses a portion of the building without rehabilitating the complete lateral-force-resisting system is termed Partial Rehabilitation.

### C1.4.3.2 Partial Rehabilitation Objective

A Partial Rehabilitation should be designed and constructed considering future completion of a Rehabilitation Objective intended to improve the performance of the entire structure.

## 1.5 TARGET BUILDING PERFORMANCE LEVELS

A target Building Performance Level shall consist of a combination of a Structural Performance Level selected from the levels specified in Section 1.5.1 and a Nonstructural Performance Level selected from the levels specified in Section 1.5.2. The target Building Performance Level shall be designated alphanumerically in accordance with Section 1.5.3.

## C1.5 TARGET BUILDING PERFORMANCE LEVELS

Building performance is a combination of the performance of both structural and nonstructural components. Table C1-2 describes the approximate limiting levels of structural and nonstructural damage that may be expected of buildings rehabilitated to the levels defined in this standard. On average, the expected damage would be less. For comparative purposes, the estimated performance of a new building subjected to the BSE-1 level of shaking is indicated. Performance descriptions in Table C1-2 are estimates rather than precise predictions, and variation among buildings of the same target Building Performance Level must be expected.

Building performance in this standard is expressed in terms of target Building Performance Levels. These target Building Performance Levels are discrete damage states selected from among the infinite spectrum of possible damage states that buildings could experience during an earthquake. The particular

damage states identified as target Building Performance Levels in this standard have been selected because they have readily identifiable consequences associated with the postearthquake disposition of the building that are meaningful to the building community. These include the ability to resume normal functions within the building, the advisability of postearthquake occupancy, and the risk to life safety.

Due to inherent uncertainties in prediction of ground motion and analytical prediction of building performance, some variation in actual performance should be expected. Compliance with this standard should not be considered a guarantee of performance. Information on the reliability of achieving various performance levels can be found in Chapter 2 of FEMA 274 (FEMA 1997).

**Table C1-2. Damage Control and Building Performance Levels**

| | Target Building Performance Levels | | | |
|---|---|---|---|---|
| | Collapse Prevention Level (5-E) | Life Safety Level (3-C) | Immediate Occupancy Level (1-B) | Operational Level (1-A) |
| Overall Damage | Severe | Moderate | Light | Very Light |
| General | Little residual stiffness and strength, but load-bearing columns and walls function. Large permanent drifts. Some exits blocked. Infills and unbraced parapets failed or at incipient failure. Building is near collapse. | Some residual strength and stiffness left in all stories. Gravity-load-bearing elements function. No out-of-plane failure of walls or tipping of parapets. Some permanent drift. Damage to partitions. Building may be beyond economical repair. | No permanent drift. Structure substantially retains original strength and stiffness. Minor cracking of facades, partitions, and ceilings as well as structural elements. Elevators can be restarted. Fire protection operable. | No permanent drift. Structure substantially retains original strength and stiffness. Minor cracking of facades, partitions, and ceilings as well as structural elements. All systems important to normal operation are functional. |
| Nonstructural components | Extensive damage. | Falling hazards mitigated but many architectural, mechanical, and electrical systems are damaged. | Equipment and contents are generally secure, but may not operate due to mechanical failure or lack of utilities. | Negligible damage occurs. Power and other utilities are available, possibly from standby sources. |
| Comparison with performance intended for buildings designed under the NEHRP Provisions, for the Design Earthquake | Significantly more damage and greater risk. | Somewhat more damage and slightly higher risk. | Less damage and lower risk. | Much less damage and lower risk. |

### 1.5.1 Structural Performance Levels and Ranges

The Structural Performance Level of a building shall be selected from four discrete Structural Performance Levels and two intermediate Structural Performance Ranges defined in this section.

The discrete Structural Performance Levels are Immediate Occupancy (S-1), Life Safety (S-3), Collapse Prevention (S-5), and Not Considered (S-6). Design procedures and acceptance criteria corresponding to these Structural Performance Levels shall be as specified in Chapters 4 through 9 or Chapter 10.

The intermediate Structural Performance Ranges are the Damage Control Range (S-2) and the Limited Safety Range (S-4). Acceptance criteria for performance within the Damage Control Structural Performance Range shall be obtained by interpolating between the acceptance criteria provided for the Immediate Occupancy and Life Safety Structural Performance Levels. Acceptance criteria for performance within the Limited Safety Structural Performance Range shall be obtained by interpolating between the acceptance criteria provided for the Life Safety and Collapse Prevention Structural Performance Levels.

### C1.5.1 Structural Performance Levels and Ranges

A wide range of structural performance requirements could be desired by individual building owners. The four Structural Performance Levels defined in this standard have been selected to correlate with the most commonly specified structural performance requirements. The two Structural Performance Ranges permit users with other requirements to customize their building Rehabilitation Objectives.

Table C1-3 relates these Structural Performance Levels to the limiting damage states for common vertical elements of lateral-force-resisting systems. Table C1-4 relates these Structural Performance Levels to the limiting damage states for common horizontal elements of building lateral-force-resisting systems. Later sections of this standard specify design parameters (such as *m*-factors, component capacities, and inelastic deformation capacities) specified as limiting values for attaining these Structural Performance Levels for a known earthquake demand.

The drift values given in Table C1-3 are typical values provided to illustrate the overall structural response associated with various Structural Performance Levels. They are not provided in these tables as drift limit requirements for this standard, and do not supersede component or element deformation limits that are specified in Chapters 4 through 9, and 11.

The expected postearthquake state of the buildings described in these tables is for comparative purposes and should not be used in the postearthquake safety evaluation process.

#### 1.5.1.1 Immediate Occupancy Structural Performance Level (S-1)

Structural Performance Level S-1, Immediate Occupancy, shall be defined as the postearthquake damage state in which a structure remains safe to occupy, essentially retains its pre-earthquake design strength and stiffness, and is in compliance with the acceptance criteria specified in this standard for this Structural Performance Level.

#### C1.5.1.1 Immediate Occupancy Structural Performance Level (S-1)

Structural Performance Level S-1, Immediate Occupancy, means the postearthquake damage state in which only very limited structural damage has occurred. The basic vertical- and lateral-force-resisting systems of the building retain nearly all of their pre-earthquake strength and stiffness. The risk of life-threatening injury as a result of structural damage is very low, and although some minor structural repairs may be appropriate, these would generally not be required prior to reoccupancy.

#### 1.5.1.2 Damage Control Structural Performance Range (S-2)

Structural Performance Range S-2, Damage Control, shall be defined as the continuous range of damage states between the Life Safety Structural Performance Level (S-3) and the Immediate Occupancy Structural Performance Level (S-1).

#### C1.5.1.2 Damage Control Structural Performance Range (S-2)

Design for the Damage Control Structural Performance Range may be desirable to minimize repair time and operation interruption, as a partial means of protecting valuable equipment and contents or to preserve important historic features when the cost of design for immediate occupancy is excessive.

#### 1.5.1.3 Life Safety Structural Performance Level (S-3)

Structural Performance Level S-3, Life Safety, shall be defined as the postearthquake damage state in which a structure has damaged components but retains a margin against onset of partial or total collapse, and

## Table C1-3. Structural Performance Levels and Damage[1,2,3]—Vertical Elements

| Elements | Type | Structural Performance Levels | | |
|---|---|---|---|---|
| | | Collapse Prevention (S-5) | Life Safety (S-3) | Immediate Occupancy (S-1) |
| Concrete Frames | Primary | Extensive cracking and hinge formation in ductile elements. Limited cracking and/or splice failure in some nonductile columns. Severe damage in short columns. | Extensive damage to beams. Spalling of cover and shear cracking (< 1/8-in. width) for ductile columns. Minor spalling in nonductile columns. Joint cracks < 1/8 in. wide. | Minor hairline cracking. Limited yielding possible at a few locations. No crushing (strains below 0.003). |
| | Secondary | Extensive spalling in columns (limited shortening) and beams. Severe joint damage. Some reinforcing buckled. | Extensive cracking and hinge formation in ductile elements. Limited cracking and/or splice failure in some nonductile columns. Severe damage in short columns. | Minor spalling in a few places in ductile columns and beams. Flexural cracking in beams and columns. Shear cracking in joints < 1/16-in. width. |
| | Drift | 4% transient or permanent. | 2% transient; 1% permanent. | 1% transient; negligible permanent. |
| Steel Moment Frames | Primary | Extensive distortion of beams and column panels. Many fractures at moment connections, but shear connections remain intact. | Hinges form. Local buckling of some beam elements. Severe joint distortion; isolated moment connection fractures, but shear connections remain intact. A few elements may experience partial fracture. | Minor local yielding at a few places. No fractures. Minor buckling or observable permanent distortion of members. |
| | Secondary | Same as primary. | Extensive distortion of beams and column panels. Many fractures at moment connections, but shear connections remain intact. | Same as primary. |
| | Drift | 5% transient or permanent. | 2.5% transient; 1% permanent. | 0.7% transient; negligible permanent. |
| Braced Steel Frames | Primary | Extensive yielding and buckling of braces. Many braces and their connections may fail. | Many braces yield or buckle but do not totally fail. Many connections may fail. | Minor yielding or buckling of braces. |
| | Secondary | Same as primary. | Same as primary. | Same as primary. |
| | Drift | 2% transient or permanent. | 1.5% transient; 0.5% permanent. | 0.5% transient; negligible permanent. |
| Concrete Walls | Primary | Major flexural and shear cracks and voids. Sliding at joints. Extensive crushing and buckling of reinforcement. Failure around openings. Severe boundary element damage. Coupling beams shattered and virtually disintegrated. | Some boundary element stress, including limited buckling of reinforcement. Some sliding at joints. Damage around openings. Some crushing and flexural cracking. Coupling beams: extensive shear and flexural cracks; some crushing, but concrete generally remains in place. | Minor hairline cracking of walls, < 1/16 in. wide. Coupling beams experience cracking < 1/8-in. width. |
| | Secondary | Panels shattered and virtually disintegrated. | Major flexural and shear cracks. Sliding at joints. Extensive crushing. Failure around openings. Severe boundary element damage. Coupling beams shattered and virtually disintegrated. | Minor hairline cracking of walls. Some evidence of sliding at construction joints. Coupling beams experience cracks < 1/8-in. width. Minor spalling. |
| | Drift | 2% transient or permanent. | 1% transient; 0.5% permanent. | 0.5% transient; negligible permanent. |

| Elements | Type | Structural Performance Levels | | |
|---|---|---|---|---|
| | | Collapse Prevention (S-5) | Life Safety (S-3) | Immediate Occupancy (S-1) |
| Unreinforced Masonry Infill Walls | Primary | Extensive cracking and crushing; portions of face course shed. | Extensive cracking and some crushing but wall remains in place. No falling units. Extensive crushing and spalling of veneers at corners of openings. | Minor (< 1/8-in. width) cracking of masonry infills and veneers. Minor spalling in veneers at a few corner openings. |
| | Secondary | Extensive crushing and shattering; some walls dislodge. | Same as primary. | Same as primary. |
| | Drift | 0.6% transient or permanent. | 0.5% transient; 0.3% permanent. | 0.1% transient; negligible permanent. |
| Unreinforced Masonry (Noninfill) Walls | Primary | Extensive cracking; face course and veneer may peel off. Noticeable in-plane and out-of-plane offsets. | Extensive cracking. Noticeable in-plane offsets of masonry and minor out-of-plane offsets. | Minor (< 1/8-in. width) cracking of veneers. Minor spalling in veneers at a few corner openings. No observable out-of-plane offsets. |
| | Secondary | Nonbearing panels dislodge. | Same as primary. | Same as primary. |
| | Drift | 1% transient or permanent. | 0.6% transient; 0.6% permanent. | 0.3% transient; 0.3% permanent. |
| Reinforced Masonry Walls | Primary | Crushing; extensive cracking. Damage around openings and at corners. Some fallen units. | Extensive cracking (< 1/4 in.) distributed throughout wall. Some isolated crushing. | Minor (< 1/8-in. width) cracking. No out-of-plane offsets. |
| | Secondary | Panels shattered and virtually disintegrated. | Crushing; extensive cracking; damage around openings and at corners; some fallen units. | Same as primary. |
| | Drift | 1.5% transient or permanent. | 0.6% transient; 0.6% permanent. | 0.2% transient; 0.2% permanent. |
| Wood Stud Walls | Primary | Connections loose. Nails partially withdrawn. Some splitting of members and panels. Veneers dislodged. | Moderate loosening of connections and minor splitting of members. | Distributed minor hairline cracking of gypsum and plaster veneers. |
| | Secondary | Sheathing sheared off. Let-in braces fractured and buckled. Framing split and fractured. | Connections loose. Nails partially withdrawn. Some splitting of members and panels. | Same as primary. |
| | Drift | 3% transient or permanent. | 2% transient; 1% permanent. | 1% transient; 0.25% permanent. |
| Precast Concrete Connections | Primary | Some connection failures but no elements dislodged. | Local crushing and spalling at connections, but no gross failure of connections. | Minor working at connections; cracks < 1/16-in. width at connections. |
| | Secondary | Same as primary. | Some connection failures but no elements dislodged. | Minor crushing and spalling at connections. |
| Foundations | General | Major settlement and tilting. | Total settlements < 6 in. and differential settlements < 1/2 in. in 30 ft. | Minor settlement and negligible tilting. |

[1]Damage states indicated in this table are provided to allow an understanding of the severity of damage that may be sustained by various structural elements where present in structures meeting the definitions of the Structural Performance Levels. These damage states are not intended for use in postearthquake evaluation of damage or for judging the safety of, or required level of repair to, a structure following an earthquake.

[2]Drift values, differential settlements, crack widths, and similar quantities indicated in these tables are not intended to be used as acceptance criteria for evaluating the acceptability of a rehabilitation design in accordance with the analysis procedures provided in this standard; rather, they are indicative of the range of drift that typical structures containing the indicated structural elements may undergo when responding within the various Structural Performance Levels. Drift control of a rehabilitated structure may often be governed by the requirements to protect nonstructural components. Acceptable levels of foundation settlement or movement are highly dependent on the construction of the superstructure. The values indicated are intended to be qualitative descriptions of the approximate behavior of structures meeting the indicated levels.

[3]For limiting damage to frame elements of infilled frames, refer to the rows for concrete or steel frames.

**Table C1-4. Structural Performance Levels and Damage[1,2]—Horizontal Elements**

| | Structural Performance Levels | | |
|---|---|---|---|
| Element | Collapse Prevention (S-5) | Life Safety (S-3) | Immediate Occupancy (S-1) |
| Metal Deck Diaphragms | Large distortion with buckling of some units and tearing of many welds and seam attachments. | Some localized failure of welded connections of deck to framing and between panels. Minor local buckling of deck. | Connections between deck units and framing intact. Minor distortions. |
| Wood Diaphragms | Large permanent distortion with partial withdrawal of nails and extensive splitting of elements. | Some splitting at connections. Loosening of sheathing. Observable withdrawal of fasteners. Splitting of framing and sheathing. | No observable loosening or withdrawal of fasteners. No splitting of sheathing or framing. |
| Concrete Diaphragms | Extensive crushing and observable offset across many cracks. | Extensive cracking (< 1/4-in. width). Local crushing and spalling. | Distributed hairline cracking. Some minor cracks of larger size (< 1/8-in. width). |
| Precast Diaphragms | Connections between units fail. Units shift relative to each other. Crushing and spalling at joints. | Extensive cracking (< 1/4-in. width). Local crushing and spalling. | Some minor cracking along joints. |

[1]Damage states indicated in this table are provided to allow an understanding of the severity of damage that may be sustained by various structural elements where present in structures meeting the definitions of the Structural Performance Levels. These damage states are not intended for use in postearthquake evaluation of damage or for judging the safety of, or required level of repair to, a structure following an earthquake.

[2]Drift values, differential settlements, crack widths, and similar quantities indicated in these tables are not intended to be used as acceptance criteria for evaluating the acceptability of a rehabilitation design in accordance with the analysis procedures provided in this standard; rather, they are indicative of the range of drift that typical structures containing the indicated structural elements may undergo when responding within the various Structural Performance Levels. Drift control of a rehabilitated structure may often be governed by the requirements to protect nonstructural components. Acceptable levels of foundation settlement or movement are highly dependent on the construction of the superstructure. The values indicated are intended to be qualitative descriptions of the approximate behavior of structures meeting the indicated levels.

Concrete Diaphragms

is in compliance with the acceptance criteria specified in this standard for this Structural Performance Level.

### C1.5.1.3 Life Safety Structural Performance Level (S-3)

Structural Performance Level S-3, Life Safety, means the postearthquake damage state in which significant damage to the structure has occurred but some margin against either partial or total structural collapse remains. Some structural elements and components are severely damaged but this has not resulted in large falling debris hazards, either inside or outside the building. Injuries may occur during the earthquake; however, the overall risk of life-threatening injury as a result of structural damage is expected to be low. It should be possible to repair the structure; however, for economic reasons this may not be practical. Although the damaged structure is not an imminent collapse risk, it would be prudent to implement structural repairs or install temporary bracing prior to reoccupancy.

### 1.5.1.4 Limited Safety Structural Performance Range (S-4)

Structural Performance Range S-4, Limited Safety, shall be defined as the continuous range of damage states between the Life Safety Structural Performance Level (S-3) and the Collapse Prevention Structural Performance Level (S-5).

### 1.5.1.5 Collapse Prevention Structural Performance Level (S-5)

Structural Performance Level S-5, Collapse Prevention, shall be defined as the postearthquake damage state in which a structure has damaged components and continues to support gravity loads but retains no margin against collapse, and is in compliance with the acceptance criteria specified in this standard for this Structural Performance Level.

### C1.5.1.5 Collapse Prevention Structural Performance Level (S-5)

Structural Performance Level S-5, Collapse Prevention, means the postearthquake damage state in which the building is on the verge of partial or total collapse. Substantial damage to the structure has occurred, potentially including significant degradation in the stiffness and strength of the lateral-force-resisting system, large permanent lateral deformation of the structure, and—to a more limited extent—degradation in vertical-load-carrying capacity. However, all significant components of the gravity-load-resisting system must continue to carry their gravity loads. Significant risk of injury due to falling hazards from structural debris may exist. The structure may not be technically practical to repair and is not safe for reoccupancy, as aftershock activity could induce collapse.

### 1.5.1.6 Structural Performance Not Considered (S-6)

A building rehabilitation that does not address the performance of the structure shall be classified as Structural Performance Not Considered (S-6).

### C1.5.1.6 Structural Performance Not Considered (S-6)

Some owners may desire to address certain nonstructural vulnerabilities in a rehabilitation program—for example, bracing parapets or anchoring hazardous materials storage containers—without addressing the performance of the structure itself. Such rehabilitation programs are sometimes attractive because they can permit a significant reduction in seismic risk at relatively low cost.

### 1.5.2 Nonstructural Performance Levels

The Nonstructural Performance Level of a building shall be selected from five discrete Nonstructural Performance Levels, consisting of Operational (N-A), Immediate Occupancy (N-B), Life Safety (N-C), Hazards Reduced (N-D), and Not Considered (N-E). Design procedures and acceptance criteria for rehabilitation of nonstructural components shall be as specified in Chapter 10 or 11.

### C1.5.2 Nonstructural Performance Levels

Nonstructural Performance Levels other than Not Considered (N-E) are summarized in Tables C1-5 through C1-7. Nonstructural components addressed in this standard include architectural components such as partitions, exterior cladding, and ceilings; and mechanical and electrical components, including HVAC systems, plumbing, fire suppression systems, and lighting.

Occupant contents and furnishings (such as inventory and computers) are included in these tables for some levels but generally are not covered with specific requirements.

### 1.5.2.1 Operational Nonstructural Performance Level (N-A)

Nonstructural Performance Level N-A, Operational, shall be defined as the postearthquake damage state in which the nonstructural components are able to support the pre-earthquake functions present in the building.

### C1.5.2.1 Operational Nonstructural Performance Level (N-A)

At this level, most nonstructural systems required for normal use of the building—including lighting, plumbing, HVAC, and computer systems—are functional, although minor cleanup and repair of some items may be required. This Nonstructural Performance Level requires considerations beyond those that are normally within the sole province of the structural engineer. In addition to assuring that nonstructural components are properly mounted and braced within the structure, it is often necessary to provide emergency standby utilities. It also may be necessary to perform rigorous qualification testing of the ability of key electrical and mechanical equipment items to function during or after strong shaking.

Specific design procedures and acceptance criteria for this Nonstructural Performance Level are not included in this standard. Although the state of the art for commercial construction does not provide a complete set of references to be used for the seismic qualification and checking of nonstructural components, the user is referred to the following documents that may be useful in seismically qualifying mechanical and electrical equipment for Operational Performance.

1. AC-156. *Acceptance Criteria for Seismic Qualification Testing of Nonstructural Components* (ICBO 2000).
2. DOE/EH-545. *Seismic Evaluation Procedure for Equipment in U.S. Department of Energy Facilities* (U.S. Department of Energy 1997).
3. IEEE 693. *IEEE Recommended Practice for Seismic Design of Substations* (IEEE 1997).
4. CERL Technical Report 97/58. *The CERL Equipment Fragility and Protection, Experimental Definition of Equipment Vulnerability to Transient Support Motions* (CERL 1997).

**Table C1-5. Nonstructural Performance Levels and Damage[1]—Architectural Components**

| Component | Nonstructural Performance Levels | | | |
| | Hazards Reduced[2] (N-D) | Life Safety (N-C) | Immediate Occupancy (N-B) | Operational (N-A) |
|---|---|---|---|---|
| Cladding | Severe distortion in connections. Distributed cracking, bending, crushing, and spalling of cladding components. Some fracturing of cladding, but panels do not fall in areas of public assembly. | Severe distortion in connections. Distributed cracking, bending, crushing, and spalling of cladding components. Some fracturing of cladding, but panels do not fall. | Connections yield; minor cracks (< 1/16-in. width) or bending in cladding. | Connections yield; minor cracks (< 1/16-in. width) or bending in cladding. |
| Glazing | General shattered glass and distorted frames in unoccupied areas. Extensive cracked glass; little broken glass in occupied areas. | Extensive cracked glass; little broken glass. | Some cracked panes; none broken. | Some cracked panes; none broken. |
| Partitions | Distributed damage; some severe cracking, crushing, and racking in some areas. | Distributed damage; some severe cracking, crushing, and racking in some areas. | Cracking to about 1/16-in. width at openings. Minor crushing and cracking at corners | Cracking to about 1/16-in. width at openings. Minor crushing and cracking at corners. |
| Ceilings | Extensive damage. Dropped suspended ceiling tiles. Moderate cracking in hard ceilings. | Extensive damage. Dropped suspended ceiling tiles. Moderate cracking in hard ceilings. | Minor damage. Some suspended ceiling tiles disrupted. A few panels dropped. Minor cracking in hard ceilings. | Generally negligible damage. Isolated suspended panel dislocations, or cracks in hard ceilings. |
| Parapets and Ornamentation | Extensive damage; some falling in unoccupied areas. | Extensive damage; some falling in unoccupied areas. | Minor damage. | Minor damage. |
| Canopies and Marquees | Moderate damage. | Moderate damage. | Minor damage. | Minor damage. |
| Chimneys and Stacks | Extensive damage. No collapse. | Extensive damage. No collapse. | Minor cracking. | Negligible damage. |
| Stairs and Fire Escapes | Extensive racking. Loss of use. | Some racking and cracking of slabs. Usable. | Minor damage. | Negligible damage. |
| Doors | Distributed damage. Many racked and jammed doors. | Distributed damage. Some racked and jammed doors. | Minor damage. Doors operable. | Minor damage. Doors operable. |

[1]Damage states indicated in this table are provided to allow an understanding of the severity of damage that may be sustained by various nonstructural components meeting the Nonstructural Performance Levels defined in this standard. These damage states are not intended for use in postearthquake evaluation of damage or for judging the safety or required level of repair following an earthquake.

[2]For the Hazards Reduced Performance Level, high-hazard nonstructural components evaluated or rehabilitated to the Life Safety criteria will have Hazards Reduced performance identical to that expected for the Life Safety Performance Level.

## Table C1-6. Nonstructural Performance Levels and Damage[1]—Mechanical, Electrical, and Plumbing Systems/Components

| System/Component | Nonstructural Performance Levels | | | |
| --- | --- | --- | --- | --- |
| | Hazards Reduced[2] (N-D) | Life Safety (N-C) | Immediate Occupancy (N-B) | Operational (N-A) |
| Elevators | Elevators out of service; counterweights off rails. | Elevators out of service; counterweights do not dislodge. | Elevators operable; can be started when power available. | Elevators operate. |
| HVAC Equipment | Most units do not operate; many slide or overturn; some suspended units fall. | Units shift on supports, rupturing attached ducting, piping, and conduit, but do not fall. | Units are secure and most operate if power and other required utilities are available. | Units are secure and operate. Emergency power and other utilities provided, if required. |
| Manufacturing Equipment | Units slide and overturn; utilities disconnected. Heavy units require reconnection and realignment. Sensitive equipment may not be functional. | Units slide, but do not overturn; utilities not available; some realignment required to operate. | Units secure, and most operable if power and utilities available. | Units secure and operable; power and utilities available. |
| Ducts | Ducts break loose of equipment and louvers; some supports fail; some ducts fall. | Ducts break loose from equipment and louvers; some supports fail; some ducts fall. | Minor damage at joints, but ducts remain serviceable. | Negligible damage. |
| Piping | Some lines rupture. Some supports fail. Some piping falls. | Minor damage at joints, with some leakage. Some supports damaged, but systems remain suspended. | Minor leaks develop at a few joints. | Negligible damage. |
| Fire Sprinkler Systems | Some sprinkler heads damaged by collapsing ceilings. Leaks develop at couplings. Some branch lines fail. | Some sprinkler heads damaged by swaying ceilings. Leaks develop at some couplings. | Minor leakage at a few heads or pipe joints. System remains operable. | Negligible damage. |
| Fire Alarm Systems | Ceiling mounted sensors damaged. May not function. | Ceiling mounted sensors damaged. May not function. | System is functional. | System is functional. |
| Emergency Lighting | Some lights fall. Power may not be available. | Some lights fall. Power may be available from emergency generator. | System is functional. | System is functional. |
| Electrical Distribution Equipment | Units slide and/or overturn, rupturing attached conduit. Uninterruptable Power Source systems fail. Diesel generators do not start. | Units shift on supports and may not operate. Generators provided for emergency power start; utility service lost. | Units are secure and generally operable. Emergency generators start, but may not be adequate to service all power requirements. | Units are functional. Emergency power is provided, as needed. |
| Light Fixtures | Many broken light fixtures. Falling hazards generally avoided in heavier fixtures (> 20 lb) in areas of public assembly. | Many broken light fixtures. Falling hazards generally avoided in heavier fixtures (> 20 lb). | Minor damage. Some pendant lights broken. | Negligible damage. |
| Plumbing | Some fixtures broken; lines broken; mains disrupted at source. | Some fixtures broken, lines broken; mains disrupted at source. | Fixtures and lines serviceable; however, utility service may not be available. | System is functional. On-site water supply provided, if required. |

[1]Damage states indicated in this table are provided to allow an understanding of the severity of damage that may be sustained by various nonstructural components meeting the Nonstructural Performance Levels defined in this standard. These damage states are not intended for use in postearthquake evaluation of damage or for judging the safety or required level of repair following an earthquake.
[2]For the Hazards Reduced Performance Level, high-hazard nonstructural components evaluated or rehabilitated to the Life Safety criteria will have Hazards Reduced performance identical to that expected for the Life Safety Performance Level.

**Table C1-7. Nonstructural Performance Levels and Damage[1]—Contents**

| Contents | Nonstructural Performance Levels | | | |
| --- | --- | --- | --- | --- |
| | Hazards Reduced[2] (N-D) | Life Safety (N-C) | Immediate Occupancy (N-B) | Operational (N-A) |
| Computer Systems | Units roll and overturn, disconnect cables. Raised access floors collapse. Power not available. | Units shift and may disconnect cables, but do not overturn. Power not available. | Units secure and remain connected. Power may not be available to operate, and minor internal damage may occur. | Units undamaged and operable; power available. |
| Desktop Equipment | Some equipment slides off desks. | Some equipment slides off desks. | Some equipment slides off desks. | Equipment secured to desks and operable. |
| File Cabinets | Cabinets overturn and spill contents. | Cabinets overturn and spill contents. | Drawers slide open, but cabinets do not tip. | Drawers slide open, but cabinets do not tip. |
| Book Shelves | Shelves overturn and spill contents. | Books slide off shelves. | Books slide on shelves. | Books remain on shelves. |
| Hazardous Materials | Minor damage; occasional materials spilled; gaseous materials contained. | Minor damage; occasional materials spilled; gaseous materials contained. | Negligible damage; materials contained. | Negligible damage; materials contained. |
| Art Objects | Objects damaged by falling, water, dust. | Objects damaged by falling, water, dust. | Some objects may be damaged by falling. | Objects undamaged. |

[1]Damage states indicated in this table are provided to allow an understanding of the severity of damage that may be sustained by various nonstructural components meeting the Nonstructural Performance Levels defined in this standard. These damage states are not intended for use in postearthquake evaluation of damage or for judging the safety or required level of repair following an earthquake.

[2]For the Hazards Reduced Performance Level, high-hazard nonstructural components evaluated or rehabilitated to the Life Safety criteria will have Hazards Reduced performance identical to that expected for the Life Safety Performance Level.

Where equipment and systems are required to be seismically qualified to achieve operational performance, it is recommended that the seismic qualification procedures, testing, evaluation, and documentation be peer reviewed. The peer review can follow the procedures found in Sections 9.2.8 and 9.3.7 for Design Reviews except that items to be reviewed are nonstructural components and systems.

### 1.5.2.2 Immediate Occupancy Nonstructural Performance Level (N-B)

Nonstructural Performance Level N-B, Immediate Occupancy, shall be defined as the postearthquake damage state in which nonstructural components are damaged but building access and life safety systems—including doors, stairways, elevators, emergency lighting, fire alarms, and fire suppression systems—generally remain available and operable, provided that power is available.

### C1.5.2.2 Immediate Occupancy Nonstructural Performance Level (N-B)

Minor window breakage and slight damage could occur to some components. Presuming that the build-

ing is structurally safe, occupants could safely remain in the building, although normal use may be impaired and some cleanup and inspection may be required. In general, components of mechanical and electrical systems in the building are structurally secured and should be able to function if necessary utility service is available. However, some components may experience misalignments or internal damage and be nonoperable. Power, water, natural gas, communications lines, and other utilities required for normal building use may not be available. The risk of life-threatening injury due to nonstructural damage is very low.

### 1.5.2.3 Life Safety Nonstructural Performance Level (N-C)

Nonstructural Performance Level N-C, Life Safety, shall be defined as the postearthquake damage state in which nonstructural components are damaged but the damage is not life-threatening.

### C1.5.2.3 Life Safety Nonstructural Performance Level (N-C)

Nonstructural Performance Level C, Life Safety, is the postearthquake damage state in which potentially significant and costly damage has occurred to nonstructural components but they have not become dislodged and fallen, threatening life safety either inside or outside the building. Egress routes within the building are not extensively blocked but may be impaired by lightweight debris. HVAC, plumbing, and fire suppression systems may have been damaged, resulting in local flooding as well as loss of function. Although injuries may occur during the earthquake from the failure of nonstructural components, overall, the risk of life-threatening injury is very low. Restoration of the nonstructural components may take extensive effort.

### 1.5.2.4 Hazards Reduced Nonstructural Performance Level (N-D)

Nonstructural Performance Level N-D, Hazards Reduced, shall be defined as the postearthquake damage state in which nonstructural components are damaged and could potentially create falling hazards, but high-hazard nonstructural components identified in Chapter 11, Table 11-1, are secured to prevent falling into areas of public assembly. Preservation of egress, protection of fire suppression systems, and similar life-safety issues are not addressed in this Nonstructural Performance Level.

### C1.5.2.4 Hazards Reduced Nonstructural Performance Level (N-D)

Nonstructural Performance Level D, Hazards Reduced, represents a postearthquake damage state in which extensive damage has occurred to nonstructural components, but large or heavy items that pose a high risk of falling hazard to a large number of people—such as parapets, cladding panels, heavy plaster ceilings, or storage racks—are prevented from falling. The hazards associated with exterior components along portions of the exterior of the building that are available for public occupancy have been reduced. Although isolated serious injury could occur from falling debris, failures that could injure large numbers of persons—either inside or outside the structure—should be avoided.

Nonstructural components that are small, lightweight, or close to the ground may fall but should not cause serious injury. Larger nonstructural components in areas that are less likely to be populated may also fall.

The intent of the Hazards Reduced Performance Level is to address significant nonstructural hazards without needing to rehabilitate all of the nonstructural components in a building. When using this performance level, it will generally be appropriate to consider Hazards Reduced Performance as equivalent to Life

Safety Performance for the most-hazardous, highest-risk subset of the nonstructural components in the building.

### 1.5.2.5 Nonstructural Performance Not Considered (N-E)

A building rehabilitation that does not address nonstructural components shall be classified as Nonstructural Performance Not Considered (N-E).

### C1.5.2.5 Nonstructural Performance Not Considered (N-E)

In some cases, the decision to rehabilitate the structure may be made without addressing the vulnerabilities of nonstructural components. It may be desirable to do this when rehabilitation must be performed without interruption of building operation. In some cases, it is possible to perform all or most of the structural rehabilitation from outside occupied building areas. Extensive disruption of normal operation may be required to perform nonstructural rehabilitation. Also, since many of the most severe hazards to life safety occur as a result of structural vulnerabilities, some municipalities may wish to adopt rehabilitation ordinances that require structural rehabilitation only.

### 1.5.3 Designation of Target Building Performance Levels

A target Building Performance Level shall be designated alphanumerically with a numeral representing the Structural Performance Level and a letter representing the Nonstructural Performance Level (such as 1-B or 3-C). If a Structural Performance Level other than Immediate Occupancy (S-1), Life Safety (S-3), Collapse Prevention (S-5), or Not Considered (S-6) is selected, the numerical designation shall represent the Structural Performance Range for Damage Control (S-2) or Limited Safety (S-4).

### C1.5.3 Designation of Target Building Performance Levels

Several common target Building Performance Levels described in this Section are shown in Fig. C1-2. Many combinations are possible as structural performance can be selected at any level in the two Structural Performance Ranges. Table C1-8 indicates the possible combinations of target Building Performance Levels and provides names for those most likely to be selected as the basis for design.

### 1.5.3.1 Operational Building Performance Level (1-A)

To attain the Operational Building Performance Level (1-A), the structural components of the building shall meet the requirements of Section 1.5.1.1 for the

Immediate Occupancy Structural Performance Level (S-1) and the nonstructural components shall meet the requirements of Section 1.5.2.1 for the Operational Nonstructural Performance Level (N-A).

### C1.5.3.1 Operational Building Performance Level (1-A)

Buildings meeting this target Building Performance Level are expected to sustain minimal or no damage to their structural and nonstructural components. The building is suitable for its normal occupancy and use, although possibly in a slightly impaired mode, with power, water, and other required utilities provided from emergency sources, and possibly with some nonessential systems not functioning. Buildings meeting this target Building Performance Level pose an extremely low risk to life safety.

Under very low levels of earthquake ground motion, most buildings should be able to meet or exceed this target Building Performance Level. Typically, however, it will not be economically practical to design for this target Building Performance Level for severe ground shaking, except for buildings that house essential services.

### 1.5.3.2 Immediate Occupancy Building Performance Level (1-B)

To attain the Immediate Occupancy Building Performance Level (1-B), the structural components of the building shall meet the requirements of Section 1.5.1.1 for the Immediate Occupancy Structural Performance Level (S-1) and the nonstructural components of the building shall meet the requirements of Section 1.5.2.2 for the Immediate Occupancy Nonstructural Performance Level (N-B).

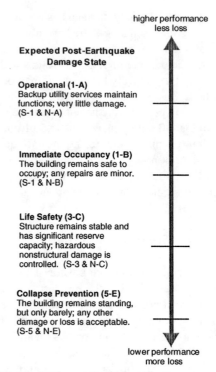

higher performance
less loss

**Expected Post-Earthquake Damage State**

**Operational (1-A)**
Backup utility services maintain functions; very little damage. (S-1 & N-A)

**Immediate Occupancy (1-B)**
The building remains safe to occupy; any repairs are minor. (S-1 & N-B)

**Life Safety (3-C)**
Structure remains stable and has significant reserve capacity; hazardous nonstructural damage is controlled. (S-3 & N-C)

**Collapse Prevention (5-E)**
The building remains standing, but only barely; any other damage or loss is acceptable. (S-5 & N-E)

lower performance
more loss

**FIGURE C1-2. Target Building Performance Levels and Ranges.**

### C1.5.3.2 Immediate Occupancy Building Performance Level (1-B)

Buildings meeting this target Building Performance Level are expected to sustain minimal or no damage to their structural elements and only minor damage to their nonstructural components. While it would be safe to reoccupy a building meeting this target Building Performance Level immediately following a major

### Table C1-8. Target Building Performance Levels and Ranges

| Nonstructural Performance Levels | Structural Performance Levels and Ranges | | | | | |
|---|---|---|---|---|---|---|
| | Immediate Occupancy (S-1) | Damage Control Range (S-2) | Life Safety (S-3) | Limited Safety Range (S-4) | Collapse Prevention (S-5) | Not Considered (S-6) |
| Operational (N-A) | Operational 1-A | 2-A | Not recommended | Not recommended | Not recommended | Not recommended |
| Immediate Occupancy (N-B) | Immediate Occupancy 1-B | 2-B | 3-B | Not recommended | Not recommended | Not recommended |
| Life Safety (N-C) | 1-C | 2-C | Life Safety 3-C | 4-C | 5-C | 6-C |
| Hazards Reduced (N-D) | Not recommended | 2-D | 3-D | 4-D | 5-D | 6-D |
| Not Considered (N-E) | Not recommended | Not recommended | Not recommended | 4-E | Collapse Prevention 5-E | Not rehabilitation |

earthquake, nonstructural systems may not function, either because of the lack of electrical power or internal damage to equipment. Therefore, although immediate reoccupancy of the building is possible, it may be necessary to perform some cleanup and repair and await the restoration of utility service before the building can function in a normal mode. The risk to life safety at this target Building Performance Level is very low.

Many building owners may wish to achieve this level of performance when the building is subjected to moderate earthquake ground motion. In addition, some owners may desire such performance for very important buildings under severe earthquake ground shaking. This level provides most of the protection obtained under the Operational Building Performance Level without the cost of providing standby utilities and performing rigorous seismic qualification of equipment performance.

### 1.5.3.3 Life Safety Building Performance Level (3-C)

To attain the Life Safety Building Performance Level (3-C), the structural components of the building shall meet the requirements of Section 1.5.1.3 for the Life Safety Structural Performance Level (S-3) and the nonstructural components shall meet the requirements of Section 1.5.2.3 for the Life Safety Nonstructural Performance Level (N-C).

### C1.5.3.3 Life Safety Building Performance Level (3-C)

Buildings meeting this level may experience extensive damage to structural and nonstructural components. Repairs may be required before reoccupancy of the building occurs, and repair may be deemed economically impractical. The risk to life safety in buildings meeting this target Building Performance Level is low.

This target Building Performance Level entails somewhat more damage than anticipated for new buildings that have been properly designed and constructed for seismic resistance when subjected to their design earthquakes. Many building owners will desire to meet this target Building Performance Level for severe ground shaking.

### 1.5.3.4 Collapse Prevention Building Performance Level (5-E)

To attain the Collapse Prevention Building Performance Level (5-E), the structural components of the building shall meet the requirements of Section 1.5.1.5 for the Collapse Prevention Structural Performance Level (S-5). Nonstructural components are not considered.

### C1.5.3.4 Collapse Prevention Building Performance Level (5-E)

Buildings meeting this target Building Performance Level may pose a significant hazard to life safety resulting from failure of nonstructural components. However, because the building itself does not collapse, gross loss of life may well be avoided. Many buildings meeting this level will be complete economic losses.

Sometimes this level has been selected as the basis for mandatory seismic rehabilitation ordinances enacted by municipalities, as it results in mitigation of the most severe life-safety hazards at relatively low cost.

## 1.6 SEISMIC HAZARD

Seismic hazard due to ground shaking shall be based on the location of the building with respect to causative faults, the regional and site-specific geologic characteristics, and a selected Earthquake Hazard Level. Assessment of seismic hazard due to earthquake-induced geologic site hazards shall be performed in accordance with Chapter 4.

Seismic hazard due to ground shaking shall be defined as acceleration response spectra or acceleration time-histories on either a probabilistic or deterministic basis. Acceleration response spectra shall be developed in accordance with either the General Procedure of Section 1.6.1 or the Site-Specific Procedure of Section 1.6.2. Acceleration time histories shall be developed in accordance with Section 1.6.2.2. The level of seismicity of the site of the building shall be determined as specified in Section 1.6.3.

Unless otherwise approved, the site-specific procedure shall be used where any of the following conditions apply:

1. The building is located on Type E soils (as defined in Section 1.6.1.4) and the mapped BSE-2 spectral response acceleration at short periods ($S_S$) exceeds 2.0;
2. The building is located on Type F soils as defined in Section 1.6.1.4.

**EXCEPTION**: Where $S_S$ determined in accordance with Section 1.6.1.1 is less than 0.20, use of a Type E soil profile shall be permitted.

## C1.6 SEISMIC HAZARD

The analysis and design procedures of this standard are primarily aimed at improving performance of buildings under loads and deformations imposed by seismic shaking. However, other seismic hazards could exist at the building site that could damage the building regardless of its ability to resist ground shaking. These hazards include fault rupture, liquefaction or other shaking-induced soil failures, landslides, and inundation from off-site effects such as dam failure or tsunami.

This standard requires hazards due to earthquake shaking to be defined on either a probabilistic or a deterministic basis. Probabilistic hazards are defined in terms of the probability that more severe demands will be experienced (probability of exceedance) in a given period (often 50 years). Deterministic demands are defined within a level of confidence in terms of a specific magnitude event on a particular major active fault.

This standard defines two basic Earthquake Hazard Levels: Basic Safety Earthquake 1 (BSE-1) and Basic Safety Earthquake 2 (BSE-2).

In addition to the BSE-1 and BSE-2 Earthquake Hazard Levels, Rehabilitation Objectives may be formed considering ground shaking due to Earthquake Hazard Levels with any defined probability of exceedance, or based on any deterministic event on a specific fault.

Site-specific procedures should be used where the Maximum Considered Earthquake (MCE) maps do not adequately characterize the local hazard. Such conditions may exist at some near-fault locations.

### 1.6.1 General Procedure for Hazard Due to Ground Shaking

The seismic hazard due to ground shaking shall be defined for any Earthquake Hazard Level using approved spectral response acceleration contour maps of 5%-damped response spectrum ordinates for short-period (0.2 sec) and long-period (1 sec) response.

The short-period spectral response acceleration parameter, $S_S$, and the long-period response acceleration parameter, $S_1$, shall be determined as follows:

1. If the desired Earthquake Hazard Level corresponds to one of the mapped Earthquake Hazard Levels, obtain spectral response acceleration parameters directly from the maps. Values between contour lines shall be interpolated in accordance with the procedure in Section 1.6.1.1;
2. If the desired Earthquake Hazard Level does not correspond to the mapped levels of hazard, then obtain the spectral response acceleration parameters from the available maps and modify them to the desired hazard level, either by logarithmic interpolation or extrapolation, in accordance with Section 1.6.1.3. It shall also be permitted to obtain the spectral response acceleration parameters by direct interpolation of the seismic hazard curves where available;

3. Obtain design spectral response acceleration parameters by adjusting the mapped or modified spectral response acceleration parameters for site class effects, in accordance with Section 1.6.1.4;
4. If the desired Earthquake Hazard Level is the Basic Safety Earthquake 2 (BSE-2), obtain spectral response acceleration parameters in accordance with Section 1.6.1.1;
5. If the desired Earthquake Hazard Level is the Basic Safety Earthquake 1 (BSE-1), obtain the spectral response acceleration parameters in accordance with Section 1.6.1.2; and
6. Using the design spectral response acceleration parameters that have been adjusted for site class effects, develop the general response spectrum in accordance with Section 1.6.1.5.

### C1.6.1 General Procedure for Hazard Due to Ground Shaking

This standard uses the latest national earthquake hazard maps developed by the United States Geological Survey (USGS) as part of a joint effort with the Building Seismic Safety Council, known as Project 97. National probabilistic maps were developed for ground motions with a 10% chance of exceedance in 50 years, a 10% chance of exceedance in 100 years (which can also be expressed as a 5% chance of exceedance in 50 years), and a 10% chance of exceedance in 250 years (which also can be expressed as a 2% chance of exceedance in 50 years). These probabilities correspond to motions that are expected to occur, on average, about once every 500, 1,000, and 2,500 years. In addition, local ground motions in regions with well-defined earthquake sources, known as deterministic motions, were used to develop MCE maps. Background information on the development of the MCE maps through Project 97 can be found in the *2003 NEHRP Provisions* Commentary (FEMA 2004).

The Rehabilitation Objective options featured in this standard allow consideration of any ground motion that may be of interest. However, for defining BSE-1 and BSE-2 Earthquake Hazard Levels, and for convenience in defining the ground motion for other Earthquake Hazard Levels, the 10%/50-year probabilistic maps and the MCE maps developed in Project 97 are referenced in this standard. This collection of maps, referred to as the National Earthquake Hazards Reduction Program (NEHRP) design map set (Maps 1 through 32), is available from the FEMA Distribution Center at 1-800-480-2520, online at http://earthquake.usgs.gov/research/hazmaps/, or on a CD-ROM from the USGS.

The MCE ground motion maps were developed by the USGS in conjunction with the Seismic Design Procedure Group (SDPG) appointed by the Building Seismic Safety Council (BSSC). The effort utilized the latest seismological information to develop design response acceleration parameters with the intent of providing a uniform margin against collapse in all areas of the United States. The MCE ground motion maps are based on seismic hazard maps, which are (1) 2%/50-year earthquake ground motion hazard maps for regions of the United States that have different ground motion attenuation relationships, and (2) deterministic ground motion maps in regions of high seismicity with the appropriate ground motion attenuation relationships for each region. The deterministic maps are used in regions of high seismicity where frequent large earthquakes are known to occur, and the rare earthquake ground motions corresponding to the 2%/50-year hazard are controlled by the large uncertainties in the hazard studies, which results in unusually high ground motions. These high ground motions were judged by the Seismic Design Procedure Group (SDGP) to be inappropriate for use in design. The use of these different maps to develop the MCE maps required SDGP to define guidelines for integrating the maps into the design ground motion maps.

The most rigorous guideline developed was for integrating the probabilistic and the deterministic maps. To integrate the probabilistic maps and the deterministic map, a transition zone set at 150% of the level of the *1994 NEHRP Provisions* was used and is extensively discussed in the *2003 NEHRP Provisions* Commentary. The goal of this guideline was to not exceed the deterministic ground motion in these areas of high seismicity where earthquake faults and maximum magnitudes are relatively well-defined. The remaining guidelines were more subjective and were related to smoothing irregular contours, joining contours in areas where closely spaced contours of equal values occurred (particularly in areas where faults are known to exist, but the hazard parameters are not well-defined), increasing the response acceleration parameters in small areas surrounded by higher parameters, and so forth.

Based on the process used to develop the MCE maps, there are some locations where the mapped acceleration response parameters in the MCE maps exceed the mapped acceleration response parameters in the 2%/50-year probabilistic maps. These locations occur primarily in the New Madrid, Missouri area; the Salt Lake City, Utah area; coastal California; and the Seattle, Washington area. This is an intended result of the process and the mapped values represent the appropriate values as determined by SDGP.

This standard requires earthquake shaking demands to be expressed in terms of ground motion response spectra or suites of ground motion time histories, depending on the analysis procedure selected. Although the maps provide a ready source for this type of information, this standard may be used with approved seismic hazard data from any source, as long as it is expressed as a response spectrum.

### 1.6.1.1 BSE-2 Spectral Response Acceleration Parameters

The design short-period spectral response acceleration parameter, $S_{XS}$, and design spectral response acceleration parameter at a 1-sec period, $S_{X1}$, for the BSE-2 Earthquake Hazard Level shall be determined using values of $S_S$ and $S_1$ taken from approved Maximum Considered Earthquake (MCE) spectral response acceleration contour maps and modified for site class in accordance with Section 1.6.1.4.

Parameters $S_S$ and $S_1$ shall be obtained by interpolating between the values shown on the map for the spectral response acceleration contour lines on either side of the site, or by using the value shown on the map for the higher contour adjacent to the site.

### C1.6.1.1 BSE-2 Spectral Response Acceleration Parameters

The latest MCE contour maps are contained in Maps 1 through 24 of the NEHRP design map set.

The BSE-2 Earthquake Hazard Level is consistent with MCE in FEMA 450 (FEMA 2004). In most areas of the United States, the BSE-2 Earthquake Hazard Level has a 2% probability of exceedance in 50 years (2%/50-year). In regions close to known faults with significant slip rates and characteristic earthquakes with magnitudes in excess of about 6.0, the BSE-2 Earthquake Hazard Level is limited by a deterministic estimate of ground motion based on 150% of the median attenuation of the shaking likely to be experienced as a result of such a characteristic event. Ground shaking levels determined in this manner will typically correspond to a probability of exceedance greater than 2% in 50 years.

### 1.6.1.2 BSE-1 Spectral Response Acceleration Parameters

The design short-period spectral response acceleration parameter, $S_{XS}$, and design spectral response acceleration parameter at a 1-sec period, $S_{X1}$, for the BSE-1 Earthquake Hazard Level shall be taken as the smaller of the following:

1. The values of $S_S$ and $S_1$ taken from approved 10%/50-year spectral response acceleration

contour maps and modified for site class in accordance with Section 1.6.1.4. Values between contour lines shall be interpolated in accordance with the procedure in Section 1.6.1.1; or

2. Two-thirds of the values of the parameters for the BSE-2 Earthquake Hazard Level, determined in accordance with Section 1.6.1.1.

### C1.6.1.2 BSE-1 Spectral Response Acceleration Parameters

The latest 10%/50-year contour maps are contained in Maps 25 through 30 of the NEHRP design map set. In determining BSE-1 parameters, the modification for site class shall be made prior to application of the two-thirds factor on BSE-2 parameters.

This standard has not directly adopted the concept of a design earthquake solely based on two-thirds of the MCE level, as in FEMA 450 (FEMA 2004). This design earthquake would have a different probability of exceedance throughout the nation (depending on the seismicity of the particular region), which would be inconsistent with the intent of this standard to permit design for specific levels of performance for hazards that have specific probabilities of exceedance. The BSE-1 Earthquake Hazard Level is similar, but not identical, to the concept of the FEMA 450 design earthquake. It is defined as ground shaking having a 10% probability of exceedance in 50 years (10%/50-year), but not exceeding values used for new buildings taken as two-thirds of the BSE-2 motion (i.e., two-thirds MCE).

### 1.6.1.3 Adjustment of Mapped Response Acceleration Parameters for Other Probabilities of Exceedance

Acceleration response spectra for earthquake hazard level corresponding to probabilities of exceedance other than 2%/50 years and 10%/50 years shall be determined using the procedures specified in Sections 1.6.1.3.1 or 1.6.1.3.2.

*1.6.1.3.1 Probabilities of Exceedance Between 2%/50 Years and 10%/50 Years* For probabilities of exceedance, $P_{EY}$, between 2%/50 years and 10%/50 years, where the mapped BSE-2 short-period spectral response acceleration parameter, $S_S$, is less than 1.5, the modified mapped short-period spectral response acceleration parameter, $S_S$, and modified mapped spectral response acceleration parameter at a 1-sec period, $S_1$, shall be determined from Eq. 1-1:

$$\ln(S_i) = \ln(S_{i10/50}) + \{[\ln(S_{iBSE-2}) - \ln(S_{i10/50})]$$
$$\cdot [0.606 \ln(P_R) - 3.73]\}$$

(Eq. 1-1)

where

$\ln(S_i)$ = natural logarithm of the spectral response acceleration parameter ("$i$" = "$S$" for short-period, or "$i$" = 1 for 1-sec period) at the desired probability of exceedance;

$\ln(S_{i10/50})$ = natural logarithm of the spectral response acceleration parameter ("$i$" = "$S$" for short-period, or "$i$" = 1 for 1-sec period) at a 10%/50-year exceedance rate;

$\ln(S_{iBSE-2})$ = natural logarithm of the spectral response acceleration parameter ("$i$" = "$S$" for short-period, or "$i$" = 1 for 1-sec period) for the BSE-2 hazard level; and

$\ln(P_R)$ = natural logarithm of the mean return period corresponding to the exceedance probability of the desired Earthquake Hazard Level.

The mean return period, $P_R$, at the desired exceedance probability shall be calculated from Eq. 1-2:

$$P_R = \frac{-Y}{\ln(1 - P_{EY})}$$

(Eq. 1-2)

where $P_{EY}$ is the probability of exceedance (expressed as a decimal) in time $Y$ (years) for the desired Earthquake Hazard Level.

Where the mapped BSE-2 short-period spectral response acceleration parameter, $S_S$, is greater than or equal to 1.5, the modified mapped short-period spectral response acceleration parameter, $S_S$, and the modified mapped spectral response acceleration parameter at a 1-sec period, $S_1$, for probabilities of exceedance between 2%/50 years and 10%/50 years shall be determined from Eq. 1-3:

$$S_i = S_{i10/50}\left(\frac{P_R}{475}\right)^n$$

(Eq. 1-3)

where $S_i$, $S_{i10/50}$, and $P_R$ are as defined above and $n$ shall be obtained from Table 1-1.

**Table 1-1. Values of Exponent $n$ for Determination of Response Acceleration Parameters at Earthquake Hazard Levels between 10%/50 Years and 2%/50 Years**

| Region | Values of Exponent $n$ for | |
|---|---|---|
| | $S_S$ | $S_1$ |
| California | 0.29 | 0.29 |
| Pacific Northwest | 0.56 | 0.67 |
| Intermountain | 0.50 | 0.60 |
| Central U.S. | 0.98 | 1.09 |
| Eastern U.S. | 0.93 | 1.05 |

Sites where mapped BSE-2 values of $S_S \geq 1.5$.

*C1.6.1.3.1 Probabilities of Exceedance Between 2%/50 Years and 10%/50 Years* Tables 1-1 through 1-3 specify five regions, three of which are not yet specifically defined, namely Intermountain, Central U.S., and Eastern U.S.

*1.6.1.3.2 Probabilities of Exceedance Greater than 10%/50 Years* For probabilities of exceedance greater than 10%/50 years, where the mapped short-period spectral response acceleration parameter, $S_S$, is less than 1.5, the modified mapped short-period spectral response acceleration parameter, $S_S$, and the modified mapped spectral response acceleration parameter at a 1-sec period, $S_1$, shall be determined from Eq. 1-3, where the exponent $n$ is obtained from Table 1-2.

For probabilities of exceedance greater than 10%/50 years, where the mapped short-period spectral response acceleration parameter, $S_S$, is greater than or equal to 1.5, the modified mapped short-period spectral response acceleration parameter, $S_S$, and the modified mapped spectral response acceleration parameter at a 1-sec period, $S_1$, shall be determined from Eq. 1-3, where the exponent $n$ is obtained from Table 1-3.

**Table 1-2. Values of Exponent *n* for Determination of Response Acceleration Parameters at Probabilities of Exceedance Greater than 10%/50 Years**

| | Values of Exponent $n$ for | |
|---|---|---|
| Region | $S_S$ | $S_1$ |
| California | 0.44 | 0.44 |
| Pacific Northwest and Intermountain | 0.54 | 0.59 |
| Central and Eastern U.S. | 0.77 | 0.80 |

Sites where mapped BSE-2 values of $S_S < 1.5$.

**Table 1-3. Values of Exponent *n* for Determination of Response Acceleration Parameters at Probabilities of Exceedance Greater than 10%/50 Years**

| | Values of Exponent $n$ for | |
|---|---|---|
| Region | $S_S$ | $S_1$ |
| California | 0.44 | 0.44 |
| Pacific Northwest | 0.89 | 0.96 |
| Intermountain | 0.54 | 0.59 |
| Central U.S. | 0.89 | 0.89 |
| Eastern U.S. | 1.25 | 1.25 |

Sites where mapped BSE-2 values of $S_S \geq 1.5$.

*1.6.1.4 Adjustment for Site Class*

The design short-period spectral response acceleration parameter, $S_{XS}$, and the design spectral response acceleration parameter at 1 sec, $S_{X1}$, shall be obtained from Eqs. 1-4 and 1-5, respectively, as follows:

$$S_{XS} = F_a S_S \qquad \text{(Eq. 1-4)}$$

$$S_{X1} = F_v S_1 \qquad \text{(Eq. 1-5)}$$

where $F_a$ and $F_v$ are site coefficients determined respectively from Tables 1-4 and 1-5, based on the site class and the values of the response acceleration parameters $S_S$ and $S_1$ for the selected return period.

**Table 1-4. Values of $F_a$ as a Function of Site Class and Mapped Short-Period Spectral Response Acceleration $S_S$**

| Site Class | Mapped Spectral Acceleration at Short-Period $S_S$[1] | | | | |
|---|---|---|---|---|---|
| | $S_S \leq 0.25$ | $S_S = 0.50$ | $S_S = 0.75$ | $S_S = 1.00$ | $S_S \geq 1.25$ |
| A | 0.8 | 0.8 | 0.8 | 0.8 | 0.8 |
| B | 1.0 | 1.0 | 1.0 | 1.0 | 1.0 |
| C | 1.2 | 1.2 | 1.1 | 1.0 | 1.0 |
| D | 1.6 | 1.4 | 1.2 | 1.1 | 1.0 |
| E | 2.5 | 1.7 | 1.2 | 0.9 | 0.9 |
| F | * | * | * | * | * |

*Site-specific geotechnical investigation and dynamic site response analyses shall be performed.
[1]Straight-line interpolation shall be used for intermediate values of $S_S$.

**Table 1-5. Values of $F_v$ as a Function of Site Class and Mapped Spectral Response Acceleration at 1-Sec Period $S_1$**

| Site Class | Mapped Spectral Acceleration at Short-Period $S_1$[1] | | | | |
|---|---|---|---|---|---|
| | $S_1 \leq 0.1$ | $S_1 = 0.2$ | $S_1 = 0.3$ | $S_1 = 0.4$ | $S_1 \geq 0.50$ |
| A | 0.8 | 0.8 | 0.8 | 0.8 | 0.8 |
| B | 1.0 | 1.0 | 1.0 | 1.0 | 1.0 |
| C | 1.7 | 1.6 | 1.5 | 1.4 | 1.3 |
| D | 2.4 | 2.0 | 1.8 | 1.6 | 1.5 |
| E | 3.5 | 3.2 | 2.8 | 2.4 | 2.4 |
| F | * | * | * | * | * |

*Site-specific geotechnical investigation and dynamic site response analyses shall be performed.
[1]Straight-line interpolation shall be used for intermediate values of $S_1$.

SEISMIC REHABILITATION OF EXISTING BUILDINGS

*1.6.1.4.1 Site Classes* Site classes shall be defined as follows:

1. **Class A**: Hard rock with average shear wave velocity, $\overline{v_s} > 5,000$ ft/sec;

2. **Class B**: Rock with 2,500 ft/sec $< \overline{v_s} < 5,000$ ft/sec;

3. **Class C**: Very dense soil and soft rock with 1,200 ft/sec $< \overline{v_s} \le 2,500$ ft/sec or with either standard blow count $\overline{N} > 50$ or undrained shear strength $\overline{s_u} > 2,000$ psf;

4. **Class D**: Stiff soil with 600 ft/sec $< \overline{v_s} \le 1,200$ ft/sec or with $15 < \overline{N} \le 50$ or 1,000 psf $\le \overline{s_u} < 2,000$ psf;

5. **Class E**: Any profile with more than 10 ft of soft clay defined as soil with plasticity index $PI > 20$, or water content $w > 40\%$, and $\overline{s_u} < 500$ psf or a soil profile with $\overline{v_s} < 600$ ft/sec; and

6. **Class F**: Soils requiring site-specific evaluations:

   6.1. Soils vulnerable to potential failure or collapse under seismic loading, such as liquefiable soils, quick and highly sensitive clays, or collapsible weakly cemented soils;
   
   6.2 Peats and/or highly organic clays ($H > 10$ ft of peat and/or highly organic clay, where $H$ = thickness of soil);
   
   6.3 Very high plasticity clays ($H > 25$ ft with $PI > 75$); or
   
   6.4 Very thick soft/medium-stiff clays ($H > 120$ ft).

The parameters $\overline{v_s}$, $\overline{N}$, and $\overline{s_u}$ are, respectively, the average values of the shear wave velocity, Standard Penetration Test (SPT) blow count, and undrained shear strength of the upper 100 ft of soils at the site. These values shall be calculated from Eq. 1-6:

$$\overline{v_s}, \overline{N}, \overline{s_u} = \frac{\sum_{i=1}^{n} d_i}{\sum_{i=1}^{n} \frac{d_i}{v_{si}}, \frac{d_i}{N_i}, \frac{d_i}{s_{ui}}} \quad \text{(Eq. 1-6)}$$

where

$N_i$ = SPT blow count in soil layer $i$;
$n$ = number of layers of similar soil materials for which data are available;
$d_i$ = depth of layer $i$;
$s_{ui}$ = undrained shear strength in layer $i$;
$v_{si}$ = shear wave velocity of the soil in layer $i$; and

$$\sum_{i=1}^{n} d_i = 100 \text{ ft.} \quad \text{(Eq. 1-7)}$$

Where $v_s$ data are available for the site, such data shall be used to classify the site. If such data are not available, $N$ data shall be used for cohesionless soil sites (sands, gravels), and $s_u$ data for cohesive soil sites (clays). For rock in profile Classes B and C, classification shall be based either on measured or estimated values of $v_s$. Classification of a site as Class A rock shall be based on measurements of $v_s$ either for material at the site itself or for rock having the same formation adjacent to the site; otherwise, Class B rock shall be assumed. Class A or B profiles shall not be assumed to be present if there is more than 10 ft of soil between the rock surface and the base of the building.

*1.6.1.4.2 Default Site Class* If there are insufficient data available to classify a soil profile as Class A, B, or C, and there is no evidence of soft clay soils characteristic of Class E in the vicinity of the site, the default site class shall be taken as Class D. If there is evidence of Class E soils in the vicinity of the site and no other data supporting selection of Class A, B, C, or D, the default site class shall be taken as Class E.

**1.6.1.5** General Response Spectrum

A general response spectrum shall be developed as specified in Sections 1.6.1.5.1 through 1.6.1.5.3.

*1.6.1.5.1 General Horizontal Response Spectrum* A general horizontal response spectrum as shown in Fig. 1-1 shall be developed using Eqs. 1-8, 1-9, and 1-10 for spectral response acceleration, $S_a$, versus structural period, $T$, in the horizontal direction.

$$S_a = S_{XS}\left[\left(\frac{5}{B_1} - 2\right)\frac{T}{T_S} + 0.4\right] \quad \text{(Eq. 1-8)}$$

for $0 < T < T_0$, and

$$S_a = S_{XS}/B_1 \quad \text{for} \quad T_0 \le T \le T_s, \quad \text{and} \quad \text{(Eq. 1-9)}$$

$$S_a = S_{X1}/(B_1 T), \text{ for } T > T_S \quad \text{(Eq. 1-10)}$$

where $T_S$ and $T_0$ are given by Eqs. 1-11 and 1-12:

$$T_S = S_{X1}/S_{XS} \quad \text{(Eq. 1-11)}$$

$$T_0 = 0.2T_S \quad \text{(Eq. 1-12)}$$

and where

$$B_1 = 4/[5.6 - \ln(100\beta)] \quad \text{(Eq. 1-13)}$$

and $\beta$ is the effective viscous damping ratio.

26

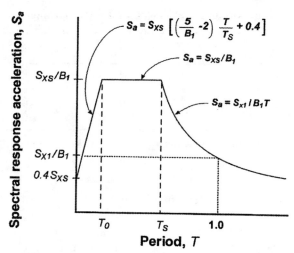

$$S_a = S_{XS}\left[\left(\frac{5}{B_1}-2\right)\frac{T}{T_S}+0.4\right]$$

$$S_a = S_{XS}/B_1$$

$$S_a = S_{x1}/B_1T$$

**FIGURE 1-1. General Horizontal Response Spectrum.**

Use of spectral response accelerations calculated using Eq. 1-8 in the extreme short-period range ($T < T_0$) shall only be permitted in dynamic analysis procedures and only for modes other than the fundamental mode.

*1.6.1.5.2 General Vertical Response Spectrum* Where a vertical response spectrum is required, it shall be developed by taking two-thirds of the spectral ordinates, at each period, obtained for the horizontal response spectrum or by alternative rational procedures approved by the code official. Alternatively, it shall be permitted to develop a site-specific vertical response spectrum in accordance with Section 1.6.2.

*C1.6.1.5.2 General Vertical Response Spectrum* Traditionally, the vertical response spectra are taken as two-thirds of the horizontal spectra developed for the site. While this is a reasonable approximation for most sites, vertical response spectra at near-field sites located within a few kilometers of the zone of fault rupture can have stronger vertical response spectra than indicated by this approximation. Development of site-specific response spectra is recommended where vertical response must be considered for buildings on such sites.

Other methods for scaling the horizontal spectrum have been proposed by Bozorgnia et al. (1996). Kehoe and Attalla (2000) present modeling considerations that should be accounted for where analyzing for vertical effects.

*1.6.1.5.3 Damping Ratios* A 5% damped response spectrum shall be used for the rehabilitation design of

all buildings and structural systems except those meeting the following criteria:

1. For structures without exterior cladding, an effective viscous damping ratio, $\beta$, equal to 2% of critical damping ($\beta = 0.02$) shall be assumed;
2. For structures with wood diaphragms and interior partitions and cross walls that interconnect the diaphragm levels at a maximum spacing of 40 ft on center transverse to the direction of motion, an effective viscous damping ratio, $\beta$, equal to 10% of critical damping ($\beta = 0.10$) shall be permitted; and
3. For structures rehabilitated using seismic isolation technology or enhanced energy dissipation technology, an equivalent effective viscous damping ratio, $\beta$, shall be calculated using the procedures specified in Chapter 9.

**1.6.2 Site-Specific Procedure for Hazard Due to Ground Shaking**
Where site-specific ground shaking characterization is used as the basis of rehabilitation design, the characterization shall be developed in accordance with this section.

*1.6.2.1 Site-Specific Response Spectra*
Development of site-specific response spectra shall be based on the geologic, seismologic, and soil characteristics associated with the specific site and as specified in Sections 1.6.2.1.1 through 1.6.2.1.4.

*C1.6.2.1 Site-Specific Response Spectra*
The code official should consider requiring an independent third-party review of the site-specific spectra by an individual with expertise in the evaluation of ground motion.

*1.6.2.1.1 Damping Ratios* Response spectra shall be developed for an effective viscous damping ratio of 5% of critical damping ($\beta = 0.05$) and for other damping ratios appropriate to the indicated structural behavior, as defined in Section 1.6.1.5.

*1.6.2.1.2 Minimum Spectral Amplitude* The 5% damped site-specific spectral amplitudes in the period range of greatest significance to the structural response shall not be specified less than 70% of the spectral amplitudes of the General Response Spectrum.

*1.6.2.1.3 Basis of the Response Spectra* Probabilistic site-specific spectra that represent the BSE-1 Earthquake Hazard Level shall be mean spectra at the 10%/50-year probability of exceedance. Probabilistic site-specific spectra that represent the BSE-2 Earthquake Hazard Level shall be mean spectra at the 2%/50-year probability of exceedance. Deterministic BSE-2 site-specific spectra shall be taken as 150% of the median spectra for the characteristic event on the controlling fault.

*1.6.2.1.4 Site-Specific BSE-2 Spectral Response Acceleration Parameters* The site-specific response acceleration parameters for the BSE-2 Earthquake Hazard Level shall be taken as the smaller of the following:

1. The values of the parameters from mean probabilistic site-specific spectra at the 2%/50-year probability of exceedance; or
2. The values of the parameters from 150% of median deterministic site-specific spectra.

*1.6.2.1.5 Site-Specific BSE-1 Spectral Response Acceleration Parameters* The site-specific response acceleration parameters for the BSE-1 Earthquake Hazard Level shall be taken as the smaller of the following:

1. The values of the parameters from mean probabilistic site-specific spectra at the 10%/50-year probability of exceedance; or
2. Two-thirds of the values of the parameters determined for the BSE-2 Earthquake Hazard Level.

*1.6.2.1.6 Site-Specific Response Acceleration Parameters* Where a site-specific response spectrum has been developed and other sections of this standard require the design response acceleration parameters, $S_{XS}$, $S_{X1}$, and $T_S$ they shall be obtained using the site-specific response spectrum in accordance with this section. Values of the design response acceleration parameter at short periods, $S_{XS}$, shall be taken as the response acceleration obtained from the site-specific spectrum at a period of 0.2 sec, except that it shall not be taken as less than 90% of the peak response acceleration at any period. In order to obtain values for the design spectral response acceleration parameter $S_{X1}$, a curve of the form $S_a = S_{X1}/T$ shall be graphically overlaid on the site-specific spectrum such that, at any period, the value of $S_a$ obtained from the curve is not less than 90% of that which would be obtained directly from the spectrum. The value of $T_S$ shall be determined in accordance with Eq. 1-11.

**Table 1-6. Level of Seismicity Definitions**

| Level of Seismicity[1] | $S_{XS}$ | $S_{X1}$ |
|---|---|---|
| Low | <0.167 | <0.067 |
| Moderate | ≥0.167 | ≥0.067 |
| | <0.500 | <0.200 |
| High | ≥0.500 | ≥0.200 |

[1]The higher level of seismicity defined by $S_{XS}$ or $S_{X1}$ shall govern.

*1.6.2.2 Acceleration Time Histories*

Time history analysis shall be performed with no fewer than three data sets (each containing two horizontal components or, if vertical motion is to be considered, two horizontal components and one vertical component) of ground motion time histories that shall be selected and scaled from no fewer than three recorded events. Time histories shall have magnitude, fault distances, and source mechanisms that are consistent with those that control the design earthquake ground motion. Where three recorded ground-motion time history data sets having these characteristics are not available, simulated time history data sets having equivalent duration and spectral content shall be used to make up the total number required. For each data set, the square root of the sum of the squares (SRSS) of the 5%-damped site-specific spectra of the scaled horizontal components shall be constructed. The data sets shall be scaled such that the average value of the SRSS spectra does not fall below 1.3 times the 5%-damped spectrum for the design earthquake for periods between 0.2T and 1.5T (where T is the fundamental period of the building).

**1.6.3 Level of Seismicity**

The level of seismicity shall be defined as High, Moderate, or Low as defined in Table 1-6.

The values of $S_{XS}$ and $S_{X1}$ used to determine the Level of Seismicity shall be two-thirds of the BSE-2 values defined in Section 1.6.1.1.

**2.0 GENERAL REQUIREMENTS**

**2.1 SCOPE**

This chapter sets forth general requirements for data collection, analysis procedures, methods, and strategies for the design of seismic rehabilitation projects.

Section 2.2 specifies data collection procedures for obtaining required as-built information on buildings. Section 2.3 outlines the Simplified and Systematic Methods for seismic rehabilitation of buildings. Section 2.4 specifies limitations on selecting analysis procedures, and defines component behavior types and corresponding acceptance criteria. Section 2.5 identifies acceptable rehabilitation strategies. Section 2.6 contains general design requirements for rehabilitation designs. Section 2.7 specifies construction quality assurance requirements. Section 2.8 specifies procedures for developing alternative modeling parameters and acceptance criteria.

## 2.2 AS-BUILT INFORMATION

The as-built information on building configuration, building components, site and foundation, and adjacent structures shall be obtained in accordance with Sections 2.2.1, 2.2.2, 2.2.3, and 2.2.4, respectively. This data shall be obtained from available drawings, specifications, and other documents for the existing construction. The data collected shall be in sufficient detail to permit classification of components as primary or secondary as specified in Section 2.2.5 and shall comply with the data collection requirements of Section 2.2.6. Data collected from available documents shall be supplemented and verified by on-site investigations including nondestructive examination and testing of building materials and components as required in Section 2.2.6.

At least one site visit shall be made to observe exposed conditions of building configuration, building components, site and foundation, and adjacent structures (made accessible by the owner) to verify that as-built information obtained from other sources is representative of the existing conditions.

## C2.2 AS-BUILT INFORMATION

Existing building characteristics pertinent to seismic performance should be obtained from the following sources, as appropriate:

1. Field observation of exposed conditions and configuration made accessible by the owner;
2. Construction documents, engineering analyses, reports, soil borings and test logs, maintenance histories, and manufacturers' literature and test data, which may be available from the owner or the code official;
3. Reference standards and codes from the period of construction as cited in Chapters 5 through 8;

4. Destructive and nondestructive examination and testing of selected building materials and components as specified in Section 2.2.6; and
5. Interviews with building owners, tenants, managers, the original architect and engineer, contractor(s), and the local building official.

The information required for an existing building may also be available from a previously conducted seismic evaluation of the building. Where seismic rehabilitation has been mandated according to building construction classification, familiarity with the building type and typical seismic deficiencies is recommended. Such information is available from several sources, including ASCE 31 (ASCE 2002). Such information may be sufficient for Simplified Rehabilitation. Additional as-built information may be needed for Systematic Rehabilitation.

Where a destructive and nondestructive testing program is necessary to obtain as-built information, it is prudent to perform preliminary calculations on select key locations or parameters prior to establishing a detailed testing program. These obtain knowledge at a reasonable cost and with as little disruption as possible of construction features and materials properties at concealed locations.

If the building is a historic structure, it is also important to identify the locations of historically significant features and fabric, which should be thoroughly investigated. Care should be taken in the design and investigation process to minimize the impact of work on these features. Refer to the *Standards for the Treatment of Historic Properties with Guidelines for Preserving, Rehabilitating, Restoring, and Reconstructing Historic Buildings* (Secretary of the Interior 1995), as discussed in Appendix A. The services of a historic preservation expert may be necessary.

### 2.2.1 Building Configuration
The as-built building configuration information shall include data on the type and arrangement of existing structural components of the vertical- and lateral-force-resisting systems, and the nonstructural components of the building that either affect the stiffness or strength of the structural components or affect the continuity of the structural load path. The as-built building configuration shall be examined to identify the vertical and lateral load paths.

### C2.2.1 Building Configuration
The as-built information on building configuration should identify the load-resisting components. Load-resisting components may include structural and

nonstructural components that participate in resisting lateral loads, whether or not they were intended to do so by the original designers. This information should identify potential seismic deficiencies in load-resisting components, which may include discontinuities in the load path, weak links, irregularities, and inadequate strength and deformation capacities.

ASCE 31 (ASCE 2002) is one example of a seismic evaluation tool that offers guidance on building configuration.

### 2.2.2 Component Properties

Sufficient as-built information shall be collected on components of the building, including their geometric and material properties and their interconnection with other components, to permit computation of their strengths and deformation capacities. To account for any uncertainty associated with component as-built information, a knowledge factor, $\kappa$, shall be used in the capacity evaluation as specified in Section 2.2.6.4.

### C2.2.2 Component Properties

Meaningful structural analysis of a building's probable seismic behavior and reliable design of rehabilitation measures require good understanding of the existing components (such as beams, columns, and diaphragms), their interconnection, and their material properties (mainly the mechanical properties, such as strength, deformability, and toughness). The strength and deformation capacity of existing components should be computed, as specified in Chapters 4 through 9 and 11, based on derived material properties and detailed component knowledge. Existing component action strengths must be determined for two basic purposes: to allow calculation of their ability to deliver load to other components, and to allow determination of their capacity to resist forces and deformations.

### 2.2.3 Site and Foundation Information

Data on foundation configuration and soil surface and subsurface conditions at the site shall be obtained from existing documentation, visual site reconnaissance, or a program of site-specific subsurface investigation in accordance with Chapter 4. A site-specific subsurface investigation shall be performed where Enhanced Rehabilitation Objectives are selected, or where insufficient data are available to quantify foundation capacities or determine the presence of geologic site hazards identified in Section 4.2.2. Where historic information indicates geologic site hazards have occurred in the vicinity of the site, a site-specific subsurface investigation shall be performed to investigate the potential for geologic site hazards at the site. Use

of applicable existing foundation capacity or geologic site hazard information available for the site shall be permitted.

A site reconnaissance shall be performed to observe variations from existing building drawings, foundation modifications not shown on existing documentation, presence of adjacent development or grading activities, and evidence of poor foundation performance.

### C2.2.3 Site and Foundation Information

Sources of applicable existing site and foundation information include original design information, foundation capacity information included on the drawings, and previous geotechnical reports for the site or for other sites in the immediate vicinity.

Adjacent building development or grading activities that impose loads on or reduce the lateral support of the structure can affect building performance in a future earthquake. Evidence of poor foundation performance can include settlement of building floor slabs and foundations, or differential movement visible at adjacent exterior sidewalks or other miscellaneous site construction.

### 2.2.4 Adjacent Buildings

Sufficient data shall be collected on the configuration of adjacent structures to permit investigation of the interaction issues identified in Sections 2.2.4.1 through 2.2.4.3. If the necessary information on adjacent structures is not available, the authority having jurisdiction shall be informed of the potential consequences of the interactions that are not being evaluated.

#### 2.2.4.1 Building Pounding

Data shall be collected to permit investigation of the effects of building pounding in accordance with Section 2.6.10, wherever a portion of an adjacent structure is located within 4% of the height above grade at the location of potential impact.

#### C2.2.4.1 Building Pounding

Building pounding can alter the basic response of the building to ground motion and impart additional inertial loads and energy to the building from the adjacent structure. Of particular concern is the potential for extreme local damage to structural elements at the zones of impact.

#### 2.2.4.2 Shared Element Condition

Data shall be collected on adjacent structures that share common vertical- or lateral-force-resisting elements with the building to permit investigation in accordance with Section 2.6.9.

### C2.2.4.2 Shared Element Condition

Buildings sharing common elements, such as party walls, have several potential problems. If the buildings attempt to move independently, one building may pull the shared element away from the other, resulting in a partial collapse. If the buildings behave as an integral unit, the additional mass and inertial loads of one structure may result in extreme demands on the lateral-force-resisting system of the other. All instances of shared elements should be reported to the building owner and the owner should be encouraged to inform adjacent building owners of hazards if identified.

### 2.2.4.3 Hazards from Adjacent Buildings

Data on hazards from adjacent buildings shall be collected to permit consideration of their potential to damage the subject building as a result of an earthquake. If there is a potential for such hazards from an adjacent building, the authority having jurisdiction over the subject building shall be informed of the effect of such hazards on achieving the selected Rehabilitation Objective.

### C2.2.4.3 Hazards from Adjacent Buildings

Hazards from adjacent buildings such as falling debris, aggressive chemical leakage, fire, or explosion that may impact building performance or the operation of the building after an earthquake should be considered and discussed with the building owner. Consideration should be given to hardening those portions of the building that may be impacted by debris or other hazards from adjacent structures. Where Immediate Occupancy of the building is desired and ingress to the building may be impaired by such hazards, consideration should be given to providing suitably resistant access to the building. Sufficient information should be collected on adjacent structures to allow preliminary evaluation of the likelihood and nature of hazards such as potential falling debris, fire, and blast pressures. Evaluations similar to those in FEMA 154 (FEMA 1988) may be adequate for this purpose.

### 2.2.5 Primary and Secondary Components

Data shall be collected to classify components as primary or secondary in accordance with Section 2.4.4.2. Data on primary and secondary components shall be collected in sufficient detail to permit modeling and analysis of such components in accordance with the requirements of this standard.

### 2.2.6 Data Collection Requirements

Data on the as-built condition of the structure, components, site, and adjacent buildings shall be collected in sufficient detail to perform the selected analysis procedure. The extent of data collected shall be consistent with minimum, usual, or comprehensive levels of knowledge as specified in Section 2.2.6.1, 2.2.6.2, or 2.2.6.3. The required level of knowledge shall be determined considering the selected Rehabilitation Objective and analysis procedure in accordance with Table 2-1.

**Table 2-1. Data Collection Requirements**

| Data | Level of Knowledge | | | | | | | |
|---|---|---|---|---|---|---|---|---|
| | Minimum | | Usual | | | | Comprehensive | |
| Rehabilitation Objective | BSO or Lower | | BSO or Lower | | Enhanced | | Enhanced | |
| Analysis Procedures | LSP, LDP | | All | | All | | All | |
| Testing | No Tests | | Usual Testing | | Usual Testing | | Comprehensive Testing | |
| Drawings | Design Drawings or Equivalent | | Design Drawings or Equivalent | | Design Drawings or Equivalent | | Construction Documents or Equivalent | |
| Condition Assessment | Visual | Comprehensive | Visual | Comprehensive | Visual | Comprehensive | Visual | Comprehensive |
| Material Properties | From drawings or default values | From default values | From drawings and tests | From usual tests | From drawings and tests | From usual tests | From documents and tests | From comprehensive tests |
| Knowledge Factor ($\kappa$) | 0.75 | 0.75 | 1.00 | 1.00 | 0.75 | 0.75 | 1.00 | 1.00 |

### 2.2.6.1 Minimum Data Collection Requirements

As a minimum, collection of as-built information shall consist of the following:

1. Information shall be obtained from design drawings with sufficient information to analyze component demands and calculate component capacities. For minimum data collection, the design drawings shall show, as a minimum, the configuration of the vertical- and lateral-force-resisting system and typical connections with sufficient detail to carry out linear analysis procedures. Where design drawings are available, information shall be verified by a visual condition assessment in accordance with Chapters 5 through 8;
2. In the absence of sufficient information from design drawings, incomplete or nonexistent information shall be supplemented by a comprehensive condition assessment, including destructive and nondestructive investigation in accordance with Chapters 5 through 8;
3. In the absence of material test records and quality assurance reports, use of default material properties in accordance with Chapters 5 through 8 shall be permitted;
4. Information needed on adjacent buildings, referenced in Section 2.2.4, shall be gained through field surveys and research of as-built information made available by the owner of the subject building; and
5. Site and foundation information shall be collected in accordance with Section 2.2.3.

### 2.2.6.2 Usual Data Collection Requirements

Usual collection of as-built information shall consist of the following:

1. Information shall be obtained from design drawings with sufficient information to analyze component demands and calculate component capacities. For usual data collection, the design drawings shall show, as a minimum, the configuration of the vertical- and lateral-force-resisting system and typical connections with sufficient detail to carry out the selected analysis procedure. Where design drawings are available, information shall be verified by a visual condition assessment in accordance with Chapters 5 through 8;
2. In the absence of sufficient information from design drawings, incomplete or nonexistent information shall be supplemented by a comprehensive condition assessment, including destructive and nondestructive investigation in accordance with Chapters 5 through 8;

3. In the absence of material test records and quality assurance reports, material properties shall be determined by usual materials testing in accordance with Chapters 5 through 8;
4. Information needed on adjacent buildings, referenced in Section 2.2.4, shall be gained through field surveys and research of as-built information made available by the owner of the subject building; and
5. Site and foundation information shall be collected in accordance with Section 2.2.3.

### 2.2.6.3 Comprehensive Data Collection Requirements

Comprehensive collection of as-built information shall consist of the following:

1. Information shall be obtained from construction documents including design drawings, specifications, material test records, and quality assurance reports covering original construction and subsequent modifications to the structure. Where construction documents are available, information shall be verified by a visual condition assessment in accordance with Chapters 5 through 8;
2. If construction documents are incomplete, missing information shall be supplemented by a comprehensive condition assessment, including destructive and nondestructive investigation in accordance with Chapters 5 through 8;
3. In the absence of material test records and quality assurance reports, material properties shall be determined by comprehensive materials testing in accordance with Chapters 5 through 8. The coefficient of variation in material test results shall be less than 20%;
4. Information needed on adjacent buildings, referenced in Section 2.2.4, shall be gained through field surveys and research of as-built information made available by the owner of the subject building; and
5. Site and foundation information shall be collected in accordance with Section 2.2.3.

### C2.2.6.3 Comprehensive Data Collection Requirements

Where materials testing results have a coefficient of variation greater than 20%, additional materials testing can be performed until the coefficient of variation is less than 20% or a knowledge factor consistent with a lesser data collection requirement can be used.

### 2.2.6.4 Knowledge Factor

*2.2.6.4.1 General* To account for uncertainty in the collection of as-built data, a knowledge factor, $\kappa$, shall be selected from Table 2-1 considering the selected

Rehabilitation Objective, analysis procedure, and data collection process. Knowledge factors shall be selected from Table 2-1 on an individual component basis as determined by the level of knowledge obtained for that component during data collection. Knowledge factors shall be applied to determine component capacities as specified in Section 2.4.4.6.

*C2.2.6.4.1 General* The $\kappa$ factor is used to express the confidence with which the properties of the building components are known, where calculating component capacities. The value of the factor is established from the knowledge obtained based on access to original construction documents, or condition assessments including destructive or nondestructive testing of representative components. The values of the factor have been established, indicating whether the level of knowledge is "minimum," "usual," or "comprehensive."

*2.2.6.4.2 Linear Procedures* Where linear procedures are used, data collection consistent with the minimum level of knowledge shall be permitted.

*2.2.6.4.3 Nonlinear Procedures* Where nonlinear procedures are used, data collection consistent with either the usual or comprehensive levels of knowledge shall be performed.

*2.2.6.4.4 Assumed Values of Knowledge Factor* It shall be permitted to perform an analysis in advance of the data collection process using an assumed value of $\kappa$, provided the value of $\kappa$ is substantiated by data collection in accordance with the requirements of Section 2.2.6 prior to implementation of the rehabilitation strategies.

If the assumed value of $\kappa$ is not supported by subsequent data collection, the analysis shall be revised to include a revised $\kappa$ consistent with the data collected in accordance with the requirements of Section 2.2.6.

If an analysis using an assumed value of $\kappa$ results in no required rehabilitation of the structure, the value of $\kappa$ shall be substantiated by data collection in accordance with the requirements of Section 2.2.6 before the analysis is finalized.

## 2.3 REHABILITATION METHODS

Seismic rehabilitation of the building shall be performed to achieve the selected Rehabilitation Objective in accordance with the requirements of the Simplified Rehabilitation Method of Section 2.3.1 or the Systematic Rehabilitation Method of Section 2.3.2.

### 2.3.1 Simplified Rehabilitation Method

The Simplified Rehabilitation Method shall be permitted for buildings that conform to one of the Model Building Types contained in Chapter 10, Table 10-1, and all limitations in that table with regard to building size and level of seismicity.

Use of the Simplified Rehabilitation Method shall be restricted to Limited Rehabilitation Objectives consisting of the Life Safety Building Performance Level (3-C) at the BSE-1 Earthquake Hazard Level, or Partial Rehabilitation as defined in Section 1.4.3.2.

The Simplified Rehabilitation Method shall be performed in accordance with the requirements of Chapters 2, 10, and 11.

### C2.3.1 Simplified Rehabilitation Method

Simplified Rehabilitation may be applied to certain buildings of regular configuration that do not require advanced analytical procedures. The primary intent of Simplified Rehabilitation is to reduce seismic risk efficiently, where possible and appropriate, by seeking Limited Objectives. Partial Rehabilitation measures, which target high-risk building deficiencies such as parapets and other exterior falling hazards, are included as Simplified Rehabilitation techniques, but their use should not be limited to buildings that conform to the limitations of Table 10-1 in Chapter 10.

The Simplified Rehabilitation Method is less complicated than the complete analytical rehabilitation design procedures found under Systematic Rehabilitation. In many cases, Simplified Rehabilitation represents a cost effective improvement in seismic performance, and it often requires less detailed evaluation or partial analysis to qualify for a specific performance level.

### 2.3.2 Systematic Rehabilitation Method

The Systematic Rehabilitation Method shall be permitted for all rehabilitation designs and shall be required for rehabilitations that do not satisfy the criteria of Section 2.3.1. The Systematic Rehabilitation Method includes the following steps:

1. An analysis procedure shall be selected in accordance with the requirements and limitations of Section 2.4;
2. A preliminary rehabilitation scheme shall be developed using one or more of the rehabilitation strategies defined in Section 2.5; and
3. An analysis of the building, including rehabilitation measures, shall be performed, and the results of the analysis shall be evaluated in accordance with the requirements of Chapters 2 through 9 and 11 to verify that the rehabilitation design meets the selected Rehabilitation Objective.

## C2.3.2 Systematic Rehabilitation Method

Systematic Rehabilitation may be applied to any building and involves thorough checking of each existing structural component, the design of new ones, and verification of acceptable overall performance represented by expected displacements and internal forces. The Systematic Rehabilitation Method focuses on the nonlinear behavior of structural response and employs procedures not previously emphasized in seismic codes.

The Systematic Rehabilitation Method is intended to be complete and contains all requirements to reach any specified performance level. Systematic Rehabilitation is an iterative process, similar to the design of new buildings, in which modifications of the existing structure are assumed for the purposes of a preliminary design and analysis, and the results of the analysis are verified as acceptable on a component basis. If either new or existing components still prove to be inadequate, the modifications are adjusted and, if necessary, a new analysis and verification cycle is performed. A preliminary design is needed to define the extent and configuration of corrective measures in sufficient detail to estimate the interaction of the stiffness, strength, and post-yield behavior of all new, modified, or existing components to be used for lateral force resistance. The designer is encouraged to include all components with significant lateral stiffness in a mathematical model to assure deformation capability under realistic seismic drifts. However, just as in the design of new buildings, it may be determined that certain components will not be considered part of the lateral-force-resisting system, as long as deformation compatibility checks are made on these components to assure their adequacy.

A mathematical model, developed for the preliminary design, must be constructed in connection with one of the analysis procedures defined in Chapter 3. These are the linear procedures (Linear Static Procedure and Linear Dynamic Procedure, LSP and LDP) and the nonlinear procedures (Nonlinear Static Procedure and Nonlinear Dynamic Procedure, NSP and NDP). With the exception of the NDP, this standard defines the analysis and rehabilitation design procedures sufficiently that compliance can be checked by an authority having jurisdiction in a manner similar to design reviews for new buildings. Modeling assumptions to be used in various situations are given in Chapters 4 through 9, and in Chapter 11 for nonstructural components. Requirements for seismic demand are given in Chapter 1. Requirements are specified for use of the NDP; however, considerable judgment is required in its application. Criteria for applying ground motion for various analysis procedures are given, but definitive rules for developing ground motion input are not included in this standard.

This standard specifies acceptance criteria for stiffness, strength, and ductility characteristics of structural components for three discrete structural performance levels in Chapters 4 though 8 for use in the Systematic Rehabilitation Method, and acceptance criteria for the performance of nonstructural components in Chapter 11 for use in Systematic and Simplified Rehabilitation Methods.

Inherent in the concept of performance levels and ranges is the assumption that performance can be measured using analytical results such as story drift ratios or strength and ductility demands on individual components. To enable structural verification at the selected performance level, the stiffness, strength, and ductility characteristics of many common components have been derived from laboratory tests and analytical studies and are presented in a standard format in Chapters 4 through 8 of this standard.

This standard specifies two new technologies in Chapter 9: seismic isolation and energy dissipation, for use in seismic rehabilitation of buildings using the Systematic Rehabilitation Method.

It is expected that testing of existing materials and components will continue and that additional corrective measures and products will be developed. It is also expected that systems and products intended to modify structural response beneficially will be advanced. The format of the analysis techniques and acceptance criteria of this standard allows rapid incorporation of such technology. Section 2.8 gives specific requirements in this regard. It is expected that this standard will have a significant impact on testing and documentation of existing materials and systems as well as on new products.

## 2.4 ANALYSIS PROCEDURES

An analysis of the building, including rehabilitation measures, shall be conducted to determine the forces and deformations induced in components of the building by ground motion corresponding to the selected Earthquake Hazard Level, or by other seismic geologic site hazards specified in Section 4.2.2.

The analysis procedure shall comply with one of the following:

1. Linear analysis subject to limitations specified in Section 2.4.1, and complying with the Linear Static Procedure (LSP) in accordance with Section 3.3.1, or the Linear Dynamic Procedure (LDP) in accordance with Section 3.3.2;

2. Nonlinear analysis subject to limitations specified in Section 2.4.2, and complying with the NSP in accordance with Section 3.3.3, or the NDP in accordance with Section 3.3.4; or
3. Alternative rational analysis in accordance with Section 2.4.3.

The analysis results shall comply with the applicable acceptance criteria selected in accordance with Section 2.4.4.

## C2.4 ANALYSIS PROCEDURES

The linear procedures maintain the traditional use of a linear stress–strain relationship, but incorporate adjustments to overall building deformations and material acceptance criteria to permit better consideration of the probable nonlinear characteristics of seismic response. The Nonlinear Static Procedure (NSP), often called "pushover analysis," uses simplified nonlinear techniques to estimate seismic structural deformations. The NDP, commonly known as nonlinear time history analysis, requires considerable judgment and experience to perform, as described in Commentary Section C2.4.2.2 of this standard.

### 2.4.1 Linear Procedures

Linear procedures shall be permitted for buildings which do not have an irregularity defined in Section 2.4.1.1. For buildings that have one or more of the irregularities defined in Section 2.4.1.1, linear procedures shall not be used unless the earthquake demands on the building comply with the demand capacity ratio (DCR) requirements in Section 2.4.1.1. For buildings incorporating base isolation systems or supplemental energy dissipation systems, the additional limitations of Section 9.2.4 or Section 9.3.4 shall apply.

### C2.4.1 Linear Procedures

The results of the linear procedures can be very inaccurate where applied to buildings with highly irregular structural systems, unless the building is capable of responding to the design earthquake(s) in a nearly elastic manner. The procedures of Section 2.4.1.1 are intended to evaluate whether the building is capable of nearly elastic response.

### 2.4.1.1 Method to Determine Limitations on Use of Linear Procedures

The methodology presented in this section shall be used to determine the applicability of linear analysis procedures based on four configurations of irregularity defined in Sections 2.4.1.1.1 through 2.4.1.1.4.

The determination of irregularity shall be based on the configuration of the rehabilitated structure. A linear analysis to determine irregularity shall be performed by either an LSP in accordance with Section 3.3.1 or an LDP in accordance with Section 3.3.2. The results of this analysis shall be used to identify the magnitude and uniformity of distribution of inelastic demands on the primary elements and components of the lateral-force-resisting system.

The magnitude and distribution of inelastic demands for existing and added primary elements and components shall be defined by DCRs and computed in accordance with Eq. 2-1:

$$DCR = \frac{Q_{UD}}{Q_{CE}}$$ (Eq. 2-1)

where

$Q_{UD}$ = force due to the gravity and earthquake loads calculated in accordance with Section 3.4.2; and
$Q_{CE}$ = expected strength of the component or element, calculated as specified in Chapters 5 through 8.

DCRs shall be calculated for each action (such as axial force, moment, or shear) of each primary component. The critical action for the component shall be the one with the largest DCR. The DCR for this action shall be termed the critical component DCR. The largest DCR for any element at a particular story is termed the critical element DCR at that story. If an element at a particular story is composed of multiple components, then the component with the largest computed DCR shall define the critical component for the element at that story.

If one or more component DCRs exceed 2.0 and any irregularity described in Section 2.4.1.1.1 through Section 2.4.1.1.4 is present, then linear procedures are not applicable and shall not be used.

### C2.4.1.1 Method to Determine Limitations on Use of Linear Procedures

The magnitude and distribution of inelastic demands are indicated by demand-capacity ratios (DCRs). Note that these DCRs are not used to determine the acceptability of component behavior. The adequacy of structural components must be evaluated using the procedures contained in Chapter 3 along with the acceptance criteria provided in Chapters 4 through 8. DCRs are used only to determine a structure's regularity. It should be noted that for complex structures, such as buildings with perforated shear walls, it may be easier to use one of the nonlinear procedures than to ensure that the building has sufficient regularity to permit use of linear procedures.

If all of the computed controlling DCRs for a component are less than or equal to 1.0, then the component is expected to respond elastically to the earthquake ground shaking being evaluated. If one or more of the computed DCRs for a component are greater than 1.0, then the component is expected to respond inelastically to the earthquake ground shaking.

*2.4.1.1.1 In-Plane Discontinuity Irregularity* An in-plane discontinuity irregularity shall be considered to exist in any primary element of the lateral-force-resisting system wherever a lateral-force-resisting element is present in one story, but does not continue or is offset within the plane of the element in the story immediately below. Figure 2-1 depicts such a condition.

*2.4.1.1.2 Out-of-Plane Discontinuity Irregularity* An out-of-plane discontinuity irregularity shall be considered to exist in any primary element of the lateral-force-resisting system where an element in one story is offset out-of-plane relative to that element in an adjacent story, as depicted in Fig. 2-2.

*2.4.1.1.3 Weak Story Irregularity* A weak story irregularity shall be considered to exist in any direction of the building if the ratio of the average shear DCR of any story to that of an adjacent story in the same direction exceeds 125%. The average DCR of a story shall be calculated by Eq. 2-2:

$$\overline{DCR} = \frac{\sum\limits_{1}^{n} DCR_i V_i}{\sum\limits_{1}^{n} V_i} \qquad \text{(Eq. 2-2)}$$

where

$\overline{DCR}$ = average DCR for the story;
$DCR_i$ = critical action DCR for element $i$ of the story;
$V_i$ = total calculated lateral shear force in an element $i$ due to earthquake response, assuming that the structure remains elastic; and
$n$ = total number of elements in the story.

For buildings with flexible diaphragms, each line of framing shall be independently evaluated.

*2.4.1.1.4 Torsional Strength Irregularity* A torsional strength irregularity shall be considered to exist in any story if the diaphragm above the story under consideration is not flexible and, for a given direction, the ratio of the critical element DCRs for primary elements on one side of the center of resistance of a story, to those on the other side of the center of resistance of the story, exceeds 1.5.

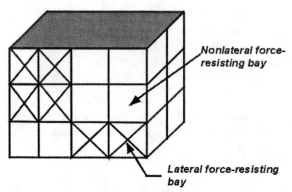

**FIGURE 2-1. In-Plane Discontinuity in Lateral System.**

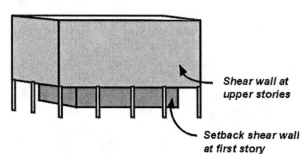

**FIGURE 2-2. Typical Building with Out-of-Plane Offset Irregularity.**

*2.4.1.2 Limitations on Use of the Linear Static Procedure*

Where Section 2.4.1.1 permits the use of linear procedures, the Linear Static Procedure shall not be used for a building with one or more of the following characteristics:

1. The fundamental period of the building, $T$, is greater than or equal to 3.5 times $T_s$;
2. The ratio of the horizontal dimension at any story to the corresponding dimension at an adjacent story exceeds 1.4 (excluding penthouses);
3. The building has a torsional stiffness irregularity in any story. A torsional stiffness irregularity exists in a story if the diaphragm above the story under consideration is not flexible and the results of the analysis indicate that the drift along any side of the structure is more than 150% of the average story drift;
4. The building has a vertical stiffness irregularity. A vertical stiffness irregularity exists where the average drift in any story (except penthouses) is more than 150% of that of the story above or below; and
5. The building has a non-orthogonal lateral-force-resisting system.

### C2.4.1.2 Limitations on Use of the Linear Static Procedure

For buildings that have irregular distributions of mass or stiffness, irregular geometries, or non-orthogonal lateral-force-resisting systems, the distribution of demands predicted by an LDP analysis will be more accurate than those predicted by the LSP. Either the response spectrum method or time history method may be used for evaluation of such structures.

## 2.4.2 Nonlinear Procedures

Nonlinear procedures shall be permitted for any of the rehabilitation strategies contained in Section 2.5. Nonlinear procedures shall be used for analysis of buildings where linear procedures are not permitted. Data collection for use with nonlinear procedures shall be in accordance with Section 2.2.6.

### 2.4.2.1 Nonlinear Static Procedure

The NSP shall be permitted for structures with all of the following characteristics:

1. The strength ratio, $R$, calculated in accordance with Eq. 3-15 in Chapter 3, is less than $R_{max}$ calculated in accordance with Eq. 3-16 in Chapter 3; and
2. Higher mode effects are not significant, as defined in this section.

To determine if higher modes are significant, a modal response spectrum analysis shall be performed for the structure using sufficient modes to produce 90% mass participation. A second response spectrum analysis shall also be performed, considering only the first mode participation. Higher mode effects shall be considered significant if the shear in any story resulting from the modal analysis considering modes required to obtain 90% mass participation exceeds 130% of the corresponding story shear considering only the first mode response.

If higher mode effects are significant, the NSP shall be permitted if an LDP analysis is also performed to supplement the NSP. Buildings with significant higher mode effects must meet the acceptance criteria of this standard for both analysis procedures, except that an increase by a factor of 1.33 shall be permitted in the LDP acceptance criteria for deformation-controlled actions (*m*-factors) provided in Chapters 5 through 9. A building analyzed using the NSP, with or without a supplementary LDP evaluation, shall meet the acceptance criteria for nonlinear procedures specified in Section 3.4.3.

If $R$ exceeds $R_{max}$, an NDP analysis shall be performed.

### C2.4.2.1 Nonlinear Static Procedure

The NSP is generally a more reliable approach to characterizing the performance of a structure than are linear procedures. However, it is not exact and cannot accurately account for changes in dynamic response as the structure degrades in stiffness, nor can it account for higher mode effects in multi-degree of freedom (MDOF) systems. Where the NSP is utilized on a structure that has significant higher mode response, the LDP is also employed to verify the adequacy of the design. Where this approach is taken, less-restrictive criteria are permitted for the LDP, recognizing the significantly improved knowledge that is obtained by performing both analysis procedures.

The strength ratio, $R$, is a measure of the extent of nonlinearity, and $R_{max}$ is a measure of the system degradation. Structures that experience nonlinear demands exceeding $R_{max}$ have significant degradation and an NDP is required to confirm the dynamic stability of the building.

### 2.4.2.2 Nonlinear Dynamic Procedure

The NDP shall be permitted for all structures. Where the NDP procedure is used, the authority having jurisdiction shall consider the requirement of review and approval by an independent third-party engineer with experience in seismic design and nonlinear procedures.

### C2.4.2.2 Nonlinear Dynamic Procedure

The NDP consists of nonlinear time-history analysis, a sophisticated approach to examining the inelastic demands produced on a structure by a specific suite of ground motion time histories. As with the NSP, the results of the NDP can be directly compared with test data on the behavior of representative structural components in order to identify the structure's probable performance when subjected to a specific ground motion. Potentially, the NDP can be more accurate than the NSP in that it avoids some of the approximations made in the more simplified analysis. Time-History Analysis automatically accounts for higher mode effects and shifts in inertial load patterns as structural softening occurs. In addition, for a given earthquake record, this approach directly solves for the maximum global displacement demand produced by the earthquake on the structure, eliminating the need to estimate this demand based on general relationships.

Despite these advantages, the NDP requires considerable judgment and experience to perform. These analyses tend to be highly sensitive to small changes in assumptions with regard to either the character of the ground motion record used in the analysis, or the nonlinear stiffness behavior of the elements. As an example, two ground motion records enveloped by the same response spectrum can produce radically different results with regard to the distribution and amount of inelasticity predicted in the structure. In order to apply this approach reliably to rehabilitation design, it is necessary to perform a number of such analyses, using varied assumptions. The sensitivity of the analysis results to the assumptions incorporated is the principal reason why this method should be used only on projects where the engineer is thoroughly familiar with nonlinear dynamic analysis techniques and limitations.

### 2.4.3 Alternative Rational Analysis

Use of an approved alternative analysis procedure that is rational and based on fundamental principles of engineering mechanics and dynamics shall be permitted. Such alternative analyses shall not adopt the acceptance criteria contained in this standard without first determining their applicability. All projects using alternative rational analysis procedures shall be reviewed and approved by an independent third-party engineer with experience in seismic design.

### 2.4.4 Acceptance Criteria

#### 2.4.4.1 General

The acceptability of force and deformation actions shall be evaluated for each component in accordance with the requirements of Section 3.4. Prior to selecting component acceptance criteria for use in Section 3.4, each component shall be classified as primary or secondary in accordance with Section 2.4.4.2, and each action shall be classified as deformation-controlled (ductile) or force-controlled (nonductile) in accordance with Section 2.4.4.3. Component strengths, material properties, and component capacities shall be determined in accordance with Sections 2.4.4.4, 2.4.4.5, and 2.4.4.6, respectively. Component acceptance criteria not specified in this standard shall be determined by qualification testing in accordance with Section 2.8.

The rehabilitated building shall be provided with at least one continuous load path to transfer seismic loads, induced by ground motion in any direction, from the point of application of the seismic load to the final point of resistance. All primary and secondary components shall be capable of resisting force and deformation actions within the applicable acceptance criteria of the selected performance level.

#### 2.4.4.2 Primary and Secondary Components

Components that affect the lateral stiffness or distribution of forces in a structure, or are loaded as a result of lateral deformation of the structure, shall be classified as primary or secondary, even if they are not intended to be part of the lateral-force-resisting system.

A structural component that is required to resist seismic forces in order for the structure to achieve the selected performance level shall be classified as primary.

A structural component that is not required to resist seismic forces in order for the structure to achieve the selected performance level shall be permitted to be classified as secondary.

#### C2.4.4.2 Primary and Secondary Components

The designation of primary and secondary components has been introduced to allow some flexibility in the rehabilitation analysis and design process. Primary components are those that the engineer relies on to resist the specified earthquake effects. Secondary components are those that the engineer does not rely on to resist the specified earthquake effects. Typically, the secondary designation will be used where a component does not add considerably or reliably to the earthquake resistance. In all cases, the engineer must verify that gravity loads are sustained by the structural system, regardless of the designation of primary and secondary components.

The secondary designation typically will be used where one or more of the following cases apply:

1. The secondary designation may be used where a nonstructural component does not contribute significantly or reliably to resist earthquake effects in any direction. A gypsum partition is a nonstructural component that might be designated secondary in a building because it does not provide significant stiffness or strength in any direction;
2. The secondary designation may be used where a structural component does not contribute significantly to resist earthquake effects. A slab-column interior frame is an element whose structural components might be designated as secondary in a building braced by much stiffer and stronger perimeter frames or shear walls. If the stronger perimeter frames or shear walls exist only in one direction, the components of the slab-column interior frame may be designated as secondary for that direction only. The connection at the base of a column that is nominally pinned where it connects to the foundation is a component that might

be designated as secondary because the moment resistance is low, relative to the entire system resistance; and

3. The secondary designation may be used where a component, intended in the original design of the building to be primary, is deformed beyond the point where it can be relied on to resist earthquake effects. For example, it is conceivable that coupling beams connecting wall piers will exhaust their deformation capacity before the entire structural system capacity is reached. In such cases, the engineer may designate these as secondary, allowing them to be deformed beyond their useful limits, provided that damage to these secondary components does not result in loss of gravity load capacity.

### 2.4.4.3 Deformation-Controlled and Force-Controlled Actions

All actions shall be classified as either deformation-controlled or force-controlled using the component force versus deformation curves shown in Fig. 2-3.

The Type 1 curve depicted in Fig. 2-3 is representative of ductile behavior where there is an elastic range (points 0 to 1 on the curve) followed by a plastic range (points 1 to 3) with non-negligible residual strength and ability to support gravity loads at point 3. The plastic range includes a strain-hardening or -softening range (points 1 to 2) and a strength-degraded range (points 2 to 3). Primary component actions exhibiting this behavior shall be classified as deformation-controlled if the strain-hardening or -softening range is such that $e > 2g$; otherwise, they shall be classified as force-controlled. Secondary com-

ponent actions exhibiting Type 1 behavior shall be classified as deformation-controlled for any $e/g$ ratio.

The Type 2 curve depicted in Fig. 2-3 is representative of ductile behavior where there is an elastic range (points 0 to 1 on the curve) and a plastic range (points 1 to 2) followed by loss of strength and loss of ability to support gravity loads beyond point 2. Primary and secondary component actions exhibiting this type of behavior shall be classified as deformation-controlled if the plastic range is such that $e \geq 2g$; otherwise, they shall be classified as force-controlled.

The Type 3 curve depicted in Fig. 2-3 is representative of a brittle or nonductile behavior where there is an elastic range (points 0 to 1 on the curve) followed by loss of strength and loss of ability to support gravity loads beyond point 1. Primary and secondary component actions displaying Type 3 behavior shall be classified as force-controlled.

### C2.4.4.3 Deformation-Controlled and Force-Controlled Actions

Acceptance criteria for primary components that exhibit Type 1 behavior typically are within the elastic or plastic ranges between points 0 and 2, depending on the performance level. Acceptance criteria for secondary components that exhibit Type 1 behavior can be within any of the performance ranges.

Acceptance criteria for primary and secondary components exhibiting Type 2 behavior will be within the elastic or plastic ranges, depending on the performance level.

Acceptance criteria for primary and secondary components exhibiting Type 3 behavior will always be within the elastic range.

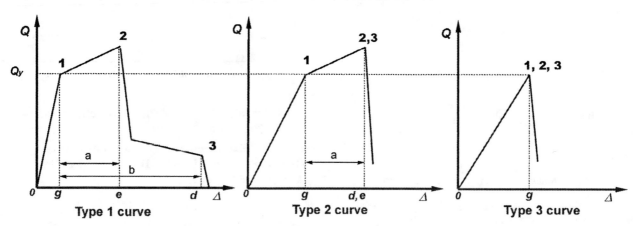

**FIGURE 2-3. Component Force Versus Deformation Curves.**

Table C2-1 provides some examples of possible deformation- and force-controlled actions in common framing systems. Classification of deformation- or force-controlled actions are specified for foundation and framing components in Chapters 4 through 8.

A given component may have a combination of both deformation- and force-controlled actions.

Classification as a deformation-controlled action is not up to the discretion of the user. Deformation-controlled actions have been defined in this standard by the designation of $m$-factors or nonlinear deformation capacities in Chapters 4 through 8. Where such values are not designated and component testing justifying Type 1 or Type 2 behavior is absent, actions are to be taken as force-controlled.

Figure C2-1 shows the generalized force versus deformation curves used throughout this standard to specify component modeling and acceptance criteria for deformation-controlled actions in any of the four basic material types. Linear response is depicted between point A (unloaded component) and an effective yield point B. The slope from point B to point C is typically a small percentage (0%–10%) of the elastic slope, and is included to represent phenomena such as strain hardening. Point C has an ordinate that represents the strength of the component, and an

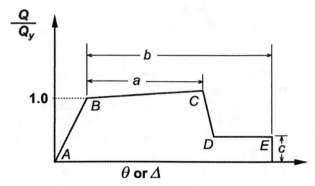

(a) Deformation

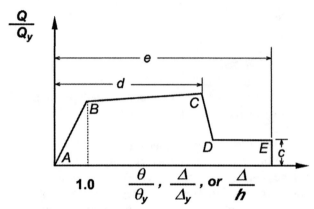

(b) Deformation ratio

**Table C2-1. Examples of Possible Deformation-Controlled and Force-Controlled Actions**

| Component | Deformation-Controlled Action | Force-Controlled Action |
|---|---|---|
| Moment Frames | | |
| • Beams | Moment ($M$) | Shear ($V$) |
| • Columns | — | Axial load ($P$), $V$ |
| • Joints | — | $V^1$ |
| Shear Walls | $M, V$ | $P$ |
| Braced Frames | | |
| • Braces | $P$ | — |
| • Beams | — | $P$ |
| • Columns | — | $P$ |
| • Shear Link | $V$ | $P, M$ |
| Connections | $P, V, M^2$ | $P, V, M$ |
| Diaphragms | $M, V^3$ | $P, V, M$ |

[1]Shear may be a deformation-controlled action in steel moment frame construction.

[2]Axial, shear, and moment may be deformation-controlled actions for certain steel and wood connections.

[3]If the diaphragm carries lateral loads from vertical seismic resisting elements above the diaphragm level, then $M$ and $V$ shall be considered force-controlled actions.

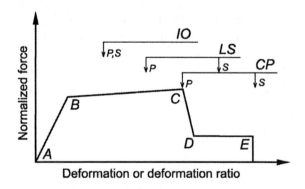

(c) Component or element deformation acceptance criteria

**FIGURE C2-1. Generalized Component Force–Deformation Relations for Depicting Modeling and Acceptance Criteria.**

abscissa value equal to the deformation at which significant strength degradation begins (line CD). Beyond point D, the component responds with substantially reduced strength to point E. At deformations greater than point E, the component strength is essentially zero.

The sharp transition as shown on idealized curves in Fig. C2-1 between points C and D can result in computational difficulty and an inability to converge where used as modeling input in nonlinear computerized analysis software. In order to avoid this computational instability, a small slope (10 vertical to 1 horizontal) may be provided to the segment of these curves between points C and D.

For some components it is convenient to prescribe acceptance criteria in terms of deformation (such as $\theta$ or $\Delta$), while for others it is more convenient to give criteria in terms of deformation ratios. To accommodate this, two types of idealized force versus deformation curves are used in Figs. C2-1 (a) and (b). Figure C2-1(a) shows normalized force ($Q/Q_y$) versus deformation ($\theta$ or $\Delta$) and the parameters $a$, $b$, and $c$. Figure C2-1(b) shows normalized force ($Q/Q_y$) versus deformation ratio ($\theta/\theta_y$, $\Delta/\Delta_y$, or $\Delta/n$) and the parameters $d$, $e$, and $c$. Elastic stiffnesses and values for the parameters $a$, $b$, $c$, $d$, and $e$ that can be used for modeling components are given in Chapters 5 through 8. Acceptance criteria for deformation or deformation ratios for primary components (P) and secondary components (S) corresponding to the target Building Performance Levels of Collapse Prevention (CP), Life Safety (LS), and Immediate Occupancy (IO) as shown in Fig. C2-1(c) are given in Chapters 5 through 8.

### 2.4.4.4 Expected and Lower-Bound Strength

In Fig. 2-3, $Q_y$ represents the yield strength of the component. Where evaluating the behavior of deformation-controlled actions, the expected strength, $Q_{CE}$, shall be used. $Q_{CE}$ is defined as the mean value of resistance of a component at the deformation level anticipated for a population of similar components, including consideration of the variability in material strength as wells as strain hardening and plastic section development. Where evaluating the behavior of force-controlled actions, a lower-bound estimate of the component strength, $Q_{CL}$, shall be used. $Q_{CL}$ is defined as the mean minus one standard deviation of the yield strengths, $Q_y$, for a population of similar components.

### C2.4.4.4 Expected and Lower-Bound Strength

In Fig. 2-3, the strength of a component is affected by inherent variability of the strength of the materials comprising the individual components as

well as differences in workmanship and physical condition. See Chapters 5 through 8 for specific direction regarding the calculation of expected and lower-bound strengths of components.

### 2.4.4.5 Material Properties

Expected material properties shall be based on mean values of tested material properties. Lower-bound material properties shall be based on mean values of tested material properties minus one standard deviation ($\sigma$).

Nominal material properties, or properties specified in construction documents, shall be taken as lower-bound material properties unless otherwise specified in Chapters 5 through 8. Corresponding expected material properties shall be calculated by multiplying lower-bound values by appropriate factors specified in Chapters 5 through 8 to translate from lower-bound to expected values.

### C2.4.4.5 Material Properties

Where calculations are used to determine expected or lower-bound strengths of components, expected or lower-bound material properties, respectively, shall be used.

### 2.4.4.6 Component Capacities

2.4.4.6.1 *General* Detailed criteria for calculation of individual component force and deformation capacities shall comply with the requirements in individual materials chapters as follows:

1. Foundations—Chapter 4;
2. Components composed of steel or cast iron—Chapter 5;
3. Components composed of reinforced concrete—Chapter 6;
4. Components composed of reinforced or unreinforced masonry—Chapter 7;
5. Components composed of timber, light metal studs, gypsum, or plaster products—Chapter 8;
6. Seismic isolation systems and energy dissipation systems—Chapter 9; and
7. Nonstructural (architectural, mechanical, and electrical) components—Chapter 11.

Elements and components composed of combinations of materials are covered in the chapters associated with each material.

2.4.4.6.2 *Linear Procedures* If linear procedures are used, capacities for deformation-controlled actions shall be defined as the product of $m$-factors and expected strengths, $Q_{CE}$. Capacities for force-controlled actions shall be defined as lower-bound strengths, $Q_{CL}$, as summarized in Table 2-2.

*2.4.4.6.3 Nonlinear Procedures* If nonlinear procedures are used, component capacities for deformation-controlled actions shall be taken as permissible inelastic deformation limits, and component capacities for force-controlled actions shall be taken as lower-bound strengths, $Q_{CL}$, as summarized in Table 2-3.

### Table 2-2. Calculation of Component Action Capacity—Linear Procedures

| Parameter | Deformation-Controlled | Force-Controlled |
|---|---|---|
| Existing Material Strength | Expected mean value with allowance for strain-hardening | Lower-bound value (approximately mean value $-1\sigma$ level) |
| Existing Action Capacity | $\kappa \cdot Q_{CE}$ | $\kappa \cdot Q_{CL}$ |
| New Material Strength | Expected material strength | Specified material strength |
| New Action Capacity | $Q_{CE}$ | $Q_{CL}$ |

### Table 2-3. Calculation of Component Action Capacity—Nonlinear Procedures

| Parameter | Deformation-Controlled | Force-Controlled |
|---|---|---|
| Deformation Capacity (Existing Component) | $\kappa \cdot$ Deformation limit | N/A |
| Deformation Capacity (New Component) | Deformation limit | N/A |
| Strength Capacity (Existing Component) | N/A | $\kappa \cdot Q_{CL}$ |
| Strength Capacity (New Component) | N/A | $Q_{CL}$ |

## 2.5 REHABILITATION STRATEGIES

A Rehabilitation Objective shall be achieved by implementing rehabilitation measures based on a strategy of addressing deficiencies identified by a prior seismic evaluation. Each rehabilitation measure shall be evaluated in conjunction with other rehabilitation measures, and the existing structure as a whole, to assure that the complete rehabilitation scheme achieves the target Building Performance Level for the selected Earthquake Hazard Level. The effects of rehabilitation on stiffness, strength, and deformability shall be taken into account in an analytical model of the rehabilitated structure. The compatibility of new and existing components shall be checked at displacements consistent with the demands produced by the selected Earthquake Hazard Level and geologic site hazards present at the site.

One or more of the following strategies shall be permitted as rehabilitation measures.

- Local modification of components
- Removal or reduction of existing irregularities
- Global structural stiffening
- Global structural strengthening
- Mass reduction
- Seismic isolation, in accordance with Chapter 9
- Supplemental energy dissipation, in accordance with Chapter 9
- Other rehabilitation strategies approved by the authority having jurisdiction.

## C2.5 REHABILITATION STRATEGIES

Although not specifically required by any of the strategies, it is very beneficial for the rehabilitated lateral-force-resisting system to have an appropriate level of redundancy so that any localized failure of a few components of the system will not result in local collapse or an instability. This should be considered when developing rehabilitation designs.

**Local Modification of Components.** Some existing buildings have substantial strength and stiffness, but some of their components may not have adequate strength, toughness, or deformation capacity to satisfy the Rehabilitation Objectives. An appropriate strategy for such structures may be to perform local modifications of components that are inadequate while retaining the basic configuration of the building's lateral-force-resisting system. Local modifications that can be considered include improvement of component connectivity, component strength, component deformation capacity, or all three. This strategy tends to be the most economical rehabilitation approach where only a few of the building's components are inadequate.

Local strengthening allows one or more understrength components or connections to resist the strength demands predicted by the analysis without affecting the overall response of the structure. This could include measures such as cover plating steel beams or columns, or adding wood structural panel sheathing to an existing timber diaphragm. Such measures increase the strength of the component and allow it to resist more earthquake-induced force before the onset of damage.

Local corrective measures that improve the deformation capacity or ductility of a component allow it to resist large deformation levels with reduced amounts of damage, without necessarily increasing the strength. One such measure is placement of a confinement jacket around a reinforced concrete column to improve its ability to deform without spalling or degrading reinforcement splices. Another measure is reduction of the cross section of selected structural components to increase their flexibility and response displacement capacity.

**Removal or Reduction of Existing Irregularities.** Removal or reduction of existing irregularities may be an effective rehabilitation strategy if a seismic evaluation shows that the irregularities result in the inability of the building to meet the selected Structural Performance Level.

The results of analysis should be reviewed to detect existing irregularities. Stiffness, mass, and strength irregularities may be detected by reviewing the results of a linear analysis, by examining the distribution of structural displacements and DCRs, or by reviewing the results of a nonlinear analysis by examining the distribution of structural displacements and inelastic deformation demands. If the distribution of values of structural displacements, DCRs, or inelastic deformation demands predicted by the analysis is nonuniform with disproportionately high values within one story relative to the adjacent story, or at one side of a building relative to the other, then an irregularity exists.

Such irregularities are often, but not always, caused by the presence of a discontinuity in the structure, such as termination of a perimeter shear wall above the first story. Simple removal of the irregularity may be sufficient to reduce demands predicted by the analysis to acceptable levels. However, removal of dis-

continuities may be inappropriate in the case of historic buildings, and the effect of such alterations on important historic features should be considered carefully.

Effective corrective measures for removal or reduction of irregularities, such as soft or weak stories, include the addition of braced frames or shear walls within the soft or weak story. Torsional irregularities can be corrected by the addition of moment frames, braced frames, or shear walls to balance the distribution of stiffness and mass within a story. Discontinuous components such as columns or walls can be extended through the zone of discontinuity.

Partial demolition can also be an effective corrective measure for irregularities, although this obviously has significant impact on the appearance and utility of the building, and this may not be an appropriate alternative for historic structures. Portions of the structure that create the irregularity, such as setback towers or side wings, can be removed. Expansion joints can be created to transform a single irregular building into multiple regular structures; however, care must be taken to avoid the potential problems associated with pounding.

**Global Structural Stiffening.** Global stiffening of the structure may be an effective rehabilitation strategy if the results of a seismic evaluation show deficiencies attributable to excessive lateral deflection of the building and critical components do not have adequate ductility to resist the resulting deformations. Construction of new braced frames or shear walls within an existing structure are effective measures for adding stiffness.

**Global Structural Strengthening.** Global strengthening of the structure may be an effective rehabilitation strategy if the results of a seismic evaluation show unacceptable performance attributable to a global deficiency in structural strength. This can be identified where the onset of global inelastic behavior occurs at levels of ground shaking that are substantially less than the selected level of ground shaking, or large DCRs (or inelastic deformation demands) are present throughout the structure. By providing supplemental strength to such a lateral-force-resisting system, it is possible to raise the threshold of ground motion at which the onset of damage occurs. Shear walls and braced frames are effective elements for this purpose, but they may be significantly stiffer than the structure to which they are added, requiring them to provide nearly all of the structure's lateral resistance. Moment-resisting frames, being more flexible, may be more compatible with existing elements in some structures; however, such flexible elements may not become effective in the building's response until existing brittle elements have already been damaged.

**Mass Reduction.** Mass reduction may be an effective rehabilitation strategy if the results of a seismic evaluation show deficiencies attributable to excessive building mass, global structural flexibility, or global structural weakness. Mass and stiffness control the amount of force and deformation induced in a structure by ground motion. Reductions in mass can result in direct reductions in both the amount of force and the deformation demand produced by earthquakes and, therefore, can be used in lieu of structural strengthening and stiffening. Mass can be reduced through demolition of upper stories, replacement of heavy cladding and interior partitions, or removal of heavy storage and equipment loads.

**Seismic Isolation.** Seismic isolation may be an effective rehabilitation strategy if the results of a seismic evaluation show deficiencies attributable to excessive seismic forces or deformation demands, or if it is desired to protect important contents and nonstructural components from damage. Where a structure is seismically isolated, compliant bearings are inserted between the superstructure and its foundations. This produces a system (structure and isolation bearings) with a nearly rigid body translation of the structure above the bearings. Most of the deformation induced in the isolated system by the ground motion occurs within the compliant bearings, which are specifically designed to resist these concentrated displacements. Most bearings also have excellent energy dissipation characteristics (damping). Together, this results in greatly reduced demands on the existing structural and nonstructural components of the building and its contents. For this reason, seismic isolation is often an appropriate strategy to achieve Enhanced Rehabilitation Objectives that include the protection of historic fabric, valuable contents, and equipment, or for buildings that contain important operations and functions. This technique is most effective for relatively stiff buildings with low profiles and large mass. It is less effective for light, flexible structures.

**Supplemental Energy Dissipation.** Installation of supplemental energy dissipation devices may be an effective rehabilitation strategy if the results of a seismic evaluation show deficiencies attributable to excessive deformations due to global structural flexibility in a building. Many available technologies allow the energy imparted to a structure by ground motion to be dissipated in a controlled manner through the action of special devices—fluid viscous dampers (hydraulic cylinders), yielding plates, or friction pads—resulting in an overall reduction in the displacements of the structure. The most commonly used devices dissipate energy through frictional, hysteretic, or viscoelastic processes. In order to dissipate substantial energy, dissipation devices typically must undergo significant deformation (or stroke), which requires that the structure experience substantial lateral displacements. Therefore, these systems are most effective in structures that are relatively flexible and have some inelastic deformation capacity. Energy dissipaters are most commonly installed in structures as components of braced frames. Depending on the characteristics of the device, either static or dynamic stiffness is added to the structure as well as energy dissipation capacity (damping). In some cases, although the structural displacements are reduced, the forces delivered to the structure can actually be increased.

## 2.6 GENERAL DESIGN REQUIREMENTS

The requirements of this section shall apply to all buildings for which the Systematic Rehabilitation Method is selected for any target Building Performance Level and any selected Earthquake Hazard Level unless specified otherwise.

### 2.6.1 Multidirectional Seismic Effects
Components shall be designed to resist seismic forces acting in any horizontal direction. Seismic forces in the vertical direction shall be considered where required by Section 2.6.11. Multidirectional seismic effects shall be considered in the analysis as specified in Section 3.2.7.

### 2.6.2 P-$\Delta$ Effects
Components of buildings shall be designed for P-$\Delta$ effects, defined as the combined effects of gravity loads acting in conjunction with lateral drifts due to seismic forces, as specified in Section 3.2.5.

### 2.6.3 Horizontal Torsion
Components of buildings shall be designed to resist the effects of horizontal torsion as specified in Section 3.2.2.2.

### 2.6.4 Overturning
Components of buildings shall be designed to resist the effects of overturning at each intermediate level as well as the base of the structure. Stability against overturning shall be evaluated as specified in Section 3.2.10. Effects of overturning on foundations shall be evaluated as specified in Section 4.4.

## 2.6.5 Continuity

All structural components shall be tied together to form a complete load path for the transfer of inertial forces generated by the dynamic response of portions of the structure to the rest of the structure. Actions resulting from the forces specified in this section shall be considered force-controlled.

1. Smaller portions of a structure, such as outstanding wings, shall be connected to the structure as a whole. Component connections shall be capable of resisting, in any direction, the horizontal force calculated using Eq. 2-3. These connections are not required if the individual portions of the structure are self-supporting and are separated by a seismic joint permitting independent movement during dynamic response.

$$F_p = 0.133 S_{XS} W \qquad \text{(Eq. 2-3)}$$

where

$F_p$ = horizontal force in any direction for the design of connections between two portions of a structure;

$S_{XS}$ = spectral response acceleration parameter at short periods for the selected Earthquake Hazard Level and damping, adjusted for site class; and

$W$ = weight of the smaller portion of the structure.

2. A positive connection for resisting horizontal force acting parallel to the member shall be provided for each beam, girder, or truss to its support. The connection shall have a minimum strength of 5% of the dead load and live load reaction.
3. Where a sliding support is provided at the end of a component, the bearing length shall be sufficient to accommodate the expected differential displacement between the component and the support.

## C2.6.5 Continuity

A continuous structural system with adequately interconnected elements is one of the most important prerequisites for acceptable seismic performance. The requirements of this section are similar to parallel provisions contained in FEMA 450 (FEMA 2004).

## 2.6.6 Diaphragms

Diaphragms shall be defined as horizontal elements that transfer earthquake-induced inertial forces to vertical elements of the lateral-force-resisting systems through the collective action of diaphragm components, including chords, collectors, and ties.

Diaphragms shall be provided at each level of the structure as necessary to connect building masses to the primary vertical elements of the lateral-force-resisting system. The analytical model of the building shall account for the behavior of the diaphragms as specified in Section 3.2.4.

Diaphragms and their connections to vertical elements providing lateral support shall comply with the design requirements specified in Section 5.8 for metal diaphragms, Section 6.10 for concrete diaphragms, Section 6.11 for precast concrete diaphragms, and Section 8.5 for wood diaphragms.

### 2.6.6.1 Diaphragm Chords

Except for diaphragms evaluated as "unchorded" as specified in Chapter 8, a boundary component shall be provided at each diaphragm edge (either at the perimeter or at an opening) to resist tension or compression resulting from the diaphragm moment. This boundary component shall be a continuous diaphragm chord, a continuous component of a wall or frame element, or a continuous combination of wall, frame, and chord components. The boundary components shall be designed to transfer accumulated lateral forces at the diaphragm boundaries. At re-entrant corners in diaphragms and at the corners of openings in diaphragms, diaphragm chords shall be extended a distance sufficient to develop the accumulated diaphragm boundary forces into the diaphragm beyond the corner.

### 2.6.6.2 Diaphragm Collectors

At each vertical element, a diaphragm collector shall be provided to transfer to the element accumulated diaphragm forces that are in excess of the forces transferred directly to the element in shear. The diaphragm collector shall be extended beyond the element and attached to the diaphragm to transfer accumulated forces.

### 2.6.6.3 Diaphragm Ties

Diaphragms shall be provided with continuous tension ties between chords or boundaries. At a minimum, ties shall be designed for axial tension as a force-controlled action calculated using Eq. 2-4.

$$F_p = 0.4 S_{XS} W \qquad \text{(Eq. 2-4)}$$

where

$F_p$ = axial tensile force for the design of ties between the diaphragm and chords or boundaries;

$S_{XS}$ = spectral response acceleration parameter at short periods for the selected hazard level and damping, adjusted for site class; and

$W$ = weight tributary to that portion of the diaphragm extending half the distance to each adjacent tie or diaphragm boundary.

Where diaphragms of timber, gypsum, or metal deck construction provide lateral support for walls of masonry or concrete construction, ties shall be designed for the wall anchorage forces specified in Section 2.6.7 for the area of wall tributary to the diaphragm tie.

### C2.6.6 Diaphragms

The concept of a diaphragm chord, consisting of an edge member provided to resist diaphragm flexural stresses through direct axial tension or compression, is not familiar to many engineers. Buildings with solid structural walls on all sides often do not require diaphragm chords. However, buildings with highly perforated perimeter walls do require these components for proper diaphragm behavior. This section of this standard requires that these components be provided where appropriate.

A common problem in buildings that nominally have robust lateral-force-resisting systems is a lack of adequate attachment between the diaphragms and the vertical elements of the lateral-force-resisting to effect shear transfer. This is particularly a problem in buildings that have discrete shear walls or frames as their vertical lateral-force-resisting elements. This section provides a reminder that it is necessary to detail a formal system of force delivery from the diaphragm to the walls and frames.

Diaphragms that support heavy perimeter walls have occasionally failed due to tension induced by out-of-plane forces generated in the walls. This section is intended to ensure that sufficient tensile ties are provided across diaphragms to prevent such failures. The design force for these tensile ties, taken as $0.4S_{XS}$ times the weight, is an extension of provisions contained in the *1994 Uniform Building Code* (ICBO 1994). In that code, parts and portions of structures are designed for a force calculated as $C_p I Z$ times the weight of the component, with typical values of $C_p$ being 0.75 and $Z$ being the effective peak ground acceleration for which the building is designed. The 1994 *Uniform Building Code* provisions use an allowable stress basis. This standard uses a strength basis. Therefore, a factor of 1.4 was applied to the $C_p$ value, and a factor of $1/(2.5)$ was applied to adjust the $Z$ value to an equivalent $S_{XS}$ value, resulting in a coefficient of 0.4.

### 2.6.7 Walls

Walls shall be evaluated for out-of-plane inertial forces as required by this section and as further required for specific structural systems in Chapters 5 through 8. Actions that result from application of the forces specified in this section shall be considered

force-controlled. Nonstructural walls shall be evaluated using the provisions of Chapter 11.

#### 2.6.7.1 Out-of-Plane Anchorage to Diaphragms

Walls shall be positively anchored to all diaphragms that provide lateral support for the wall or are vertically supported by the wall. Walls shall be anchored to diaphragms at horizontal distances not exceeding 8 ft, unless it can be demonstrated that the wall has adequate capacity to span horizontally between the supports for greater distances. Anchorage of walls to diaphragms shall be designed for forces calculated using Eq. 2-5, which shall be developed in the diaphragm. If sub-diaphragms are used, each sub-diaphragm shall be capable of transmitting the shear forces due to wall anchorage to a continuous diaphragm tie. Sub-diaphragms shall have length-to-depth ratios not exceeding 3:1. Where wall panels are stiffened for out-of-plane behavior by pilasters or similar components, anchors shall be provided at each such component and the distribution of out-of-plane forces to wall anchors and diaphragm ties shall consider the stiffening effect and accumulation of forces at these components.

$$F_p = \chi S_{XS} W \qquad \text{(Eq. 2-5)}$$

where

$F_p$ = design force for anchorage of walls to diaphragms;

$\chi$ = factor from Table 2-4 for the selected Structural Performance Level. Increased values of $\chi$ shall be used where anchoring to flexible diaphragms;

$S_{XS}$ = spectral response acceleration parameter at short periods for the selected hazard level and damping, adjusted for site class; and

$W$ = weight of the wall tributary to the anchor.

#### Table 2-4. Coefficient $\chi$ for Calculation of Out-of-Plane Wall Forces

| Structural Performance Level | Flexible Diaphragms | Other Diaphragms |
|---|---|---|
| Collapse Prevention | 0.9 | 0.3 |
| Life Safety | 1.2 | 0.4 |
| Immediate Occupancy | 1.8 | 0.6 |

[1]Value of $\chi$ for flexible diaphragms need not be applied to out-of-plane strength of walls in Section 2.6.7.2.

**EXCEPTION:**

1. $F_p$ shall not be less than the minimum of 400 lb/ft or 400 $S_{XS}$ (lb/ft) for concrete or masonry walls.

### 2.6.7.2 Out-of-Plane Strength

Wall components shall have adequate strength to span between locations of out-of-plane support when subjected to out-of-plane forces calculated using Eq. 2-6.

$$F_p = \chi S_{XS} W \qquad \text{(Eq. 2-6)}$$

where

$F_p$ = out-of-plane force per unit area for design of a wall spanning between two out-of-plane supports;

$\chi$ = factor from Table 2-4 for the selected performance level. Values of $\chi$ for flexible diaphragms need not be applied to out-of-plane strength of wall components;

$S_{XS}$ = spectral response acceleration at short periods for the selected hazard level and damping, adjusted for site class; and

$W$ = weight of the wall per unit area.

### C2.6.7.2 Out-of-Plane Strength

Application of these requirements for unreinforced masonry walls and infills is further defined in Chapter 7.

### 2.6.8 Nonstructural Components

Nonstructural components, including architectural, mechanical, and electrical components, shall be anchored and braced to the structure in accordance with the provisions of Chapter 11.

### 2.6.9 Structures Sharing Common Elements

Buildings sharing common vertical- or lateral-force-resisting elements shall be rehabilitated considering interconnection of the two structures, or they shall be separated as specified in this section.

### 2.6.9.1 Interconnection

Buildings sharing common elements, other than foundation elements, shall be thoroughly tied together so as to behave as an integral unit. Ties between the structures at each level shall be designed for the forces specified in Section 2.6.5. Analyses of the combined response of the buildings shall account for the interconnection of the structures and shall evaluate the structures as one integral unit.

If the shared common elements are foundation elements and the superstructures meet the separation requirements of Section 2.6.10, the structures need not be tied together. Shared foundation elements shall be designed considering an analysis of the combined response of the two buildings.

### 2.6.9.2 Separation

Buildings sharing common elements shall be completely separated by introducing seismic joints between the structures meeting the requirements of Section 2.6.10. Independent lateral-force-resisting systems shall be provided for each structure. Independent vertical support shall be provided on each side of the seismic joint, unless slide bearings are used and adequate bearing length is provided to accommodate the expected independent lateral movement of each structure. It shall be assumed for such purposes that the structures move out of phase with each other in opposite directions simultaneously. The original shared element shall be either completely removed, or anchored to one of the structures in accordance with the applicable requirements of Section 2.6.5.

### 2.6.10 Building Separation

### 2.6.10.1 Minimum Separation

Buildings shall be separated from adjacent structures to prevent pounding by a minimum distance $s_i$ at any level $i$ given by Eq. 2-7 unless exempted as specified in Section 2.6.10.2.

$$s_i = \sqrt{\Delta_{i1}^2 + \Delta_{i2}^2} \qquad \text{(Eq. 2-7)}$$

where

$\Delta_{i1}$ = lateral deflection of the building under consideration, at level $i$, relative to the ground, calculated in accordance with the provisions of this standard for the selected hazard level; and

$\Delta_{i2}$ = lateral deflection of an adjacent building, at level $i$, relative to the ground, estimated using the provisions of this standard or other approved approximate procedure. Alternatively, it shall be permitted to assume $\Delta_{i2} = (0.03)(h_i)$ for any structure in lieu of a more detailed analysis, where $h_i$ is the height of level $i$ above grade.

The value of $s_i$ need not exceed 0.04 times the height of the level under consideration above grade at the location of potential impact.

### 2.6.10.2 Exceptions

For Structural Performance Levels of Life Safety or lower, buildings adjacent to structures that have diaphragms located at the same elevation, and differ in height by less than 50% of the height of the shorter building, need not meet the minimum separation distance specified in Section 2.6.10.1.

Where an approved analysis procedure that accounts for the change in dynamic response of the structures due to impact is used, the rehabilitated buildings need not meet the minimum separation distance specified in Section 2.6.10.1. Such an analysis shall demonstrate that:

1. The structures are capable of transferring forces resulting from impact, for diaphragms located at the same elevation; or
2. The structures are capable of resisting all required vertical and lateral forces considering the loss of any elements or components damaged by impact of the structures.

### C2.6.10.2 Exceptions

This standard permits rehabilitated buildings to experience pounding as long as the effects are adequately considered by analysis methods that account for the transfer of momentum and energy between the structures as they impact.

Approximate methods of accounting for these effects can be obtained by performing nonlinear time-history analyses of both structures (Johnson 1992). Approximate elastic methods for evaluating these effects have also been developed and are presented in the literature (Kasai 1990).

Buildings that are likely to experience significant pounding should not be considered capable of meeting Enhanced Rehabilitation Objectives. This is because significant local crushing of components is likely to occur at points of impact. Furthermore, the very nature of the impact is such that high-frequency shocks can be transmitted through the structures and potentially be very damaging to architectural components and mechanical and electrical systems. Such damage is not consistent with the performance expected of buildings designed to Enhanced Rehabilitation Objectives.

### 2.6.11 Vertical Seismic Effects

The effects of the vertical response of a structure to earthquake ground motion shall be considered for the following cases:

1. Cantilever components of structures;
2. Prestressed components of structures; and
3. Structural components in which demands due to gravity loads specified in Section 3.2.8 exceed 80% of the nominal capacity of the component.

## 2.7 CONSTRUCTION QUALITY ASSURANCE

Construction of seismic rehabilitation work shall be checked for quality of construction and general compliance with the intent of the plans and specifications of the rehabilitation design. Construction quality assurance shall conform to the requirements of this section and the additional testing and inspection requirements of the building code and reference standards of Chapters 5 through 11.

## C2.7 CONSTRUCTION QUALITY ASSURANCE

The design professional responsible for the seismic rehabilitation of a specific building may find it appropriate to specify more stringent or more detailed requirements. Such additional requirements may be particularly appropriate for those buildings having Enhanced Rehabilitation Objectives.

### 2.7.1 Construction Quality Assurance Plan

A Quality Assurance Plan (QAP) shall be prepared by the design professional and approved by the authority having jurisdiction. The QAP shall identify components of the work that are subject to quality assurance procedures and identify special inspection, testing, and observation requirements to confirm construction quality. The QAP shall also include a process for modifying the rehabilitation design to reflect unforeseen conditions discovered during construction.

### C2.7.1 Construction Quality Assurance Plan

The quality assurance plan (QAP) should, as a minimum, include the following:

1. Required contractor quality control procedures; and
2. Required design professional construction quality assurance services, including but not limited to the following:
   2.1 Review of required contractor submittals;
   2.2 Monitoring of required inspection reports and test results;
   2.3 Construction consultation as required by the contractor on the intent of the construction documents; and
   2.4 Construction observation in accordance with Section 2.7.2.1.

## 2.7.2 Construction Quality Assurance Requirements

### 2.7.2.1 Requirements for the Design Professional

The design professional shall be responsible for preparing the QAP applicable to the portion of the work for which they are in responsible charge, overseeing the implementation of the plan, and reviewing special inspection and testing reports.

The design professional shall be responsible for performing periodic structural observation of the rehabilitation work. Structural observation shall be performed at significant stages of construction, and shall include visual observation of the work for substantial conformance with the construction documents and confirmation of conditions assumed during design. Structural observation shall be performed in addition to any special inspection and testing that is otherwise required for the work.

The design professional shall be responsible for modifying the rehabilitation design to reflect conditions discovered during construction.

### C2.7.2.1 Requirements for the Design Professional

Following structural observations, the design professional should report any observed deficiencies in writing to the owner's representative, the special inspector, the contractor, and the code official. Upon completion of the work, the design professional should submit to the authority having jurisdiction a written statement attesting that the site visits have been made, and identifying any reported deficiencies that, to the best of the structural construction observer's knowledge, have not been resolved or rectified.

### 2.7.2.2 Special Inspection

The owner shall engage the services of a special inspector to observe construction of the following rehabilitation work:

1. Items designated in Section A.9.3.3 of Appendix A of ASCE 7 (ASCE 2005); and
2. Other work designated for such special inspection by the design professional or the authority having jurisdiction.

### 2.7.2.3 Testing

The special inspector shall be responsible for verifying that special test requirements, as described in the QAP, are performed by an approved testing agency for the following rehabilitation work:

1. Work described in Section A.9.3.4 of Appendix A of ASCE 7 (ASCE 2005);

2. Other work designated for such testing by the design professional or the authority having jurisdiction.

### 2.7.2.4 Reporting and Compliance Procedures

The special inspector shall furnish copies of progress reports to the owner's representative and the design professional, noting any uncorrected deficiencies and corrections of previously reported deficiencies. All observed deficiencies shall be brought to the immediate attention of the contractor for correction.

Upon completion of construction, the special inspector shall submit a final report to the owner's representative and the design professional, indicating the extent to which inspected work was completed in accordance with approved construction documents. Noncompliant work shall have been corrected prior to completion of construction.

### C2.7.2 Construction Quality Assurance Requirements

The special inspector should be a qualified person who should demonstrate competence, to the satisfaction of the authority having jurisdiction, for inspection of the particular type of construction or operation requiring special inspection.

### 2.7.3 Responsibilities of the Authority Having Jurisdiction

The authority having jurisdiction shall be responsible for reviewing and approving the QAP and specifying minimum special inspection, testing, and reporting requirements.

### C2.7.3 Responsibilities of the Authority Having Jurisdiction

The authority having jurisdiction should act to enhance and encourage the protection of the public that is represented by such rehabilitation. These actions should include those described in the following subsections.

### C2.7.3.1 Construction Document Submittals— Permitting

As part of the permitting process, the authority having jurisdiction should require that construction documents be submitted for a permit to construct the proposed seismic rehabilitation measures. The documents should include a statement of the design basis for the rehabilitation, drawings (or adequately detailed sketches), structural/seismic calculations, and a QAP as recommended by Section 2.7.1. Appropriate structural construction specifications are also recommended if structural requirements are not adequately defined by notes on drawings.

The authority having jurisdiction should require that it be demonstrated (in the design calculations, by third-party review, or by other means) that the design of the seismic rehabilitation measures has been performed in conformance with local building regulations, the stated design basis, the intent of this standard, and/or accepted engineering principles. The authority having jurisdiction should be aware that compliance with the building code provisions for new structures is often not possible and is not required by th's standard. It is not intended that the authority having jurisdiction assure compliance of the submittals w th the structural requirements for new construction.

The authority having jurisdiction should maintain a permanent public file of the construction documents submitted as part of the permitting process for construction of the seismic rehabilitation measures.

### C2.7.3.2 Construction Phase Role

The authority having jurisdiction should monitor the implementation of the QAP. In particular, the following actions should be taken:

1. Files of inspection reports should be maintained for a defined length of time following completion of construction and issuance of a certificate of occupancy. These files should include both reports submitted by special inspectors employed by the owner, as in Section 2.7.2.2, and those submitted by inspectors employed by the authority having jurisdiction;
2. Prior to issuance of a certificate of occupancy, the authority having jurisdiction should ascertain that either all reported noncompliant aspects of construction have been rectified, or such noncompliant aspects have been accepted by the design professional in responsible charge as acceptable substitutes that are consistent with the general intent of the construction documents; and
3. Files of test reports prepared in accordance with Section 2.7.2.4 should be maintained for a defined length of time following completion of construction and issuance of a certificate of occupancy.

## 2.8 ALTERNATIVE MODELING PARAMETERS AND ACCEPTANCE CRITERIA

For elements, components, systems, and materials for which structural modeling parameters and acceptance criteria are not provided in this standard, it shall be permitted to derive the required parameters and acceptance criteria using the experimentally obtained cyclic response characteristics of the subassembly, determined in accordance with this section. Approved independent review of this process shall be conducted.

### 2.8.1 Experimental Setup

Where relevant data on the inelastic force-deformation behavior for a structural subassembly are not available, such data shall be obtained from experiments consisting of physical tests of representative subassemblies as specified in this section. Each subassembly shall be an identifiable portion of the structural element or component, the stiffness of which is required to be modeled as part of the structural analysis process. The objective of the experiment shall be to estimate the lateral-force-displacement relationships (stiffness) for the subassemblies at different loading increments, together with the strength and deformation capacities for the desired Structural Performance Levels. These properties shall be used in developing an analytical model of the structure to calculate its response to earthquake ground shaking and other hazards, and in developing acceptance criteria for strength and deformations. The limiting strength and deformation capacities shall be determined from the experimental program using the average values of a minimum of three tests performed for the same design configuration and test conditions.

The experimental setup shall replicate the construction details, support and boundary conditions, and loading conditions expected in the building. The loading shall consist of fully reversed cyclic loading at increasing displacement levels with the number of cycles and displacement levels based on expected response of the structure to the design earthquake. Increments shall be continued until the subassembly exhibits complete failure, characterized by the loss of lateral- and vertical-load resistance.

### 2.8.2 Data Reduction and Reporting

A report shall be prepared for each experiment. The report shall include the following:

1. Description of the subassembly being tested.
2. Description of the experimental setup, including:
    2.1. Details on fabrication of the subassembly;
    2.2. Location and date of testing;
    2.3. Description of instrumentation employed;
    2.4. Name of the person in responsible charge of the test; and
    2.5. Photographs of the specimen, taken prior to testing.

3. Description of the loading protocol employed, including:

   3.1. Increment of loading (or deformation) applied;

   3.2. Rate of loading application; and

   3.3. Duration of loading at each stage.

4. Description (including photographic documentation) and limiting deformation value for all important behavior states observed during the test, including the following, as applicable:

   4.1. Elastic range with effective stiffness reported;

   4.2. Plastic range;

   4.3. Onset of visible damage;

   4.4. Loss of lateral-force-resisting capacity;

   4.5. Loss of vertical-force-resisting capacity;

   4.6. Force–deformation plot for the subassembly (noting the various behavior states); and

   4.7. Description of limiting behavior states defined as the onset of specific damage mode, change in stiffness or behavior (such as initiation of cracking or yielding), and failure modes.

## 2.8.3 Design Parameters and Acceptance Criteria

The following procedure shall be followed to develop structural modeling parameters and acceptance criteria for subassemblies based on experimental data:

1. An idealized lateral-force–deformation pushover curve shall be developed from the experimental data for each experiment and for each direction of loading with unique behavior. The curve shall be plotted in a single quadrant (positive force versus positive deformation, or negative force versus negative deformation). The curve shall be constructed as follows:

   1.1. The appropriate quadrant of data shall be taken from the lateral-force–deformation plot from the experimental report.

   1.2. A smooth "backbone" curve shall be drawn through the intersection of the first cycle curve for the $i$-th deformation step with the second cycle curve of the $(i - 1)$th deformation step, for all $i$ steps, as indicated in Fig. 2-4.

   1.3. The backbone curve so derived shall be approximated by a series of linear segments, drawn to form a multisegmented curve conforming to one of the types indicated in Fig. 2-3.

2. The multilinear curves derived for all experiments involving the subassembly shall be compared and an average multilinear representation of the subassembly behavior shall be derived based on these curves. Each segment of the composite curve shall be assigned the average stiffness (either positive or negative) of the similar segments in the multilinear curves for the various experiments. Each segment on the composite curve shall terminate at the average of the deformation levels at which the similar segments of the multilinear curves for the various experiments terminate.

3. The stiffness of the subassembly for use in linear procedures shall be taken as the slope of the first segment of the composite curve. The composite multilinear force–deformation curve shall be used for modeling in nonlinear procedures.

4. For the purpose of determining acceptance criteria, subassembly actions shall be classified as being either force-controlled or deformation-controlled. Subassembly actions shall be classified as force-controlled unless any of the following apply:

   4.1. The full backbone curve, including strength degradation and residual strength, is modeled; the composite multilinear force–deformation curve for the subassembly, determined in accordance with requirements in paragraph 2 above, conforms to either Type 1 or Type 2, as indicated in Fig. 2-3; and the deformation parameter $d$ is at least twice the deformation parameter $g$.

   4.2. Bilinear modeling is performed in accordance with the simplified NSP procedure of Section 3.3.3.2.2; the composite multilinear force–deformation curve for the subassembly, determined in accordance with requirements in paragraph 2 above, conforms to either Type 1 or Type 2, as indicated in Fig. 2-3; and the deformation parameter $e$ is at least twice the deformation parameter $g$.

   4.3. Secondary components in which the composite multilinear force–deformation curve for the subassembly, determined in accordance with requirements in paragraph 2 above, conforms to Type 1, as indicated in Fig. 2-3.

5. The strength capacity, $Q_{CL}$, for force-controlled actions evaluated using either the linear or nonlinear procedures shall be taken as the mean minus one standard deviation strength $Q_y$ determined from the series of representative subassembly tests.

6. The acceptance criteria for deformation-controlled actions used in nonlinear procedures shall be the deformations corresponding with the following points on the curves of Fig. 2-3:

   6.1. Immediate Occupancy

      6.1.1. The deformation at which permanent, visible damage occurred in the experiments but not greater than 0.67 times the deformation limit for Life Safety specified in 6.2.1.

6.2. Primary Components
    6.2.1. Life Safety: 0.75 times the deformation at point 2 on the curves; and
    6.2.2. Collapse Prevention: The deformation at point 2 on the curves but not greater than 0.75 times the deformation at point 3.
6.3. Secondary Components
    6.3.1. Life Safety: 0.75 times the deformation at point 3; and
    6.3.2. Collapse Prevention: 1.0 times the deformation at point 3 on the curve.
7. The *m*-factors used as acceptance criteria for deformation-controlled actions in linear procedures shall be determined as follows: (a) obtain the deformation acceptance criteria given in paragraph 6 above; (b) then obtain the ratio of this deformation to the deformation at yield, represented by the deformation parameter *g* in the curves shown in Fig. 2-3; (c) then multiply this ratio by a factor 0.75 to obtain the acceptable *m*-factor.

## C2.8 ALTERNATIVE MATERIALS AND METHODS OF CONSTRUCTION

This section provides guidance for developing appropriate data to evaluate construction materials and detailing systems not specifically covered by this standard. This standard specifies stiffnesses, *m*-factors, strengths, and deformation capacities for a wide range of components. To the extent practical, this standard has been formatted to provide broad coverage of the various common construction types present in the national inventory of buildings. However, it is fully anticipated that in the course of evaluating and rehabilitating existing buildings, construction systems and component detailing practices that are not specifically covered by this standard will be encountered. Furthermore, it is anticipated that new methods and materials, not currently in use, will be developed that may have direct application to building rehabilitation. This section provides a method for obtaining the needed design parameters and acceptance criteria for elements, components, and construction details not specifically included in this standard.

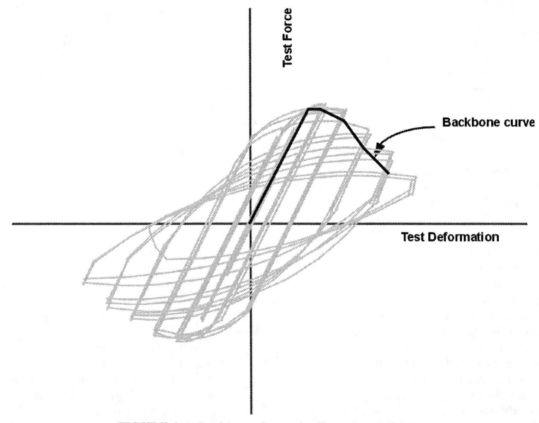

**FIGURE 2-4. Backbone Curve for Experimental Data.**

The approach taken in this section is similar to that used to derive the basic design parameters and acceptance criteria contained in this standard for various components, except that no original experimentation was performed. The required story–force deformation curves were derived by this standard's developers, either directly from research testing available in the literature or based on the judgment of engineers knowledgeable about the behavior of the particular materials and systems.

### C2.8.1 Experimental Setup

This standard requires performing a minimum of three separate tests of each unique subassembly. This is because there can be considerable variation in the results of testing performed on "identical" specimens, just as there is inherent variability in the behavior of actual components in buildings. The use of multiple test data allows some of the uncertainty with regard to actual behavior to be defined.

A specific testing protocol has not been recommended, as selection of a suitable protocol is dependent on the anticipated failure mode of the subassembly as well as the character of excitation it is expected to experience in the real structure. In one widely used protocol, the Applied Technology Council's *Guidelines for Seismic Testing of Components of Steel Structures* (ATC 1992), the specimen is subjected to a series of quasi-static, fully reversed cyclic displacements that are incremented from displacement levels corresponding to elastic behavior, to those at which failure of the specimen occurs. Other protocols that entail fewer or more cycles of displacement, and more rapid loading rates, have also been employed. In selecting an appropriate test protocol, it is important that sufficient increments of loading be selected to characterize adequately the force-deformation behavior of the subassembly throughout its expected range of performance. In addition, the total energy dissipated by the test specimen should be similar to that which the subassembly is anticipated to experience in the real structure. Tests should always proceed to a failure state, so that the margin against failure of the subassembly in service can be judged.

If the structure is likely to be subjected to strong impulsive ground motions, such as those that are commonly experienced within a few kilometers of the fault rupture, consideration should be given to using a protocol that includes one or more very large displacements at the initiation of the loading, to simulate the large initial response induced by impulsive motion. Alternatively, a single monotonic loading to failure may be useful as a performance measure for subassemblies representing components in structures subject to impulsive motion.

### C2.8.2 Data Reduction and Reporting

It is important that data from experimental programs be reported in a uniform manner so that the performance of different subassemblies may be compared. The data reporting requirements specified are the minimum thought to be adequate to allow development of the required design parameters and acceptance criteria for the various Systematic Rehabilitation Procedures. Some engineers and researchers may desire additional data from the experimentation program to allow calibration of their analytical models and to permit improved understanding of the probable behavior of the subassemblies in the real structure.

### C2.8.3 Design Parameters and Acceptance Criteria

A multistep procedure for developing design parameters and acceptance criteria for use with both the linear and nonlinear procedures is provided. The basic approach consists of the development of an approximate story lateral-force-deformation curve for the subassembly, based on the experimental data.

In developing the representative story lateral-force-deformation curve from the experimentation, use of the "backbone" curve is required. This takes into account, in an approximate manner, the strength and stiffness deterioration commonly experienced by structural components. The backbone curve is defined by points given by the intersection of an unloading branch and the loading curve of the next load cycle that goes to a higher level of displacement, as illustrated in Fig. 2-4.

## 3.0 ANALYSIS PROCEDURES

### 3.1 SCOPE

This chapter sets forth requirements for analysis of buildings using the Systematic Rehabilitation Method. Section 3.2 specifies general analysis requirements for the mathematical modeling of buildings including basic assumptions, consideration of torsion, diaphragm flexibility, P-$\Delta$ effects, soil–structure interaction (SSI), multidirectional effects, and overturning. Section 3.3 defines four analysis procedures included in this standard. Section 3.4 defines component acceptance criteria.

Analysis of buildings with seismic isolation or energy dissipation systems shall comply with the requirements of Chapter 9. Analysis of buildings using the Simplified Rehabilitation Method shall comply with the requirements of Chapter 10.

## C3.1 SCOPE

The relationship of the analysis procedures described in this chapter with specifications in other chapters of this standard is as follows.

- Information on Rehabilitation Objectives, including Earthquake Hazard Levels and target Building Performance Levels, is provided in Chapter 1.
- The provisions set forth in this chapter are intended for Systematic Rehabilitation. Provisions for Simplified Rehabilitation are presented in Chapter 10.
- Guidelines for selecting an appropriate analysis procedure are provided in Chapter 2. Chapter 3 describes the loading requirements, mathematical model, and detailed analytical procedures required to estimate seismic force and deformation demands on components of a building. Information on the calculation of appropriate stiffness and strength characteristics for components is provided in Chapters 4 through 9.
- General design requirements are specified in Section 2.6 for multidirectional excitation effects, P-Δ effects, horizontal torsion, overturning, continuity of the framing system, diaphragms, walls, nonstructural components, building separation, structures sharing common components, and vertical seismic effects.
- Component strength and deformation demands obtained from analysis using procedures described in this chapter, based on component acceptance criteria outlined in this chapter, are compared with permissible values provided in Chapters 4 through 9 for the desired performance level.
- Design methods for walls subjected to out-of-plane seismic forces are addressed in Chapter 2. Analysis and design methods for nonstructural components (including mechanical and electrical equipment) are presented in Chapter 11.

## 3.2 GENERAL ANALYSIS REQUIREMENTS

An analysis of the building, as specified in Section 2.4, shall be conducted in accordance with the requirements of this section and Section 2.6.

### 3.2.1 Analysis Procedure Selection

An analysis of the building shall be performed using the Linear Static Procedure (LSP), the Linear Dynamic Procedure (LDP), the Nonlinear Static Procedure (NSP), or the Nonlinear Dynamic Procedure (NDP), selected based on the limitations specified in Section 2.4. Use of alternative rational analysis procedures as described in Section 2.4.3 shall also be permitted.

### C3.2.1 Analysis Procedure Selection

Four procedures are presented for seismic analysis of buildings: two linear procedures and two nonlinear procedures. The two linear procedures are termed the Linear Static Procedure (LSP) and the Linear Dynamic Procedure (LDP). The two nonlinear procedures are termed the Nonlinear Static Procedure (NSP) and the Nonlinear Dynamic Procedure (NDP).

Either the linear procedures of Sections 3.3.1 and 3.3.2 or the nonlinear procedures of Sections 3.3.3 and 3.3.4 may be used to analyze a building, subject to the limitations set forth in Section 2.4.

Linear procedures are appropriate where the expected level of nonlinearity is low. This is measured by component demand capacity ratios (DCRs) of less than 2.0.

Static procedures are appropriate where higher mode effects are not significant. This is generally true for short, regular buildings. Dynamic procedures are required for tall buildings and for buildings with torsional irregularities or nonorthogonal systems.

The NSP is acceptable for most buildings, but should be used in conjunction with the LDP if mass participation in the first mode is low.

The term "linear" in linear analysis procedures implies "linearly elastic." The analysis procedure, however, may include geometric nonlinearity of gravity loads acting through lateral displacements and implicit material nonlinearity of concrete and masonry components using properties of cracked sections. The term "nonlinear" in nonlinear analysis procedures implies explicit material nonlinearity or inelastic material response, but geometric nonlinearity may also be included.

### 3.2.2 Mathematical Modeling

#### 3.2.2.1 Basic Assumptions

A building shall be modeled, analyzed, and evaluated as a three-dimensional assembly of components. Alternatively, use of a two-dimensional model shall be permitted if the building meets one of the following conditions:

1. The building has rigid diaphragms as defined in Section 3.2.4 and horizontal torsion effects do not exceed the limits specified in Section 3.2.2.2, or

horizontal torsion effects are accounted for as specified in Section 3.2.2.2; or
2. The building has flexible diaphragms as defined in Section 3.2.4.

If two-dimensional models are used, the three-dimensional nature of components and elements shall be considered when calculating stiffness and strength properties.

If the building contains out-of-plane offsets in vertical lateral-force-resisting elements, the model shall explicitly account for such offsets in the determination of diaphragm demands.

Modeling stiffness of structural components shall be based on the stiffness requirements of Chapters 4 through 8.

For nonlinear procedures, a connection shall be explicitly modeled if the connection is weaker than or has less ductility than the connected components or if the flexibility of the connection results in a change in the connection forces or deformations of more than 10%.

### C3.2.2.1 Basic Assumptions

For two-dimensional models, the three-dimensional nature of components and elements should be recognized in calculating their stiffness and strength properties. For example, shear walls and other bracing systems may have "L" or "T" or other three-dimensional cross sections where contributions of both the flanges and webs should be accounted for in calculating stiffness and strength properties.

In this standard, component stiffness is generally taken as the effective stiffness based on the secant stiffness to yield level forces. Specific direction on calculating effective stiffness is provided in each material chapter for each type of structural system.

Examples of where connection flexibility may be important to model include the panel zone of steel moment-resisting frames and the "joint" region of perforated masonry or concrete walls.

### 3.2.2.2 Horizontal Torsion

The effects of horizontal torsion shall be considered in accordance with this section. Torsion need not be considered in buildings with flexible diaphragms as defined in Section 3.2.4.

#### 3.2.2.2.1 Total Torsional Moment
The total horizontal torsional moment at a story shall be equal to the sum of the actual torsional moment and the accidental torsional moment calculated as follows:

1. The actual torsional moment at a story shall be calculated by multiplying the seismic story shear force

by the eccentricity between the center of mass and the center of rigidity measured perpendicular to the direction of the applied load. The center of mass shall be based on all floors above the story under consideration. The center of rigidity of a story shall include all vertical seismic elements in the story; and
2. The accidental torsional moment at a story shall be calculated as the seismic story shear force multiplied by a distance equal to 5% of the horizontal dimension at the given floor level measured perpendicular to the direction of the applied load.

#### 3.2.2.2.2 Consideration of Torsional Effects
Effects of horizontal torsion shall be considered in accordance with the following requirements:

1. Increased forces and displacements due to actual torsion shall be calculated for all buildings;
2. The displacement multiplier, $\eta$, at each floor shall be calculated as the ratio of the maximum displacement at any point on the floor diaphragm to the average displacement ($\delta_{max}/\delta_{avg}$). Displacements shall be calculated for the applied loads;
3. Increased forces and displacements due to accidental torsion shall be considered unless the accidental torsional moment is less than 25% of the actual torsional moment, or the displacement multiplier $\eta$ due to the applied load and accidental torsion is less than 1.1 at every floor;
4. For linear analysis procedures, forces and displacements due to accidental torsion shall be amplified by a factor, $A_x$, as defined by Eq. 3-1, where the displacement multiplier $\eta$ due to total torsional moment exceeds 1.2 at any level;

$$A_x = \left(\frac{\eta}{1.2}\right)^2 \le 3.0 \qquad \text{(Eq. 3-1)}$$

5. If the displacement modifier $\eta$ due to total torsional moment at any floor exceeds 1.5, two-dimensional models shall not be permitted and three-dimensional models that account for the spatial distribution of mass and stiffness shall be used;
6. Where two-dimensional models are used, the effects of horizontal torsion shall be calculated as follows:
   6.1. For the LSP and the LDP, forces and displacements shall be amplified by the maximum value of $\eta$ calculated for the building;
   6.2. For the NSP, the target displacement shall be amplified by the maximum value of $\eta$ calculated for the building;
   6.3. For the NDP, the amplitude of the ground acceleration record shall be amplified by the

maximum value of $\eta$ calculated for the building; and

7. The effects of accidental torsion shall not be used to reduce force and deformation demands on components.

### C3.2.2.2 Horizontal Torsion

Actual torsion is due to the eccentricity between the centers of mass and stiffness. Accidental torsion is intended to cover the effects of the rotational component of the ground motion, differences between computed and actual stiffness, and unfavorable distributions of dead and live load masses.

The 10% threshold on additional displacement due to accidental torsion is based on judgment. The intent is to reward those building frames that are torsionally redundant and possess high torsional stiffness. Such structures are likely to be much less susceptible to torsional response than those framing systems possessing low redundancy and low torsional stiffness.

### 3.2.2.3 Primary and Secondary Components

Components shall be classified as primary or secondary as defined in Section 2.4.4.2. Primary components shall be evaluated for earthquake-induced forces and deformations in combination with gravity load effects. Secondary components shall be evaluated for earthquake-induced deformations in combination with gravity load effects.

Mathematical models for use with linear analysis procedures shall include the stiffness and resistance of only the primary components. If the total lateral stiffness of secondary components in a building exceeds 25% of the total initial stiffness of primary components, some secondary components shall be reclassified as primary to reduce the total stiffness of secondary components to less than 25% of primary. If the inclusion of a secondary component will increase the force or deformation demands on a primary component, the secondary component shall be reclassified as primary and included in the model.

Mathematical models for use with nonlinear procedures shall include the stiffness and resistance of primary and secondary components. The strength and stiffness degradation of primary and secondary components shall be modeled explicitly. For the simplified NSP of Section 3.3.3.2.2, only primary components shall be included in the model and degradation shall not be modeled.

Nonstructural components shall be classified as structural components and shall be included in mathematical models if their lateral stiffness exceeds 10% of the total initial lateral stiffness of a story.

Components shall not be selectively designated primary or secondary to change the configuration of a building from irregular to regular.

### C3.2.2.3 Primary and Secondary Components

Due to limitations inherent in each analysis method, the manner in which primary and secondary components are handled differs for linear and nonlinear procedures. Since strength and stiffness degradation of secondary components is likely, their resistance is unreliable. Linear procedures cannot account for this degradation, so only primary components are included in linear analysis models. This is conservative in linear analyses because it will result in the highest demands placed on the primary components that remain. Secondary components, however, must still be checked against the acceptance criteria given in Chapters 5 through 8.

In nonlinear procedures, strength and stiffness degradation can be modeled. Since degradation of the overall system can increase displacement demands, inclusion of both primary and secondary components is conservative in nonlinear analyses.

For linear procedures, this standard limits the amount of lateral resistance that can be provided by secondary components. The main reason for this limitation is to minimize the potential for sudden loss of lateral-force-resisting components to produce irregular structural response that is difficult to detect. The contribution of secondary components can be checked by temporarily including them in the analysis model and examining the change in response.

### 3.2.2.4 Stiffness and Strength Assumptions

Stiffness and strength properties of components shall be determined in accordance with the requirements of Chapters 4 through 9, and 11.

### 3.2.2.5 Foundation Modeling

The foundation system shall be modeled considering the degree of fixity provided at the base of the structure. Rigid or flexible base assumptions shall be permitted in accordance with the requirements for soil–structure interaction in Section 3.2.6 and foundation acceptability in Section 4.4.3. Foundation modeling shall consider movement due to geologic site hazards specified in Section 4.2.2, and load-deformation characteristics specified in Section 4.4.2.

### C3.2.2.5 Foundation Modeling

Methods for modeling foundations and estimation of ground movements due to seismic geologic site hazards are referenced in Chapter 4, and may require the expertise of a geotechnical engineer or a geologist.

The decision to model foundation flexibility must consider impacts on the behavior of structural components in the building. Rigid base models for concrete shear walls on independent spread footings may maximize deformation demands on the walls themselves, but could underestimate the demands on other secondary components in the building, such as beams and columns in moment frames, which may be sensitive to additional building movement.

## 3.2.3 Configuration

Building irregularities defined in Section 2.4.1.1 shall be based on the plan and vertical configuration of the rehabilitated structure. Irregularity shall be determined, both with and without the contribution of secondary components.

## C3.2.3 Configuration

One objective of seismic rehabilitation should be the improvement of the regularity of a building through the judicious placement of new framing elements.

Adding seismic framing elements at certain locations will improve the regularity of the building and should be considered as a means to improve seismic performance of the building.

Secondary components can lose significant strength and stiffness after initial earthquake shaking and may no longer be effective. Therefore, regularity of the building should be determined both with and without the contribution of secondary components.

## 3.2.4 Diaphragms

### 3.2.4.1 General

Diaphragms shall be classified as flexible, stiff, or rigid in accordance with Section 3.2.4.2.

### 3.2.4.2 Classification of Diaphragms

Diaphragms shall be classified as flexible where the maximum horizontal deformation of the diaphragm along its length is more than twice the average story drift of the vertical lateral-force-resisting elements of the story immediately below the diaphragm.

Diaphragms shall be classified as rigid where the maximum lateral deformation of the diaphragm is less than half the average story drift of the vertical lateral-force-resisting elements of the associated story.

Diaphragms that are neither flexible nor rigid shall be classified as stiff.

For the purpose of classifying diaphragms, story drift and diaphragm deformations shall be calculated using the pseudo-lateral force specified in Eq. 3-10. The in-plane deflection of the diaphragm shall be calculated for an in-plane distribution of lateral force

consistent with the distribution of mass, and all in-plane lateral forces associated with offsets in the vertical seismic framing at that diaphragm level.

### 3.2.4.3 Mathematical Modeling

Mathematical modeling of buildings with rigid diaphragms shall account for the effects of horizontal torsion as specified in Section 3.2.2.2. Mathematical models of buildings with stiff or flexible diaphragms shall account for the effects of diaphragm flexibility by modeling the diaphragm as an element with in-plane stiffness consistent with the structural characteristics of the diaphragm system. Alternatively, for buildings with flexible diaphragms at each floor level, each lateral-force-resisting element in a vertical plane shall be permitted to be designed independently, with seismic masses assigned on the basis of tributary area.

## C3.2.4 Diaphragms

Evaluation of diaphragm demands should be based on the likely distribution of horizontal inertial forces. For flexible diaphragms, such a distribution may be given by Eq. C3-1 and is illustrated in Fig. C3-1.

$$f_d = \frac{1.5F_d}{L_d}\left[1 - \left(\frac{2x}{L_d}\right)^2\right] \qquad \text{(Eq. C3-1)}$$

where

$f_d$ = inertial load per foot;
$F_d$ = total inertial load on a flexible diaphragm;
$x$ = distance from the center line of flexible diaphragm; and
$L_d$ = distance between lateral support points for diaphragm.

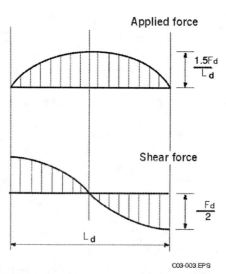

C03-003.EPS

**FIGURE C3-1. Plausible Force Distribution in a Flexible Diaphragm.**

### 3.2.5 P-Δ Effects

P-Δ effects shall be included in linear and nonlinear analysis procedures.

For nonlinear procedures, static P-Δ effects shall be incorporated in the analysis by including in the mathematical model the nonlinear force–deformation relationship of all components subjected to axial forces.

### C3.2.5 P-Δ Effects

Static P-Δ effects are caused by gravity loads acting through the deformed configuration of a building and result in an increase in lateral displacements.

Dynamic P-Δ effects are caused by a negative post-yield stiffness that increases story drift and the target displacement. The degree by which dynamic P-Δ effects increase displacements depends on the following:

1. The ratio of the negative post-yield stiffness to the effective elastic stiffness;
2. The fundamental period of the building;
3. The strength ratio, $R$;
4. The hysteretic load–deformation relations for each story;
5. The frequency characteristics of the ground motion; and
6. The duration of the strong ground motion.

Because of the number of parameters involved, it is difficult to capture dynamic P-Δ effects in linear and nonlinear static analysis procedures. For the NSP, dynamic instability is measured by the strength ratio, $R$. For the NDP, dynamic P-Δ effects are captured explicitly in the analysis.

### 3.2.6 Soil–Structure Interaction

The effects of soil–structure interaction (SSI) shall be evaluated for those buildings in which an increase in fundamental period due to SSI effects will result in an increase in spectral accelerations. For other buildings, the effects of SSI need not be evaluated.

SSI effects shall be calculated using the explicit modeling procedure, or other approved rational procedure. Where the LSP is used, the simplified procedure shall be permitted.

### C3.2.6 Soil–Structure Interaction

Interaction between the structure and the supporting soil consists of the following:

- Foundation flexibility—introduction of flexibility at the foundation–soil interface;

- Kinematic effects—filtering of the ground motions transmitted to the structure based on the geometry and properties of the foundation;
- Foundation damping effects—dissipation of energy through radiation and hysteretic soil damping.

Foundation flexibility is covered in Section 4.4. Consideration of soil–structure interaction (SSI) effects caused by kinematic interaction or foundation damping, which serve to reduce the shaking input to the structure relative to the free-field motion, is covered in Section 4.5.

SSI may modify the seismic demands on a building. It can reduce spectral accelerations and lateral forces, but can increase lateral displacements and secondary forces due to P-Δ effects. Reductions in seismic demand due to explicit modeling of foundation flexibility, foundation damping, or kinematic effects can be significant, and should be used where applicable. Where SSI effects are not required to be evaluated, use of all three effects alone or in combination is permitted.

For those rare cases (such as near-field and soft soil sites) in which the increase in period due to SSI increases spectral accelerations, the effects of SSI on building response must be evaluated. Further discussion of SSI effects can be found in FEMA 440 (FEMA 2005).

#### 3.2.6.1 Simplified Procedure

Calculation of SSI effects using the simplified procedure shall comply with the procedure in ASCE 7 (ASCE 2005) utilizing the effective fundamental period and effective fundamental damping ratio of the foundation–structure system. Combination of these effects with kinematic interaction effects calculated in accordance with Section 4.5.1 shall be permitted.

#### 3.2.6.2 Explicit Modeling Procedure

Calculation of SSI effects using the explicit modeling procedure shall be based on a mathematical model that includes the flexibility and damping of individual foundation elements. Foundation stiffness parameters shall comply with the requirements of Section 4.4.2. Damping ratios for individual foundation elements shall not exceed the value used for the elastic superstructure. In lieu of explicitly modeling damping, use of the effective damping ratio of the structure–foundation system, $\beta_0$, calculated in accordance with Section 4.5.2, shall be permitted.

For the NSP, the effective damping ratio of the foundation–structure system, $\beta_0$, calculated in accordance with Section 4.5.2, shall be used to modify spectral demands.

Combination of damping effects with kinematic interaction effects calculated in accordance with Section 4.5.1 shall be permitted.

## 3.2.7 Multidirectional Seismic Effects

Buildings shall be designed for seismic motion in any horizontal direction. Multidirectional seismic effects shall be considered to act concurrently as specified in Section 3.2.7.1 for buildings meeting the following criteria:

1. The building has plan irregularities as defined in Section 2.4.1.1; or
2. The building has one or more primary columns which form a part of two or more intersecting frame or braced frame elements.

All other buildings shall be permitted to be designed for seismic motions acting nonconcurrently in the direction of each principal axis of the building.

### 3.2.7.1 Concurrent Seismic Effects

Where concurrent multidirectional seismic effects must be considered, horizontally oriented, orthogonal x- and y-axes shall be established. Components of the building shall be designed for combinations of forces and deformations from separate analyses performed for ground motions in X and Y directions as follows:

1. Where the LSP or LDP is used as the basis for design, elements and components shall be designed for (a) forces and deformations associated with 100% of the design forces in the X direction plus the forces and deformations associated with 30% of the design forces in the Y direction; and for (b) forces and deformations associated with 100% of the design forces in the Y direction plus the forces and deformations associated with 30% of the design forces in the X direction. Other combination rules shall be permitted where verified by experiment or analysis; and
2. Where the NSP or NDP is used as the basis for design, elements and components of the building shall be designed for (a) forces and deformations associated with 100% of the design displacement in the X direction only, plus the forces (not deformations) associated with 30% of the design displacements in the Y direction only; and for (b) forces and deformations associated with 100% of the design displacements in the Y direction only, plus the forces (not deformations) associated with 30% of the design displacements in the X direction only. Design displacements shall be determined in accordance with Section 3.3.3 for NSP and Section 3.3.4

for NDP. Other combination rules shall be permitted where verified by experiment or analysis.

### 3.2.7.2 Vertical Seismic Effects

For components in which Section 2.6.11 requires consideration of vertical seismic effects, the vertical response of a structure to earthquake ground motion need not be combined with the effects of the horizontal response.

## 3.2.8 Component Gravity Loads for Load Combinations

The following actions due to gravity loads, $Q_G$, shall be considered for combination with actions due to seismic loads.

Where the effects or actions of gravity and seismic loads are additive, the action due to design gravity loads, $Q_G$, shall be obtained in accordance with Eq. 3-2:

$$Q_G = 1.1 \ (Q_D + Q_L + Q_S) \qquad \text{(Eq. 3-2)}$$

where

$Q_D$ = action due to design dead loads;
$Q_L$ = action due to design live load, equal to 25% of the unreduced design live load, but not less than the actual live load; and
$Q_S$ = action due to effective snow load contribution.

Where the effects or actions of gravity and seismic loads are counteracting, the action due to design gravity loads, $Q_G$, shall be obtained in accordance with Eq. 3-3:

$$Q_G = 0.9 Q_D \qquad \text{(Eq. 3-3)}$$

where

$Q_D$ = action due to design dead loads.

Where the design flat roof snow load calculated in accordance with ASCE 7 (ASCE 2005) exceeds 30 psf, the effective snow load shall be taken as 20% of the design snow load. Where the design flat roof snow load is less than 30 psf, the effective snow load shall be permitted to be zero.

## C3.2.8 Component Gravity Loads for Load Combinations

Evaluation of components for gravity and wind forces, in the absence of earthquake forces, is beyond the scope of this document.

## 3.2.9 Verification of Design Assumptions

Each component shall be evaluated to verify that locations of inelastic deformations assumed in the analysis are consistent with strength and equilibrium requirements along the component length. Each component shall also be evaluated for postearthquake residual gravity load capacity by a rational analysis procedure approved by the authority having jurisdiction that accounts for potential redistribution of gravity loads and reduction of strength or stiffness caused by earthquake damage to the structure.

### C3.2.9 Verification of Design Assumptions

It is important that assumptions about locations of potential inelastic activity in the structure are verified. In linear procedures, the potential for inelastic flexural action is restricted to the beam ends because flexural yielding along the span length can lead to unconservative results. In nonlinear procedures, potential inelastic activity should occur only where specifically modeled. Where demands due to gravity load combinations of Section 3.2.8 exceed 50% of the capacity of the component at any location along its length, the potential for inelastic activity exists and should be investigated. Sample procedures for verifying design assumptions are contained in Section C3.2.9 of FEMA 274 (FEMA 1997).

### 3.2.10 Overturning

Structures shall be designed to resist overturning effects caused by seismic forces. Each vertical-force-resisting element receiving earthquake forces due to overturning shall be investigated for the cumulative effects of seismic forces applied at and above the level under consideration. The effects of overturning shall be evaluated at each level of the structure as specified in Section 3.2.10.1 for linear procedures, or Section 3.2.10.2 for nonlinear procedures. The effects of overturning on foundations and geotechnical components shall be considered in the evaluation of foundation strength and stiffness as specified in Chapter 4.

### C3.2.10 Overturning

Response to earthquake ground motion results in a tendency for structures and individual vertical elements of structures to overturn about their bases. Although actual overturning failure is very rare, overturning effects can result in significant stresses, as demonstrated in some local and global failures. In new building design, earthquake effects, including overturning, are evaluated for lateral forces that are significantly reduced (by an $R$-factor) from those that may actually develop in the structure.

For elements with positive attachment between levels that behave as single units, such as reinforced concrete walls, the overturning effects are resolved into component forces (e.g., flexure and shear at the base of the wall). The element is then proportioned with adequate strength using $m$-factors, where appropriate, to resist overturning effects resulting from these force levels.

Some elements, such as wood shear walls and foundations, may not be designed with positive attachment between levels. An overturning stability check is typically performed for such elements when designed using codes for new buildings. If the element has sufficient dead load to remain stable under the overturning effects of the design lateral forces and has sufficient shear connection to the level below, then the design is deemed adequate. However, if dead load is inadequate to provide stability, then tie-downs, piles, or other types of uplift anchors are provided to resist the residual overturning caused by the design forces.

In the linear and nonlinear procedures of this standard, lateral forces are not reduced by an $R$-factor, as they are for new buildings, so computed overturning effects are larger than typically calculated for new buildings. Although the procedure used for new buildings is not completely rational, it has resulted in successful performance. Therefore, it may not be appropriate to require that structures and elements of structures remain stable for the pseudo-lateral forces used in the linear procedures in this standard. Instead, the designer must determine if positive direct attachment will be used to resist overturning effects or if dead loads will be used. If positive direct attachment will be used, then the overturning effect at this attachment is treated just as any other component action.

However, if dead loads alone are used to resist overturning, then overturning is treated as a force-controlled behavior. The real overturning demands can be estimated by considering the overall limiting strength of the component.

There is no simple rational method available, shown to be consistent with observed behavior, to design or evaluate elements for overturning effects. The method described in this standard is rational but inconsistent with procedures used for new buildings. To improve damage control, the full lateral forces used in the linear procedures of this standard are required for checking acceptability for performance levels higher than life safety.

Additional studies are needed on the parameters that control overturning in seismic rehabilitation. Information regarding consideration of rocking behavior can be found in Commentary Section C4.4.2.

### 3.2.10.1 Linear Procedures

Where linear procedures are used, overturning effects shall be resisted through the stabilizing effect of dead loads acting alone or in combination with positive connection of structural components to components below the level under consideration.

Where dead loads alone are used to resist the effects of overturning, Eq. 3-4 shall be satisfied:

$$M_{ST} > M_{OT}/(C_1 C_2 J) \qquad \text{(Eq. 3-4)}$$

where

$M_{OT}$ = total overturning moment induced on the element by seismic forces applied at and above the level under consideration. Overturning moment shall be determined based on design seismic forces calculated in accordance with Section 3.3.1 for LSP and 3.3.2 for LDP;

$M_{ST}$ = stabilizing moment produced by dead loads acting on the element;

$C_1$ and $C_2$ = coefficients defined in Section 3.3.1.3; and

$J$ = coefficient defined in Section 3.4.2.1.2.

The quantity $M_{OT}/(C_1 C_2 J)$ need not exceed the overturning moment on the element, as limited by the expected strength of the structure. The element shall be evaluated for the effects of increased compression at the end about which it is being overturned. For this purpose, compression at the end of the element shall be considered a force-controlled action.

Alternatively, the load combination represented by Eq. 3-5 shall be permitted for evaluating the adequacy of dead loads alone to resist the effects of overturning.

$$0.9 M_{ST} > M_{OT}/(C_1 C_2 R_{OT}) \qquad \text{(Eq. 3-5)}$$

where

$R_{OT}$ = 10.0 for Collapse Prevention;
= 8.0 for Life Safety; and
= 4.0 for Immediate Occupancy.

Where Eq. 3-4 or 3-5 for dead load stability against the effects of overturning is not satisfied, positive attachment between elements of the structure at and immediately above and below the level under consideration shall be provided. If the level under consideration is the base of the structure, positive attachment shall be provided between the structure and the supporting soil, unless nonlinear procedures are used to rationalize overturning stability. Positive attachments shall be capable of resisting earthquake forces in combination with gravity loads as force- or deformation-controlled actions in accordance with Eq. 3-16 or 3-17 and applicable acceptance criteria of Eq. 3-18 or 3-19.

### C3.2.10.1 Linear Procedures

For evaluating the adequacy of dead loads to provide stability against overturning, the alternative procedure of Section 3.2.10.1 is intended to provide a method that is consistent with prevailing practice specified in current codes for new buildings.

### 3.2.10.2 Nonlinear Procedures

Where nonlinear procedures are used, the effects of earthquake-induced uplift on the tension side of an element, or rocking, shall be included in the analytical model as a nonlinear degree of freedom. The adequacy of elements above and below the level at which uplift or rocking occurs, including the foundations, shall be evaluated for any redistribution of forces or deformations that occurs as a result of this rocking.

## 3.3 ANALYSIS PROCEDURES

Selection of an appropriate analysis procedure shall comply with Section 3.2.1.

### 3.3.1 Linear Static Procedure

#### 3.3.1.1 Basis of the Procedure

If the LSP is selected for seismic analysis of the building, the design seismic forces, their distribution over the height of the building, and the corresponding internal forces and system displacements shall be determined using a linearly elastic, static analysis in accordance with this section.

Buildings shall be modeled with linearly elastic stiffness and equivalent viscous damping values consistent with components responding at or near yield level, as defined in Section 2.4.4. The pseudo-lateral force defined in Section 3.3.1.3 shall be used to calculate internal forces and system displacements due to the design earthquake.

Results of the LSP shall be checked using the acceptance criteria of Section 3.4.2.

#### C3.3.1.1 Basis of the Procedure

The magnitude of the pseudo-lateral force has been selected with the intention that, when applied to the linearly elastic model of the building, it will result in design displacement amplitudes approximating maximum displacements expected during the design earthquake. The procedure is keyed to the displacement response of the building because displacements are a better indicator of damage in the nonlinear range of building response than are forces. In this range, relatively small changes in force demand correspond to large changes in displacement demand. If the building responds essentially elastically to the design earthquake, the calculated internal forces will be reasonable approximations of those expected during the design earthquake. If the building responds inelastically to the design earthquake, as commonly will be the case, the actual internal forces that would develop in the building will be less than the internal forces calculated using a pseudo-lateral force.

Calculated internal forces typically will exceed those that the building can develop because of anticipated inelastic response of components. These design forces are evaluated through the acceptance criteria of Section 3.4.2, which include modification factors and alternative analysis procedures to account for anticipated inelastic response demands and capacities.

### 3.3.1.2 Period Determination

The fundamental period of a building shall be calculated for the direction under consideration using one of the following analytical, empirical, or approximate methods specified in this section.

*3.3.1.2.1 Method 1—Analytical* Eigenvalue (dynamic) analysis of the mathematical model of the building shall be performed to determine the fundamental period of the building.

*3.3.1.2.2 Method 2—Empirical* The fundamental period of the building shall be determined in accordance with Eq. 3-6:

$$T = C_t h_n^\beta \qquad \text{(Eq. 3-6)}$$

where

$T$ = fundamental period (in sec) in the direction under consideration;
$C_t$ = 0.035 for steel moment-resisting frame systems;
= 0.018 for concrete moment-resisting frame systems;
= 0.030 for steel eccentrically-braced frame systems;

= 0.020 for wood buildings (Types 1 and 2 in Table 10-2, Chapter 10);
= 0.020 for all other framing systems;
$h_n$ = height (in ft) above the base to the roof level; and
$\beta$ = 0.80 for steel moment-resisting frame systems;
= 0.90 for concrete moment-resisting frame systems;
= 0.75 for all other framing systems.

*3.3.1.2.3 Method 3—Approximate*

1. For any building, use of Rayleigh's method to approximate the fundamental period shall be permitted.
2. For one-story buildings with single-span flexible diaphragms, use of Eq. 3-7 to approximate the fundamental period shall be permitted.

$$T = (0.1\Delta_w + 0.078\Delta_d)^{0.5} \qquad \text{(Eq. 3-7)}$$

where $\Delta_w$ and $\Delta_d$ are in-plane wall and diaphragm displacements in inches, due to a lateral load in the direction under consideration, equal to the weight of the diaphragm.

3. For one-story buildings with multiple-span diaphragms, use of Eq. 3-7 shall be permitted as follows: a lateral load equal to the weight tributary to the diaphragm span under consideration shall be applied to calculate a separate period for each diaphragm span. The period that maximizes the pseudo-lateral force shall be used for design of all walls and diaphragm spans in the building.
4. For unreinforced masonry buildings with single-span flexible diaphragms, six stories or less in height, use of Eq. 3-8 to approximate the fundamental period shall be permitted.

$$T = (0.078\Delta_d)^{0.5} \qquad \text{(Eq. 3-8)}$$

where $\Delta_d$ is the maximum in-plane diaphragm displacement in inches, due to a lateral load in the direction under consideration, equal to the weight tributary to the diaphragm.

### C3.3.1.2 Period Determination

*C3.3.1.2.1 Method 1—Analytical* For many buildings, including multistory buildings with well-defined framing systems, the preferred approach to obtaining the period for design is Method 1. By this method, the building is modeled using the modeling procedures of Chapters 4 through 8 and 11, and the period is obtained by Eigenvalue analysis. Flexible diaphragms may be modeled as a series of lumped masses and diaphragm finite elements.

Contrary to procedures in codes for new buildings, there is no maximum limit on period calculated using Method 1. This omission is intended to encourage the use of more advanced analyses. It is felt that sufficient controls on analyses and acceptance criteria are present within this standard to provide appropriately conservative results using calculated periods.

*C3.3.1.2.2 Method 2—Empirical* Empirical equations for period, such as that used in Method 2, intentionally underestimate the actual period and will generally result in conservative estimates of pseudo-lateral force. Studies have shown that depending on actual mass or stiffness distributions in a building, the results of Method 2 may differ significantly from those of Method 1. The $C_t$ values specified for Method 2 are generally consistent with FEMA 302 (FEMA 1997) but have been modified based on recent published research on measured building response to earthquakes.

*C3.3.1.2.3 Method 3—Approximate* Rayleigh's method for approximating the fundamental period of vibration of a building is presented in Eq. C3-2. The equation uses the shape function given by the static deflections of each floor due to the applied lateral forces.

$$T = 2\pi \left[ \frac{\sum_{i=1}^{n} w_i \delta_i^2}{g \sum_{i=1}^{n} F_i \delta_i} \right]^{1/2} \quad \text{(Eq. C3-2)}$$

where

$w_i$ = portion of the effective seismic weight located on or assigned to floor level $i$;
$\delta_i$ = displacement at floor $i$ due to lateral load $F_i$;
$F_i$ = lateral load applied at floor level $i$; and
$n$ = total number of stories in the vertical seismic framing.

Equations 3-7 and 3-8 of Method 3 are appropriate for systems with rigid vertical elements and flexible diaphragms in which the dynamic response of the system is concentrated in the diaphragm. Use of Method 2 on these systems to calculate the period

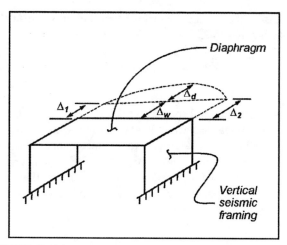

**FIGURE C3-2. Diaphragm and Wall Displacement Terminology.**

based on the stiffness of the vertical elements will substantially underestimate the period of actual dynamic response and overestimate the pseudo-lateral force.

Equation 3-8 is a special case developed specifically for unreinforced masonry (URM) buildings. In this method, wall deformations are assumed negligible compared to diaphragm deflections.

For illustration of wall and diaphragm displacements, see Fig. C3-2. Where calculating diaphragm displacements for the purpose of estimating period using Eq. 3-7 or 3-8, the diaphragm shall be considered to remain elastic under the prescribed lateral loads.

### 3.3.1.3 Determination of Forces and Deformations

Forces and deformations in elements and components shall be calculated for the pseudo-lateral force of Section 3.3.1.3.1, using component stiffnesses calculated in accordance with Chapters 4 through 8. Pseudo-lateral forces shall be distributed throughout the building in accordance with Sections 3.3.1.3.2 through 3.3.1.3.4. Alternatively, for unreinforced masonry buildings in which the fundamental period is calculated using Eq. 3-8, pseudo-lateral forces shall be permitted to be distributed in accordance with Section 3.3.1.3.5. Actions and deformations shall be modified to consider the effects of horizontal torsion in accordance with Section 3.2.2.2.

*3.3.1.3.1 Pseudo-Lateral Force* The pseudo-lateral force in a given horizontal direction of a building shall be determined using Eq. 3-9. This load shall be used to design the vertical elements of the lateral-force-resisting system.

$$V = C_1 C_2 C_m S_a W \qquad \text{(Eq. 3-9)}$$

where

$V$ = pseudo-lateral force;

$C_1$ = modification factor to relate expected maximum inelastic displacements to displacements calculated for linear elastic response. For periods less than 0.2 sec, $C_1$ need not be taken greater than the value at $T = 0.2$ sec. For periods greater than 1.0 sec, $C_1 = 1.0$.

$$C_1 = 1 + \frac{R - 1}{aT^2}$$

where

$a$ = site class factor;
 = 130 site Class A, B;
 = 90 site Class C;
 = 60 site Class D, E, F;

$R$ = strength ratio calculated in accordance with Eq. 3-15 with the elastic base shear capacity substituted for shear yield strength, $V_y$;

$T$ = fundamental period of the building in the direction under consideration, calculated in accordance with Section 3.3.1.2, including modification for SSI effects of Section 3.2.6, if applicable;

$C_2$ = modification factor to represent the effect of pinched hysteresis shape, cyclic stiffness degradation, and strength deterioration on maximum displacement response. For periods greater than 0.7 sec, $C_2 = 1.0$.

$$C_2 = 1 + \frac{1}{800} \left( \frac{R - 1}{T} \right)^2$$

$C_m$ = effective mass factor to account for higher mode mass participation effects obtained

from Table 3-1. $C_m$ shall be taken as 1.0 if the fundamental period, $T$, is greater than 1.0 sec;

$S_a$ = response spectrum acceleration, at the fundamental period and damping ratio of the building in the direction under consideration. The value of $S_a$ shall be obtained from the procedure specified in Section 1.6; and

$W$ = effective seismic weight of the building including the total dead load and applicable portions of other gravity loads listed below:

1. In areas used for storage, a minimum 25% of the floor live load shall be applicable. The live load shall be permitted to be reduced for tributary area as approved by the authority having jurisdiction. Floor live load in public garages and open parking structures is not applicable.
2. Where an allowance for partition load is included in the floor load design, the actual partition weight or a minimum weight of 10 psf of floor area, whichever is greater, shall be applicable.
3. Total operating weight of permanent equipment.
4. Where the design flat roof snow load calculated in accordance with ASCE 7 exceeds 30 psf, the effective snow load shall be taken as 20% of the design snow load. Where the design flat roof snow load is less than 30 psf, the effective snow load shall be permitted to be zero.

*C3.3.1.3.1 Pseudo-Lateral Force* **Coefficient $C_1$.** This modification factor is to account for the difference in maximum elastic and inelastic displacement amplitudes in structures with relatively stable and full hysteretic loops. The values of the coefficient are based on analytical and experimental investigations of the earthquake response of yielding structures. The quantity, $R$, is the ratio of the required elastic strength to

### Table 3-1. Values for Effective Mass Factor $C_m$[1]

| No. of Stories | Concrete Moment Frame | Concrete Shear Wall | Concrete Pier-Spandrel | Steel Moment Frame | Steel Concentric Braced Frame | Steel Eccentric Braced Frame | Other |
|---|---|---|---|---|---|---|---|
| 1–2 | 1.0 | 1.0 | 1.0 | 1.0 | 1.0 | 1.0 | 1.0 |
| 3 or more | 0.9 | 0.8 | 0.8 | 0.9 | 0.9 | 0.9 | 1.0 |

[1]$C_m$ shall be taken as 1.0 if the fundamental period, $T$, is greater than 1.0 sec.

the yielding strength of the structure. For linear analyses, $R$ may be determined using:

$$R = \frac{DCR_{max}}{1.5} C_m \geq 1.0$$

where $DCR_{max}$ is the largest $DCR$ computed for any primary component, taking $C_1 = C_2 = C_m = 1.0$.

The expression above is obtained by substituting Eq. 3-9 into Eq. 3-15 and assuming that the elastic base shear capacity (fully yielded strength, $V_y$) is mobilized at a shear that is 1.5 times the shear at first yield (as indicated by the largest primary component $DCR$). The latter assumption is based on representative values for system overstrength. As is indicated in Fig. C4.2-1 of FEMA 450 (FEMA 2004), the factor relating design force level to fully yielded strength is $\Omega_0$. Sources of overstrength are design $\phi$ factors, expected material properties in excess of nominal material properties, and global system response. As this standard prescribes use of $\phi = 1$ and expected material properties, the only additional source of overstrength is global system response. Using representative values for these contributions to overstrength ($\Omega_0 = 2.5$, $\phi = 0.75$, and expected/nominal $= 1.25$), the factor relating shear at first yield to elastic base shear capacity is 1.5. Additional commentary regarding this coefficient is provided in C.3.3.3.3.2.

**Coefficient $C_2$.** This coefficient adjusts design values based on component hysteresis characteristics, cyclic stiffness degradation, and strength deterioration. For buildings with systems that do not exhibit degradation of stiffness and/or strength, the $C_2$ coefficient can be assumed to be 1.0. This would include buildings with modern concrete or steel special moment-resisting frames, steel eccentrically braced frames, and buckling-restrained braced frames as either the original system or the system added during seismic rehabilitation. See Section C3.3.3.3.2 and FEMA 274 (FEMA 1997) for additional discussion.

**Coefficient, $C_m$.** The effective mass factor was developed to reduce the conservatism of the LSP for buildings where higher mode mass participation reduces lateral forces up to 20% depending on building type. See FEMA 357 (FEMA 2000), Appendix E for more information on the development of $C_m$.

*3.3.1.3.2 Vertical Distribution of Seismic Forces* The vertical distribution of the pseudo-lateral force shall be as specified in this section for all buildings except unreinforced masonry buildings, for which the pseudo-lateral force shall be permitted to be distributed in accordance with Section 3.3.1.3.5. The lateral load $F_x$

applied at any floor level $x$ shall be determined in accordance with Eqs. 3-10 and 3-11:

$$F_x = C_{vx} V \quad \text{(Eq. 3-10)}$$

$$C_{vx} = \frac{w_x h_x^k}{\sum_{i=1}^{n} w_i h_i^k} \quad \text{(Eq. 3-11)}$$

where

$C_{vx}$ = vertical distribution factor;
$k$ = 2.0 for $T \geq 2.5$ sec;
   = 1.0 for $T \leq 0.5$ sec (linear interpolation shall be used to calculate values of $k$ for intermediate values of $T$);
$V$ = pseudo-lateral force from Eq. 3-9;
$w_i$ = portion of the effective seismic weight $W$ located on or assigned to floor level $i$;
$w_x$ = portion of the effective seismic weight $W$ located on or assigned to floor level $x$;
$h_i$ = height (in ft) from the base to floor level $i$; and
$h_x$ = height (in ft) from the base to floor level $x$.

*3.3.1.3.3 Horizontal Distribution of Seismic Forces*
The seismic forces at each floor level of the building calculated using Eq. 3-10 shall be distributed according to the distribution of mass at that floor level.

*3.3.1.3.4 Diaphragms* Diaphragms shall be designed to resist the combined effects of the inertial force, $F_{px}$, calculated in accordance with Eq. 3-12, and horizontal forces resulting from offsets in, or changes in the stiffness of, the vertical seismic framing elements above and below the diaphragm. Actions resulting from offsets in or changes in the stiffness of the vertical seismic framing elements shall be taken as force-controlled, unless smaller forces are justified by other rational analysis, and shall be added directly to the diaphragm inertial forces.

$$F_{px} = \frac{\sum_{i=x}^{n} F_i}{\sum_{i=x}^{n} w_i} w_x \quad \text{(Eq. 3-12)}$$

where

$F_{px}$ = total diaphragm inertial force at level $x$;
$F_i$ = lateral load applied at floor level $i$ given by Eq. 3-10;
$w_i$ = portion of the effective seismic weight $W$ located on or assigned to floor level $i$; and
$w_x$ = portion of the effective seismic weight $W$ located on or assigned to floor level $x$.

The seismic load on each flexible diaphragm shall be distributed along the span of that diaphragm, proportional to its displaced shape.

Diaphragms receiving horizontal forces from discontinuous vertical elements shall be taken as force-controlled. Actions on other diaphragms shall be considered force- or deformation-controlled as specified for diaphragm components in Chapters 5 through 8.

*C3.3.1.3.4 Diaphragms* Further information on load distribution in flexible diaphragms is given in Section C3.2.4.

*3.3.1.3.5 Distribution of Seismic Forces for Unreinforced Masonry Buildings with Flexible Diaphragms* For unreinforced masonry buildings with flexible diaphragms for which the fundamental period is calculated using Eq. 3-8, it shall be permitted to calculate and distribute the pseudo-lateral force as follows:

1. For each span of the building and at each level, calculate period from Eq. 3-8;
2. Using Eq. 3-9, calculate pseudo-lateral force for each span;
3. Apply the lateral loads calculated for all spans and calculate forces in vertical seismic-resisting elements using tributary loads;
4. Diaphragm forces for evaluation of diaphragms shall be determined from the results of step 3 above and distributed along the diaphragm span considering its deflected shape; and
5. Diaphragm deflection shall not exceed 6 in. for this method of distribution of pseudo-lateral force to be applicable.

*C3.3.1.3.5 Distribution of Seismic Forces in Unreinforced Masonry Buildings with Flexible Diaphragms* These provisions are based on Appendix Chapter 1 of the *1997 Uniform Code for Building Conservation* (ICBO 1997). See FEMA 357 (FEMA 2000), Appendix D for more information.

### 3.3.2 Linear Dynamic Procedure

#### 3.3.2.1 Basis of the Procedure

If the LDP is selected for seismic analysis of the building, the design seismic forces, their distribution over the height of the building, and the corresponding internal forces and system displacements shall be determined using a linearly elastic, dynamic analysis in compliance with the requirements of this section.

Buildings shall be modeled with linearly elastic stiffness and equivalent viscous damping values consistent with components responding at or near yield level, as defined in Section 2.4.4. Modeling and analysis procedures to calculate forces and deformations shall be in accordance with Section 3.3.2.2.

Results of the LDP shall be checked using the acceptance criteria of Section 3.4.2.

#### C3.3.2.1 Basis of the Procedure

Modal spectral analysis is carried out using linearly elastic response spectra that are not modified to account for anticipated nonlinear response. As with the LSP, it is expected that the LDP will produce displacements that approximate maximum displacements expected during the design earthquake, but will produce internal forces that exceed those that would be obtained in a yielding building.

Calculated internal forces typically will exceed those that the building can sustain because of anticipated inelastic response of components. These design forces are evaluated through the acceptance criteria of Section 3.4.2, which include modification factors and alternative analysis procedures to account for anticipated inelastic response demands and capacities.

#### 3.3.2.2 Modeling and Analysis Considerations

*3.3.2.2.1 General* The ground motion characterized for dynamic analysis shall comply with the requirements of Section 3.3.2.2.2. The dynamic analysis shall be performed using the response spectrum method in accordance with Section 3.3.2.2.3 or the time-history method in accordance with Section 3.3.2.2.4.

*3.3.2.2.2 Ground Motion Characterization* The horizontal ground motion shall be characterized for design by the requirements of Section 1.6 and shall be one of the following:

1. A response spectrum as specified in Section 1.6.1.5;
2. A site-specific response spectrum as specified in Section 1.6.2.1; or
3. Ground acceleration time histories as specified in Section 1.6.2.2.

*3.3.2.2.3 Response Spectrum Method* Dynamic analysis using the response spectrum method shall calculate peak modal responses for sufficient modes to capture at least 90% of the participating mass of the building in each of two orthogonal principal horizontal directions of the building. Modal damping ratios shall reflect the damping in the building at deformation levels less than the yield deformation.

Peak member forces, displacements, story forces, story shears, and base reactions for each mode of response shall be combined by either the square root sum of squares (SRSS) rule or the complete quadratic combination (CQC) rule.

Multidirectional seismic effects shall be considered in accordance with the requirements of Section 3.2.7.

### 3.3.2.2.4 Time-History Method

Dynamic analysis using the time-history method shall calculate building response at discrete time steps using discretized recorded or synthetic time histories as base motion. The damping matrix associated with the mathematical model shall reflect the damping in the building at deformation levels near the yield deformation.

Response parameters shall be calculated for each time-history analysis. If fewer than seven time-history analyses are performed, the maximum response of the parameter of interest shall be used for design. If seven or more time-history analyses are performed, the average value of each response parameter shall be permitted to be used for design.

Multidirectional seismic effects shall be considered in accordance with the requirements of Section 3.2.7. Alternatively, an analysis of a three-dimensional mathematical model using simultaneously imposed consistent pairs of earthquake ground motion records along each of the horizontal axes of the building shall be permitted.

### C3.3.2.2 Modeling and Analysis Considerations

The LDP includes two analysis methods, namely, the Response Spectrum Method and the Time-History Method. The Response Spectrum Method uses peak modal responses calculated from dynamic analysis of a mathematical model. Only those modes contributing significantly to the response need to be considered. Modal responses are combined using rational methods to estimate total building response quantities. The Time-History Method (also termed Response-History Analysis) involves a time-step-by-time-step evaluation of building response, using discretized recorded or synthetic earthquake records as base motion input. Pairs of ground motion records for simultaneous analysis along each horizontal axis of the building should be consistent. Consistent pairs are the orthogonal motions expected at a given site based on the same earthquake. Guidance for correlation between two sets

of time histories is provided in the U.S. Nuclear Regulatory Commission *Regulatory Guide 1.92* (USNRC 1976).

### 3.3.2.3 Determination of Forces and Deformations

*3.3.2.3.1 Modification of Demands* All forces and deformations calculated using either the Response Spectrum or the Time-History Method shall be multiplied by the product of the modification factors $C_1$ and $C_2$ defined in Section 3.3.1.3, and further modified to consider the effects of torsion in accordance with Section 3.2.2.2.

*3.3.2.3.2 Diaphragms* Diaphragms shall be designed to resist the combined effects of the seismic forces calculated by the LDP, and the horizontal forces resulting from offsets in, or changes in stiffness of, the vertical seismic framing elements above and below the diaphragm. The seismic forces calculated by the LDP shall be taken as not less than 85% of the forces calculated using Eq. 3-12. Actions resulting from offsets in, or changes in stiffness of, the vertical seismic framing elements shall be taken as force-controlled, unless smaller forces are justified by a rational analysis approved by the authority having jurisdiction.

Diaphragms receiving horizontal forces from discontinuous vertical elements shall be taken as force-controlled. Actions on other diaphragms shall be considered force- or deformation-controlled as specified for diaphragm components in Chapters 5 through 8.

### 3.3.3 Nonlinear Static Procedure

#### 3.3.3.1 Basis of the Procedure

If the NSP is selected for seismic analysis of the building, a mathematical model directly incorporating the nonlinear load-deformation characteristics of individual components of the building shall be subjected to monotonically increasing lateral loads representing inertia forces in an earthquake until a target displacement is exceeded. Mathematical modeling and analysis procedures shall comply with the requirements of Section 3.3.3.2. The target displacement shall be calculated by the procedure in Section 3.3.3.3.

#### C3.3.3.1 Basis of the Procedure

The target displacement is intended to represent the maximum displacement likely to be experienced during the design earthquake. Because the mathematical model accounts directly for effects of material inelastic response, the calculated internal forces will be reasonable approximations of those expected during the design earthquake.

### 3.3.3.2 Modeling and Analysis Considerations

*3.3.3.2.1 General* Selection of a control node, selection of lateral load patterns, determination of the fundamental period, and application of the analysis procedure shall comply with the requirements of this section.

The relation between base shear force and lateral displacement of the control node shall be established for control node displacements ranging between zero and 150% of the target displacement, $\delta_t$.

The component gravity loads shall be included in the mathematical model for combination with lateral loads as specified in Section 3.2.8. The lateral loads shall be applied in both the positive and negative directions, and the maximum seismic effects shall be used for design.

The analysis model shall be discretized to represent the load-deformation response of each component along its length to identify locations of inelastic action.

Primary and secondary components of lateral-force-resisting elements shall be included in the model, as specified in Section 3.2.2.3.

The force–displacement behavior of all components shall be explicitly included in the model using full backbone curves that include strength degradation and residual strength, if any.

The NSP shall be used in conjunction with the acceptance criteria of Sections 3.4.3.2.1. and 3.4.3.2.3.

*C3.3.3.2.1 General* The requirement to carry out the analysis to at least 150% of the target displacement is meant to encourage the engineer to investigate likely building performance and behavior of the model under extreme load conditions that exceed the design values. The engineer should recognize that the target displacement represents a mean displacement value for the design earthquake loading, and that there is considerable scatter about the mean. Estimates of the target displacement may be unconservative for buildings with low strength compared with the elastic spectral demands.

*3.3.3.2.2 Simplified NSP Analysis* The use of a simplified NSP analysis shall be permitted as follows:

1. Only primary components are modeled;
2. The force–displacement characteristics of components are bilinear, and the degrading portion of the backbone curve is not explicitly modeled; and

3. Components not meeting the acceptance criteria for primary components are designated as secondary, and removed from the mathematical model.

A simplified NSP analysis shall be used with the acceptance criteria of Sections 3.4.3.2.2 and 3.4.3.2.3.

*C3.3.3.2.2 Simplified Nonlinear Static Procedure Analysis* The simplified NSP differs from the NSP in that component degradation is not explicitly included in the mathematical model. Therefore, more stringent acceptance criteria are used and component demands must be within the acceptance criteria limits for primary components. Where using the simplified NSP analysis, care should be taken to make sure that removal of degraded components from the model does not result in changes to the regularity of the structure that would significantly alter the dynamic response. In pushing with a static load pattern, the NSP does not capture changes in the dynamic characteristics of the structure as yielding and degradation take place.

In order to explicitly evaluate deformation demands on secondary components that are to be excluded from the model, one might consider including them in the model, but with negligible stiffness, to obtain deformation demands without significantly affecting the overall response.

*3.3.3.2.3 Control Node Displacement* The control node shall be located at the center of mass at the roof of a building. For buildings with a penthouse, the floor of the penthouse shall be regarded as the level of the control node. The displacement of the control node in the mathematical model shall be calculated for the specified lateral loads.

*3.3.3.2.4 Lateral Load Distribution* Lateral loads shall be applied to the mathematical model in proportion to the distribution of inertia forces in the plane of each floor diaphragm. The vertical distribution of these forces shall be proportional to the shape of the fundamental mode in the direction under consideration.

*C3.3.3.2.4 Lateral Load Distribution* The distribution of lateral inertial forces determines relative magnitudes of shears, moments, and deformations within the structure. The actual distribution of these forces is expected to vary continuously during earthquake response as portions of the structure yield and stiffness characteristics change. The extremes of this distribution will depend on the severity of the earthquake shaking and the degree of nonlinear response of the structure. Use of more than one lateral load pattern has been used in the past as a way to bound the range of design actions that may occur during actual dynamic response. Recent research [FEMA 440 (FEMA 2005)] has shown that multiple load patterns do little to improve the accuracy of nonlinear static procedures and that a single pattern based on the first mode shape is recommended.

*3.3.3.2.5 Idealized Force–Displacement Curve* The nonlinear force–displacement relationship between base shear and displacement of the control node shall be replaced with an idealized relationship to calculate the effective lateral stiffness, $K_e$, and effective yield strength, $V_y$, of the building as shown in Fig. 3-1.

The first line segment of the idealized force–displacement curve shall begin at the origin and have a slope equal to the effective lateral stiffness, $K_e$. The effective lateral stiffness, $K_e$, shall be taken as the secant stiffness calculated at a base shear force equal to 60% of the effective yield strength of the structure. The effective yield strength, $V_y$, shall not be taken as greater than the maximum base shear force at any point along the force–displacement curve.

The second line segment shall represent the positive post-yield slope ($\alpha_1 K_e$), determined by a point ($V_d$, $\Delta_d$) and a point at the intersection with the first line segment such that the areas above and below the actual curve are approximately balanced. ($V_d$, $\Delta_d$) shall be a point on the actual force–displacement curve at the calculated target displacement, or at the displacement corresponding to the maximum base shear, whichever is least.

The third line segment shall represent the negative post yield slope ($\alpha_2 K_e$), determined by the point at the end of the positive post-yield slope ($V_d$, $\Delta_d$) and the point at which the base shear degrades to 60% of the effective yield strength.

*C3.3.3.2.5 Idealized Force–Displacement Curve* The idealized force–displacement curve is developed using an iterative graphical procedure to balance the areas below the actual and idealized curves up to $\Delta_d$ such that the idealized curve has the properties defined in this section. The definition of the idealized force–displacement curve was modified from the definition in FEMA 356 (FEMA 2000) based on the recommendations of FEMA 440 (FEMA 2005).

*3.3.3.2.6 Period Determination* The effective fundamental period in the direction under consideration shall be based on the idealized force–displacement curve defined in Section 3.3.3.2.5. The effective fundamental period, $T_e$, shall be calculated in accordance with Eq. 3-13:

$$T_e = T_i \sqrt{\frac{K_i}{K_e}} \qquad \text{(Eq. 3-13)}$$

where

$T_i$ = elastic fundamental period (in seconds) in the direction under consideration calculated by elastic dynamic analysis;

$K_i$ = elastic lateral stiffness of the building in the direction under consideration calculated using the modeling requirements of Section 3.2.2.4; and

$K_e$ = effective lateral stiffness of the building in the direction under consideration.

*3.3.3.2.7 Analysis of Mathematical Models* Separate mathematical models representing the framing along two orthogonal axes of the building shall be developed for two-dimensional analysis. A mathematical model representing the framing along two orthogonal axes of the building shall be developed for three-dimensional analysis.

The effects of horizontal torsion shall be evaluated in accordance with Section 3.2.2.2.

Independent analysis along each of the two orthogonal principal axes of the building shall be permitted unless concurrent evaluation of multidirectional effects is required by Section 3.2.7.

### 3.3.3.3 Determination of Forces and Deformations

*3.3.3.3.1 General* For buildings with rigid diaphragms at each floor level, the target displacement, $\delta_t$, shall be calculated in accordance with Eq. 3-14 or by an approved procedure that accounts for the nonlinear response of the building.

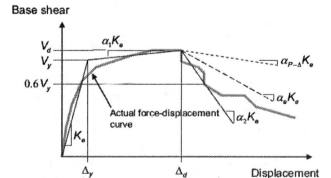

**FIGURE 3-1. Idealized Force–Displacement Curves.**

For buildings with nonrigid diaphragms at each floor level, diaphragm flexibility shall be explicitly included in the model. The target displacement shall be calculated as specified for rigid diaphragms, except that it shall be amplified by the ratio of the maximum displacement at any point on the roof to the displacement at the center of mass of the roof ($\delta_{max}/\delta_{cm}$). $\delta_{max}$ and $\delta_{cm}$ shall be based on a response spectrum analysis of a three-dimensional model of the building. The target displacement so calculated shall be no less than that displacement given by Eq. 3-14. No line of vertical seismic framing shall be evaluated for displacements smaller than the target displacement.

Alternatively, for buildings with flexible diaphragms at each floor level, a target displacement shall be calculated for each line of vertical seismic framing. The target displacement for an individual line of vertical seismic framing shall be as specified for buildings with rigid diaphragms, except that the masses shall be assigned to each line on the basis of tributary area.

Forces and deformations corresponding to the control node displacement equaling or exceeding the target displacement shall comply with acceptance criteria of Section 3.4.3.

*3.3.3.3.2 Target Displacement* The target displacement, $\delta_t$, at each floor level shall be calculated in accordance with Eq. 3-14 and as specified in Section 3.3.3.3.1.

$$\delta_t = C_0 C_1 C_2 S_a \frac{T_e^2}{4\pi^2} g \qquad \text{(Eq. 3-14)}$$

where

$C_0$ = modification factor to relate spectral displacement of an equivalent single-degree of freedom (SDOF) system to the roof displacement of the building

multi-degree of freedom (MDOF) system calculated using one of the following procedures:

- The first mode mass participation factor multiplied by the ordinate of the first mode shape at the control node;
- The mass participation factor calculated using a shape vector corresponding to the deflected shape of the building at the target displacement multiplied by ordinate of the shape vector at the control node; or
- The appropriate value from Table 3-2;

$C_1$ = modification factor to relate expected maximum inelastic displacements to displacements calculated for linear elastic response. For periods less than 0.2 sec, $C_1$ need not be taken greater than the value at $T = 0.2$ sec. For periods greater than 1.0 sec, $C_1 = 1.0$.

$$C_1 = 1 + \frac{R - 1}{aT_e^2}$$

where

$a$ = site class factor:
= 130 site Class A, B;
= 90 site Class C;
= 60 site Class D, E, F;

$T_e$ = effective fundamental period of the building in the direction under consideration, in seconds;

$T_s$ = characteristic period of the response spectrum, defined as the period associated with the transition from the constant acceleration segment of the spectrum to the constant velocity segment of the spectrum per Sections 1.6.1.5 and 1.6.2.1;

$R$ = ratio of elastic strength demand to yield strength coefficient calculated in

**Table 3-2. Values for Modification Factor $C_0$[1]**

| Number of Stories | Shear Buildings[2] | | Other Buildings |
| | Triangular Load Pattern (1.1, 1.2, 1.3) | Uniform Load Pattern (2.1) | Any Load Pattern |
|---|---|---|---|
| 1 | 1.0 | 1.0 | 1.0 |
| 2 | 1.2 | 1.15 | 1.2 |
| 3 | 1.2 | 1.2 | 1.3 |
| 5 | 1.3 | 1.2 | 1.4 |
| 10+ | 1.3 | 1.2 | 1.5 |

[1]Linear interpolation shall be used to calculate intermediate values.
[2]Buildings in which, for all stories, story drift decreases with increasing height.

accordance with Eq. 3-15. Use of the NSP is not permitted where $R$ exceeds $R_{max}$, per Section 2.4.2.1;

$C_2$ = modification factor to represent the effect of pinched hysteresis shape, cyclic stiffness degradation, and strength deterioration on maximum displacement response. For periods greater than 0.7 sec, $C_2$=1.0;

$$C_2 = 1 + \frac{1}{800}\left(\frac{R-1}{T_e}\right)^2$$

$S_a$ = response spectrum acceleration, at the effective fundamental period and damping ratio of the building in the direction under consideration, as calculated in Sections 1.6.1.5 and 1.6.2.1; and

$g$ = acceleration of gravity.

The strength ratio $R$ shall be calculated in accordance with Eq. 3-15:

$$R = \frac{S_a}{V_y/W} \cdot C_m \qquad \text{(Eq. 3-15)}$$

where $S_a$ is defined above, and

$V_y$ = yield strength calculated using results of the NSP for the idealized nonlinear force–displacement curve developed for the building in accordance with Section 3.3.3.2.5;

$W$ = effective seismic weight, as calculated in Section 3.3.1.3.1; and

$C_m$ = effective mass factor from Table 3-1. Alternatively, $C_m$ taken as the effective modal mass participation factor calculated for the fundamental mode using an Eigenvalue analysis shall be permitted. $C_m$ shall be taken as 1.0 if the fundamental period, $T$, is greater than 1.0 sec.

For buildings with negative post-yield stiffness, the maximum strength ratio, $R_{max}$, shall be calculated in accordance with Eq. 3-16.

$$R_{max} = \frac{\Delta_d}{\Delta_y} + \frac{|\alpha_e|^{-h}}{4} \qquad \text{(Eq. 3-16)}$$

where

$\Delta_d$ = lesser of target displacement, $\delta_t$, or displacement at maximum base shear defined in Fig. 31;

$\Delta_y$ = displacement at effective yield strength defined in Fig. 3-1;

$h$ = $1 + 0.15 \cdot \ln T_e$; and

$\alpha_e$ = effective negative post-yield slope ratio defined in Eq. 3-17.

The effective negative post-yield slope ratio, $\alpha_e$, shall be calculated in accordance with Eq. 3-17:

$$\alpha_e = \alpha_{P-\Delta} + \lambda(\alpha_2 - \alpha_{P-\Delta}) \qquad \text{(Eq. 3-17)}$$

where

$\alpha_2$ = negative post-yield slope ratio defined in Fig. 3-1. This includes P-$\Delta$ effects, in-cycle degradation, and cyclic degradation;

$\alpha_{P-\Delta}$ = negative slope ratio caused by P-$\Delta$ effects; and

$\lambda$ = near field effect factor:
= 0.8 if $S_1 \geq 0.6$ (Maximum Considered Earthquake, MCE);
= 0.2 if $S_1 < 0.6$ (MCE).

*C3.3.3.3.2 Target Displacement* This standard presents the Coefficient Method for calculating target displacement. Other procedures can also be used. Section C3.3.3.3 of FEMA 274 (FEMA 1997) presents additional background information on the Coefficient Method and another acceptable procedure referred to as the Capacity Spectrum Method.

The $C_0$ coefficient accounts for the difference between the roof displacement of a multi-degree of freedom (MDOF) building and the displacement of the equivalent single-degree of freedom (SDOF) system. Using only the first mode shape ($\phi_1$) and elastic behavior, coefficient $C_0$ is equal to:

$$C_0 = \phi_{1,r}\frac{\{\phi_1\}^T[M]\{1\}}{\{\phi_1\}^T[M]\{\phi_1\}} \qquad \text{(Eq. C3-3)}$$

$$= \phi_{1,r}\Gamma_1$$

where

$\phi_{1,r}$ = the ordinate of mode shape 1 at the roof (control node);

$[M]$ = a diagonal mass matrix; and

$\Gamma_1$ = the first mode mass participation factor.

Since the mass matrix is diagonal, Eq. C3-3 can be rewritten as:

$$C_0 = \phi_{1,r}\frac{\sum\limits_1^N m_i\phi_{i,n}}{\sum\limits_1^N m_i\phi_{i,n}^2} \qquad \text{(Eq. C3-4)}$$

where

$m_i$ = the mass at level $i$; and

$\phi_{i,n}$ = the ordinate of mode shape $i$ at level $n$.

If the absolute value of the roof (control node) ordinate of each mode shape is set equal to unity, the value of coefficient $C_0$ is equal to the first mode mass participation factor.

Explicit calculation of $C_0$ using the actual deflected shape may be beneficial in terms of lower amplification of target displacement. The actual shape vector may take on any form, particularly since it is intended to simulate the time-varying deflection profile of the building responding inelastically to the ground motion and will likely be different from the elastic first-mode shape. If this method is used, the mass participation factor, $\Gamma_1$, must be calculated using the actual deflected shape as the shape vector in lieu of the mode shape.

Use of the tabulated values, which are based on a straight-line vector with equal masses at each floor level, is approximate (particularly if masses vary much over the height of the building) and may be overly conservative.

Coefficients for estimating the target displacement have been modified based on the recommendations contained in FEMA 440 (FEMA 2005).

FEMA 440 concluded that the previous cap on the $C_1$ factor was not appropriate and a simplified equation was recommended based on $R$, effective period, $T_e$, and the site class factor, $a$, with a revised cap at $T = 0.2$ sec. FEMA 440 recommended site class factors for site classes B, C, and D only. The site class factor for site class A was set equal to that for B and the site class factor for site classes E and F was set equal to that for D. The use of the simplified $C_1$ equation to estimate displacements for soft soil sites, including classes E and F, will have higher uncertainty due to high dispersions of the results in studies of SDOF oscillators on soft soils. See FEMA 440 for more discussion on uncertainties related to the $C_1$ equation.

The $C_2$ factor was revised to better account for the effects of cyclic degradation of stiffness as recommended in FEMA 440. For buildings with systems that do not exhibit degradation of stiffness and/or strength, the $C_2$ coefficient can be assumed to be 1.0. This would include buildings with modern concrete or steel special moment-resisting frames, steel eccentrically braced frames, and buckling-restrained braced frames as either the original system or the system added during seismic rehabilitation.

The $C_3$ coefficient has been eliminated and replaced with a maximum strength ratio, $R_{max}$, which is intended to measure dynamic instability. Where the value for $R_{max}$ is exceeded, an NDP analysis is required to capture strength degradation and dynamic P-$\Delta$ effects to confirm dynamic stability of the building. As recommended in FEMA 440, the NDP analysis

should include the in-cycle or cyclic strength or stiffness degradation in the hysteretic models of the components as required. The effective negative post-yield slope ratio, $\alpha_e$, was introduced in FEMA 440 as a variable necessary to determine the maximum strength ratio, $R_{max}$, that a building can have before dynamic instability is a concern. The negative slope caused by P-$\Delta$ effects, $\alpha_{P-\Delta}$, is based on the restoring force needed to balance the overturning moment caused by the weight of the structure displaced an amount $\Delta$, acting at the effective height of the first mode. It can be determined using structural analysis software by comparing the stiffness results of an analysis run with P-$\Delta$ effects to one run without P-$\Delta$ effects considered.

*3.3.3.3.3 Modification of Demands* The target displacement shall be modified to consider the effects of horizontal torsion in accordance with Section 3.2.2.2.

*3.3.3.3.4 Diaphragms* Diaphragms shall be designed to resist the combined effects of the horizontal forces resulting from offsets in, or changes in stiffness of, the vertical seismic framing elements above and below the diaphragm, and the diaphragm forces determined using either Section 3.3.1.3.4 or Section 3.3.2.3.2.

### 3.3.4 Nonlinear Dynamic Procedure

#### 3.3.4.1 Basis of the Procedure

If the NDP is selected for seismic analysis of the building, a mathematical model directly incorporating the nonlinear load-deformation characteristics of individual components of the building shall be subjected to earthquake shaking represented by ground motion time histories in accordance with Section 1.6.2.2 to obtain forces and displacements.

Calculated displacements and forces shall be compared directly with acceptance criteria specified in Section 3.4.3.

#### C3.3.4.1 Basis of the Procedure

The basis, modeling approaches, and acceptance criteria of the NDP are similar to those for the NSP. The main exception is that the response calculations are carried out using time-history analysis. With the NDP, the design displacements are not established using a target displacement but, instead, are determined directly through dynamic analysis using ground motion time histories. Calculated response can be highly sensitive to characteristics of individual ground motions; therefore, the analysis should be carried out with more than one ground motion record. Because the numerical model accounts directly for effects of material inelastic response, the calculated internal forces will be reasonable approximations of those expected during the design earthquake.

### 3.3.4.2 Modeling and Analysis Considerations

*3.3.4.2.1 General* The modeling and analysis requirements specified in Section 3.3.3.2 for the NSP shall apply to the NDP, excluding considerations of control node and target displacements.

*3.3.4.2.2 Ground Motion Characterization* For the NDP, earthquake shaking shall be characterized by discretized recorded or synthetic earthquake records as base motion meeting the requirements of Section 1.6.2.2.

*3.3.4.2.3 Time-History Method* For the NDP, time-history analysis shall be performed using horizontal ground motion time histories prepared according to the requirements of Section 1.6.2.2.

Multidirectional seismic effects shall be accounted for in accordance with Section 3.2.7. Alternatively, an analysis of a three-dimensional mathematical model using simultaneously imposed consistent pairs of earthquake ground motion records along each of the horizontal axes of the building shall be permitted.

*C3.3.4.2.3 Time-History Method* Guidance for correlation between sets of time histories is provided in the U.S. Nuclear Regulatory Commission *Regulatory Guide 1.92* (USNRC 1976).

### 3.3.4.3 Determination of Forces and Deformations
Forces and deformations shall be determined in accordance with Section 3.3.2.2.4.

*3.3.4.3.1 Modification of Demands* The effects of torsion shall be considered in accordance with Section 3.2.2.2.

*3.3.4.3.2 Diaphragms* Diaphragms shall be designed to resist the effects of the seismic forces calculated by dynamic analysis, including the effects of the horizontal forces resulting from offsets in, or changes in stiffness of, the vertical seismic framing elements above and below the diaphragm.

## 3.4 ACCEPTANCE CRITERIA

### 3.4.1 General Requirements
Components analyzed using the linear procedures of Section 3.3.1 and Section 3.3.2 shall satisfy the requirements of Section 3.4.2. Components analyzed

using the nonlinear procedures of Section 3.3.3 and Section 3.3.4 shall satisfy the requirements of Section 3.4.3.

Prior to selecting component acceptance criteria, components shall be classified as primary or secondary, and actions shall be classified as deformation-controlled or force-controlled, as defined in Section 2.4.4.

Foundations shall satisfy the criteria specified in Chapter 4.

### C3.4.1 General Requirements
The linear analysis procedures are intended to provide a conservative estimate of building response and performance in an earthquake. Since the actual response of buildings to earthquakes is typically nonlinear, nonlinear analysis procedures should provide a more accurate representation of building response and performance. In recognition of the improved estimates of nonlinear analysis, the acceptance criteria for nonlinear procedures are more accurate and less conservative than those for linear procedures. Buildings that do not comply with the linear analysis acceptance criteria may comply with nonlinear acceptance criteria. Therefore, performing a nonlinear analysis is recommended to minimize or eliminate unnecessary seismic rehabilitation. Users are urged to report to the building owner limitations on the use of linear procedures and to pursue nonlinear analyses where linear acceptance criteria are not met.

### 3.4.2 Linear Procedures

#### 3.4.2.1 Design Forces and Deformations
Component design forces and deformations shall be calculated in accordance with linear analysis procedures of Sections 3.3.1 or 3.3.2.

*3.4.2.1.1 Deformation-Controlled Actions* Deformation-controlled design actions, $Q_{UD}$, shall be calculated in accordance with Eq. 3-18:

$$Q_{UD} = Q_G \pm Q_E \qquad \text{(Eq. 3-18)}$$

where

$Q_E$ = action due to design earthquake loads calculated using forces and analysis models described in either Section 3.3.1 or Section 3.3.2;

$Q_G$ = action due to design gravity loads as defined in Section 3.2.8; and

$Q_{UD}$ = deformation-controlled design action due to gravity loads and earthquake loads.

*C3.4.2.1.1 Deformation-Controlled Actions* Because of possible anticipated nonlinear response of the structure, the design actions as represented by Eq. 3-18 may exceed the actual strength of the component to resist these actions. The acceptance criteria of Section 3.4.2.2.1 take this overload into account through use of a factor, *m*, which is an indirect measure of the nonlinear deformation capacity of the component.

*3.4.2.1.2 Force-Controlled Actions* Force-controlled design actions, $Q_{UF}$, shall be calculated using one of the following methods:

1. $Q_{UF}$ shall be taken as the maximum action that can be developed in a component based on a limit-state analysis considering the expected strength of the components delivering load to the component under consideration, or the maximum action developed in the component as limited by the nonlinear response of the building.
2. Alternatively, $Q_{UF}$ shall be calculated in accordance with Eq. 3-19.

$$Q_{UF} = Q_G \pm \frac{Q_E}{C_1 C_2 J} \qquad \text{(Eq. 3-19)}$$

where

$Q_{UF}$ = force-controlled design action due to gravity loads in combination with earthquake loads; and

$J$ = force-delivery reduction factor, greater than or equal to 1.0, taken as the smallest demand capacity ratio (DCR) of the components in the load path delivering force to the component in question, calculated in accordance with Eq. 2-1. Alternatively, values of $J$ equal to 2.0 for a High Level of Seismicity, 1.5 for a Moderate Level of Seismicity, and 1.0 for a Low Level of Seismicity shall be permitted where not based on calculated DCRs. $J$ shall be taken as 1.0 for the Immediate Occupancy Structural Performance Level. In any case where the forces contributing to $Q_{UF}$ are delivered by components of the lateral force resisting system that remain elastic, $J$ shall be taken as 1.0.

*C3.4.2.1.2 Force-Controlled Actions* The basic approach for calculating force-controlled actions for design differs from that used for deformation-controlled actions because nonlinear deformations associated with forced-controlled actions are not permitted. Therefore, force demands for force-controlled actions should not exceed the force capacity (strength).

Ideally, an inelastic mechanism for the structure will be identified and the force-controlled actions, $Q_{UF}$, for design will be determined by limit analysis using that mechanism. This approach will always produce a conservative estimate of the design actions, even if an incorrect mechanism is selected. Where it is not possible to use limit (or plastic) analysis, or in cases where design forces do not produce significant nonlinear response in the building, it is acceptable to determine the force-controlled actions for design using Eq. 3-19.

Coefficients $C_1$ and $C_2$ were introduced in Eq. 3-9 to amplify the design base shear to achieve a better estimate of the maximum displacements expected for buildings responding in the inelastic range. Displacement amplifiers, $C_1$ and $C_2$, are divided out of Eq. 3-19 when seeking an estimate of the force level present in a component where the building is responding inelastically.

Since $J$ is included for force-controlled actions, it may appear to be more advantageous to treat an action as force-controlled where *m*-factors are less than $J$. However, proper application of force-controlled criteria requires a limit state analysis of demand and lower-bound calculation of capacity that will yield a safe result whether an action is treated as force- or deformation-controlled.

### 3.4.2.2 Acceptance Criteria for Linear Procedures

*3.4.2.2.1 Deformation-Controlled Actions* Deformation-controlled actions in primary and secondary components shall satisfy Eq. 3-20.

$$m\kappa Q_{CE} \geq Q_{UD} \qquad \text{(Eq. 3-20)}$$

where

$m$ = component demand modification factor to account for expected ductility associated with this action at the selected Structural Performance Level. *m*-factors are specified in Chapters 4 through 8;

$Q_{CE}$ = expected strength of the component at the deformation level under consideration for deformation-controlled actions. $Q_{CE}$, the expected strength, shall be determined considering all coexisting actions on the component under the design loading condition by procedures specified in Chapters 4 through 8; and

$\kappa$ = knowledge factor defined in Section 2.2.6.4.

*3.4.2.2.2 Force-Controlled Actions* Force-controlled actions in primary and secondary components shall satisfy Eq. 3-21:

$$\kappa Q_{CL} \geq Q_{UF} \qquad \text{(Eq. 3-21)}$$

where

$Q_{CL}$ = lower-bound strength of a component at the deformation level under consideration for force-controlled actions. $Q_{CL}$, the lower-bound strength, shall be determined considering all coexisting actions on the component under the design loading condition by procedures specified in Chapters 4 through 8.

*3.4.2.2.3 Verification of Design Assumptions* In addition to the requirements in Section 3.2.9, the following verification of design assumptions shall be made.

Where moments due to gravity loads in horizontally spanning primary components exceed 75% of the expected moment strength at any location, the possibility for inelastic flexural action at locations other than member ends shall be specifically investigated by comparing flexural actions with expected member strengths. Where linear procedures are used, formation of flexural plastic hinges away from member ends shall not be permitted.

### 3.4.3 Nonlinear Procedures

#### 3.4.3.1 Design Forces and Deformations

Component design forces and deformations shall be calculated in accordance with nonlinear analysis procedures of Sections 3.3.3 or 3.3.4.

#### 3.4.3.2 Acceptance Criteria for Nonlinear Procedures

*3.4.3.2.1 Deformation-Controlled Actions* Primary and secondary components shall have expected deformation capacities not less than maximum deformation demands calculated at the target displacement. Primary and secondary component demands shall be within the acceptance criteria for secondary components at the selected Structural Performance Level. Expected deformation capacities shall be determined considering all coexisting forces and deformations in accordance with Chapters 4 through 8.

Acceptance criteria for the simplified NSP analysis of Section 3.3.3.2.1 shall be as specified in Section 3.4.3.2.2.

*C3.4.3.2.1 Deformation-Controlled Actions* Where all components are explicitly modeled with full backbone curves, the NSP can be used to evaluate the full contribution of all components to the lateral force resistance of the structure as they degrade to residual strength values. Where degradation is explicitly evaluated in the NSP, components can be relied upon for lateral-force resistance out to the secondary component limits of response.

Studies on the effects of different types of strength degradation are presented in FEMA 440 (FEMA 2005).

As components degrade, the post-yield slope of the force–displacement curve becomes negative. The strength ratio, $R_{max}$, limits the extent of degradation based on the degree of negative post-yield slope.

*3.4.3.2.2 Deformation-Controlled Actions for the Simplified Nonlinear Static Analysis* Primary and secondary components modeled using the simplified NSP analysis of Section 3.3.3.2.2 shall meet the requirements of this section. Expected deformation capacities shall not be less than maximum deformation demands calculated at the target displacement. Primary component demands shall be within the acceptance criteria for primary components at the selected Structural Performance Level. Demands on other components shall be within the acceptance criteria for secondary components at the selected Structural Performance Level. Expected deformation capacities shall be determined considering all coexisting forces and deformations by procedures specified in Chapters 4 through 8.

*C3.4.3.2.2 Deformation-Controlled Actions for the Simplified Nonlinear Static Analysis* In the simplified NSP analysis, primary components are not modeled with full backbone curves. Degradation cannot be explicitly evaluated and degraded components cannot be reliably used to the secondary component limits of response. For this reason, the lateral-force resistance of the structure consists of primary components measured against primary component acceptance criteria.

*3.4.3.2.3 Force-Controlled Actions* Primary and secondary components shall have lower bound strengths not less than the maximum design forces. Lower-bound strengths shall be determined considering all coexisting forces and deformations by procedures specified in Chapters 4 through 8.

*3.4.3.2.4 Verification of Design Assumptions* In addition to the requirements in Section 3.2.9, the following verification of design assumptions shall be made:

• Flexural plastic hinges shall not form away from component ends unless they are explicitly accounted for in modeling and analysis.

### 4.0 FOUNDATIONS AND GEOLOGIC SITE HAZARDS

### 4.1 SCOPE

This chapter sets forth general requirements for consideration of foundation load–deformation characteristics, seismic rehabilitation of foundations, and mitigation of geologic site hazards in the Systematic Rehabilitation of buildings.

Section 4.2 specifies data collection for site characterization and defines geologic site hazards. Section 4.3 outlines procedures for mitigation of geologic site hazards. Section 4.4 provides soil strength and stiffness parameters for consideration of foundation load–deformation characteristics. Section 4.5 specifies procedures for consideration of soil–structure (SSI) effects. Section 4.6 specifies seismic earth pressures on building walls. Section 4.7 specifies requirements for seismic rehabilitation of foundations.

## C4.1 SCOPE

This chapter provides geotechnical engineering provisions for building foundations and seismic-geologic site hazards. Acceptability of the behavior of the foundation system and foundation soils for a given performance level cannot be determined apart from the context of the behavior of the superstructure.

Geotechnical requirements for buildings that are suitable for Simplified Rehabilitation are included in Chapter 10. Structural engineering issues of foundation systems are discussed in the chapters on Steel (Chapter 5), Concrete (Chapter 6), Masonry (Chapter 7), and Wood and Light Metal Framing (Chapter 8).

## 4.2 SITE CHARACTERIZATION

Site characterization shall include collection of information on the building foundation as specified in Section 4.2.1, and on seismic geologic site hazards as specified in Section 4.2.2.

## C4.2 SITE CHARACTERIZATION

The guidance of the authorities having jurisdiction over historical matters should be obtained if historic or archeological resources are present at the site.

### 4.2.1 Foundation Information

Information on the foundation supporting the building to be rehabilitated, nearby foundation conditions, design foundation loads, and load–deformation characteristics of the foundation soils shall be obtained as specified in Sections 4.2.1.1 through 4.2.1.3.

### 4.2.1.1 Foundation Conditions

*4.2.1.1.1 Structural Foundation Information* The following structural information shall be obtained for the foundation of the building to be rehabilitated in accordance with the data collections requirements of Section 2.2.6:

1. Foundation type;
2. Foundation configuration, including dimensions and locations; and
3. Material composition and details of construction.

*C4.2.1.1.1 Structural Foundation Information* Foundation types may consist of shallow isolated or continuous spread footings, mat foundations, deep foundations of driven piles, cast-in-place concrete piers, or drilled shafts of concrete.

Foundation configuration information includes dimensions and locations, depths of embedment of shallow foundations, pile tip elevations, and variations in cross section along the length of the pile for tapered piles and belled caissons.

Foundation material types include concrete, steel, and wood. Foundation installation methods include cast-in-place and open/closed-end driving.

With this minimum amount of information, presumptive or prescriptive procedures may be used to determine the bearing capacity of the foundations. However, additional information is required for site-specific assessments of foundation bearing capacity and stiffness. Acquiring this additional information involves determining unit weights, shear strength, friction angle, compressibility characteristics, soil moduli, and Poisson's ratio.

*4.2.1.1.2 Subsurface Soil Conditions* The following information on subsurface soil conditions shall be obtained as required for the selected rehabilitation objectives:

1. For rehabilitation objectives that include Collapse Prevention and Life Safety Performance Levels, the type, composition, consistency, relative density, and layering of soils shall be determined to a depth at which the stress imposed by the building is less than or equal to 10% of the building weight divided by the total foundation area. For buildings with friction piles, the depth so calculated shall be increased by two-thirds of the pile length. For end bearing piles, the depth of investigation shall be the pile length plus 10 ft. The location of the water table and its seasonal fluctuations beneath the building shall be determined.

2. For enhanced rehabilitation objectives, the soil unit weight, $\gamma$; soil cohesion, $c$; soil friction angle, $\phi$; soil compressibility characteristics, soil shear modulus, $G$; and Poisson's ratio, $\nu$, for each type of soil, shall be determined.

*C4.2.1.1.2 Subsurface Soil Conditions* Specific foundation information developed for an adjacent or nearby building may be useful if subsurface soils and ground water conditions in the site region are known to be uniform. However, less confidence will result if subsurface data are developed from anywhere but the site of the building being rehabilitated. Adjacent sites where construction has been done recently may provide a guide for evaluation of subsurface conditions at the site being considered. Sources of existing geotechnical information are discussed in C2.2.3.

### 4.2.1.2 Design Foundation Loads

Information on the design foundation loads shall be obtained, including separate information on dead loads and live loads. Alternatively, the design foundation loads shall be calculated where information on the design foundation loads is not available.

### C4.2.1.2 Design Foundation Loads

Design drawings may indicate information regarding the allowable bearing capacity of the foundation components. This information can be used directly in a presumptive or prescriptive evaluation of the foundation capacity. Construction records may also be available indicating ultimate pile capacities if load tests were performed. Information on the existing loads on the structure is relevant to determining the amount of overload that the foundations may be capable of resisting during an earthquake.

### 4.2.1.3 Load–Deformation Characteristics Under Seismic Loading

Load–deformation characteristics of foundations shall be obtained from geotechnical reports, or shall be determined in accordance with the requirements of Section 4.4.

### C4.2.1.3 Load–Deformation Characteristics Under Seismic Loading

Traditional geotechnical engineering treats load–deformation characteristics for long-term dead loads plus frequently applied live loads only. In most cases, long-term settlement governs foundation design. Short-term (earthquake) load–deformation characteristics have not traditionally been used for design; consequently, such relationships are not generally found in the geotechnical reports for existing buildings.

### 4.2.2 Seismic Geologic Site Hazards

Seismic rehabilitation shall include an assessment of earthquake-induced hazards at the site due to fault rupture, liquefaction, differential compaction, landsliding, and an assessment of earthquake-induced flooding or inundation in accordance with Sections 4.2.2.1 through 4.2.2.5.

If the resulting ground movements cause unacceptable performance in the building for the selected performance level, then the hazards shall be mitigated in accordance with Section 4.3.

### 4.2.2.1 Fault Rupture

A geologic fault shall be defined as a plane or zone along which earth materials on opposite sides have moved differentially in response to tectonic forces.

Geologic site information shall be obtained to determine if an active geologic fault is present under the building foundation. If a fault is present, the following information shall be obtained:

1. The degree of activity based on the age of most recent movement;
2. The fault type—whether it is a strike-slip, normal-slip, reverse-slip, or thrust fault;
3. The sense of slip with respect to building geometry;
4. Magnitudes of vertical and/or horizontal displacements consistent with the selected earthquake hazard level; and
5. The width and distribution of the fault-rupture zone.

### C4.2.2.1 Fault Rupture

Buildings found to straddle active faults should be assessed to determine if any rehabilitation is warranted, possibly to reduce the collapse potential of the structure given the likely amount and direction of fault displacement.

### 4.2.2.2 Liquefaction

Liquefaction shall be defined as an earthquake-induced process in which saturated, loose, granular soils lose shear strength and liquefy as a result of an increase in pore-water pressure during earthquake shaking.

Subsurface soil and ground water information shall be obtained to determine if liquefiable materials are present under the building foundation. If liquefiable soils are present, the following information shall be obtained: soil type, soil density, depth to water table, ground surface slope, proximity of free-face conditions, and lateral and vertical differential displacements.

A site shall be regarded as free from liquefaction hazard if the site soils, or similar soils in the site vicinity, have not experienced historical liquefaction, and if any of the following criteria are met:

1. The geologic materials underlying the site are either bedrock or have very low liquefaction susceptibility according to the relative susceptibility ratings based upon the type of deposit and geologic age of the deposit, as shown in Table 4-1;
2. The soils underlying the site are stiff clays or clayey silts;
3. The soils are not highly sensitive, based on local experience;
4. The soils are cohesionless with a minimum normalized Standard Penetration Test (SPT) resistance, $(N_1)_{60}$, value of 30 blows/0.3m (30 blows/ft) as defined in ASTM D1586-99 (ASTM 1999), for depths below the ground water table, or with clay content greater than 20%; or
5. The ground water table is at least 35 ft below the deepest foundation depth, or 50 ft below the ground surface, whichever is shallower, including considerations for seasonal and historic groundwater level rises, and any slopes or free-face conditions in the site vicinity do not extend below the ground water elevation at the site.

If a liquefaction hazard is determined to exist at the site, then a more detailed evaluation of potential ground movements due to liquefaction shall be performed using procedures approved by the authority having jurisdiction.

**Table 4-1. Estimated Susceptibility to Liquefaction of Surficial Deposits During Strong Ground Shaking**

| Type of Deposit | General Distribution of Cohesionless Sediments in Deposits | Likelihood that Cohesionless Sediments, When Saturated, Would Be Susceptible to Liquefaction (by Age of Deposit) | | | |
|---|---|---|---|---|---|
| | | Modern < 500 years | Holocene < 11,000 years | Pleistocene < 2 million years | Pre-Pleistocene > 2 million years |
| (a) Continental Deposits | | | | | |
| River Channel | Locally variable | Very high | High | Low | Very low |
| Flood Plain | Locally variable | High | Moderate | Low | Very low |
| Alluvial Fan, Plain | Widespread | Moderate | Low | Low | Very low |
| Marine Terrace | Widespread | — | Low | Very low | Very low |
| Delta, Fan Delta | Widespread | High | Moderate | Low | Very low |
| Lacustrine, Playa | Variable | High | Moderate | Low | Very low |
| Colluvium | Variable | High | Moderate | Low | Very low |
| Talus | Widespread | Low | Low | Very low | Very low |
| Dune | Widespread | High | Moderate | Low | Very low |
| Loess | Variable | High | High | High | Unknown |
| Glacial Till | Variable | Low | Low | Very low | Very low |
| Tuff | Rare | Low | Low | Very low | Very low |
| Tephra | Widespread | High | Low | Unknown | Unknown |
| Residual Soils | Rare | Low | High | Very low | Very low |
| Sebka | Locally variable | High | Moderate | Low | Very low |
| (b) Coastal Zone Deposits | | | | | |
| Delta | Widespread | Very high | High | Low | Very low |
| Esturine | Locally variable | High | Moderate | Low | Very low |
| Beach, High Energy | Widespread | Moderate | Low | Very low | Very low |
| Beach, Low Energy | Widespread | High | Moderate | Low | Very low |
| Lagoon | Locally variable | High | Moderate | Low | Very low |
| Foreshore | Locally variable | High | Moderate | Low | Very Low |
| (c) Fill Materials | | | | | |
| Uncompacted Fill | Variable | Very high | — | — | — |
| Compacted Fill | Variable | Low | — | — | — |

*C4.2.2.2 Liquefaction*

Soil liquefaction is a phenomenon in which a soil below the groundwater table loses a substantial amount of strength due to strong earthquake ground shaking. Recently deposited (i.e., geologically young) and relatively loose natural soils and uncompacted or poorly compacted fill soils are potentially susceptible to liquefaction. Loose sands and silty sands are particularly susceptible; loose silts and gravels also have potential for liquefaction. Dense natural soils and well-compacted fills have low susceptibility to liquefaction. Clay soils are generally not susceptible, except for highly sensitive clays found in some geographic regions.

The following information may be necessary for evaluating the liquefaction potential of soils:

**Soil type:** Whether liquefiable soils [i.e., granular (sand, silty sand, nonplastic silt) soils] are present;

**Soil density:** Whether liquefiable soils are loose to medium dense;

**Depth to water table:** Whether liquefiable soils are saturated at any time during seasonal fluctuations of the water table;

**Ground surface slope and proximity of free-face conditions:** Whether liquefiable soils are at a gently sloping site or in the proximity of free-surface conditions; and

**Lateral and vertical differential displacement:** Amount and direction at the building foundation should be calculated.

**Seed-Idriss Procedure for Evaluating Liquefaction Potential.** The potential for liquefaction to occur may be assessed by a variety of available approaches (National Research Council 1985). The most commonly utilized approach is the Seed-Idriss simplified empirical procedure, presented by Seed and Idriss (1971; 1982) and subsequently updated by many researchers, that utilizes Standard Penetration Test (SPT) blow count data. Using SPT data to assess liquefaction potential due to an earthquake is considered a reasonable engineering approach (Seed and Idriss 1982; Seed et al. 1985; NRC 1985) because many of the factors affecting penetration resistance affect the liquefaction resistance of sandy soils in a similar way, and because these liquefaction potential evaluation procedures are based on actual performance of soil deposits during worldwide historical earthquakes. Section C4.2.2.2 of FEMA 274 (FEMA 1997) provides more guidance for evaluating liquefaction potential, but readers should note that Youd et al. (2001) includes an update of the methods described in FEMA 274.

**Evaluating Potential for Lateral Spreading.** Lateral spreads are ground-failure phenomena that can occur on gently sloping ground underlain by liquefied soil. Earthquake ground shaking affects the stability of sloping ground containing liquefiable materials due to seismic inertia forces within the slope and shaking-induced strength reductions in the liquefiable materials. Temporary instability due to seismic inertia forces is manifested by lateral downslope movement that can potentially involve large land areas. For the duration of ground shaking associated with moderate to large earthquakes, there could be many such occurrences of temporary instability, producing an accumulation of downslope movement. The resulting movements can range from a few inches or less to tens of feet, and are characterized by breaking up of the ground and horizontal and vertical offsets.

Various relationships for estimating lateral spreading displacement have been proposed, including the Liquefaction Severity Index (LSI) by Youd and Perkins (1978), a relationship incorporating slope and liquefied soil thickness by Hamada et al. (1986), a modified LSI approach presented by Baziar et al. (1992), and a relationship by Bartlett and Youd (1992), in which they characterize displacement potential as a function of earthquake and local site characteristics (e.g., slope, liquefaction thickness, and grain size distribution). The relationship of Bartlett and Youd (1992), which is empirically based on analysis of case histories where lateral spreading did and did not occur, is relatively widely used, especially for initial assessments of the hazard. More site-specific analyses can also be made based on slope stability and deformation analysis procedures using undrained residual strengths for liquefied sand (Seed and Harder 1990; Stark and Mesri 1992), along with either Newmark-type simplified displacement analyses (Newmark 1965; Franklin and Chang 1977; Makdisi and Seed 1978; Yegian et al. 1991) or more complex deformation analysis approaches.

**Evaluating Potential for Flow Slides.** Flow generally occurs in liquefied materials found on steeper slopes and may involve ground movements of hundreds of feet or more. As a result, flow slides can be the most catastrophic of the liquefaction-related ground-failure phenomena. Fortunately, flow slides occur much less commonly than do lateral spreads. Whereas lateral spreading requires earthquake inertia forces to create instability for movement to occur, flow movements occur when the gravitational forces acting on a ground slope exceed the strength of the liquefied materials within the slope. The potential for flow sliding can be assessed by carrying out static slope stability analyses using undrained residual strengths for the liquefied materials.

**Evaluating Potential for Bearing Capacity Failure.** The occurrence of liquefaction in soils supporting foundations can result in bearing capacity failures and large, plunging-type settlements. In fact, the buildup of pore-water pressures in a soil to less than a complete liquefaction condition will still reduce soil strength and may threaten bearing capacity if the strength is reduced sufficiently.

The potential for bearing capacity failure beneath a spread footing depends on the depth of the liquefied (or partially liquefied) layer below the footing, the size of the footing, and the load. If lightly loaded small footings are located sufficiently above the depth of liquefied materials, bearing capacity failure may not occur. The foundation bearing capacity for a case where a footing is located some distance above a liquefied layer can be assessed by evaluating the strength of the liquefied (excess pore pressure ratio = 1.0), partially liquefied, and nonliquefied strata, then applying bearing capacity formulations for layered systems (Meyerhof 1974; Hanna and Meyerhof 1980; Hanna 1981). The capacity of friction pile or pier foundations can be similarly assessed, based on the strengths of the liquefied, partially liquefied, and nonliquefied strata penetrated by the foundations.

**Evaluating Potential for Liquefaction-Induced Settlements.** Following the occurrence of liquefaction, over time the excess pore-water pressures built up in the soil will dissipate, drainage will occur, and the soil will densify, manifesting at the ground surface as settlement. Differential settlements occur due to lateral variations in soil stratigraphy and density. Typically, such settlements are much smaller and tend to be more uniform than those due to bearing capacity failure. They may range from a few inches to a few feet at the most where thick, loose soil deposits liquefy.

One approach to estimating the magnitude of such ground settlement, analogous to the Seed-Idriss simplified empirical procedure for liquefaction potential evaluation (i.e., using SPT blow count data and cyclic stress ratio), has been presented by Tokimatsu and Seed (1987). Relationships presented by Ishihara and Yoshimine (1992) are also available for assessing settlement.

**Evaluating Increased Lateral Earth Pressures on Building Walls.** Liquefaction of soils adjacent to building walls increases lateral earth pressures which can be approximated as a fluid pressure having a unit weight equal to the saturated unit weight of the soil plus the inertial forces on the soil equal to the hydrodynamic pressure.

**Evaluating Potential for Flotation of Buried Structures.** A common phenomenon accompanying liquefaction is the flotation of tanks or structures that are embedded in liquefied soil. A building with a basement surrounded by liquefied soil can be susceptible to either flotation or bearing capacity failure, depending on the building weight and the structural continuity (i.e., whether the basement acts as an integral unit). The potential for flotation of a buried or embedded structure can be evaluated by comparing the total weight of the buried or embedded structure with the increased uplift forces occurring due to the buildup of liquefaction-induced pore-water pressures.

### 4.2.2.3 Differential Compaction

Differential compaction shall be defined as an earthquake-induced process in which foundation soils compact and the foundation settles in a nonuniform manner across a site.

Subsurface soil information shall be obtained to determine if soils susceptible to differential compaction are present under the building foundation.

A site shall be regarded as free of a differential compaction hazard if the soil conditions meet both of the following criteria:

1. Geologic materials below the ground water table do not pose a liquefaction hazard, based on the criteria in Section 4.2.2.2; and
2. Geologic deposits above the ground water table are either Pleistocene in geologic age (older than 11,000 years), stiff clays or clayey silts, or cohesionless sands, silts, and gravels with a minimum $(N_1)_{60}$ of 20 blows/0.3 m (20 blows/ft).

If a differential compaction hazard is determined to exist at the site, then a more detailed evaluation shall be performed using approved procedures.

### C4.2.2.3 Differential Compaction

Differential compaction or densification of soils may accompany strong ground shaking. The resulting differential settlements can be damaging to structures. Types of soil susceptible to liquefaction (i.e., relatively loose natural soils, or uncompacted or poorly compacted fill soils) are also susceptible to compaction. Compaction can occur in soils above and below the groundwater table.

Situations most susceptible to differential compaction include heavily graded areas where deep fills have been placed to create building sites for development. If the fills are not well compacted, they may be susceptible to significant settlements, and differential settlements may occur above variable depths of fill placed in canyons and near the transitions of cut and filled areas.

### 4.2.2.4 Landsliding

A landslide shall be defined as the down-slope mass movement of earth resulting from any cause. Subsurface soil information shall be obtained to determine if soils susceptible to a landslide that will cause differential movement of the building foundation are present at the site.

Slope stability shall be evaluated at sites with:

1. Existing slopes exceeding 18 degrees (three horizontal to one vertical); and/or
2. Prior histories of instability (rotational or translational slides, or rock fall).

Use of pseudo-static analyses shall be permitted to determine slope stability if the soils are not susceptible to liquefaction based on Section 4.2.2.2 or otherwise expected to lose shear strength during deformation. If soils are susceptible to liquefaction based on Section 4.2.2.2 or are otherwise expected to lose shear strength during deformation, dynamic analyses shall be performed to determine slope stability.

Pseudo-static analyses shall use a seismic coefficient equal to $S_{XS}/5$, to approximate one-half the peak ground acceleration at the site associated with the selected Rehabilitation Objective. Sites with a static factor of safety equal to or greater than 1.0 shall be judged to have adequate stability, and require no further stability analysis.

A sliding-block displacement analysis shall be performed for sites with a static factor of safety of less than 1.0. The displacement analysis shall determine the magnitude of ground movement and its effect upon the performance of the structure.

In addition to the effects of landslides that directly undermine the building foundation, the effects of rock fall or slide debris from adjacent slopes shall be evaluated using approved procedures.

### C4.2.2.4 Landsliding

If no blocks of rock are present at the site but a cliff or steep slope is located nearby, then the likely performance of the cliff under earthquake loading should be evaluated. The earthquake loading condition for cliff performance must be compatible with the earthquake loading condition selected for the Rehabilitation Objective for the building.

Some sites may be exposed to hazards from major landslides moving onto the site from upslope, or retrogressive removal of support from downslope. Such conditions should be identified during site characteri-

zation, and may pose special challenges if adequate investigation requires access to adjacent property.

### 4.2.2.5 Flooding or Inundation

For seismic rehabilitation of buildings for performance levels higher than Life Safety, site information shall be obtained to determine if the following sources of earthquake-induced flooding or inundation are present:

1. Dams located upstream, subject to damage by earthquake shaking or fault rupture;
2. Pipelines, aqueducts, and water storage tanks located upstream, subject to damage by fault rupture, earthquake-induced landslides, or strong shaking;
3. Coastal areas within tsunami zones or areas adjacent to bays or lakes, subject to seiche waves; and/or
4. Low-lying areas with shallow groundwater, subject to regional subsidence and surface ponding of water, resulting in inundation of the site.

Damage to buildings from earthquake-induced flooding or inundation shall be evaluated for its effect upon the performance of the structure.

In addition to the effects of earthquake-induced flooding or inundation, scour of building foundation soils from swiftly flowing water shall be evaluated using procedures approved by the authority having jurisdiction.

## 4.3 MITIGATION OF SEISMIC-GEOLOGIC SITE HAZARDS

Mitigation of seismic-geologic hazards identified in Section 4.2 shall be accomplished through modification of the structure, foundation, or soil conditions, or by other methods approved by the authority having jurisdiction. The structure, foundation, and soil for the rehabilitated structure shall meet the acceptance criteria for the appropriate chapters of this standard for the selected Rehabilitation Objective.

## C4.3 MITIGATION OF SEISMIC-GEOLOGIC SITE HAZARDS

Opportunities exist to improve seismic performance under the influence of some site hazards at reasonable cost; however, some site hazards may be so severe that they are economically impractical to include in risk-reduction measures. The discussions presented in this section are based on the concept that the extent of site hazards is discovered after the decision for seismic rehabilitation of a building has been made; however, the decision to rehabilitate a building and the selection of a Rehabilitation Objective may have been made with full knowledge that significant site hazards exist and must be mitigated as part of the rehabilitation.

Possible mitigation strategies for seismic geologic site hazards are presented in the following sections.

# 1. Fault Rupture

If the structural performance of a building evaluated for the calculated ground movement due to fault rupture during earthquake fails to comply with the requirements for the selected performance level, mitigation schemes should be employed that include one or more of the following measures to achieve acceptable performance: stiffening of the structure and/or its foundation; strengthening of the structure and/or its foundation; and modifications to the structure and/or its foundation to distribute the effects of differential vertical movement over a greater horizontal distance to reduce angular distortion.

Large movements caused by fault rupture generally cannot be mitigated economically. If the structural consequences of the estimated horizontal and vertical displacements are unacceptable for any performance level, either the structure, its foundation, or both, might be stiffened or strengthened to reach acceptable performance. Measures are highly dependent on specific structural characteristics and inadequacies. Grade beams and reinforced slabs are effective in increasing resistance to horizontal displacement. Horizontal forces are sometimes limited by sliding friction capacity of spread footings or mats. Vertical displacements are similar in nature to those caused by long-term differential settlement.

# 2. Liquefaction

If the structural performance of a building evaluated for the calculated ground movement due to liquefaction during an earthquake fails to comply with the requirements for the selected performance level, then one or more of the following mitigation measures should be implemented to achieve acceptable performance.

## 2.1 Modification of the Structure

The structure should be strengthened to improve resistance against the predicted liquefaction-induced

ground deformation. This solution may be feasible for small ground deformations.

## 2.2 Modification of the Foundation

The foundation system should be modified to reduce or eliminate the differential foundation displacements by underpinning existing shallow foundations to achieve bearing on deeper, nonliquefiable strata or by stiffening a shallow foundation system by a system of grade beams between isolated footings, or any other approved method.

## 2.3 Modification of the Soil Conditions

One or more of the following ground improvement techniques should be implemented to reduce or eliminate the liquefaction under existing buildings: soil grouting (either throughout the entire liquefiable strata beneath a building, or locally beneath foundation components); installation of drains; or installation of permanent dewatering systems.

Other types of ground improvement widely used for new construction are less applicable to existing buildings because of the effects of the procedures on the building. Thus, removal and replacement of liquefiable soil or in-place densification of liquefiable soil by various techniques are not applicable beneath an existing building.

## 2.4 Mitigation of the Lateral Spreading

Large soil volumes should be stabilized and/or buttressing structures should be constructed.

If the potential for significant liquefaction-induced lateral spreading movements exists at a site, then the mitigation of the liquefaction hazard may be more difficult. This is because the potential for lateral spreading movements beneath a building may depend on the behavior of the soil mass at distances well beyond the building as well as immediately beneath it.

# 3. Differential Compaction

If the structural performance of a building evaluated for the calculated differential compaction during earthquake fails to comply with the requirements for the selected performance level, then one or more mitigation measures similar to those recommended for liquefaction should be implemented to achieve acceptable performance.

# 4. Landslide

If the structural performance of a building evaluated for the calculated ground movement due to landslide during earthquake fails to comply with the requirements for the selected performance level, then one or more of the following mitigation measures should be implemented to achieve acceptable performance:

> Regrading;
> Drainage;
> Buttressing;
> Structural improvements:
>> Gravity walls;
>> Tieback/soil nail walls;
>> Mechanically stabilized earth walls;
>> Barriers for debris torrents or rock fall;
>> Building strengthening to resist deformation;
>> Grade beams; and
>> Shear walls.
> Soil modification/replacement:
>> Grouting; and
>> Densification.

## 5. Flooding or Inundation

If the structural performance of a building evaluated for the effects of earthquake-induced flooding and inundation fails to comply with the requirements for the selected performance level, then one or more of the following mitigating measures should be implemented to achieve acceptable performance:

1. Improvement of nearby dam, pipeline, or aqueduct facilities independent of the rehabilitated building;
2. Diversion of anticipated peak flood flows;
3. Installation of pavement around the building to reduce scour; and
4. Construction of sea wall or breakwater for tsunami or seiche protection.

## 4.4 FOUNDATION STRENGTH AND STIFFNESS

Foundation strength and stiffness shall be determined in accordance with this section.

## C4.4 FOUNDATION STRENGTH AND STIFFNESS

It is assumed that foundation soils are not susceptible to significant strength loss due to earthquake loading. In general, soils have considerable ductility unless they degrade significantly in stiffness and strength under cyclic loading. With this assumption, the provisions of this section provide an overview of the requirements and procedures for evaluating the ability of foundations to withstand the imposed seismic loads without excessive deformations.

The amount of acceptable deformations for foundations in such soils depends primarily on the effect of the deformation on the structure, which in turn depends on the desired Structural Performance Level. However, foundation yield associated with mobilization at upper-bound expected capacity during earthquake loading may be accompanied by progressive permanent foundation settlement during continued cyclic loading, albeit in most cases this settlement probably would be less than a few inches. In general, if the real loads transmitted to the foundation during earthquake loading do not exceed upper-bound expected soil capacities, it can be assumed that foundation deformations will be relatively small.

Parametric analyses to cover uncertainties in soil load–deformation characteristics are required. One alternative is to perform the NSP or NDP because the nonlinear load–deformation characteristics of the foundations can be directly incorporated in these analyses (Section 4.4.2). In static analyses, a somewhat conservative interpretation of the results is recommended because cyclic loading effects cannot be incorporated directly.

### 4.4.1 Expected Capacities of Foundations

The expected capacity of foundation components shall be determined by presumptive, prescriptive, or site-specific methods as specified in Sections 4.4.1.1 through 4.4.1.3. Capacities shall be based on foundation information obtained as specified in Section 4.2.1.

### C4.4.1 Expected Capacities of Foundations

Design values recommended by geotechnical engineers are generally consistent with lower-bound values. It is important to obtain information on the actual factor of safety applied to arrive at design values so that soil capacities are understood and expected values can be properly derived.

#### 4.4.1.1 Presumptive Capacities

*4.4.1.1.1 Presumptive Capacities of Shallow Foundations* Calculation of presumptive expected capacities for spread footings and mats shall be permitted using the parameters specified in Table 4-2.

*4.4.1.1.2 Presumptive Capacities of Deep Foundations*
It shall be permitted to determine pile and pier capacity parameters using Table 4-3, 4-4, 4-5, and 4-6.

**Table 4-3** Typical Pile and Pier Capacity Parameters: Bearing Capacity Factors, $N_q$

**Table 4-4** Typical Pile and Pier Capacity Parameters: Effective Horizontal Stress Factors, $F_{di}$ and $F_{ui}$

**Table 4-5** Typical Pile and Pier Capacity Parameters: Friction Angle, $\delta$ (degrees)

**Table 4-6** Typical Pile and Pier Capacity Parameters: Cohesion, $c_t$, and Adhesion, $c_a$ (psf)

**Capacities of Piles or Piers in Granular Soils**
Calculation of presumptive expected capacities of piles or piers in granular soils shall be permitted using the procedure shown in Fig. 4-1.

**Table 4-2. Parameters for Calculating Presumptive Expected Foundation Load Capacities of Spread Footings and Mats**

| Class of Materials[2] | Vertical Foundation Pressure[3] Lbs/Sq. Ft ($q_c$) | Lateral Bearing Pressure Lbs/Sq. Ft/Ft of Depth Below Natural Grade[4] | Lateral Sliding[1] Coefficient[5] | Resistance[6] Lbs/Sq. Ft |
|---|---|---|---|---|
| Massive Crystalline Bedrock | 8,000 | 2,400 | 0.80 | — |
| Sedimentary and Foliated Rock | 4,000 | 800 | 0.70 | — |
| Sandy Gravel and/or Gravel (GW and GP) | 4,000 | 400 | 0.70 | — |
| Sand, Silty Sand, Clayey Sand, Silty Gravel, and Clayey Gravel (SW, SP, SM, SC, GM, and GC) | 3,000 | 300 | 0.50 | — |
| Clay, Sandy Clay, Silty Clay, and Clayey Silt (CL, ML, MH, and CH) | 2,000[7] | 200 | — | 260 |

[1]Lateral bearing and lateral sliding resistance shall be permitted to be combined.
[2]For soil classifications OL, OH, and PT (i.e., organic clays and peat), a foundation investigation shall be required.
[3]All values of expected bearing capacities are for footings having a minimum width of 12 in. and a minimum depth of 12 in. into natural grade. Except where Footnote 7 applies, an increase of 20% is allowed for each additional foot of width or depth to a maximum value of three times the designated value.
[4]Shall be permitted to be increased by the amount of the designated value for each additional foot of depth to a maximum of 15 times the designated value.
[5]Coefficient applied to the dead load.
[6]Lateral sliding resistance value to be multiplied by the contact area. In no case shall the lateral sliding resistance exceed one-half of the dead load.
[7]No increase for width shall be permitted.

**Table 4-3. Typical Pile and Pier Capacity Parameters: Bearing Capacity Factors, $N_q$**

| | Angle of Shearing Resistance for Soil, $\phi$ (degrees) | | | | | | | | | | | | |
|---|---|---|---|---|---|---|---|---|---|---|---|---|---|
| Placement | 26 | 28 | 30 | 31 | 32 | 33 | 34 | 35 | 36 | 37 | 38 | 39 | 40+ |
| Driven Pile | 10 | 15 | 21 | 24 | 29 | 35 | 42 | 50 | 62 | 77 | 86 | 120 | 145 |
| Drilled Pier | 5 | 8 | 10 | 12 | 14 | 17 | 21 | 25 | 30 | 38 | 43 | 60 | 72 |

**Table 4-4. Typical Pile and Pier Capacity Parameters: Effective Horizontal Stress Factors, $F_{di}$ and $F_{ui}$**

| Pile or Pier Type | Downward $F_{di}$[1] | | Upward $F_{ui}$[1] | |
|---|---|---|---|---|
| | Low | High | Low | High |
| Driven H-Pile | 0.5 | 1.0 | 0.3 | 0.5 |
| Driven Straight Prismatic Pile | 1.0 | 1.5 | 0.6 | 1.0 |
| Driven Tapered Pile | 1.5 | 2.0 | 1.0 | 1.3 |
| Driven Jetted Pile | 0.4 | 0.9 | 0.3 | 0.6 |
| Drilled Pier | 0.7 | 0.7 | 0.4 | 0.4 |

[1]Expected values that are selected on the basis of conditions established in accordance with Section 4.4.1.1 shall not fall outside the range of values indicated by Low and High.

**Table 4-5. Typical Pile and Pier Capacity Parameters: Friction Angle, $\delta$ (degrees)**

| Pile or Pier Material | $\delta$ |
|---|---|
| Steel | 20 |
| Concrete | $0.75\,\phi$ |
| Timber | $0.75\,\phi$ |

**Table 4-6. Typical Pile and Pier Capacity Parameters: Cohesion, $c_t$, and Adhesion, $c_a$ (psf)**

| Pile Material | Consistency of Soil (Approximate SPT Blow Count) | Cohesion, $c_t$[1] | | Adhesion, $c_a$[1] | |
|---|---|---|---|---|---|
| | | Low | High | Low | High |
| Timber and Concrete | Very Soft (< 2) | 0 | 250 | 0 | 250 |
| | Soft (2–4) | 250 | 500 | 250 | 480 |
| | Medium Stiff (4–8) | 500 | 1,000 | 480 | 750 |
| | Stiff (8–15) | 1,000 | 2,000 | 750 | 950 |
| | Very Stiff (> 15) | 2,000 | 4,000 | 950 | 1,300 |
| Steel | Very Soft (< 2) | 0 | 250 | 0 | 250 |
| | Soft (2–4) | 250 | 500 | 250 | 460 |
| | Medium Stiff (4–8) | 500 | 1,000 | 460 | 700 |
| | Stiff (8–15) | 1,000 | 2,000 | 700 | 720 |
| | Very Stiff (> 15) | 2,000 | 4,000 | 720 | 750 |

[1]Expected values that are selected on the basis of conditions established in accordance with Section 4.4.1.1 shall not fall outside the range of values indicated by Low and High

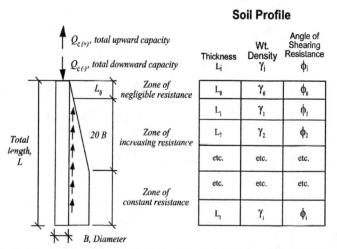

**FIGURE 4-1. Presumptive Expected Capacities of Piles or Piers in Granular Soils.**

**Capacities of Piles or Piers in Cohesive Soils**

Calculation of presumptive expected capacities of piles or piers in cohesive soils shall be permitted using the procedure shown in Fig. 4-2

*C4.4.1.1.2 Presumptive Capacities of Deep Foundations* The calculation procedures for presumptive expected capacities of piles or piers specified in this section are adapted from ATC-40 (ATC 1996), NAVFAC DM-7.01 (NAVFAC 1986a), and NAVFAC DM-7.02 (NAVFAC 1986b).

### 4.4.1.2 Prescriptive Expected Capacities

Prescriptive expected capacities shall be used where construction documents or previous geotechnical reports for the existing building are available and provide information on foundation soil design parameters. Calculation of prescriptive expected capacities by the following methods shall be permitted:

1. The prescriptive expected bearing capacity, $q_c$, for a spread footing shall be calculated using Eq. 4-1:

$$q_c = 3q_{allow} \qquad \text{(Eq. 4-1)}$$

where

$q_{allow}$ = allowable bearing pressure specified in available documents for the gravity load design of shallow foundations (dead plus live loads);

2. For deep foundations, the prescriptive expected vertical capacity, $Q_c$, of individual piles or piers shall be calculated using Eq. 4-2:

$$Q_c = 3Q_{allow} \qquad \text{(Eq. 4-2)}$$

where

$Q_{allow}$ = allowable vertical capacity specified in available documents for the gravity load design of deep foundations (dead plus live loads); and

3. Alternatively, the prescriptive expected capacity, $q_c$ or $Q_c$, of any foundation, shallow or deep, shall be calculated using Eq. 4-3:

$$q_c \text{ or } Q_c = 1.5Q_G \qquad \text{(Eq. 4-3)}$$

where $Q_G$ = gravity load action as specified in Section 3.2.8, expressed in terms of pressure or load.

### 4.4.1.3 Site-Specific Capacities

For buildings where the methods specified in Sections 4.4.1.1 and 4.4.1.2 do not apply, a subsurface geotechnical investigation shall be conducted to determine expected ultimate foundation capacities based on the specific characteristics of the building site.

### 4.4.2 Load–Deformation Characteristics for Foundations

If building foundations are explicitly modeled in the mathematical model of the building, the load–deformation characteristics shall be calculated in accordance with Section 4.4.2.1 for shallow bearing foundations, Section 4.4.2.2 for pile foundations, and Section 4.4.2.3 for drilled shafts.

For explicit modeling of other types of foundations, load–deformation characteristics shall be calculated by an approved method.

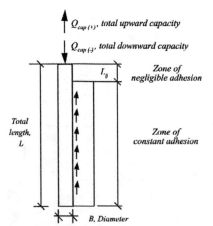

**FIGURE 4-2. Presumptive Expected Capacities of Piles or Piers in Cohesive Soils.**

Nonlinear behavior of foundations shall be represented by an equivalent elasto-plastic load–deformation relationship unless another approved relationship is available.

Where foundation components are modeled explicitly, the analysis shall be performed using upper- and lower-bound load–deformation characteristics of foundations as illustrated in Fig. 4-3(a) and defined in this section. Where foundation components are not modeled explicitly, the analysis shall be bounded by the upper- and lower-bound foundation capacity as defined in this section. In lieu of explicit evaluation of uncertainties in foundation characteristics, it shall be permitted to take the upper-bound stiffness and capacity values as two times the values given in this section and the lower-bound stiffness and capacity values as one-half of the values given in this section.

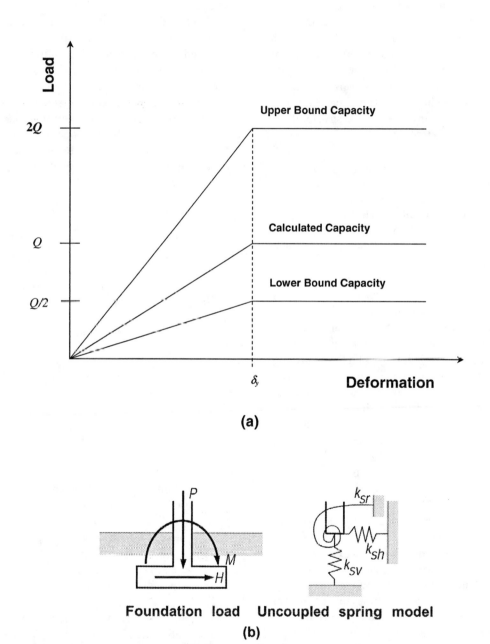

(a)

Foundation load    Uncoupled spring model

(b)

**FIGURE 4-3. (a) Idealized Elasto-Plastic Load–Deformation Behavior for Soils; (b) Uncoupled Spring Model for Rigid Footings.**

**C4.4.2 Load–Deformation Characteristics for Foundations**

Load–deformation characteristics are required where the effects of foundations are to be taken into account in LSPs or LDPs, NSPs (pushover), or NDPs (time history). Foundation load–deformation parameters characterized by both stiffness and capacity can have a significant effect on both structural response and load distribution among structural components.

While it is recognized that the load–deformation behavior of foundations is nonlinear, an equivalent elasto-plastic representation of load–deformation behavior is recommended because of the difficulties in determining soil properties and the likely variability of soils supporting foundations. In addition, to allow for such variability or uncertainty, an upper- and lower-bound approach to defining stiffness and capacity is required to evaluate the sensitivity of the structural response to these parameters.

The sources of this uncertainty include variations due to rate of loading, assumed elasto-plastic soil behavior, level of strain, cyclic loading, and variability of soil properties. These sources of variability produce results that are generally within a factor of two above or below the expected value. It is conceivable that certain conditions will fall outside the bounds prescribed in this standard. However, it is not the objective to guarantee that the answer is always within the applied factor. Instead, the intent is (1) that solution sensitivity be identified; and (2) that the bounds, considered reasonably, capture the expected behavior. Current practice (both conventional and within the nuclear industry) has suggested that variation by a factor of two is generally appropriate. Geotechnical engineers often use a safety factor of two to establish lower-bound values for use in design. Consistent with the approach taken in ASCE 4 (ASCE 1998), if additional testing is performed, the range could be narrowed to that defined by multiplying and dividing by $(1 + C_v)$, where the coefficient of variation, $C_v$, is defined as the standard deviation divided by the mean. In no case should $C_v$ be taken to be less than 0.5.

It is important that geotechnical engineers report the average results obtained and the actual factor of safety applied to arrive at design values. The design values recommended by geotechnical engineers are generally consistent with the lower bound. If such reduced values were used by the structural engineer as expected values, the application of the prescribed upper- and lower-bound variations would not achieve the intended aim.

**Consideration of Foundation Rocking.** Buildings may rock on their foundations in an acceptable manner provided the structural components can accommodate the resulting displacements and deformations. Consideration of rocking can be used to limit the force input to a building; however, rocking should not be considered simultaneously with the effects of soil flexibility.

The design professional is directed to FEMA 274 (FEMA 1997) and the work of Yim and Chopra (1985), Housner (1963), Makris and Roussos (1998), Priestley et al. (1978), and Makris and Konstantinidis (2001) for additional information on rocking behavior. In using those references two points of caution should be noted: (1) Makris and Konstantinidis report that the simple response-spectrum-based design method proposed by Priestley should not be used as it is based on an erroneous dynamic characterization of the rocking problem; and (2) physical experiments conducted by Priestley show that the common theoretical assumption of perfectly inelastic collisions during rocking overestimates the actual energy reduction.

### 4.4.2.1 Shallow Bearing Foundations

*4.4.2.1.1 Stiffness Parameters* The initial shear modulus, $G_o$, shall be calculated in accordance with Eq. 4-4 or 4-5 where $v_s$ is the shear wave velocity at low strains, $\gamma$ is the weight density of the soil, and $g$ is the acceleration due to gravity:

$$G_0 = \frac{\gamma v_s^2}{g} \qquad \text{(Eq. 4-4)}$$

$$G_0 \cong 20{,}000(N_1)_{60}^{1/3}\sqrt{\sigma_0'} \qquad \text{(Eq. 4-5)}$$

where

$(N_1)_{60}$ = Standard Penetration Test blow count normalized for an effective stress of 1.0 ton psf confining pressure and corrected to an equivalent hammer energy efficiency of 60%;
$\sigma_o'$ = effective vertical stress in psf;
$\sigma_o'$ = $\gamma_t d - \gamma_w(d - d_w)$;
$\gamma_t$ = total unit weight of soil;
$\gamma_w$ = unit weight of water;
$d$ = depth to sample; and
$d_w$ = depth to groundwater level.

$G_o$ in Eq. 4-5 is expressed in lbs/psf, as is $\sigma_o'$.
The effective shear modulus, $G$, shall be calculated in accordance with Table 4-7.

**Table 4-7. Effective Shear Modulus Ratio ($G/G_0$)**

| Site Class | Effective Peak Acceleration, $S_{XS}/2.5$[1] | | | |
|---|---|---|---|---|
| | $S_{XS}/2.5 = 0$ | $S_{XS}/2.5 = 0.1$ | $S_{XS}/2.5 = 0.4$ | $S_{XS}/2.5 = 0.8$ |
| A | 1.00 | 1.00 | 1.00 | 1.00 |
| B | 1.00 | 1.00 | 0.95 | 0.90 |
| C | 1.00 | 0.95 | 0.75 | 0.60 |
| D | 1.00 | 0.90 | 0.50 | 0.10 |
| E | 1.00 | 0.60 | 0.05 | * |
| F | * | * | * | * |

[1]Use straight-line interpolation for intermediate values of $S_{XS}/2.5$.
*Site-specific geotechnical investigation and dynamic site response analyses shall be performed.

Based on relative stiffnesses of the foundation structure and the supporting soil, the foundation stiffness shall be calculated using one of the following three methods:

*C4.4.2.1.1 Stiffness Parameters* Table 4-7 is consistent with the site classification Tables 1-4 and 1-5 in that the layout and level of complexity are identical, and the indication of problem soils that require site-specific investigation (site Class F) is consistent. The following observations on the relationship between shear modulus reduction and peak ground acceleration can be made:

1. As the peak ground acceleration approaches zero, the modulus reduction factor approaches unity;
2. Modulus reduction effects are significantly more pronounced for softer soils; and

The modulus reduction factors given in both FEMA 273 (FEMA 1997) and the FEMA 302 *NEHRP Recommended Provisions* (FEMA 1997) overestimate the modulus reduction effects for Site Classes A, B, and C.

The shears and moments in foundation components are conservative where such components are considered rigid. However, soil pressures may be significantly underestimated where foundation flexibility is ignored. The flexibility and nonlinear response of soil and of foundation structures should be considered where the results would change.

For beams on elastic supports (e.g., strip footings and grade beams) with a point load at midspan, the beam may be considered rigid where:

$$\frac{EI}{L^4} > \frac{2}{3}k_{sv}B \qquad \text{(Eq. C4-1)}$$

The preceeding equation is generally consistent with traditional beam-on-elastic foundation limits (NAVFAC 1986b; Bowles 1988). The resulting soil bearing pressures are within 3% of the results, including foundation flexibility.

For rectangular plates (with plan dimensions $L$ and $B$, and thickness $t$, and mechanical properties $E_f$ and $v_f$) on elastic supports (for instance, mat foundations or isolated footings) subjected to a point load in the center, the foundation may be considered rigid where:

$$4k_{sv}\sum_{m=1}^{5}\sum_{n=1}^{5}\frac{\sin^2\left(\dfrac{m\cdot\pi}{2}\right)\sin^2\left(\dfrac{n\cdot\pi}{2}\right)}{\left[\pi^4 D_f\left(\dfrac{m^2}{L^2}+\dfrac{n^2}{B^2}\right)^2\right]+k_{sv}} < 0.03$$

$$\text{(Eq. C4-2)}$$

where

$$D_f = \frac{E_f t^3}{12(1-v_f)^2} \qquad \text{(Eq. C4-3)}$$

The above equation is based on Timoshenko's solutions for plates on elastic foundations (Timeshenko and Woinowsky-Krieger 1959). The general solution has been simplified by restriction to a center load. Only the first five values of $m$ and $n$ (in the infinite series) are required to achieve reasonable accuracy.

*4.4.2.1.2 Method 1* For shallow bearing footings that are rigid with respect to the supporting soil, an uncoupled spring model, as shown in Fig. 4-3(b), shall represent the foundation stiffness.

The equivalent spring constants shall be calculated as specified in Fig. 4-4.

*C4.4.2.1.2 Method 1* Researchers have developed spring stiffness solutions that are applicable to any solid basemat shape on the surface of, or partially or fully embedded in, a homogeneous halfspace. Such solutions are reported in Gazetas (1991). The equations in Fig. 4-4 reflect the most common conditions—rectangular foundations and rectangular strip footings. Rather than taking the approach of ATC 40 (ATC 1996), in which equations for foundations of arbitrary shape were adapted to the case of rectangular foundations, the surface stiffness equations that appear

| Degree of Freedom | Stiffness of Foundation at Surface | Note |
|---|---|---|
| Translation along x-axis | $K_{x,sur} = \dfrac{GB}{2-\nu}\left[3.4\left(\dfrac{L}{B}\right)^{0.65} + 1.2\right]$ | |
| Translation along y-axis | $K_{y,sur} = \dfrac{GB}{2-\nu}\left[3.4\left(\dfrac{L}{B}\right)^{0.65} + 0.4\dfrac{L}{B} + 0.8\right]$ | |
| Translation along z-axis | $K_{z,sur} = \dfrac{GB}{1-\nu}\left[1.55\left(\dfrac{L}{B}\right)^{0.75} + 0.8\right]$ | |
| Rocking about x-axis | $K_{xx,sur} = \dfrac{GB^3}{1-\nu}\left[0.4\left(\dfrac{L}{B}\right) + 0.1\right]$ | |
| Rocking about y-axis | $K_{yy,sur} = \dfrac{GB^3}{1-\nu}\left[0.47\left(\dfrac{L}{B}\right)^{2.4} + 0.034\right]$ | Orient axes such that $L>B$. If $L=B$, use x-axis equations for both x-axis and y-axis. |
| Torsion about z-axis | $K_{zz,sur} = GB^3\left[0.53\left(\dfrac{L}{B}\right)^{2.45} + 0.51\right]$ | |

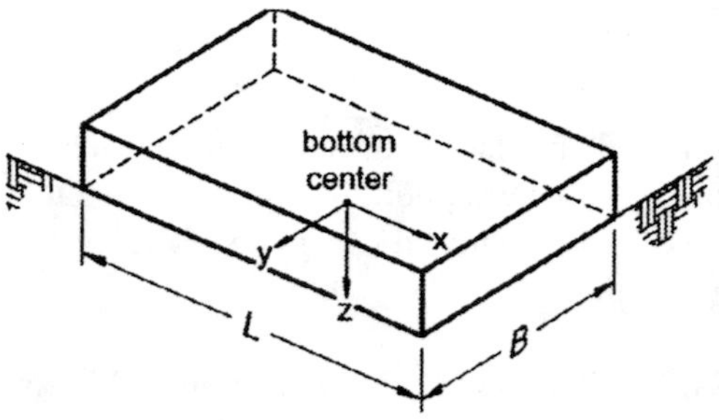

| Degree of Freedom | Correction Factor for Embedment | Note |
|---|---|---|
| Translation along x-axis | $\beta_x = \left(1 + 0.21\sqrt{\dfrac{D}{B}}\right)\cdot\left[1 + 1.6\left(\dfrac{hd(B+L)}{BL^2}\right)^{0.4}\right]$ | |
| Translation along y-axis | $\beta_y = \left(1 + 0.21\sqrt{\dfrac{D}{L}}\right)\cdot\left[1 + 1.6\left(\dfrac{hd(B+L)}{LB^2}\right)^{0.4}\right]$ | |
| Translation along z-axis | $\beta_z = \left[1 + \dfrac{1}{21}\dfrac{D}{B}\left(2 + 2.6\dfrac{B}{L}\right)\right]\cdot\left[1 + 0.32\left(\dfrac{d(B+L)}{BL}\right)^{2/3}\right]$ | |
| Rocking about x-axis | $\beta_{xx} = 1 + 2.5\dfrac{d}{B}\left[1 + \dfrac{2d}{B}\left(\dfrac{d}{D}\right)^{-0.2}\sqrt{\dfrac{B}{L}}\right]$ | d = height of effective sidewall contact (may be less than total foundation height) |
| Rocking about y-axis | $\beta_{yy} = 1 + 1.4\left(\dfrac{d}{L}\right)^{0.6}\left[1.5 + 3.7\left(\dfrac{d}{L}\right)^{1.9}\left(\dfrac{d}{D}\right)^{-0.6}\right]$ | h = depth to centroid of effective sidewall contact |
| Torsion about z-axis | $\beta_{zz} = 1 + 2.6\left(1 + \dfrac{B}{L}\right)\left(\dfrac{d}{B}\right)^{0.9}$ | For each degree of freedom, calculate $K_{emb} = \beta K_{sur}$ |

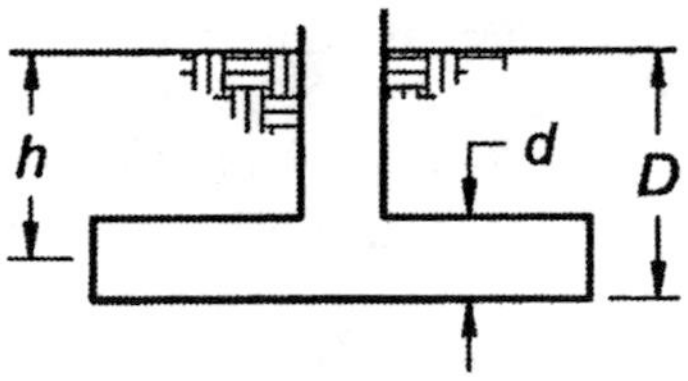

**FIGURE 4-4. Elastic Solutions for Rigid Footing Spring Constraints.**

in Fig. 4-4 are those reported by Pais and Kausel (1988) for the specific case of rectangular foundations. These equations are used because they are somewhat simpler than those that would result from an adaptation of the equations in Gazetas (1991) and because they are expected to be more accurate; Pais and Kausel report that the largest error for these shape-specific equations is expected to be "less than a few percent." Because Pais and Kausel report that "only scarce data are available for the stiffnesses of embedded rectangular foundations," the embedment correction factors shown in Fig. 4-4 are based on an adaptation of the general solutions presented in Gazetas. Concerning these embedment factors, Gazetas reports that "the errors that may result from their use will be well within an acceptable 15 percent."

Using Fig. 4-4, a two-step calculation process is required. First, the stiffness terms are calculated for a foundation at the surface. Then, an embedment correction factor is calculated for each stiffness term. The stiffness of the embedded foundation is the product of these two terms. Figure C4-1 illustrates the effects of foundation aspect ratio and embedment.

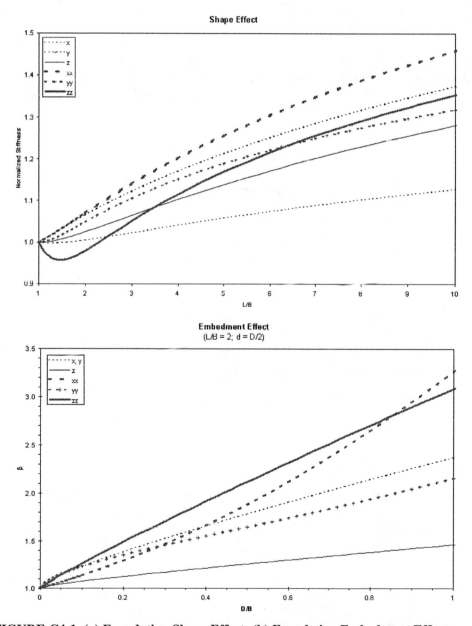

**FIGURE C4-1. (a) Foundation Shape Effect; (b) Foundation Embedment Effect.**

According to Gazetas, the height of effective sidewall contact, $d$, should be taken as the average height of the sidewall that is in good contact with the surrounding soil. It should, in general, be smaller than the nominal height of contact to account for such phenomena as slippage and separation that may occur near the ground surface. Note that $d$ will not necessarily attain a single value for all modes of oscillation. Where $d$ is taken larger than zero, the resulting stiffness includes sidewall friction and passive pressure contributions.

Although frequency-dependent solutions are available, results are reasonably insensitive to loading frequencies within the range of parameters of interest for buildings subjected to earthquakes. It is sufficient to use static stiffnesses as representative of repeated loading conditions.

*4.4.2.1.3 Method 2* For shallow bearing foundations that are not rigid with respect to the supporting soils, a finite element representation of linear or nonlinear

foundation behavior using Winkler models shall be used. Distributed vertical stiffness properties shall be calculated by dividing the total vertical stiffness by the area. Uniformly distributed rotational stiffness properties shall be calculated by dividing the total rotational stiffness of the footing by the moment of inertia of the footing in the direction of loading. Vertical and rotational stiffnesses shall be decoupled for a Winkler model. It shall be permitted to use the procedure illustrated in Fig. 4-5 to decouple these stiffnesses.

*C4.4.2.1.3 Method 2* The stiffness per unit length in these end zones is based on the vertical stiffness of a $B \times B/6$ isolated footing. The stiffness per unit length in the middle zone is equivalent to that of an infinitely long strip footing.

*4.4.2.1.4 Method 3* For shallow bearing foundations that are flexible relative to the supporting soil, the relative stiffness of foundations and supporting soil shall

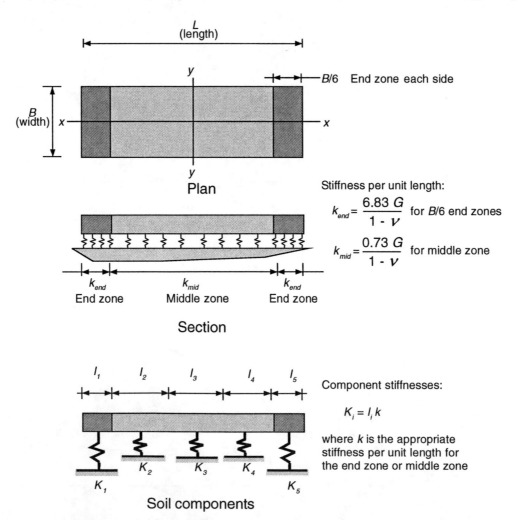

FIGURE 4-5. Vertical Stiffness Modeling for Shallow Bearing Footings.

be evaluated using theoretical solutions for beams and plates on elastic supports, approved by the authority having jurisdiction. The foundation stiffness shall be permitted to be calculated by a decoupled Winkler model using a unit subgrade spring coefficient. For flexible foundation systems, the unit subgrade spring coefficient, $k_{sv}$, shall be calculated by Eq. 4-6.

$$k_{sv} = \frac{1.3G}{B(1-v)} \qquad \text{(Eq. 4-6)}$$

where

$G$ = shear modulus;
$B$ = width of footing; and
$v$ = Poisson's ratio.

*4.4.2.1.5 Capacity Parameters* The vertical expected capacity of shallow bearing foundations shall be determined using the procedures of Section 4.4.1.

In the absence of moment loading, the expected vertical load capacity, $Q_c$, of a rectangular footing shall be calculated by Eq. 4-7.

$$Q_c = q_c BL \qquad \text{(Eq. 4-7)}$$

where

$q_c$ = expected bearing capacity determined in Section 4.4.1;
$B$ = width of footing; and
$L$ = length of footing.

The moment capacity of a rectangular footing shall be calculated by Eq. 4-8:

$$M_c = \frac{LP}{2}\left(1 - \frac{q}{q_c}\right) \qquad \text{(Eq. 4-8)}$$

where

$P$ = vertical load on footing;
$q = \dfrac{P}{BL}$ = vertical bearing pressure;
$B$ = width of footing (parallel to the axis of bending);
$L$ = length of footing in the direction of bending; and
$q_c$ = expected bearing capacity determined in Section 4.4.1.

The lateral capacity of shallow foundations shall be calculated using established principles of soil mechanics and shall include the contributions of traction at the bottom and passive pressure resistance on the leading face. Mobilization of passive pressure shall be calculated using Fig. 4-6.

*C4.4.2.1.5 Capacity Parameters* For rigid footings subject to moment and vertical load, contact stresses become concentrated at footing edges, particularly as uplift occurs. The ultimate moment capacity, $M_c$, is dependent upon the ratio of the vertical load stress, $q$, to the expected bearing capacity, $q_c$. Assuming that contact stresses are proportional to vertical displacement and remain elastic up to the expected bearing

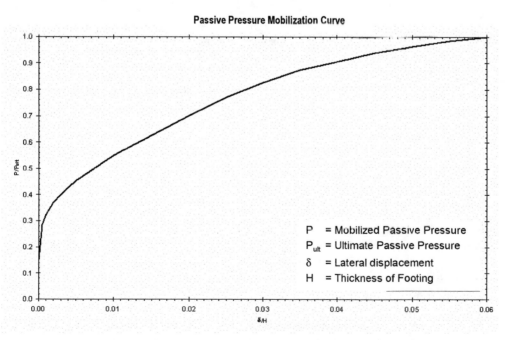

**Passive Pressure Mobilization Curve**

P = Mobilized Passive Pressure
$P_{ult}$ = Ultimate Passive Pressure
$\delta$ = Lateral displacement
H = Thickness of Footing

**FIGURE 4-6. Passive Pressure Mobilization Curve.**

capacity, $q_c$, it can be shown that uplift will occur prior to plastic yielding of the soil where $q/q_c$ is less than 0.5. If $q/q_c$ is greater than 0.5, then the soil at the toe will yield prior to uplift. This is illustrated in Fig. C4-2.

For footings subjected to lateral loads, the base traction strength is given by $V = C + N\mu$, where $C$ is the effective cohesion force (effective cohesion stress, $c$, times footing base area), $N$ is the normal (compressive) force, and $\mu$ is the coefficient of friction. If included, side traction is calculated in a similar manner. The coefficient of friction is often specified by the geotechnical consultant. In the absence of such a recommendation, $\mu$ may be based on the minimum of the effective internal friction angle of the soil and the friction coefficient between soil and foundation from published foundation references. The ultimate passive pressure strength is often specified by the geotechnical consultant in the form of passive pressure coefficients or equivalent fluid pressures. The passive pressure problem has been extensively investigated for more than 200 years. As a result, countless solutions and recommendations exist. The method used should, at a minimum, include the contributions of internal friction and cohesion, as appropriate.

As shown in Fig. 4-6, the force–displacement response associated with passive pressure resistance is highly nonlinear. However, for shallow foundations, passive pressure resistance generally accounts for much less than half of the total strength. Therefore, it is adequate to characterize the nonlinear response of shallow foundations as elastic-perfectly plastic using the initial, effective stiffness and the total expected strength. The actual behavior is expected to fall within the upper and lower bounds prescribed in this standard.

### 4.4.2.2 Pile Foundations

A pile foundation shall be defined as a deep foundation system composed of one or more driven or cast-in-place piles and a pile cap cast-in-place over the piles, which together form a pile group supporting one or more load-bearing columns, or a linear sequence of pile groups supporting a shear wall.

The requirements of this section shall apply to piles less than or equal to 24 in. in diameter. The stiffness characteristics of single large-diameter piles or drilled shafts larger than 24 in. in diameter shall comply with the requirements of Section 4.4.2.3.

*4.4.2.2.1 Stiffness Parameters* The uncoupled spring model shown in Fig. 4-3(b) shall be used to represent the stiffness of a pile foundation where the footing in the figure represents the pile cap. In calculating the vertical and rocking springs, the contribution of the soil immediately beneath the pile cap shall be neglected. The total lateral stiffness of a pile group shall

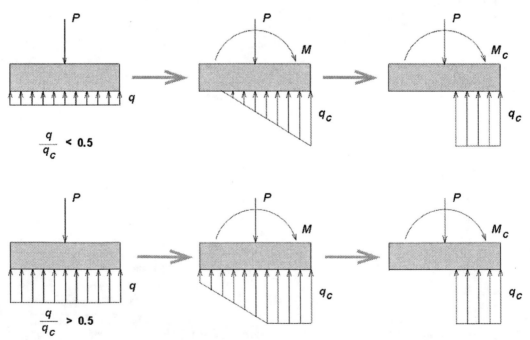

**FIGURE C4-2. Idealized Concentration of Stress at Edge of Rigid Footings Subjected to Overturning Moment.**

include the contributions of the piles (with an appropriate modification for group effects) and the passive resistance of the pile cap. The lateral stiffness of piles shall be based on classical methods or on analytical solutions using approved beam-column pile models. The lateral stiffness contribution of the pile cap shall be calculated using the passive pressure mobilization curve in Fig. 4-6.

Pile group axial spring stiffness values, $k_{sv}$, shall be calculated using Eq. 4-9.

$$k_{sv} = \sum_{n=1}^{N} \frac{A\,E}{L} \qquad \text{(Eq. 4-9)}$$

where

$A$ = cross-sectional area of a pile;
$E$ = modulus of elasticity of piles;
$L$ = length of piles; and
$N$ = number of piles in group.

The rocking spring stiffness values about each horizontal pile cap axis shall be computed by modeling each pile axial spring as a discrete Winkler spring. The rotational spring constant, $k_{sr}$, (moment per unit rotation) shall be calculated using Eq. 4-10:

$$k_{sr} = \sum_{n=1}^{N} k_{vn} S_n^2 \qquad \text{(Eq. 4-10)}$$

where

$k_{vn}$ = axial stiffness of the $n$-th pile; and
$S_n$ = distance between $n$-th pile and axis of rotation.

*C4.4.2.2.1 Stiffness Parameters* As the passive pressure resistance may be a significant part of the total strength, and deep foundations often require larger lateral displacements than shallow foundations to mobilize the expected strength, it may not be appropriate to base the force–displacement response on the initial, effective stiffness alone. Instead, the contribution of passive pressure should be based on the passive pressure mobilization curve provided in Fig. 4-6.

Although the effects of group action and the influence of pile batter are not directly accounted for in the form of the above equations, it can be reasonably assumed that the latter effects are accounted for in the range of uncertainties that must be considered in accordance with Section 4.4.1.

*4.4.2.2.2 Capacity Parameters* The expected axial capacity of piles in compression and tension shall be determined using the procedures in Section 4.4.1. The expected axial capacity in tension shall not exceed the lower-bound capacity of the foundation structural components.

The moment capacity of a pile group shall be determined assuming a rigid pile cap. Lower-bound moment capacity shall be based on triangular distribution of axial pile loading and lower-bound axial capacity of the piles. Upper-bound moment capacity shall be based on a rectangular distribution of axial pile load using full, upper-bound axial capacity of the piles.

The lateral capacity of a pile group shall include the contributions of the piles (with an appropriate modification for group effects) and the passive resistance of the pile cap. The lateral capacity of the piles shall be calculated using the same method used to calculate the stiffness. The lateral capacity of the pile cap, due to passive pressure, shall be calculated using established principles of soil mechanics. Passive pressure mobilization shall be calculated using Fig. 4-6.

*C4.4.2.2.2 Capacity Parameters* The lateral capacity of a pile cap should be calculated in the same way that the capacity of a shallow foundation is computed, except that the contribution of base traction should be neglected. Section C4.4.2.1.5 provides a more detailed description of the calculation procedure.

### 4.4.2.3 Drilled Shafts

The stiffness and capacity of drilled shaft foundations and piers of diameter less than or equal to 24 in. shall be calculated using the requirements for pile foundations specified in 4.4.2.2. For drilled shaft foundations and piers of diameter greater than 24 in., the capacity shall be calculated based on the interaction of the soil and shaft where the soil shall be represented using Winkler type models specified in Section 4.4.2.2.

### C4.4.2.3 Drilled Shafts

Where the diameter of the shaft becomes large (>24 in.), the bending and the lateral stiffness and strength of the shaft itself may contribute to the overall capacity. This is obviously necessary for the case of individual shafts supporting isolated columns.

### 4.4.3 Foundation Acceptance Criteria

The foundation soil shall comply with the acceptance criteria specified in this section. The structural components of foundations shall meet the appropriate requirements of Chapters 5 through 8. The foundation soil shall be evaluated to support all actions, including vertical loads, moments, and lateral forces applied to the soil by the foundation.

### 4.4.3.1 Simplified Rehabilitation

The foundation soil of buildings for which the Simplified Rehabilitation Method is selected in accordance with Section 2.3.1 shall comply with the requirements of Chapter 10.

### 4.4.3.2 Linear Procedures

The acceptance criteria for foundation soil analyzed by linear procedures shall be based on the modeling assumptions for the base of the structure specified in Section 4.4.3.2.1 or 4.4.3.2.2.

*4.4.3.2.1 Fixed Base Assumption* If the base of the structure is assumed to be completely rigid, the foundation soil shall be classified as deformation-controlled. Component actions shall be determined by Eq. 3-18. Acceptance criteria shall be based on Eq. 3-20, *m*-factors for foundation soil shall be 3, and the use of upper-bound component capacities shall be permitted. A fixed base assumption shall not be used for buildings being rehabilitated to the Immediate Occupancy Performance Level that are sensitive to base rotations or other types of foundation movement that would cause the structural components to exceed their acceptance criteria.

If the alternative overturning method described in Section 3.2.10.1 is used, the foundation soil shall be classified as force-controlled. Component actions shall not exceed the calculated capacities and upper-bound component capacities shall not be used.

*4.4.3.2.2 Flexible Base Assumption* If the base of the structure is assumed to be flexible and is modeled using linear foundation soil, then the foundation soil shall be classified as deformation-controlled. Component actions shall be determined by Eq. 3-18. Soil strength need not be evaluated. Acceptability of soil displacements shall be based on the ability of the structure to accommodate these displacements within the acceptance criteria for the selected Rehabilitation Objective.

### 4.4.3.3 Nonlinear Procedures

The acceptance criteria for foundation soil analyzed by nonlinear procedures shall be based on the modeling assumptions for the base of the structure specified in Section 4.4.3.3.1 or 4.4.3.3.2.

*4.4.3.3.1 Fixed Base Assumption* If the base of the structure is assumed to be completely rigid, then the base reactions for all foundation soil shall be classified as force-controlled, as determined by Eq. 3-19, and shall not exceed upper-bound component capacities.

A fixed base assumption shall not be used for buildings being rehabilitated for the Immediate Occupancy Performance Level that are sensitive to base rotations or other types of foundation movement that would cause the structural components to exceed their acceptance criteria.

*4.4.3.3.2 Flexible Base Assumption* If the base of the structure is assumed to be flexible and is modeled using flexible nonlinear foundation soil, then the foundation soil shall be classified as deformation-controlled and the displacements at the base of the structure shall not exceed the acceptance criteria of this section. For the Life Safety and Collapse Prevention Structural Performance Levels, acceptability of soil displacements shall be based on the ability of the structure to accommodate these displacements within the acceptance criteria for the selected Rehabilitation Objective. For the Immediate Occupancy Structural Performance Level, the permanent, nonrecoverable displacement of the foundation soil shall be calculated by an approved method based on the maximum total displacement, foundation and soil type, thickness of soil layers, and other pertinent factors. The acceptability of these displacements shall be based upon the ability of the structure to accommodate them within the acceptance criteria for the Immediate Occupancy Structural Performance Level.

## 4.5 KINEMATIC INTERACTION AND FOUNDATION DAMPING SOIL-STRUCTURE INTERACTION EFFECTS

Where required by Section 3.2.6., soil–structure interaction effects shall be calculated in accordance with Section 4.5.1 for kinematic interaction effects and Section 4.5.2 for foundation damping effects.

## C4.5 KINEMATIC INTERACTION AND FOUNDATION DAMPING SOIL–STRUCTURE INTERACTION EFFECTS

Foundation flexibility is covered in Section 4.4. SSI effects that serve to reduce the shaking input to the structure relative to the free-field motion (kinematic interaction and damping) are covered in this section. Procedures for calculating kinematic and damping effects were taken from recommendations in FEMA 440 (FEMA 2005) and have been included in the FEMA 368 (FEMA 2001) and FEMA 450 *NEHRP Recommended Provisions for Seismic Regulations for*

*New Buildings* (FEMA 2004) for a number of years. Further discussion of SSI effects can be found in FEMA 440.

### 4.5.1 Kinematic Interaction

Kinematic interaction effects shall be represented by ratio of response spectra (RRS) factors $RRS_{bsa}$ for base slab averaging, and $RRS_e$ for embedment, which are multiplied by the spectral acceleration ordinates on the response spectrum calculated in accordance with Section 1.6. Reduction of the response spectrum for kinematic interaction effects shall be permitted subject to the limitations in Sections 4.5.1.1 and 4.5.1.2.

#### 4.5.1.1 Base Slab Averaging

The RRS factor for base slab averaging, $RRS_{bsa}$, shall be determined using Eq. 4-11 for each period of interest. Alternatively, the RRS factor for base slab averaging shall be determined from Fig. 4-7. Reductions for base slab averaging shall not be permitted for buildings with the following characteristics:

1. Located on soft clay sites (site Classes E and F);
2. Floor and roof diaphragms classified as flexible, and foundation components that are not laterally connected;

$$RRS_{bsa} = 1 - \frac{1}{14,100}\left(\frac{b_e}{T}\right)^{1.2} \geq \text{the value for } T = 0.2 \text{ sec}$$

(Eq. 4-11)

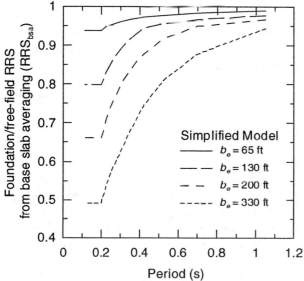

**FIGURE 4-7. Ratio of Response Spectra for Base Slab Averaging, $RRS_{bsa}$.**

where

$b_e$ = effective foundation size, ft;
$b_e = \sqrt{ab}$;
$T$ = fundamental period of the building, sec;
$a$ = longitudinal dimension of full footprint of building foundation, ft; and
$b$ = transverse dimension of full footprint of building foundation, ft.

#### C4.5.1.1 Base Slab Averaging

For base slab averaging effects to occur, foundation components must be interconnected with grade beams or concrete slabs. The method has not been rigorously studied for buildings on piles; however, it is considered reasonable to extend the application to pile-supported structures in which the pile caps are in contact with the soil and are laterally connected to one another.

#### 4.5.1.2 Embedment

The RRS factor for embedment, $RRS_e$, shall be determined using Eq. 4-12 for each period of interest. Reductions for embedment shall not be permitted for buildings with the following characteristics:

1. Located on firm rock sites (site Classes A and B), or soft clay sites (site Classes E and F); and
2. Foundation components that are not laterally connected.

$$RRS_e = \cos\left(\frac{2\pi e}{Tnv_s}\right) \geq \text{the larger of 0.453 or the}$$

$RRS_e$ value for $T = 0.2$ sec    (Eq. 4-12)

where

$e$ = foundation embedment depth, feet. A minimum of 75% of the foundation footprint shall be present at the embedment depth. The foundation embedment for buildings located on sloping sites shall be the shallowest embedment;

$v_s$ = shear wave velocity for site soil conditions, taken as average value of velocity to a depth of $b_e$ below foundation, ft/sec;

$n$ = shear wave velocity reduction factor;
$n = \sqrt{G/G_o}$; and

$G/G_o$ = effective shear modulus ratio from Table 4-7.

#### C4.5.1.2 Embedment

The embedment effect model was largely based on studies of buildings with basements. The recommendations can also be applied to buildings with embedded foundations without basements where the foundation is laterally connected. However, the embedment effect factor is not applicable to embedded individual spread footings.

### 4.5.2 Foundation Damping Soil–Structure Interaction Effects

The effects of foundation damping for nonlinear analyses shall be represented by the effective damping ratio of the structure-foundation system, $\beta_0$, determined in accordance with Eq. 4-13. Modification of the acceleration response spectrum calculated in accordance with Section 1.6 using $\beta_0$ in lieu of the effective viscous damping ratio, $\beta$, shall be permitted except where:

1. Vertical lateral-force-resisting elements are spaced at a distance less than the larger dimension of either component in the direction under consideration;
2. $v_s T / r_x / 2\pi$ (where $v_s$ = average shear wave velocity to a depth of $r_x$) and the shear stiffness of foundation soils increases with depth;
3. The soil profile consists of a soft layer overlying a very stiff material, and the system period is greater that the first-mode period of the layer.

$$\beta_o = \beta_f + \frac{\beta}{(\tilde{T}_{eff}/T_{eff})^3} \qquad \text{(Eq. 4-13)}$$

where

> $\beta_f$ = foundation-soil interaction damping ratio defined in Eq. 4-14;
> $\beta$ = effective viscous damping ratio of the building; and
> $\tilde{T}_{eff}/T_{eff}$ = effective period lengthening ratio defined in Eq. 4-15.

The foundation damping due to radiation damping, $\beta_f$, shall be determined in accordance with Eq. 4-14. Alternatively, foundation damping due to radiation damping shall be approximated using Fig. 4-8.

$$\beta_f = a_1\left(\frac{\tilde{T}_{eff}}{T_{eff}} - 1\right) + a_2\left(\frac{\tilde{T}_{eff}}{T_{eff}} - 1\right)^2 \qquad \text{(Eq. 4-14)}$$

where

$a_1 = c_e \exp(4.7 - 1.6 h/r_\theta)$;
$a_2 = c_e[25 \ln(h/r_\theta) - 16]$;
$c_e = 1.5(e/r_x) + 1$;
$h$ = effective structure height taken as the vertical distance from the foundation to the centroid of the first mode shape for multistory structures. Alternatively, $h$ shall be permitted to be approximated as 70% of the total structure height for

multistory structures or as the full height of the building for one-story structures;

$$h = \frac{\sum\limits_{i=1}^{N} w_i \phi_{i1} h_i}{\sum\limits_{i=1}^{N} w_i \phi_{i1}}$$

$r_\theta$ = equivalent foundation radius for rotation;

$$r_\theta = \left(\frac{3(1 - \nu)K_\theta}{8G}\right)^{1/3}$$

$K_\theta$ = effective rotational stiffness of the foundation;

$$K_\theta = \frac{K^*_{fixed}(h)^2}{\left(\dfrac{\tilde{T}}{T}\right)^2 - 1 - \dfrac{K^*_{fixed}}{K_x}}$$

$$K^*_{fixed} = M^*\left(\frac{2\pi}{T}\right)^2;$$

$M^*$ = effective mass for the first mode. Alternatively, it shall be permitted to take the effective mass as 70% of the total building mass, except where the mass is concentrated at a single level, it shall be taken as the total building mass;

$$M^* = \frac{W}{g}\left\{\frac{\left[\sum\limits_{i=1}^{N}(w_i \phi_{i1})/g\right]^2}{\left[\sum\limits_{i=1}^{N} w_i/g\right]\left[\sum\limits_{i=1}^{N}(w_i \phi_{i1}^2)/g\right]}\right\}$$

$W$ = total building weight;
$w_i$ = portion of the effective seismic weight located on or assigned to floor level $i$;
$\phi_{i1}$ = first mode displacement at level $i$;
$K_x$ = effective translational stiffness of the foundation;

$$K_x = \frac{8}{2 - \nu} G r_x$$

$G$ = effective shear modulus;
$\nu$ = Poisson's ratio; it shall be permitted to use 0.3 for sand and 0.45 for clay soils;
$e$ = foundation embedment depth, ft;
$r_x$ = equivalent foundation radius for translation;

$$r_x = \sqrt{A_f/\pi}$$

$A_f$ = area of the foundation footprint if the foundation components are interconnected laterally;

$T$ = fundamental period of the building using a model with a fixed base, sec; and

$\tilde{T}$ = fundamental period of the building using a model with a flexible base, sec.

The effective period lengthening ratio shall be determined in accordance with Eq. 4-15.

$$\frac{\tilde{T}_{eff}}{T_{eff}} = \left\{ 1 + \frac{1}{\mu} \left[ \left( \frac{\tilde{T}}{T} \right)^2 - 1 \right] \right\}^{0.5} \quad \text{(Eq. 4-15)}$$

where

$\mu$ = expected ductility demand. For nonlinear procedures, $\mu$ is the maximum displacement divided by the yield displacement ($\delta_t / \Delta_y$ for NSP). For linear procedures, $\mu$ is the maximum base shear divided by the elastic base shear capacity.

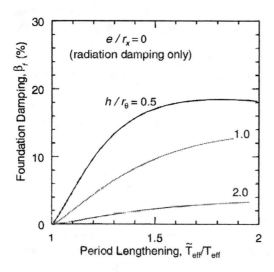

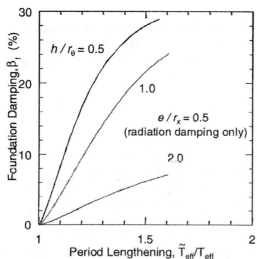

**FIGURE 4-8. Approximations of Foundation Damping, $\beta_f$.**

### C4.5.2 Foundation Damping Soil–Structure Interaction Effects

Foundation damping effects tend to be important for stiff structural systems such as shear walls and braced frames, particularly where they are supported on relatively soft soil sites such as site Classes D and E. The procedure is conservative where foundation aspect ratios exceed 2:1, and where foundations are deeply embedded ($e/r_x > 0.5$), but is potentially unconservative where wall and frame elements are close enough so that waves emanating from distinct foundations components destructively interfere with each other across the period range of interest.

The damping ratios determined in accordance with this section represent radiation damping effects only. See FEMA 440 (FEMA 2005) for further discussion of foundation damping SSI effects, including limitations.

### 4.6 SEISMIC EARTH PRESSURE

Building walls retaining soil shall be evaluated to resist additional earth pressure due to seismic forces. Unless otherwise determined from a site-specific geotechnical investigation, the seismic earth pressure acting on a building wall retaining nonsaturated, level soil above the ground water table shall be calculated using Eq. 4-16:

$$\Delta p = 0.4 k_h \gamma_t H_{rw} \quad \text{(Eq. 4-16)}$$

where

$\Delta p$ = additional earth pressure due to seismic shaking, which is assumed to be a uniform pressure;

$k_h$ = horizontal seismic coefficient in the soil, which may be assumed equal to $S_{XS}/2.5$;

$\gamma_t$ = total unit weight of soil;

$H_{rw}$ = height of the retaining wall; and

$S_{XS}$ = spectral response acceleration parameter as specified in Section 1.6.

The seismic earth pressure shall be added to the unfactored static earth pressure to obtain the total earth pressure on the wall. The wall shall be evaluated as a force-controlled component using acceptance criteria based on the type of wall construction and approved methods.

### C4.6 SEISMIC EARTH PRESSURE

Past earthquakes have not caused extensive damage to building walls below grade. In some cases, however, it is advisable to verify the adequacy of retaining walls to resist increased pressure due to seismic loading. These situations include walls of poor construction quality, unreinforced or lightly reinforced walls, walls of archaic materials, unusually tall or thin walls, damaged walls, or other conditions implying a sensitivity to increased loads.

The expression in Eq. 4-16 is a simplified approximation of the Mononobe-Okabe formulation. The actual magnitude and distribution of pressure on walls during earthquakes is very complex. If walls do not have the apparent capacity to resist the pressures estimated from the previously described approximate procedures, detailed investigation by a qualified geotechnical engineer is recommended. The seismic earth pressure from this equation is added to the unfactored static earth pressure, which includes pressure due to soil, water, and surcharge loads.

Seismic earth pressures calculated in accordance with this section are intended for use in checking acceptability of local wall components and should not be used to increase total base shear on the building.

## 4.7 FOUNDATION REHABILITATION

Foundation rehabilitation schemes shall be evaluated in conjunction with any rehabilitation of the superstructure and according to the general principles and requirements of this standard to assure that the complete rehabilitation achieves the selected building performance level for the selected earthquake hazard level. Where new rehabilitation components are used in conjunction with existing components, the effects of differential foundation stiffness on the modified structure shall be demonstrated to meet the acceptance criteria. If existing loads are not redistributed to all the components of the rehabilitated foundation by shoring and/or jacking, the effects of differential strengths and stiffnesses among individual foundation components shall be included in the analysis of the rehabilitated foundation. The effects of rehabilitation on stiffness, strength, and deformability shall be taken into account in an analytical model of the rehabilitated structure. The compatibility of new and existing components shall be checked at displacements consistent with the performance level chosen.

## C4.7 FOUNDATION REHABILITATION

Guidance for modification of foundations to improve seismic performance is provided as follows:

**Soil Material Improvements.** Improvement in existing soil materials may be effective in the rehabilitation

of foundations by achieving one or more of the following results: (1) improvement in vertical bearing capacity of footing foundations; (2) increase in the lateral frictional resistance at the base of footings; and (3) increase in the passive resistance of the soils adjacent to foundations or grade beams.

Soil improvement options to increase the vertical bearing capacity of footing foundations are limited. Soil removal and replacement and soil vibratory densification usually are not feasible because they would induce settlements beneath the footings or be expensive to implement without causing settlement. Grouting may be considered to increase bearing capacity. Different grouting techniques are discussed in FEMA 274 Section C4.3.2 (FEMA 1997). Compaction grouting can achieve densification and strengthening of a variety of soil types and/or extend foundation loads to deeper, stronger soils. The technique requires careful control to avoid causing uplift of foundation components or adjacent floor slabs during the grouting process. Permeation grouting with chemical grouts can achieve substantial strengthening of sandy soils, but the more fine-grained or silty the sand, the less effective the technique becomes. Jet grouting could also be considered. These same techniques also may be considered to increase the lateral frictional resistance at the base of footings.

Soil improvement by the following methods may be effective in increasing the passive resistance of soils adjacent to foundations or grade beams: removal and replacement of existing soils with stronger, well-compacted soils or with treated (e.g., cement-stabilized) soils; in-place mixing of existing soils with strengthening materials (e.g., cement); grouting, including permeation grouting and jet grouting; and in-place densification by impact or vibratory compaction. In-place densification by impact or vibratory compaction should be used only if the soil layers to be compacted are not too thick and vibration effects on the structure are tolerable.

**Shallow Foundation Rehabilitation.** The following measures may be effective in the rehabilitation of shallow foundations:

1. New isolated or spread footings may be added to existing structures to support new structural elements such as shear walls or frames.
2. Existing isolated or spread footings may be enlarged to increase bearing or uplift capacity. Consideration of existing contact pressures on the strength and stiffness of the modified footing may

be required unless uniform distribution is achieved by shoring and/or jacking.

3. Existing isolated or spread footings may be underpinned to increase bearing or uplift capacity. Underpinning improves bearing capacity by lowering the contact horizon of the footing. Consideration of the effects of jacking and load transfer may be required.
4. Uplift capacity may be improved by increasing the resisting soil mass above the footing.
5. Mitigation of differential lateral displacement of different portions of a building foundation may be carried out by provision of interconnection with grade beams, reinforced grade slab, or ties.

**Deep Foundation Rehabilitation.** The following measures may be effective in the rehabilitation of deep foundation consisting of driven piles made of steel, concrete, or wood, or cast-in-place concrete piers, or drilled shafts of concrete.

Shallow foundation of spread footings or mats may be provided to support new shear walls or frames or other new elements of the lateral-force-resisting system, provided the effects of differential foundation stiffness on the modified structure are analyzed and meet the acceptance criteria.

New wood piles may be provided for an existing wood pile foundation. A positive connection should be provided to transfer the uplift forces from the pile cap or foundation above to the new wood piles. Existing wood piles should be inspected for deterioration caused by decay, insect infestation, or other signs of distress prior to undertaking evaluation of existing wood pile foundation.

Driven piles made of steel, concrete, or wood, or cast-in-place concrete piers or drilled shafts of concrete, may be provided to support new structural elements such as shear walls or frames.

Driven piles made of steel, concrete, or wood, or cast-in-place concrete piers or drilled shafts of concrete, may be provided to supplement the vertical and lateral capacities of existing pile and pier foundation groups.

# 5.0 STEEL

## 5.1 SCOPE

This chapter sets forth requirements for the Systematic Rehabilitation of steel components of the lateral-force-resisting system of an existing building. The requirements of this chapter shall apply to existing steel

components of a building system, rehabilitated steel components of a building system, and new steel components added to an existing building system.

Section 5.2 specifies data collection procedures for obtaining material properties and performing condition assessments. Section 5.3 specifies general requirements. Sections 5.4, 5.5, 5.6, and 5.7 provide modeling procedures, component strengths, acceptance criteria, and rehabilitation measures for steel moment-resisting frames, steel braced frames, steel plate shear walls, and steel frames with infills. Section 5.8 provides modeling procedures, strengths, acceptance criteria, and rehabilitation measures for diaphragms used in steel structures. Section 5.9 specifies requirements for steel piles. Section 5.10 specifies requirements for components of cast or wrought iron.

## C5.1 SCOPE

Techniques for repair of earthquake-damaged steel components are not included in this standard. The design professional is referred to SAC Joint Venture publications FEMA 351 (FEMA 2000) and FEMA 353 (FEMA 2000) for information on design, evaluation, and repair of damaged steel moment-resisting frame structures.

## 5.2 MATERIAL PROPERTIES AND CONDITION ASSESSMENT

### 5.2.1 General

Mechanical properties for steel materials and components shall be based on available construction documents and as-built conditions for the particular structure as specified in Section 2.2. Where such information fails to provide adequate information to quantify material properties or document the condition of the structure, such information shall be supplemented by material tests and assessments of existing conditions as required in Section 2.2.6.

Material properties of existing steel components shall be determined in accordance with Section 5.2.2. A condition assessment shall be conducted in accordance with Section 5.2.3. The extent of materials testing and condition assessment performed shall be used to determine the knowledge factor as specified in Section 5.2.4.

Use of default material properties shall be permitted in accordance with Section 5.2.2.5. Use of material properties based on historical information as default values shall be permitted as specified in Section 5.2.2.5.

### C5.2.1 General

The extent of in-place materials testing and condition assessment that must be accomplished is related to availability and accuracy of construction and as-built records, the quality of materials used and construction performed, and the physical condition of the structure. Data such as the properties and grades of material used in component and connection fabrication may be effectively used to reduce the amount of in-place testing required. The design professional is encouraged to research and acquire all available records from original construction.

Steel components of buildings include columns, beams, braces, connections, link beams, and di..phragms. Columns, beams, and braces may be built up with plates, angles, and/or channels connected to gether with rivets, bolts, or welds. The material used in older construction is likely to be mild steel with a specified yield strength between 30 ksi and 36 ksi. Cast iron was often used for columns in much older construction, from before 1900 through the 1920s. Cast iron was gradually replaced by wrought iron and then by steel. The connectors in older construction were usually mild steel rivets or bolts. These were later replaced by high-strength bolts and welds. The seismic performance of these components will depend heavily on the condition of the in-place material. A more detailed historical perspective is given in Section C5.2 of FEMA 274 (FEMA 1997).

Great care should be exercised in selecting the appropriate rehabilitation approaches and techniques for application to historic buildings in order to preserve their unique characteristics.

### 5.2.2 Properties of In-Place Materials and Components

#### 5.2.2.1 Material Properties

*5.2.2.1.1 General* The following component and connection material properties shall be obtained for the as-built structure:

1. Yield and tensile strength of the base material.
2. Yield and tensile strength of the connection material.
3. Carbon equivalent of the base and connection material.

Structural steel components constructed after 1900 shall be classified based on ASTM specification and material grade and, if applicable, shape group in accordance with Table 5-2. Lower-bound material properties shall be taken in accordance with Table 5-2 for material conforming to the specifications listed therein. For material grades not listed in Table 5-2, lower-bound material properties shall be taken as nominal or specified properties, or shall be based on tests where the material grade or specified value is not known.

Where materials testing is required by Section 2.2.6, test methods to determine ASTM designation and material grade or to quantify material properties shall be as specified in Section 5.2.2.3.

The minimum number of tests shall comply with the requirements of Section 5.2.2.4.

The carbon equivalent of the existing components shall be determined to establish weldability of the material, unless it is confirmed that either the existing material conforms with a weldable material specification or welding to existing components will not be performed as part of the rehabilitation. The welding procedures shall be determined based on the chemistry of the base material and filler material as specified in Section 8 of AWS D1.1 [American Welding Society (AWS) 2002]. Material conforming to ASTM A36/A36M-04 (ASTM 2004), ASTM A242/A242M-03 (ASTM 2003), ASTM A307-02 (ASTM 2002), ASTM A572/572M-04 (ASTM 2004), ASTM A913/A913M-01 (ASTM 2001), ASTM A972/A972M-00 (ASTM 2000), and ASTM A992/A992M-04 (ASTM 2004) shall be deemed to be weldable.

*5.2.2.1.2 Nominal Properties* Nominal material properties specified in the American Institute of Steel Construction (AISC) *Load and Resistance Factor Design Specification for Structural Steel Buildings (LRFD)* (AISC 1999), or properties specified in construction documents, shall be taken as lower-bound material properties. Corresponding expected material properties shall be calculated by multiplying lower-bound values by an appropriate factor taken from Table 5-3 to translate from lower-bound to expected values.

Where construction documents indicate the ultimate tensile strength of weld metal, the lower-bound strength of welds shall be taken as indicated in AWS D1.1 (AWS 2002). For construction predating 1970, use of a nominal ultimate tensile strength of 60 ksi shall be permitted.

### C5.2.2.1 Material Properties

Mechanical properties of component and connection material dictate the structural behavior of the component under load. Mechanical properties of greatest interest include the expected and lower-bound estimates of yield ($F_{ye}$) and tensile ($F_{te}$) strengths of base and connection material, modulus of elasticity, ductility, toughness, elongational characteristics, and weldability.

Expected material properties should be used for deformation-controlled actions. Lower-bound material properties should be used for force-controlled actions in lieu of nominal strengths specified in the *Load and Resistance Factor Design Specification for Structural Steel Buildings (LRFD)* (AISC 1999).

### 5.2.2.2 Component Properties

The following properties of components and their connections shall be obtained for the structure:

1. Size and thickness of connected materials, including cover plates, bracing, and stiffeners;
2. Cross-sectional area, section moduli, moments of inertia, and torsional properties of components at critical sections;
3. As-built configuration of intermediate, splice, and end connections; and
4. Current physical condition of base metal and connector materials, including presence of deformation and extent of deterioration.

Review of available construction documents shall be performed to identify primary vertical- and lateral-load-carrying elements and systems, critical components and connections, and any modifications to components or overall configuration of the structure.

In the absence of deterioration, use of the nominal cross-sectional dimensions of components published by the AISC, the American Iron and Steel Institute (AISI), and other approved trade associations shall be permitted.

### 5.2.2.3 Test Methods to Quantify Properties

Laboratory testing of samples to determine in-place mechanical properties of materials and components shall be performed in compliance with consensus standards published by ASTM, the American National Standards Institute (ANSI), and other approved organizations.

The extent of in-place materials testing required to determine material properties shall be based on the data collection requirements in Section 2.2.6.

The determination of material properties shall be accomplished through removal of samples and labora-

tory testing. Sampling shall take place in regions where the decreased section strength due to the sampling remains higher than the capacity required at the reduced section to resist the design loads. Alternately, where the reduced section strength due to sampling becomes lower than the required capacity, the lost section shall be temporarily supported and restored by repairs to the section.

If a connector such as a bolt or rivet is removed for testing, a comparable bolt shall be reinstalled at the time of sampling. Destructive removal of a welded connection sample shall be accompanied by repair of the connection.

Expected material properties shall be based on mean test values. Lower-bound material properties shall be based on mean test values minus one standard deviation, except that where the material is positively identified as conforming to a defined standard material specification, lower-bound properties need not be taken less than the nominal properties for that specification.

### C5.2.2.3 Test Methods to Quantify Properties

FEMA 274 (FEMA 1997) provides information and references for several test methods.

Sampling should take place in regions where the calculated stresses (considering the lost section due to sampling) for the applied loads is less than the allowable stress, where using allowable stress design (ASD), and less than the capacity where using load and resistance factor design (LRFD).

Of greatest interest to steel building system performance are the expected yield and tensile strength of the installed materials. Notch toughness of structural steel and weld material is also important for connections that undergo cyclic loadings and deformations during earthquakes. Chemical and metallurgical properties can provide information on properties such as compatibility of welds with parent metal and potential lamellar tearing due to through-thickness stresses. Virtually all steel component elastic and inelastic limit states are related to yield and tensile strengths. Past research and accumulation of data by industry groups have resulted in published material mechanical properties for most primary metals and their date of fabrication. Section 5.2.2.5 provides default properties. This information may be used, together with tests from recovered samples, to rapidly establish expected strength properties for use in component strength and deformation analyses.

Review of other properties derived from laboratory tests, such as hardness, impact, fracture, and fatigue, is generally not needed for steel component capacity determination, but may be required for archaic materials and connection evaluation. These properties may not be needed in the analysis phase if significant rehabilitative measures are already known to be required.

To quantify material properties and analyze the performance of welded moment connections, more extensive sampling and testing may be necessary. This testing may include base and weld material chemical and metallurgical evaluation, expected strength determination, hardness, and Charpy V-notch testing of the heat-affected zone and neighboring base metal, and other tests depending on connection configuration.

Recommendations given in FEMA 351 (FEMA 2000) may also be followed to select welding procedures for welding of rehabilitative measures to existing components.

### 5.2.2.4 Minimum Number of Tests

Materials testing is not required if material properties are available from original construction documents that include material test records or material test reports. If such properties differ from default material properties given in Tables 5-1 and 5-2, material properties for rehabilitation shall be selected such that the largest demands on components and connections are generated.

*5.2.2.4.1 Usual Testing* The minimum number of tests to determine the yield and tensile strengths of steel materials for usual data collection shall be based on the following criteria:

1. If design drawings are incomplete or not available, at least one strength coupon from each steel component type shall be removed for testing, and one weld metal sample for each component type shall be obtained for testing. The sample shall consist of both local base and weld metal to determine composite strength of the connection.
2. If design drawings containing ASTM specification and material grade information are available, use of Table 5-2 to determine material properties shall be permitted without additional testing.
3. If design drawings containing material property information are available but the material properties are not listed in Table 5-2, use of nominal or specified material properties shall be permitted without additional testing.

*5.2.2.4.2 Comprehensive Testing* The minimum number of tests to determine the yield and tensile strengths of steel materials for comprehensive data collection shall be based on the following criteria:

1. If original construction documents defining material properties are inconclusive or do not exist, but the date of construction is known and the material used is confirmed to be carbon steel, at least three strength coupons and three bolts and rivets shall be randomly removed from each component type.
2. If no knowledge of the structural system and materials used exists, at least two tensile strength coupons and two bolts and rivets shall be removed from each component type for every four floors or every 200,000 sf. If it is determined from testing that more than one material grade exists, additional sampling and testing shall be performed until the extent of each grade in component fabrication has been established.
3. In the absence of construction records defining welding filler metals and processes used, at least one weld metal sample for each component type shall be obtained for laboratory testing. The sample shall consist of both local base and weld metal to determine composite strength of the connection.
4. For archaic materials, at least three strength coupons shall be extracted for each component type for every four floors or 200,000 sf of construction. If initial tests provide material properties that are consistent with properties given in Table 5-1, tests shall be required for every six floors or 300,000 sf of construction only. If these tests provide material properties that are nonuniform, additional tests shall be performed until the extent of different materials is established.

For other material properties, a minimum of three tests shall be conducted.

The results of any material testing performed shall be compared to the default values in Tables 5-1 and 5-2 for the particular era of building construction. The amount of testing shall be doubled if the expected and lower-bound yield and tensile strengths determined from testing are lower than the default values.

### C5.2.2.4 Minimum Number of Tests

In order to quantify expected strength and other properties accurately, a minimum number of tests

**Table 5-1. Default Lower-Bound Material Strengths for Archaic Materials[1,2]**

| Year | Material | Lower-Bound Yield Strength, ksi | Lower-Bound Tensile Strength, ksi |
|------|----------|---------------------------------|-----------------------------------|
| Pre-1900 | Cast Iron | 18 | — |
| Pre-1900 | Steel | 24 | 36 |

[1]Modified from unit stress values in AISC *Iron and Steel Beams from 1873 to 1952* (AISC 1983).
[2]Properties based on tables of allowable loads as published in mill catalogs.

may be required to be conducted on representative components.

The evaluating engineer should exercise judgment to determine how much variability of component sizes will constitute a significant change in material properties. It is likely that most of the sections of the same size within a building have similar material properties. Differences in material properties are more likely to occur due to differences in size groups, differences in specified material properties (36 ksi versus 50 ksi), and differences in section shapes. At a minimum, one coupon should be removed from each nominal size of each wide-flange, angle, channel, hollow structural section (HSS), and other structural shape used as part of the lateral-force-resisting system. Additional sampling should be done where large variations in member sizes occur within the building and where the building was constructed in phases or over extended time periods where members may have come from different mills or batches.

Material properties of structural steel vary much less than those of other construction materials. In fact, the expected yield and tensile stresses are usually considerably higher than the nominal specified values. As a result, testing for material properties of structural steel may not be required. The properties of wrought iron are more variable than those of steel. The strength of cast iron components cannot be determined from small sample tests, since component behavior is usually governed by inclusions and other imperfections.

If ductility and toughness are required at or near the weld, the design professional may conservatively assume that no ductility is available, in lieu of testing. In this case the joint would have to be modified if inelastic demands are anticipated and the possibility of fractures cannot be tolerated. Special requirements for welded moment frames are given in FEMA 351 (FEMA 2000).

If a higher degree of confidence in results is desired, either the sample size shall be determined using ASTM E22 (ASTM 1955) criteria, or the prior knowledge of material grades from Section 5.2.2.5 should be used in conjunction with approved statistical procedures.

Design professionals may consider using Bayesian statistics and other statistical procedures contained in FEMA 274 (FEMA 1997) to gain greater confidence in the test results obtained from the sample sizes specified in this section.

### 5.2.2.5 Default Properties

The default lower-bound material properties for steel components shall be as specified in Tables 5-1 and 5-2. Default expected strength material properties shall be determined by multiplying lower-bound values by an appropriate factor taken from Table 5-3.

Use of default material properties to determine component and connection strengths shall be permitted in conjunction with the linear analysis procedures of Chapter 3.

### 5.2.3 Condition Assessment

### 5.2.3.1 General

A condition assessment of the existing building and site shall be performed as specified in this section. A condition assessment shall include the following:

1. The physical condition of primary and secondary components shall be examined and the presence of any degradation shall be noted;
2. Verification of the presence and configuration of structural elements and components and their connections, and the continuity of load paths between components, elements, and systems; and
3. Identification of other conditions including the presence of nonstructural components that influence building performance and impose limitations on rehabilitation.

**Table 5-2. Default Lower-Bound Material Strengths[1,2]**

| Date | Specification | Remarks | Tensile Strength[3], ksi | Yield Strength[3], ksi |
|---|---|---|---|---|
| 1900 | ASTM A9 | Rivet Steel | 50 | 30 |
|  | Buildings | Medium Steel | 60 | 35 |
| 1901–1908 | ASTM A9 | Rivet Steel | 50 | 25 |
|  | Buildings | Medium Steel | 60 | 30 |
| 1909–1923 | ASTM A9 | Structural Steel | 55 | 28 |
|  | Buildings | Rivet Steel | 46 | 23 |
| 1924–1931 | ASTM A7 | Structural Steel | 55 | 30 |
|  |  | Rivet Steel | 46 | 25 |
|  | ASTM A9 | Structural Steel | 55 | 30 |
|  |  | Rivet Steel | 46 | 25 |
| 1932 | ASTM A140-32T issued as a tentative revision to ASTM A9 (Buildings) | Plates, Shapes, Bars | 60 | 33 |
|  |  | Eyebar Flats (Unannealed) | 67 | 36 |
| 1933 | ASTM A140-32T discontinued and ASTM A9 (Buildings) revised Oct. 30, 1933 | Structural Steel | 55 | 30 |
|  | ASTM A9 tentatively revised to ASTM A9-33T (Buildings) | Structural Steel | 60 | 33 |
|  | ASTM A141-32T adopted as a standard | Rivet Steel | 52 | 28 |
| 1934–Present | ASTM A9 | Structural Steel | 60 | 33 |
|  | ASTM A141 | Rivet Steel | 52 | 28 |
| 1961–1990 | ASTM A36/A36M-04 | Structural Steel |  |  |
|  | Group 1 |  | 62 | 44 |
|  | Group 2 |  | 59 | 41 |
|  | Group 3 |  | 60 | 39 |
|  | Group 4 |  | 62 | 37 |
|  | Group 5 |  | 70 | 41 |
| 1961–Present | ASTM A572/A572M-04, Grade 50 | Structural Steel |  |  |
|  | Group 1 |  | 65 | 50 |
|  | Group 2 |  | 66 | 50 |
|  | Group 3 |  | 68 | 51 |
|  | Group 4 |  | 72 | 50 |
|  | Group 5 |  | 77 | 50 |
| 1990–Present | ASTM A36/A36M-04 and Dual Grade | Structural Steel |  |  |
|  | Group 1 |  | 66 | 49 |
|  | Group 2 |  | 67 | 50 |
|  | Group 3 |  | 70 | 52 |
|  | Group 4 |  | 70 | 49 |
| 1998–Present | ASTM A992/A992M-04 | Structural Steel | 65 | 50 |

[1]Lower-bound values for material prior to 1960 are based on minimum specified values. Lower-bound values for material after 1960 are mean minus one standard deviation values from statistical data.
[2]Properties based on ASTM and AISC Structural Steel Specification Stresses.
[3]The indicated values are representative of material extracted from the flanges of wide flange shapes.

**Table 5-3. Factors to Translate Lower-Bound Steel Properties to Expected-Strength Steel Properties**

| Property | Year | Specification | Factor |
|---|---|---|---|
| Tensile Strength | Prior to 1961 | | 1.10 |
| Yield Strength | Prior to 1961 | | 1.10 |
| Tensile Strength | 1961–1990 | ASTM A36/A36M-04 | 1.10 |
| | 1961–Present | ASTM A572/A572M-04, Group 1 | 1.10 |
| | | ASTM A572/A572M-04, Group 2 | 1.10 |
| | | ASTM A572/A572M-04, Group 3 | 1.05 |
| | | ASTM A572/A572M-04, Group 4 | 1.05 |
| | | ASTM A572/A572M-04, Group 5 | 1.05 |
| | 1990–Present | ASTM A36/A36M-04 and Dual Grade, Group 1 | 1.05 |
| | | ASTM A36/A36M-04 and Dual Grade, Group 2 | 1.05 |
| | | ASTM A36/A36M-04 and Dual Grade, Group 3 | 1.05 |
| | | ASTM A36/A36M-04 and Dual Grade, Group 4 | 1.05 |
| | 1998–Present | ASTM A992/A992M-04 | 1.10 |
| Yield Strength | 1961–1990 | ASTM A36/A36M-04 | 1.10 |
| | 1961–Present | ASTM A572/A572M-04, Group 1 | 1.10 |
| | | ASTM A572/A572M-04, Group 2 | 1.10 |
| | | ASTM A572/A572M-04, Group 3 | 1.05 |
| | | ASTM A572/A572M-04, Group 4 | 1.10 |
| | | ASTM A572/A572M-04, Group 5 | 1.05 |
| | 1990–Present | ASTM A36/A36M-04, Plates | 1.10 |
| | | ASTM A36/A36M-04 and Dual Grade, Group 1 | 1.05 |
| | | ASTM A36/A36M-04 and Dual Grade, Group 2 | 1.10 |
| | | ASTM A36/A36M-04 and Dual Grade, Group 3 | 1.05 |
| | | ASTM A36/A36M-04 and Dual Grade, Group 4 | 1.05 |
| | 1998–Present | ASTM A992/A992M-04 | 1.10 |
| Tensile Strength | All | Not Listed [1] | 1.10 |
| Yield Strength | All | Not Listed [1] | 1.10 |

[1]For materials not conforming to one of the listed specifications.

### C5.2.3.1 General

The physical condition of existing components and elements and their connections must be examined for degradation. Degradation may include environmental effects (e.g., corrosion, fire damage, chemical attack) or past or current loading effects (e.g., overload, damage from past earthquakes, fatigue, fracture). The condition assessment should also examine for configurational problems observed in recent earthquakes, including effects of discontinuous components, improper welding, and poor fit-up.

Component orientation, plumbness, and physical dimensions should be confirmed during an assessment. Connections in steel components, elements, and systems require special consideration and evaluation. The load path for the system must be determined, and each connection in the load path(s) must be evaluated. This includes diaphragm-to-component and component-to-component connections. FEMA 351 (FEMA 2000) provides recommendations for inspection of welded steel moment frames.

The condition assessment also affords an opportunity to review other conditions that may influence steel elements and systems and overall building performance. Of particular importance is the identification of other elements and components that may contribute to or impair the performance of the steel system in question, including infills, neighboring buildings, and equipment attachments. Limitations posed by existing coverings, wall and ceiling space, infills, and other conditions shall also be defined such that prudent rehabilitation measures may be planned.

### 5.2.3.2 Scope and Procedures

The condition assessment shall include visual inspection of accessible structural elements and components involved in lateral-load resistance to verify information shown on available documents.

If coverings or other obstructions exist, either partial visual inspection through use of drilled holes and a fiberscope shall be used, or complete visual inspection shall be performed by local removal of covering materials. Where required by Section 2.2.6, the following shall be performed for visual and comprehensive condition assessments:

### C5.2.3.2 Scope and Procedures

For steel elements encased in concrete, it may be more cost-effective to provide an entirely new lateral-load-resisting system than undertaking a visual inspection by removal of concrete encasement and repair.

Physical condition of components and connectors may also dictate the use of certain destructive and nondestructive test methods. If steel elements are covered by well-bonded fireproofing materials or are er·:ased in durable concrete, it is likely that their condition will be suitable. However, local removal of th·se materials at connections should be performed as part of the assessment. The scope of this removal effort is dictated by the component and element design. For example, in a braced frame, exposure of several key connections may suffice if the physical condition is acceptable and the configuration matches the design drawings. However, for moment frames, it may be necessary to expose more connection points because of varying designs and the critical nature of the connections. See FEMA 351 (FEMA 2000) for inspection of welded moment frames.

*5.2.3.2.1 Visual Condition Assessment* If detailed design drawings exist, at least one connection of each connection type shall be exposed. If no deviations from the drawings exist, the sample shall be considered representative. If deviations from the existing drawings exist, then removal of additional coverings from connections of that type shall be done until the extent of deviations is determined.

*5.2.3.2.2 Comprehensive Condition Assessment* In the absence of construction drawings, at least three connections of each type shall be exposed for the primary structural components. If no deviations within a connection group are observed, the sample shall be considered representative. If deviations within a connection group are observed, then additional connections shall be exposed until the extent of deviations is determined.

### 5.2.3.3 Basis for the Mathematical Building Model

The results of the condition assessment shall be used to create a mathematical building model.

If no damage, alteration, or degradation is observed in the condition assessment, component section properties shall be taken from design drawings. If some sectional material loss or deterioration has occurred, the loss shall be quantified by direct measurement and section properties shall be reduced accordingly using principles of structural mechanics.

### 5.2.4 Knowledge Factor

A knowledge factor ($\kappa$) for computation of steel component capacities and permissible deformations shall be selected in accordance with Section 2.2.6.4 with the following additional requirements specific to steel components.

A knowledge factor of 0.75 shall be used if the components and their connectors are composed of cast or wrought iron.

## 5.3 GENERAL ASSUMPTIONS AND REQUIREMENTS

### 5.3.1 Stiffness

Component stiffnesses shall be calculated in accordance with Sections 5.4 through 5.10.

### 5.3.2 Design Strengths and Acceptance Criteria

#### 5.3.2.1 General

Classification of steel component actions as deformation- or force-controlled, and calculation of design strengths, shall be as specified in Sections 5.4 through 5.9.

#### 5.3.2.2 Deformation-Controlled Actions

Design strengths for deformation-controlled actions, $Q_{CE}$, shall be taken as expected strengths obtained experimentally or calculated using accepted principles of mechanics. Expected strength shall be defined as the mean maximum resistance expected over the range of deformations to which the component is likely to be subjected. Where calculations are used to determine mean expected strength, expected material properties (including strain hardening) shall be used. Unless other procedures are specified in this standard, procedures contained in *Load and Resistance Factor Design Specification for Structural Steel Buildings* (AISC 1999) to calculate design strength shall be permitted, except that the strength reduction factor, $\phi$, shall be taken as unity. Deformation capacities for acceptance of deformation-controlled actions shall be as specified in Sections 5.4 through 5.10.

### 5.3.2.3 Force-Controlled Actions

Design strengths for force-controlled actions, $Q_{CL}$, shall be taken as lower-bound strengths obtained experimentally or calculated using established principles of mechanics. Lower-bound strength shall be defined as mean strength minus one standard deviation. Where calculations are used to determine lower-bound strength, lower-bound material properties shall be used. Unless other procedures are specified in this standard, procedures contained in *Load and Resistance Factor Design Specification for Structural Steel Buildings (LRFD)* (AISC 1999) to calculate design strength shall be permitted, except that the strength reduction factor, $\phi$, shall be taken as unity. Where alternative definitions of design strength are used, they shall be justified by experimental evidence.

### 5.3.2.4 Anchorage to Concrete

Connections of steel components to concrete components shall comply with the provisions of this chapter and Chapter 6 for determination of strength and classification of actions as deformation-controlled or force-controlled.

The strength of connections between steel components and concrete components shall be the lowest value obtained for the limit states of the strength of the steel components, strength of the connection plates, and strength of the anchor bolts.

The strength of column base plates shall be the lowest strength calculated based on the following limit states: expected strength of welds or bolts; expected bearing stress of the concrete; and expected yield strength of the base plate.

The strength of the anchor bolt connection between the column base plate and the concrete shall be the lowest strength calculated based on the following limit states: shear or tension yield strength of the anchor bolts; loss of bond between the anchor bolts and the concrete; or failure of the concrete. Anchor bolt strengths for each failure type or limit state shall be calculated in accordance with ACI 318 (ACI 2002), using $\phi = 1.0$, or other procedures approved by the authority having jurisdiction.

For column base plate yielding, bolt yielding, and weld failure, the use of *m*-factors from Table 5-5, based on the respective limit states for partially restrained end plates, shall be permitted. Column base connection limit states controlled by anchor bolt failure modes governed by the concrete shall be considered force-controlled.

### 5.3.3 Rehabilitation Measures

Upon determining that steel elements in an existing building are deficient for the selected Rehabilitation Objective, these elements shall be rehabilitated or replaced so they are no longer deficient. If replacement of the element is selected, the new element shall be designed in accordance with this standard and detailed and constructed in accordance with a building code approved by the authority having jurisdiction.

## 5.4 STEEL MOMENT FRAMES

### 5.4.1 General

The behavior of steel moment-resisting frames is generally dependent on the connection configuration and detailing. Table 5-4 identifies the various connection types for which acceptance criteria are provided. Modeling procedures, acceptance criteria, and rehabilitation measures for Fully Restrained (FR) Moment Frames and Partially Restrained (PR) Moment Frames shall be as defined in Sections 5.4.2 and 5.4.3, respectively.

### C5.4.1 General

Steel moment frames are those frames that develop their seismic resistance through bending of steel beams and columns, and moment-resisting beam—column connections. A moment-resisting beam–column connection is one that is designed to develop moment resistance at the joint between the beam and the column and also designed to develop the shear resistance at the panel zone of the column. Beams and columns consist of either hot-rolled steel sections or cold-formed steel sections or built-up members from hot-rolled or cold-formed plates and sections. Built-up members are assembled by riveting, bolting, or welding. The components are either bare steel or steel with a nonstructural coating for protection from fire or corrosion, or both, or steel with either concrete or masonry encasement.

Following the 1994 Northridge earthquake, the SAC Joint Venture undertook a major program to address the issue of the seismic performance of moment-resisting steel frame structures. This program produced several documents which provide recommended criteria for the evaluation and upgrade of this building type. However, the design professional should be cautioned that there are some differences in the methodologies and specifics of this standard and the SAC procedures. While both methodologies utilize similar analysis procedures, there are some variations in the factors used to compute the pseudo-lateral load

## Table 5-4. Steel Moment Frame Connection Types

| Connection | Description[1,2] | Type |
|---|---|---|
| Welded Unreinforced Flange (WUF) | Full-penetration welds between beam and columns, flanges, bolted or welded web, designed prior to code changes following the Northridge earthquake | FR |
| Bottom Haunch in WUF with Slab | Welded bottom haunch added to existing WUF connection with composite slab[3] | FR |
| Bottom Haunch in WUF without Slab | Welded bottom haunch added to existing WUF connection without composite slab[3] | FR |
| Welded Cover Plate in WUF | Welded cover plates added to existing WUF connection[3] | FR |
| Improved WUF—Bolted Web | Full-penetration welds between beam and column flanges, bolted web[4] | FR |
| Improved WUF—Welded Web | Full-penetration welds between beam and column flanges, welded web[4] | FR |
| Free Flange | Web is coped at ends of beam to separate flanges, welded web tab resists shear and bending moment due to eccentricity due to coped web[4] | FR |
| Welded Flange Plates | Flange plate with full-penetration weld at column and fillet welded to beam flange[4] | FR |
| Reduced Beam Section | Connection in which net area of beam flange is reduced to force plastic hinging away from column face[4] | FR |
| Welded Bottom Haunch | Haunched connection at bottom flange only[4] | FR |
| Welded Top and Bottom Haunches | Haunched connection at top and bottom flanges[4] | FR |
| Welded Cover—Plated Flanges | Beam flange and cover-plate are welded to column flange[4] | FR |
| Top and Bottom Clip Angles | Clip angle bolted or riveted to beam flange and column flange | PR |
| Double Split Tee | Split tees bolted or riveted to beam flange and column flange | PR |
| Composite Top and Clip Angle Bottom | Clip angle bolted or riveted to column flange and beam bottom flange with composite slab | PR |
| Bolted Flange Plates | Flange plate with full-penetration weld at column and bolted to beam flange[4] | PR[5] |
| Bolted End Plate | Stiffened or unstiffened end plate welded to beam and bolted to column flange | PR[5] |
| Shear Connection with Slab | Simple connection with shear tab, composite slab | PR |
| Shear Connection without Slab | Simple connection with shear tab, no composite slab | PR |

[1]Where not indicated otherwise, definition applies to connections with bolted or welded web.
[2]Where not indicated otherwise, definition applies to connections with or without composite slab.
[3]Full-penetration welds between haunch or cover plate to column flange conform to the requirements of the AISC 341 *Seismic Provisions for Structural Steel Buildings* (AISC 2002).
[4]Full-penetration welds conform to the requirements of the AISC 341 *Seismic Provisions for Structural Steel Buildings* (AISC 2002).
[5]For purposes of modeling, the connection may be considered FR if it meets the strength and stiffness requirements of Section 5.4.2.1.

**Table 5-5. Acceptance Criteria for Linear Procedures—Structural Steel Components**

| Component/Action | IO | Primary LS | Primary CP | Secondary LS | Secondary CP |
|---|---|---|---|---|---|
| | | *m*-Factors for Linear Procedures | | | |

Beams—Flexure

a. $\dfrac{b_f}{2t_f} \le \dfrac{52}{\sqrt{F_{ye}}}$ and $\dfrac{h}{t_w} \le \dfrac{418}{\sqrt{F_{ye}}}$ — 2, 6, 8, 10, 12

b. $\dfrac{b_f}{2t_f} \ge \dfrac{65}{\sqrt{F_{ye}}}$ or $\dfrac{h}{t_w} \ge \dfrac{640}{\sqrt{F_{ye}}}$ — 1.25, 2, 3, 3, 4

c. Other — Linear interpolation between the values on lines a and b for both flange slenderness (first term) and web slenderness (second term) shall be performed, and the lowest resulting value shall be used.

Columns—Flexure [11,12]

For $P/P_{CL} < 0.2$

a. $\dfrac{b_f}{2t_f} \le \dfrac{52}{\sqrt{F_{ye}}}$ and $\dfrac{h}{t_w} \le \dfrac{300}{\sqrt{F_{ye}}}$ — 2, 6, 8, 10, 12

b. $\dfrac{b_f}{2t_f} \ge \dfrac{65}{\sqrt{F_{ye}}}$ or $\dfrac{h}{t_w} \ge \dfrac{460}{\sqrt{F_{ye}}}$ — 1.25, 1.25, 2, 2, 3

c. Other — Linear interpolation between the values on lines a and b for both flange slenderness (first term) and web slenderness (second term) shall be performed, and the lowest resulting value shall be used.

For $0.2 \le P/P_{CL} \le 0.5$

a. $\dfrac{b_f}{2t_f} \le \dfrac{52}{\sqrt{F_{ye}}}$ and $\dfrac{h}{t_w} \le \dfrac{260}{\sqrt{F_{ye}}}$ — 1.25, —[1], —[2], —[3], —[4]

b. $\dfrac{b_f}{2t_f} \ge \dfrac{65}{\sqrt{F_{ye}}}$ or $\dfrac{h}{t_w} \ge \dfrac{400}{\sqrt{F_{ye}}}$ — 1.25, 1.25, 1.5, 2, 2

c. Other — Linear interpolation between the values on lines a and b for both flange slenderness (first term) and web slenderness (second term) shall be performed, and the lowest resulting value shall be used.

| Component/Action | IO | Primary LS | Primary CP | Secondary LS | Secondary CP |
|---|---|---|---|---|---|
| Column Panel Zones—Shear | 1.5 | 8 | 11 | 12 | 12 |
| Fully Restrained Moment Connections[14] | | | | | |
| WUF[14] | 1.0 | $4.3 - 0.083d$ | $3.9 - 0.043d$ | $4.3 - 0.048d$ | $5.5 - 0.064d$ |
| Bottom Haunch in WUF with Slab | 2.3 | 2.7 | 3.4 | 3.8 | 4.7 |
| Bottom Haunch in WUF without Slab | 1.8 | 2.1 | 2.5 | 2.8 | 3.3 |
| Welded Cover Plate in WUF[13] | $3.9 - 0.059d$ | $4.3 - 0.067d$ | $5.4 - 0.090d$ | $5.4 - 0.090d$ | $6.9 - 0.118d$ |
| Improved WUF—Bolted Web[13] | $2.0 - 0.016d$ | $2.3 - 0.021d$ | $3.1 - 0.032d$ | $4.9 - 0.048d$ | $6.2 - 0.065d$ |
| Improved WUF—Welded Web | 3.1 | 4.2 | 5.3 | 5.3 | 6.7 |
| Free Flange[13] | $4.5 - 0.065d$ | $6.3 - 0.098d$ | $8.1 - 0.129d$ | $8.4 - 0.129d$ | $11.0 - 0.172d$ |
| Reduced Beam Section[13] | $3.5 - 0.016d$ | $4.9 - 0.025d$ | $6.2 - 0.032d$ | $6.5 - 0.025d$ | $8.4 - 0.032d$ |
| Welded Flange Plates<br>a. Flange Plate Net Section<br>b. Other Limit States | 2.5<br>Force-controlled | 3.3 | 4.1 | 5.7 | 7.3 |
| Welded Bottom Haunch | 2.3 | 3.1 | 3.8 | 4.6 | 5.9 |
| Welded Top and Bottom Haunch | 2.4 | 3.1 | 3.9 | 4.7 | 6.0 |
| Welded Cover—Plated Flanges | 2.5 | 2.8 | 3.4 | 3.4 | 4.2 |

*continued*

**Table 5-5. (Continued)**

| Component/Action | IO | m-Factors for Linear Procedures Primary LS | Primary CP | Secondary LS | Secondary CP |
|---|---|---|---|---|---|

| | | *m*-Factors for Linear Procedures | | | |
| | | Primary | | Secondary | |
| Component/Action | IO | LS | CP | LS | CP |
|---|---|---|---|---|---|
| **Partially Restrained Moment Connections** | | | | | |
| Top and Bottom Clip Angle[7] | | | | | |
| a. Shear Failure of Rivet or Bolt (Limit State 1)[8] | 1.5 | 4 | 6 | 6 | 8 |
| b. Tension Failure of Horizontal Leg of Angle (Limit State 2) | 1.25 | 1.5 | 2 | 1.5 | 2 |
| c. Tension Failure of Rivet or Bolt (Limit State 3)[8] | 1.25 | 1.5 | 2.5 | 4 | 4 |
| d. Flexural Failure of Angle (Limit State 4) | 2 | 5 | 7 | 7 | 14 |
| Double Split Tee[7] | | | | | |
| a. Shear Failure of Rivet or Bolt (Limit State 1)[8] | 1.5 | 4 | 6 | 6 | 8 |
| b. Tension Failure of Rivet or Bolt (Limit State 2)[8] | 1.25 | 1.5 | 2.5 | 4 | 4 |
| c. Tension Failure of Split Tee Stem (Limit State 3) | 1.25 | 1.5 | 2 | 1.5 | 2 |
| d. Flexural Failure of Split Tee (Limit State 4) | 2 | 5 | 7 | 7 | 14 |
| Bolted Flange Plate[7] | | | | | |
| a. Failure in Net Section of Flange Plate or Shear Failure of Bolts or Rivets[8] | 1.5 | 4 | 5 | 4 | 5 |
| b. Weld Failure or Tension Failure on Gross Section of Plate | 1.25 | 1.5 | 2 | 1.5 | 2 |
| Bolted End Plate | | | | | |
| a. Yield of End Plate | 2 | 5.5 | 7 | 7 | 7 |
| b. Yield of Bolts | 1.5 | 2 | 3 | 4 | 4 |
| c. Failure of Weld | 1.25 | 1.5 | 2 | 3 | 3 |
| Composite Top and Clip Angle Bottom[7] | | | | | |
| a. Failure of Deck Reinforcement | 1.25 | 2 | 3 | 4 | 6 |
| b. Local Flange Yielding and Web Crippling of Column | 1.5 | 4 | 6 | 5 | 7 |
| c. Yield of Bottom Flange Angle | 1.5 | 4 | 6 | 6 | 7 |
| d. Tensile Yield of Rivets or Bolts at Column Flange | 1.25 | 1.5 | 2.5 | 2.5 | 3.5 |
| e. Shear Yield of Beam Flange Connections | 1.25 | 2.5 | 3.5 | 3.5 | 4.5 |
| Shear Connection with Slab[13] | $2.4 - 0.011 d_{bg}$ | — | — | $13.0 - 0.290 d_{bg}$ | $17.0 - 0.387 d_{bg}$ |
| Shear Connection without Slab[13] | $8.9 - 0.193 d_{bg}$ | — | — | $13.0 - 0.290 d_{bg}$ | $17.0 - 0.387 d_{bg}$ |
| **EBF Link Beam[6,9]** | | | | | |
| a. $e \leq \dfrac{1.6 M_{CE}}{V_{CE}}$ | 1.5 | 9 | 13 | 13 | 15 |
| b. $e \geq \dfrac{2.6 M_{CE}}{V_{CE}}$ | Same as for beams. | | | | |
| c. $\dfrac{1.6 M_{CE}}{V_{CE}} < e < \dfrac{2.6 M_{CE}}{V_{CE}}$ | Linear interpolation shall be used. | | | | |

| | | m-Factors for Linear Procedures | | | |
|---|---|---|---|---|---|
| | | Primary | | Secondary | |
| Component/Action | IO | LS | CP | LS | CP |
| **Braces in Compression (except EBF braces)** | | | | | |
| a. Slender[16] | | | | | |
| $\dfrac{Kl}{r} \geq 4.2\sqrt{E/Fy}$ | | | | | |
| 1. W, I, 2L in-plane[16], 2C in-plane[16] | 1.25 | 6 | 8 | 7 | 9 |
| 2. 2L out-of-plane[16], 2C out-of-plane[16] | 1.25 | 5 | 7 | 6 | 8 |
| 3. HSS, pipes, tubes | 1.25 | 5 | 7 | 6 | 8 |
| b. Stocky[15,17] | | | | | |
| $\dfrac{Kl}{r} \leq 2.1\sqrt{E/Fy}$ | | | | | |
| 1. W, I, 2L in-plane[16], 2C in-plane[16] | 1.25 | 5 | 7 | 6 | 8 |
| 2. 2L out-of-plane[16], 2C out-of-plane[16] | 1.25 | 4 | 6 | 5 | 7 |
| 3. HSS, pipes, tubes | 1.25 | 4 | 6 | 5 | 7 |
| c. Intermediate | Linear interpolation between the values for slender and stocky braces (after application of all applicable modifiers) shall be used. | | | | |
| **Braces in Tension** (except EBF Braces)[19] | 1.25 | 6 | 8 | 8 | 10 |
| **Beams, Columns in Tension** (except EBF Beams, Columns) | 1.25 | 3 | 5 | 6 | 7 |
| **Steel Plate Shear Walls** [10] | 1.5 | 8 | 12 | 12 | 14 |
| **Diaphragm Components** | | | | | |
| Diaphragm Shear Yielding or Panel or Plate Buckling | 1.25 | 2 | 3 | 2 | 3 |
| Diaphragm Chords and Collectors— Full Lateral Support | 1.25 | 6 | 8 | 6 | 8 |
| Diaphragm Chords and Collectors— Limited Lateral Support | 1.25 | 2 | 3 | 2 | 3 |

[1] $m = 9(1 - 5/3\, P/P_{CL})$.

[2] $m = 12(1 - 5/3\, P/P_{CL})$.

[3] $m = 15(1 - 5/3\, P/P_{CL})$.

[4] $m = 18(1 - 5/3\, P/P_{CL})$.

[5] Not used.

[6] Values are for link beams with three or more web stiffeners. If no stiffeners, divide values by 2.0, but values need not be less than 1.25. Linear interpolation shall be used for one or two stiffeners.

[7] Web plate or stiffened seat shall be considered to carry shear. Without shear connection, action shall not be classified as secondary. If $d_b > 18$ in., multiply m-factors by $18/d_b$, but values need not be less than 1.0.

[8] For high-strength bolts, divide values by 2.0, but values need not be less than 1.25.

[9] Assumes ductile detailing for flexural link, in accordance with AISC *LRFD Specifications* (AISC 1999).

[10] Applicable if stiffeners, or concrete backing, is provided to prevent buckling.

[11] Columns in moment or braced frames shall be permitted to be designed for the maximum force delivered by connecting members. For rectangular or square columns, replace $b_t/2t_f$ with $b/t$, replace 52 with 110, and replace 65 with 190.

[12] Columns with $P/P_{CL} > 0.5$ shall be considered force-controlled.

[13] $d$ is the beam depth; $d_{bg}$ is the depth of the bolt group.

[14] Tabulated values shall be modified as indicated in Section 5.4.2.4.2, Item 4.

[15] In addition to consideration of connection capacity in accordance with Section 5.4.2.4.1, values for braces shall be modified for connection robustness as follows: Where brace connections do not satisfy the requirements of AISC 341, Section 13.3c (AISC 2002) , the acceptance criteria shall be multiplied by 0.8.

[16] Stitches for built-up members: Where the stitches for built-up braces do not satisfy the requirements of AISC 341, Section 13.2e (AISC 2002) , the acceptance criteria shall be multiplied by 0.5.

[17] Section compactness: Acceptance criteria applies to brace sections that are concrete-filled or seismically compact according to Table I-8-1 of AISC 341 (AISC 2002). Where the brace section is noncompact according to Table B5.1 of AISC *LRFD Specifications* (AISC 1999), the acceptance criteria shall be multiplied by 0.5. For intermediate compactness conditions, the acceptance criteria shall be multiplied by a value determined by linear interpolation between the seismically compact and the noncompact cases.

[18] Regardless of the modifiers applied, $m$ need never be taken less than 1.0.

[19] For tension-only bracing, m-factors shall be divided by 2.0, but need not be less than 1.25.

in the LSP and NSP. Where using the acceptance criteria of this section, the design professional should follow the procedures set forth in Chapter 3 of this standard without modification. The procedures in this standard and the SAC procedures are judged to result in comparable levels of drift demand.

Connections between the members shall be classified as fully restrained (FR) or partially restrained (PR), based on the strength and stiffness of the connection assembly. The connection types and definitions contained in Table 5-4, as well as the acceptance criteria for these connections, has been adopted from the referenced SAC documents, FEMA 350 (FEMA 2000), 351 (FEMA 2000), 355D (FEMA 2000), and 355F (FEMA 2000). The number of connections identified is based on research that has shown behavior to be highly dependent on connection detailing. The design professional should refer to those guidelines for more detailed descriptions of these connections as well as a methodology for determining acceptance criteria for other connection types not included in this standard.

FEMA 351 (FEMA 2000) provides an alternate methodology for determining column demands that has not been adopted into this standard.

## 5.4.2 Fully Restrained Moment Frames

### 5.4.2.1 General

FR moment frames shall be those moment frames with connections identified as FR in Table 5-4.

Moment frames with connections not included in Table 5-4 shall be defined as FR if the joint deformations (not including panel zone deformation) do not contribute more than 10% to the total lateral deflection of the frame, and the connection is at least as strong as the weaker of the two members being joined. If either of these conditions is not satisfied, the frame shall be characterized as PR.

FR moment frames encompass both Special Moment Frames and Ordinary Moment Frames, defined in AISC 341 (AISC 2002). These terms are not used in this standard, but the requirements for these systems and for general or seismic design of steel componentks specified in *Load and Resistance Factor Design Specification for Structural Steel Buildings (LRFD)* (AISC 1999) or ASCE 7 (ASCE 2005) shall be followed for new elements designed as part of the seismic rehabilitation, unless superseded by provisions in this standard.

### C5.4.2.1 General

FEMA 351 (FEMA 2000) identifies two types of connections—Type 1 (ductile) and Type 2 (brittle). These definitions are not used in this standard since the distinction is reflected in the acceptance criteria for the connections.

The most common beam-to-column connection used in steel FR moment frames since the late 1950s required the beam flange to be welded to the column flange using complete joint penetration groove welds. Many of these connections have fractured during recent earthquakes. The design professional is referred to FEMA 274 (FEMA 1997) and to FEMA 351 (FEMA 1997).

### 5.4.2.2 Stiffness

*5.4.2.2.1 Linear Static and Dynamic Procedures* The stiffness of steel members (columns and beams) and connections (joints and panel zones used with the linear procedures of Chapter 3) shall be based on principles of structural mechanics and as specified in the *Load and Resistance Factor Design Specification for Structural Steel Buildings (LRFD)* (AISC 1999) unless superseded by provisions of this section.

1. **Axial Area and Shear Area.** For components fully encased in concrete, calculation of the stiffness using full composite action shall be permitted if confining reinforcement is provided to allow the concrete to remain in place during an earthquake. Concrete confined on at least three sides, or over 75% of its perimeter, by elements of the structural steel member shall be permitted to be considered adequately confined to provide composite action.

2. **Moment of Inertia.** For components fully encased in concrete, calculation of the stiffness using full composite action shall be permitted, but the width of the composite section shall be taken as equal to the width of the flanges of the steel member and shall not include parts of the adjoining concrete floor slab, unless there is an identifiable shear transfer mechanism between the concrete slab and the steel flange which is shown to meet the applicable acceptance criteria for the selected performance level.

3. **Panel Zone Modeling.** Inclusion of panel zone flexibility shall be permitted in a frame analysis by adding a panel zone element to the mathematical model. Alternatively, adjustment of the beam flexural stiffness to account for panel zone flexibility shall be

permitted. Where the expected shear strength of panel zones exceeds the flexural strength of the beams at a beam–column connection, and the stiffness of the panel zone is at least 10 times larger than the flexural stiffness of the beam, direct modeling of the panel zone shall not be required. In such cases, rigid offsets from the center of the column shall be permitted to represent the effective span of the beam. Use of center-line analysis shall be permitted for other cases.

4. **Joint Modeling.** Modeling of connection stiffness for FR moment frames shall not be required except for joints that are intentionally reinforced to force formation of plastic hinges within the beam span, remote from the column face. For such joints, rigid elements shall be used between the column and the beam to represent the effective span of the beam.

5. **Connections.** Requirements of this section shall apply to connections identified as FR in Table 5-4 and those meeting the requirements of Section 5.4.2.1.

*5.4.2.2.2 Nonlinear Static Procedure* If the Nonlinear Static Procedure (NSP) of Chapter 3 is used, the following criteria shall apply:

1. Elastic component properties shall be modeled as specified in Section 5.4.2.2.1;
2. Plastification shall be represented by nonlinear moment-curvature and interaction relationships for beams and beam–columns derived from experiment or analysis; and
3. Linear or nonlinear behavior of panel zones shall be included in the mathematical model except as indicated in Section 5.4.2.2.1, Item 3.

In lieu of relationships derived from experiment or analysis, the generalized load–deformation curve shown in Fig. 5-1, with parameters $a$, $b$, and $c$ as defined in Tables 5-6 and 5-7, shall be used for components of steel moment frames. Modification of this curve shall be permitted to account for strain-hardening of components as follows: (1) a strain-hardening slope of 3% of the elastic slope shall be permitted for beams and columns unless a greater strain-hardening slope is justified by test data; and (2) where panel zone yielding occurs, a strain-hardening slope of 6% shall be used for the panel zone unless a greater strain-hardening slope is justified by test data.

The parameters $Q$ and $Q_y$ in Fig. 5-1 are generalized component load and generalized component expected strength, respectively. For beams and columns, $\theta$ is the total elastic and plastic rotation of the beam or column, $\theta_y$ is the rotation at yield, $\Delta$ is total elastic and plastic displacement, and $\Delta_y$ is yield displacement. For panel zones, $\theta_y$ is the angular shear deformation in radians. Figure 5-2 defines chord rotation for beams. The chord rotation shall be calculated either by adding the yield rotation, $\theta_y$, to the plastic rotation or taken to be equal to the story drift. Use of Eqs. 5-1 and 5-2 shall be permitted to calculate the yield rotation, $\theta_y$, where the point of contraflexure is anticipated to occur at the mid-length of the beam or column, respectively.

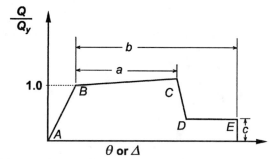

**FIGURE 5-1. Generalized Force–Deformation Relation for Steel Elements or Components.**

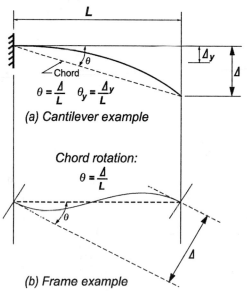

$$\theta = \frac{\Delta}{L} \qquad \theta_y = \frac{\Delta_y}{L}$$

*(a) Cantilever example*

Chord rotation:

$$\theta = \frac{\Delta}{L}$$

*(b) Frame example*

**FIGURE 5-2. Definition of Chord Rotation.**

**Table 5-6. Modeling Parameters and Acceptance Criteria for Nonlinear Procedures—Structural Steel Components**

| | Modeling Parameters | | | Acceptance Criteria[14] | | | | |
| | Plastic Rotation Angle, Radians | | Residual Strength Ratio | Plastic Rotation Angle, Radians | | | | |
| | | | | | Primary | | Secondary | |
| Component/Action | $a$ | $b$ | $c$ | IO | LS | CP | LS | CP |
|---|---|---|---|---|---|---|---|---|
| **Beams—Flexure** | | | | | | | | |
| a. $\dfrac{b_f}{2t_f} \le \dfrac{52}{\sqrt{F_{ye}}}$ and $\dfrac{h}{t_w} \le \dfrac{418}{\sqrt{F_{ye}}}$ | $9q_y$ | $11q_y$ | 0.6 | $1q_y$ | $6q_y$ | $8q_y$ | $9q_y$ | $11q_y$ |
| b. $\dfrac{b_f}{2t_f} \ge \dfrac{65}{\sqrt{F_{ye}}}$ or $\dfrac{h}{t_w} \ge \dfrac{640}{\sqrt{F_{ye}}}$ | $4q_y$ | $6q_y$ | 0.2 | $0.25q_y$ | $2q_y$ | $3q_y$ | $3q_y$ | $4q_y$ |
| c. Other | Linear interpolation between the values on lines a and b for both flange slenderness (first term) and web slenderness (second term) shall be performed, and the lower resulting value shall be used | | | | | | | |
| **Columns—Flexure[2,7]** For $P/P_{CL} < 0.2$ | | | | | | | | |
| a. $\dfrac{b_f}{2t_f} \le \dfrac{52}{\sqrt{F_{ye}}}$ and $\dfrac{h}{t_w} \le \dfrac{300}{\sqrt{F_{ye}}}$ | $9q_y$ | $11q_y$ | 0.6 | $1q_y$ | $6q_y$ | $8q_y$ | $9q_y$ | $11q_y$ |
| b. $\dfrac{b_f}{2t_f} \ge \dfrac{65}{\sqrt{F_{ye}}}$ or $\dfrac{h}{t_w} \ge \dfrac{460}{\sqrt{F_{ye}}}$ | $4q_y$ | $6q_y$ | 0.2 | $0.25q_y$ | $2q_y$ | $3q_y$ | $3q_y$ | $4q_y$ |
| c. Other | Linear interpolation between the values on lines a and b for both flange slenderness (first term) and web slenderness (second term) shall be performed, and the lower resulting value shall be used | | | | | | | |
| For $0.2 \le P/P_{CL} \le 0.5$ | | | | | | | | |
| a. $\dfrac{b_f}{2t_f} \le \dfrac{52}{\sqrt{F_{ye}}}$ and $\dfrac{h}{t_w} \le \dfrac{260}{\sqrt{F_{ye}}}$ | —[3] | —[4] | 0.2 | $0.25q_y$ | —[5] | —[3] | —[6] | —[4] |
| b. $\dfrac{b_f}{2t_f} \ge \dfrac{65}{\sqrt{F_{ye}}}$ or $\dfrac{h}{t_w} \ge \dfrac{400}{\sqrt{F_{ye}}}$ | $1q_y$ | $1.5q_y$ | 0.2 | $0.25q_y$ | $0.5q_y$ | $0.8q_y$ | $1.2q_y$ | $1.2q_y$ |
| c. Other | Linear interpolation between the values on lines a and b for both flange slenderness (first term) and web slenderness (second term) shall be performed, and the lower resulting value shall be used | | | | | | | |
| **Column Panel Zones** | $12q_y$ | $12q_y$ | 1.0 | $1q_y$ | $8q_y$ | $11q_y$ | $12q_y$ | $12q_y$ |
| **Fully Restrained Moment Connections[13]** | | | | | | | | |
| WUF[12] | $0.051 - 0.0013d$ | $0.043 - 0.00060d$ | 0.2 | $0.026 - 0.00065d$ | $0.0337 - 0.00086d$ | $0.0284 - 0.00040d$ | $0.0323 - 0.00045d$ | $0.043 - 0.00060d$ |
| Bottom Haunch in WUF with Slab | 0.026 | 0.036 | 0.2 | 0.013 | 0.0172 | 0.0238 | 0.0270 | 0.036 |
| Bottom Haunch in WUF without Slab | 0.018 | 0.023 | 0.2 | 0.009 | 0.0119 | 0.0152 | 0.0180 | 0.023 |
| Welded Cover Plate in WUF[12] | $0.056 - 0.0011d$ | $0.056 - 0.0011d$ | 0.2 | $0.028 - 0.00055d$ | $0.0319 - 0.00063d$ | $0.0426 - 0.00084d$ | $0.0420 - 0.00083d$ | $0.056 - 0.0011d$ |
| Improved WUF—Bolted Web[12] | $0.021 - 0.00030d$ | $0.050 - 0.00060d$ | 0.2 | $0.010 - 0.00015d$ | $0.0139 - 0.00020d$ | $0.0210 - 0.00030d$ | $0.0375 - 0.00045d$ | $0.050 - 0.00060d$ |
| Improved WUF—Welded Web | 0.041 | 0.054 | 0.2 | 0.020 | 0.0312 | 0.0410 | 0.0410 | 0.054 |
| Free Flange[12] | $0.067 - 0.0012d$ | $0.094 - 0.0016d$ | 0.2 | $0.034 - 0.00060d$ | $0.0509 - 0.00091d$ | $0.0670 - 0.0012d$ | $0.0705 - 0.0012d$ | $0.094 - 0.0016d$ |
| Reduced Beam Section[12] | $0.050 - 0.00030d$ | $0.070 - 0.00030d$ | 0.2 | $0.025 - 0.00015d$ | $0.0380 - 0.00023d$ | $0.0500 - 0.00030d$ | $0.0525 - 0.00023d$ | $0.07 - 0.00030d$ |
| Welded Flange Plates a. Flange Plate Net Section | 0.03 | 0.06 | 0.2 | 0.015 | 0.0228 | 0.0300 | 0.0450 | 0.06 |
| b. Other Limit States | Force-controlled | | | | | | | |
| Welded Bottom Haunch | 0.027 | 0.047 | 0.2 | 0.014 | 0.0205 | 0.0270 | 0.0353 | 0.047 |

| Component/Action | Modeling Parameters | | | Acceptance Criteria[14] | | | | |
|---|---|---|---|---|---|---|---|---|
| | Plastic Rotation Angle, Radians | | Residual Strength Ratio | Plastic Rotation Angle, Radians | | | | |
| | | | | | Primary | | Secondary | |
| | $a$ | $b$ | $c$ | IO | LS | CP | LS | CP |
| Welded Top and Bottom Haunches | 0.028 | 0.048 | 0.2 | 0.014 | 0.0213 | 0.0280 | 0.0360 | 0.048 |
| Welded Cover—Plated Flanges | 0.031 | 0.031 | 0.2 | 0.016 | 0.0177 | 0.0236 | 0.0233 | 0.031 |
| **Partially Restrained Moment Connections** | | | | | | | | |
| Top and Bottom Clip Angle[9] | | | | | | | | |
| a. Shear Failure of Rivet or Bolt (Limit State 1)[8] | 0.036 | 0.048 | 0.200 | 0.008 | 0.020 | 0.030 | 0.030 | 0.040 |
| b. Tension Failure of 0.012 Horizontal Leg of Angle (Limit State 2) | 0.018 | 0.800 | 0.003 | 0.008 | 0.010 | 0.010 | 0.015 | |
| c. Tension Failure of Rivet or Bolt (Limit State 3)[8] | 0.016 | 0.025 | 1.000 | 0.005 | 0.008 | 0.013 | 0.020 | 0.020 |
| d. Flexural Failure of Angle (Limit State 4) | 0.042 | 0.084 | 0.200 | 0.010 | 0.025 | 0.035 | 0.035 | 0.070 |
| Double Split Tee[9] | | | | | | | | |
| a. Shear Failure of Rivet or Bolt (Limit State 1)[8] | 0.036 | 0.048 | 0.200 | 0.008 | 0.020 | 0.030 | 0.030 | 0.040 |
| b. Tension Failure of Rivet or Bolt (Limit State 2)[8] | 0.016 | 0.024 | 0.800 | 0.005 | 0.008 | 0.013 | 0.020 | 0.020 |
| c. Tension Failure of Split Tee Stem (Limit State 3) | 0.012 | 0.018 | 0.800 | 0.003 | 0.008 | 0.010 | 0.010 | 0.015 |
| d. Flexural Failure of Split Tee (Limit State 4) | 0.042 | 0.084 | 0.200 | 0.010 | 0.025 | 0.035 | 0.035 | 0.070 |
| Bolted Flange Plate[9] | | | | | | | | |
| a. Failure in Net Section of Flange Plate or Shear Failure of Bolts or Rivets[8] | 0.030 | 0.030 | 0.800 | 0.008 | 0.020 | 0.025 | 0.020 | 0.025 |
| b. Weld Failure or Tension Failure on Gross Section of Plate | 0.012 | 0.018 | 0.800 | 0.003 | 0.008 | 0.010 | 0.010 | 0.015 |
| Bolted End Plate | | | | | | | | |
| a. Yield of End Plate | 0.042 | 0.042 | 0.800 | 0.010 | 0.028 | 0.035 | 0.035 | 0.035 |
| b. Yield of Bolts | 0.018 | 0.024 | 0.800 | 0.008 | 0.010 | 0.015 | 0.020 | 0.020 |
| c. Failure of Weld | 0.012 | 0.018 | 0.800 | 0.003 | 0.008 | 0.010 | 0.015 | 0.015 |
| Composite Top Clip Angle Bottom[9] | | | | | | | | |
| a. Failure of Deck Reinforcement | 0.018 | 0.035 | 0.800 | 0.005 | 0.010 | 0.015 | 0.020 | 0.030 |
| b. Local Flange Yielding and Web Crippling of Column | 0.036 | 0.042 | 0.400 | 0.008 | 0.020 | 0.030 | 0.025 | 0.035 |
| c. Yield of Bottom Flange Angle | 0.036 | 0.042 | 0.200 | 0.008 | 0.020 | 0.030 | 0.025 | 0.035 |
| d. Tensile Yield of Rivets or Bolts at Column Flange | 0.015 | 0.022 | 0.800 | 0.005 | 0.008 | 0.013 | 0.013 | 0.018 |
| e. Shear Yield of Beam Flange Connection | 0.022 | 0.027 | 0.200 | 0.005 | 0.013 | 0.018 | 0.018 | 0.023 |
| Shear Connection with Slab[12] | $0.029 - 0.00020d_{bg}$ | $0.15 - 0.0036d_{bg}$ | 0.400 | $0.014 - 0.00010d_{bg}$ | — | — | $0.1125 - 0.0027d_{bg}$ | $0.15 - 0.0036d_{bg}$ |
| Shear Connection without Slab[12] | $0.15 - 0.0036d_{bg}$ | $0.15 - 0.0036d_{bg}$ | 0.400 | $0.075 - 0.0018d_{bg}$ | — | — | $0.1125 - 0.0027d_{bg}$ | $0.15 - 0.0036d_{bg}$ |

*Continued*

**Table 5-6. (Continued)**

| Component/Action | Modeling Parameters | | | Acceptance Criteria[14] | | | | |
|---|---|---|---|---|---|---|---|---|
| | Plastic Rotation Angle, Radians | | Residual Strength Ratio | Plastic Rotation Angle, Radians | | | | |
| | | | | | Primary | | Secondary | |
| | $a$ | $b$ | $c$ | IO | LS | CP | LS | CP |
| **EBF Link Beam[10,11]** | | | | | | | | |
| a. $e \leq \dfrac{1.6\,M_{CE}}{V_{CE}}$ | 0.15 | 0.17 | 0.8 | 0.005 | 0.11 | 0.14 | 0.14 | 0.16 |
| b. $e \geq \dfrac{2.6\,M_{CE}}{V_{CE}}$ | Same as for beams | | | | | | | |
| c. $\dfrac{1.6\,M_{CE}}{V_{CE}} < e < \dfrac{2.6\,M_{CE}}{V_{CE}}$ | Linear interpolation shall be used | | | | | | | |
| **Steel Plate Shear Walls[1]** | $14q_y$ | $16q_y$ | 0.7 | $0.5q_y$ | $10q_y$ | $13q_y$ | $13q_y$ | $15q_y$ |

[1]Values are for shear walls with stiffeners to prevent shear buckling.

[2]Columns in moment or braced frames shall be permitted to be designed for the maximum force delivered by connecting members. For rectangular or square columns, replace $b_t/2t_f$ with $b/t$, replace 52 with 110, and replace 65 with 190.

[3]Plastic rotation = $11\,(1 - 5/3\,P/P_{CL})\,\theta_y$.

[4]Plastic rotation = $17\,(1 - 5/3\,P/P_{CL})\,\theta_y$.

[5]Plastic rotation = $8\,(1 - 5/3\,P/P_{CL})\,\theta_y$.

[6]Plastic rotation = $14\,(1 - 5/3\,P/P_{CL})\,\theta_y$.

[7]Columns with $P/P_{CL} > 0.5$ shall be considered force-controlled.

[8]For high-strength bolts, divide values by 2.0.

[9]Web plate or stiffened seat shall be considered to carry shear. Without shear connection, action shall not be classified as secondary. If beam depth, $d_b > 18$ in., multiply $m$-factors by $18/d_b$.

[10]Deformation is the rotation angle between link and beam outside link or column.

[11]Values are for link beams with three or more web stiffeners. If no stiffeners, divide values by 2.0. Linear interpolation shall be used for one or two stiffeners.

[12]$d$ is the beam depth; $d_{bg}$ is the depth of the bolt group. Where plastic rotations are a function of $d$ or $d_{bg}$, they need not be taken as less than 0.0.

[13]Tabulated values shall be modified as indicated in Section 5.4.2.4.3, Item 4.

[14]Primary and secondary component demands shall be within secondary component acceptance criteria where the full backbone curve is explicitly modeled including strength degradation and residual strength in accordance with Section 3.4.3.2.

**Table 5-7. Modeling Parameters and Acceptance Criteria for Nonlinear Procedures—Structural Steel Components—Axial Actions**

| Component/Action | Modeling Parameters | | | Acceptance Criteria[6] | | | | |
| --- | --- | --- | --- | --- | --- | --- | --- | --- |
| | Plastic Deformation | | Residual Strength Ratio | Plastic Deformation | | | | |
| | | | | | Primary | | Secondary | |
| | $a$ | $b$ | $c$ | IO | LS | CP | LS | CP |
| Braces in Compression (except EBF braces)[1,2] | | | | | | | | |
| a. Slender | | | | | | | | |
| $\dfrac{Kl}{r} \geq 4.2\sqrt{E/F_y}$ | | | | | | | | |
| 1. W, I, 2L In-Plane[3], 2C In-Plane[3] | $0.5\Delta_c$ | $10\Delta_c$ | 0.3 | $0.25\Delta_c$ | $6\Delta_c$ | $8\Delta_c$ | $8\Delta_c$ | $10\Delta_c$ |
| 2. 2L Out-of-Plane[3], 2C Out-of-Plane[3] | $0.5\Delta_c$ | $9\Delta_c$ | 0.3 | $0.25\Delta_c$ | $5\Delta_c$ | $7\Delta_c$ | $7\Delta_c$ | $9\Delta_c$ |
| 3. HSS, Pipes, Tubes | $0.5\Delta_c$ | $9\Delta_c$ | 0.3 | $0.25\Delta_c$ | $5\Delta_c$ | $7\Delta_c$ | $7\Delta_c$ | $9\Delta_c$ |
| b. Stocky[4] | | | | | | | | |
| $\dfrac{Kl}{r} \leq 2.1\sqrt{E/F_y}$ | | | | | | | | |
| 1. W, I, 2L In-Plane[3], 2C In-Plane[3] | $1\Delta_c$ | $8\Delta_c$ | 0.5 | $0.25\Delta_c$ | $5\Delta_c$ | $7\Delta_c$ | $7\Delta_c$ | $8\Delta_c$ |
| 2. 2L Out-of-Plane[3], 2C Out-of-Plane[3] | $1\Delta_c$ | $7\Delta_c$ | 0.5 | $0.25\Delta_c$ | $4\Delta_c$ | $6\Delta_c$ | $6\Delta_c$ | $7\Delta_c$ |
| 3. HSS, Pipes, Tubes | $1\Delta_c$ | $7\Delta_c$ | 0.5 | $0.25\Delta_c$ | $4\Delta_c$ | $6\Delta_c$ | $6\Delta_c$ | $7\Delta_c$ |
| c. Intermediate | Linear interpolation between the values for slender and stocky braces (after application of all applicable modifiers) shall be used. | | | | | | | |
| Braces in Tension (except EBF braces)[5] | $11\Delta_T$ | $14\Delta_I$ | 0.8 | $0.25\Delta_T$ | $7\Delta_T$ | $9\Delta_T$ | $11\Delta_T$ | $13\Delta_T$ |
| Beams, Columns in Tension (except EBF beams, columns)[5] | $5\Delta_T$ | $7\Delta_T$ | 1.0 | $0.25\Delta_T$ | $3\Delta_T$ | $5\Delta_T$ | $6\Delta_T$ | $7\Delta_T$ |

[1] $\Delta_c$ is the axial deformation at expected buckling load.

[2] In addition to consideration of connection capacity in accordance with Section 5.5.2.4.1, values for braces shall be modified for connection robustness as follows: Where brace connections do not satisfy the requirements of AISC 341, Section 13.3c (AISC 2002), the acceptance criteria shall be multiplied by 0.8.

[3] Stitches for built-up members: Where the stitches for built-up braces do not satisfy the requirements of AISC 341, Section 13.2e (AISC 2002), the values of $a$, $b$, and all acceptance criteria shall be multiplied by 0.5.

[4] Section compactness: Modeling parameters and acceptance criteria apply to brace sections that are concrete-filled or seismically compact according to Table I-8-1 of AISC 341 (AISC 2002). Where the brace section is noncompact according to Table B5.1 of AISC *LRFD Specifications* (AISC 1999), the acceptance criteria shall be multiplied by 0.5. For intermediate compactness conditions, the acceptance criteria shall be multiplied by a value determined by linear interpolation between the seismically compact and the noncompact cases.

[5] $\Delta_T$ is the axial deformation at expected tensile yielding load.

[6] Primary and secondary component demands shall be within secondary component acceptance criteria where the full backbone curve is explicitly modeled including strength degradation and residual strength in accordance with Section 3.4.3.2.

Beams: $\quad \theta_y = \dfrac{ZF_{ye}l_b}{6EI_b}$ (Eq. 5-1)

Columns: $\quad \theta_y = \dfrac{ZF_{ye}l_c}{6EI_c}\left(1 - \dfrac{P}{P_{ye}}\right)$ (Eq. 5-2)

$Q_{CE}$ is the component expected strength. For flexural actions of beams and columns, $Q_{CE}$ refers to the plastic moment capacity, which shall be calculated using Eqs. 5-3 and 5-4: k

Beams: $\quad Q_{CE} = M_{CE} = ZF_{ye}$ (Eq. 5-3)

Columns: $Q_{CE} = M_{CE} = 1.18ZF_{ye}\left(1 - \dfrac{P}{P_{ye}}\right) \leq ZF_{ye}$

(Eq. 5-4)

For panel zones, $Q_{CE}$ refers to the plastic shear capacity of the panel zone, which shall be calculated using Eq. 5-5:

Panel Zones: $\quad Q_{CE} = V_{CE} = 0.55F_{ye}d_ct_p$ (Eq. 5-5)

where

$d_c$ = column depth;
$E$ = modulus of elasticity;
$F_{ye}$ = expected yield strength of the material;
$I$ = moment of inertia;
$l_b$ = beam length;
$l_c$ = column length;
$M_{CE}$ = expected flexural strength;
$P$ = axial force in the member at the target displacement for nonlinear static analyses, or at the instant of computation for nonlinear dynamic analyses. For linear analyses, $P$ shall be taken as $Q_{UF}$, calculated in accordance with Section 3.4.2.1.2;
$P_{ye}$ = expected axial yield force of the member = $A_gF_{ye}$;
$t_p$ = total thickness of panel zone, including doubler plates;
$\theta$ = chord rotation;
$\theta_y$ = yield rotation;
$V_{CE}$ = expected shear strength; and
$Z$ = plastic section modulus.

*C5.4.2.2.2 Nonlinear Static Procedure* Strain hardening should be considered for all components. FEMA 355D (FEMA 2000) is a useful reference for information concerning nonlinear behavior of various tested connection configurations.

*5.4.2.2.3 Nonlinear Dynamic Procedure* The complete hysteretic behavior of each component shall be determined experimentally or by other procedures approved by the authority having jurisdiction.

*C5.4.2.2.3 Nonlinear Dynamic Procedure* FEMA 355D (FEMA 2000) is a useful reference for information concerning nonlinear behavior of various tested connection configurations.

### 5.4.2.3 Strength

*5.4.2.3.1 General* Component strengths shall be computed in accordance with the general requirements of Section 5.3.2 and the specific requirements of this section.

*5.4.2.3.2 Linear Static and Dynamic Procedures*

1. **Beams.** The strength of structural steel elements under flexural actions shall be calculated in accordance with this section if the calculated axial load does not exceed 10% of the axial strength.

The expected flexural strength, $Q_{CE}$, of beam components shall be determined using equations for design strength, $M_n$, given in AISC 341 (AISC 2002), except that $\phi$ shall be taken as 1.0 and $F_{ye}$ shall be substituted for $F_y$. The component expected strength, $Q_{CE}$, of beams and other flexural deformation-controlled members shall be the lowest value obtained for the limit states of yielding, lateral-torsional buckling, local flange buckling, or shear yielding of the web.

For fully concrete-encased beams where confining reinforcement is provided to allow the concrete to remain in place during the earthquake, the values of $b_f = 0$ and $L_b = 0$ shall be permitted to be used. For bare beams bent about their major axes and symmetric about both axes, satisfying the requirements of compact sections, and $L_b < L_p$, $Q_{CE}$ shall be computed in accordance with Eq. 5-6:

$$Q_{CE} = M_{CE} = M_{pCE} = ZF_{ye} \qquad \text{(Eq. 5-6)}$$

where

$b_f$ = width of the compression flange;
$L_b$ = distance between points braced against lateral displacement of the compression flange, or between points braced to prevent twist of the cross section, per *Load and Resistance Factor Design Specification for Structural Steel Buildings (LRFD)* (AISC 1999) ;
$L_p$ = limiting lateral unbraced length for full plastic bending capacity for uniform bending from *Load and Resistance Factor Design*

*Specification for Structural Steel Buildings (LRFD)* (AISC 1999);

$M_{pCE}$ = expected plastic moment capacity; and
$F_{ye}$ = expected yield strength of the material.

The limit states of local and lateral torsional buckling shall not be considered for components either subjected to bending about their minor axes or fully encased in concrete where confining reinforcement is provided to allow the concrete to remain in place during an earthquake.

If the beam strength is governed by the shear strength of the unstiffened web and $\frac{h}{t_w} \leq \frac{418}{\sqrt{F_y}}$, then $V_{CE}$ shall be calculated in accordance with Eq. 5-7:

$$Q_{CE} = V_{CE} = 0.6 F_{ye} A_w \qquad \text{(Eq. 5-7)}$$

where

$V_{CE}$ = expected shear strength;
$A_w$ = nominal area of the web = $d_b t_w$;
$t_w$ = web thickness;
$h$ = distance from inside of compression flange to inside of tension flange;
$F_{ye}$ = expected yield strength of the material; and
$F_y$ = specified minimum of yield strength; must be in ksi where used to determine applicability per Eq. 5-7.

If $\frac{h}{t_w} > \frac{418}{\sqrt{F_y}}$, then the value of $V_{CE}$ shall be calculated from AISC 341 (AISC 2002).

2. **Columns.** This section shall be used to evaluate flexural and axial strengths of structural steel elements if the calculated axial load exceeds 10% of the axial strength.

The lower-bound strength, $Q_{CL}$, of steel columns under axial compression shall be the lowest value obtained for the limit states of column buckling, local flange buckling, or local web buckling. The effective design strength or the lower-bound axial compressive strength, $P_{CL}$, shall be calculated in accordance with AISC 341 (AISC 2002), taking $\phi = 1.0$ and using the lower-bound strength, $F_{yLB}$, for yield strength.

The expected axial strength of a column in tension, $Q_{CE}$, shall be computed in accordance with Eq. 5-8:

$$Q_{CE} = T_{CE} = A_c F_{ye} \qquad \text{(Eq. 5-8)}$$

where

$A_c$ = area of column;

$F_{ye}$ = expected yield strength of the material; and
$T_{CE}$ = expected tensile strength of column.

3. **Panel Zone.** The strength of the panel zone shall be calculated using Eq. 5-5.

4. **FR Beam–Column Connections.** The strength of connections shall be based on the controlling mechanism considering all potential modes of failure.

*C5.4.2.3.2 Linear Static and Dynamic Procedures* **FR Beam–Column Connections.** The design professional is directed to FEMA 351 (FEMA 2000) for guidance in determining the strength of various FR connection configurations.

*5.4.2.3.3 Nonlinear Static Procedure* The complete load–deformation relationship of each component as depicted in Fig. 5-1 shall be determined in accordance with Section 5.4.2.2.2. The values for expected strength, $Q_{CE}$, shall be the same as those used for linear procedures as specified in Section 5.4.2.3.2.

*5.4.2.3.4 Nonlinear Dynamic Procedures* The complete hysteretic behavior of each component shall be determined experimentally or by other procedures approved by the authority having jurisdiction.

*C5.4.2.3.4 Nonlinear Dynamic Procedures* FEMA 355D (FEMA 2000) is a useful reference for information concerning nonlinear behavior of various tested connection configurations.

### 5.4.2.4 Acceptance Criteria

*5.4.2.4.1 General* Component acceptance criteria shall be computed in accordance with the general requirements of Section 5.3.2 and the specific requirements of this section.

*C5.4.2.4.1 General* The strength and behavior of steel moment-resisting frames is typically governed by the connections. The design professional is urged to determine the controlling limit state of the system where selecting the corresponding acceptance criterion.

*5.4.2.4.2 Linear Static and Dynamic Procedures*

1. **Beams.** The acceptance criteria of this section shall apply to flexural actions of structural steel elements that have a calculated axial load that does not exceed

10% of the axial strength. Beam flexure and shear shall be considered deformation-controlled.

For built-up shapes, the adequacy of lacing plates shall be evaluated using the provisions for tension braces in Section 5.5.2.4.

Values for the $m$-factor used in Eq. 3-20 shall be as specified in Table 5-5. For fully concrete-encased beams where confining reinforcement is provided to allow the concrete to remain in place during an earthquake, the values of $b_f = 0$ and $L_b = 0$ shall be used for the purpose of determining $m$. If $Q_{CE} < M_{pCE}$ due to lateral torsional buckling, then $m$ in Eq. 3-20 shall be replaced by $m_e$, calculated in accordance with Eq. 5-9:

$$m_e = m - (m - 1)\left(\frac{M_p - M_n}{M_p - M_r}\right) \geq 1.0 \quad \text{(Eq. 5-9)}$$

where

$M_n$ = nominal flexural capacity determined in accordance with AISC 341 (AISC 2002);

$M_p$ = plastic moment capacity determined in accordance with *Load and Resistance Factor Design Specification for Structural Steel Buildings (LRFD)* (AISC 1999) ;

$M_r$ = limiting buckling moment determined in accordance with *Load and Resistance Factor Design Specification for Structural Steel Buildings (LRFD)* (AISC 1999);

$m$ = value of $m$ given in Table 5-5; and

$m_e$ = effective $m$ computed in accordance with Eq. 5-9.

For built-up shapes, where the strength is governed by the strength of the lacing plates that carry component shear, the $m$-factor shall be taken as 0.5 times the applicable value in Table 5-5, unless larger values are justified by tests or analysis; however, $m$ need not be taken less than 1.0. For built-up laced beams and columns fully encased in concrete, local buckling of the lacing need not be considered where confining reinforcement is provided to allow the encasement to remain in place during a design earthquake.

2. **Columns.** For steel columns under combined axial compression and bending stress, where the axial column load is less than 50% of the lower-bound axial column strength, $P_{CL}$, the column shall be considered deformation-controlled for flexural behavior and force-controlled for compressive behavior and the combined strength shall be evaluated by Eq. 5-10 or 5-11.

For $0.2 \leq \dfrac{P_{UF}}{P_{CL}} \leq 0.5$,

$$\frac{P_{UF}}{P_{CL}} + \frac{8}{9}\left[\frac{M_x}{m_x M_{CEx}} + \frac{M_y}{m_y M_{CEy}}\right] \leq 1.0 \quad \text{(Eq. 5-10)}$$

For $\dfrac{P_{UF}}{P_{CL}} < 0.2$,

$$\frac{P_{UF}}{2P_{CL}} + \frac{M_x}{m_x M_{CEx}} + \frac{M_y}{m_y M_{CEy}} \leq 1.0 \quad \text{(Eq. 5-11)}$$

where

$P_{UF}$ = axial force in the member computed in accordance with Section 3.4.2.1.2;

$P_{CL}$ = lower-bound compression strength of the column;

$M_x$ = bending moment in the member for the $x$-axis computed in accordance with Section 3.4.2.1.1;

$M_y$ = bending moment in the member for the $y$-axis computed in accordance with Section 3.4.2.1.1;

$M_{CEx}$ = expected bending strength of the column for the $x$-axis;

$M_{CEy}$ = expected bending strength of the column for the $y$-axis;

$m_x$ = value of $m$ for the column bending about the $x$-axis in accordance with Table 5-5; and

$m_y$ = value of $m$ for the column bending about the $y$-axis in accordance with Table 5-5.

Steel columns with axial compressive forces exceeding 50% of the lower-bound axial compressive strength, $P_{CL}$, shall be considered force-controlled for both axial loads and flexure and shall be evaluated using Eq. 5-12:

$$\frac{P_{UF}}{P_{CL}} + \frac{M_{UFx}}{M_{CLx}} + \frac{M_{UFy}}{M_{CLy}} \leq 1 \quad \text{(Eq. 5-12)}$$

where

$P_{UF}$ = axial load in the member, calculated in accordance with Section 3.4.2.1.2;

$M_{UFx}$ = bending moment in the member about the $x$-axis, calculated in accordance with Section 3.4.2.1.2;

$M_{UFy}$ = bending moment in the member about the $y$-axis, calculated in accordance with Section 3.4.2.1.2;

$M_{CLx}$ = lower-bound flexural strength of the member about the $x$-axis; and

$M_{CLy}$ = lower-bound flexural strength of the member about the $y$-axis.

Flexural strength shall be calculated in accordance with AISC 341 (AISC 2002), taking $\phi = 1.0$ and using the lower-bound value for yield strength.

For columns under combined compression and bending, lateral bracing to prevent torsional buckling shall be provided as required by the *Load and Resistance Factor Design Specification for Structural Steel Buildings (LRFD)* (AISC 1999) .

Steel columns under axial tension shall be considered deformation-controlled and shall be evaluated using Eq. 3-20.

Steel columns under combined axial tension and bending stress shall be considered deformation-controlled and shall be evaluated using Eq. 5-13:

$$\frac{T}{m_t T_{CE}} + \frac{M_x}{m_x M_{CEx}} + \frac{M_y}{m_y M_{CEy}} \le 1.0 \quad (Eq.\ 5\text{-}13)$$

where

$M_x$ = bending moment in the member for the x-axis;
$M_y$ = bending moment in the member for the y-axis;
$M_{CEx}$ = expected bending strength of the column for the x-axis;
$M_{CEy}$ = expected bending strength of the column for the y-axis;
$m_t$ = value of $m$ for the column in tension based on Table 5-5;
$m_x$ = value of $m$ for the column bending about the x-axis based on Table 5-5;
$m_y$ = value of $m$ for the column bending about the y-axis based on Table 5-5;
$T$ = tensile load in column; and
$T_{CE}$ = expected tensile strength of column computed in accordance with Eq. 5-8.

**3. Panel Zone.** Shear behavior of panel zones shall be considered deformation-controlled and shall be evaluated using Eq. 3-20, with the expected panel zone shear strength, $Q_{CE}$, calculated according to Eq. 5-5 and m-factors taken from Table 5-5.

**4. FR Beam–Column Connections.** FR connections identified in Table 5-4 shall be considered deformation-controlled and evaluated in accordance with Eq. 3-20, with $Q_{UD}$ and $Q_{CE}$ taken as the computed demand and capacity of the critical connection component respectively, and m-factors taken from Table 5-5 as modified below.

Connection acceptance criteria are dependent on the detailing of continuity plates (column stiffeners that align with the beam flanges), the strength of the panel zone, the beam span-to-depth ratio, and the slenderness of the beam web and flanges. Tabulated

m-factors in Table 5-5 shall be modified as determined by the following four conditions. The modifications shall be cumulative, but m-factors need not be taken as less than 1.0.

4.1 If the connection does not satisfy at least one of the following conditions, the tabulated m-factors in Table 5-5 shall be multiplied by 0.8.

$$t_{cf} \ge \frac{b_{bf}}{5.2}$$

or

$$\frac{b_{bf}}{7} \le t_{cf} < \frac{b_{bf}}{5.2} \text{ and continuity plates with } t \ge \frac{t_{bf}}{2}$$

or

$$t_{cf} < \frac{b_{bf}}{7} \text{ and continuity plates with } t \ge t_{bf}$$

where

$t_{cf}$ = thickness of column flange;
$b_{bf}$ = width of beam flange;
$t$ = thickness of continuity; and
$t_{bf}$ = thickness of beam flange.

4.2 If one of the following conditions is not met, the tabulated m-factors in Table 5-5 shall be multiplied by 0.8.

$$0.6 \le \frac{V_{PZ}}{V_y} \le 0.9$$

where $V_y = 0.55 F_{ye(col)} d_c t_{cw}$ and $V_{PZ}$ is the computed panel zone shear at the development of a hinge at the critical location of the connection. For $M_y$ at the face of the column,

$$V_{PZ} = \frac{\sum M_{y(beam)}}{d_b} \left(\frac{L}{L - d_c}\right)\left(\frac{h - d_b}{h}\right)$$

where

$F_{ye(col)}$ = expected yield strength of column;
$d_c$ = column depth;
$t_{cw}$ = thickness of column web;
$M_{y(beam)}$ = yield moment of beam;
$d_b$ = depth of beam;
$L$ = length of beam, center-to-center of columns; and
$h$ = average story height of columns.

123

4.3 If the clear span-to-depth ratio, $L_c/d$, is greater than 10, the tabulated $m$-factors in Table 5-5 shall be multiplied by:

$$1.4 - 0.04 \frac{L_c}{d}$$

where

$L_c$ = length of beam, clear span between columns; and
$d$ = depth of member.

4.4 If the beam flange and web meet the following conditions, the tabulated $m$-factors in Table 5-5 need not be modified for flange and web slenderness.

$$\frac{b_f}{2t_f} < \frac{52}{\sqrt{F_{ye}}} \quad \text{and} \quad \frac{h}{t_w} < \frac{418}{\sqrt{F_{ye}}}$$

If the beam flange or web slenderness values exceed either of the following limits, the tabulated $m$-factors in Table 5-5 shall be multiplied by 0.5.

$$\frac{b_f}{2t_f} > \frac{65}{\sqrt{F_{ye}}} \quad \text{or} \quad \frac{h}{t_w} > \frac{640}{\sqrt{F_{ye}}}$$

where

$b_f$ = width of beam flange;
$t_f$ = thickness of beam flange;
$h$ = height of beam web;
$t_w$ = thickness of beam web; and
$F_{ye}$ = expected yield strength of column.

Straight-line interpolation, based on the case that results in the lower modifier, shall be used for intermediate values of beam flange or web slenderness.

Type FR connections designed to promote yielding of the beam remote from the column face shall be considered force-controlled and shall be designed using Eq. 5-14:

$$Q_{CLc} \geq Q_{CEb} \quad\quad \text{(Eq. 5-14)}$$

where

$Q_{CLc}$ = the lower-bound strength of the connection; and
$Q_{CEb}$ = expected bending strength of the beam.

*C5.4.2.4.2 Linear Static and Dynamic Procedures* **FR Beam–Column Connections.** The continuity plate modifier is based on recommendations FEMA 355F (FEMA 2000) for continuity plate detailing in relationship to column flange thickness.

The panel zone modifier is based on research in FEMA 355F indicating that connection performance is less ductile where the strength of the panel zone is either too great or too small compared to the flexural strength of the beam. The panel zone strength range between 60%–90% of the beam strength is considered to provide balanced yielding between the beam and panel zone, which results in more desirable performance.

The clear span-to-depth ratio modifier for linear acceptance criteria reflects the decreased apparent ductility that arises due to increased elastic rotations for longer beams. The decreased plastic rotation capacity of beams with very small $L_c/d$ ratios is not reflected directly. However, the modifier for linear criteria was developed so that it would be appropriate for the predominant case of $L_c/d$ ratios greater than about 5.

The beam flange and web slenderness modifier is based on the same modifications to beam acceptance criteria contained in Table 5-5. While not an aspect of the connection itself, beam flange and web slenderness affect the behavior of the connection assembly.

Type FR connections designed to promote yielding of the beam in the span, remote from the column face, are discussed in FEMA 350 (FEMA 2000).

*5.4.2.4.3 Nonlinear Static and Dynamic Procedures* Calculated component actions shall satisfy the requirements of Section 3.4.3. Maximum permissible inelastic deformations shall be taken from Tables 5-6 and 5-7.

1. **Beams.** Flexural actions of beams shall be considered deformation-controlled. Permissible plastic rotation deformation shall be as indicated in Tables 5-6 and 5-7, where $q_y$ shall be calculated in accordance with Section 5.4.2.2.2.

2. **Columns.** Axial compressive loading of columns shall be considered force-controlled, with the lower-bound axial compression capacity, $P_{CL}$, computed in accordance with Section 5.4.2.4.2.

Flexural loading of columns, with axial loads at the target displacement less than 50% of $P_{CL}$, computed in accordance with Section 5.4.2.4.2, shall be considered deformation-controlled and maximum permissible plastic rotation demands on columns, in radians, shall be as indicated in Tables 5-6 and 5-7, dependent on the axial load present and the compactness of the section.

Flexural loading of columns, with axial loads at the target displacement greater than or equal to 50% of $P_{CL}$, computed in accordance with Section 5.4.2.4.2, shall be considered force-controlled and shall conform to Eq. 5-12.

**3. FR Connection Panel Zones.** Plastic rotation demands on panel zones shall be evaluated using the acceptance criteria provided in Tables 5-6 and 5-7.

**4. FR Beam–Column Connections.** FR connections identified in Table 5-4 shall be considered deformation-controlled and the plastic rotation predicted by analysis shall be compared with the acceptance criteria in Tables 5-6 and 5-7 as modified below. Connection acceptance criteria are dependent on the detailing of continuity plates, the strength of the panel zone, the beam span-to-depth ratio, and the slenderness of the beam web and flanges as determined by the following four conditions. The modifications shall be cumulative.

4.1 If the connection does not satisfy at least one of the following conditions, the tabulated plastic rotation in Tables 5-6 and 5-7 shall be multiplied by 0.8.

$$t_{cf} \geq \frac{b_{bf}}{5.2}$$

or

$$\frac{b_{bf}}{7} \leq t_{cf} < \frac{b_{bf}}{5.2} \text{ and continuity plates with } t \geq \frac{t_{bf}}{2}$$

or

$$t_{cf} < \frac{b_{bf}}{7} \text{ and continuity plates with } t \geq t_{bf}$$

where

$t_{cf}$ = thickness of column flange;
$b_{bf}$ = width of beam flange;
$t$ = thickness of continuity plate; and
$t_{bf}$ = thickness of beam flange.

4.2 If the following condition is not met, the tabulated plastic rotations in Tables 5-6 and 5-7 shall be multiplied by 0.8.

$$0.6 \leq \frac{V_{PZ}}{V_y} \leq 0.9$$

where $V_y = 0.55F_{ye(col)} \, d_c t_{cw}$ and $V_{PZ}$ is the computed panel zone shear at the development of a hinge at the critical location of the connection. For $M_{ye}$ at the face of the column,

$$V_{PZ} = \frac{\sum M_{y(beam)}}{d_b} \left( \frac{L}{L - d_c} \right) \left( \frac{h - d_b}{h} \right)$$

where

$F_{ye(col)}$ = expected yield strength of column;
$d_c$ = column depth;
$t_{cw}$ = thickness of column web;
$M_{y(beam)}$ = yield moment of beam;
$d_b$ = depth of beam;
$L$ = length of beam, center-to-center of columns; and
$h$ = average story height of columns.

4.3 If the clear span-to-depth ratio, $L/d$, is less than 8, the tabulated plastic rotations in Tables 5-6 and 5-7 shall be multiplied by:

$$(0.5)^{[(8 - L_c/d)/3]}$$

$L_c$ = length of beam, clear span between columns; and
$d$ = depth of member.

4.4 If the beam flange and web meet the following conditions, the tabulated plastic rotations in Tables 5-6 and 5-7 need not be modified for flange and web slenderness.

$$\frac{b_f}{2t_f} < \frac{52}{\sqrt{F_{ye}}} \quad \text{and} \quad \frac{h}{t_w} < \frac{418}{\sqrt{F_{ye}}}$$

If the beam flange or web slenderness values exceed either of the following limits, the tabulated plastic rotations Tables 5-6 and 5-7 shall be multiplied by 0.5.

$$\frac{b_f}{2t_f} > \frac{65}{\sqrt{F_{ye}}} \quad \text{or} \quad \frac{h}{t_w} > \frac{640}{\sqrt{F_{ye}}}$$

where

$b_f$ = width of beam flange;
$t_f$ = thickness of beam flange;
$h$ = height of beam web;
$t_w$ = thickness of beam web; and
$F_{ye}$ = expected yield strength.

Straight-line interpolation, based on the case that results in the lower modifier, shall be used for intermediate values of beam flange or web slenderness.

Type FR connections designed to promote yielding of the beam in the span remote from the column face shall be considered force-controlled and shall be evaluated to ensure that the lower-bound strength of the connection exceeds the expected flexural strength of the beam at the connection.

*C5.4.2.4.3 Nonlinear Static and Dynamic Procedures*
**FR Beam–Column Connections.** The continuity plate modifier is based on recommendations in FEMA 355F (FEMA 2000) for continuity plate detailing in relationship to column flange thickness.

The panel zone modifier is based on research in FEMA 355F indicating that connection performance is less ductile where the strength of the panel zone is either too great or too small compared to the flexural strength of the beam. The panel zone strength range between 60% and 90% of the beam strength is considered to provide balanced yielding between the beam and panel zone, which results in more desirable performance.

The clear span-to-depth ratio modifier for nonlinear modeling and acceptance criteria reflects decreased plastic rotation capacity for beams with hinging occurring over a shorter length. This modifier is based on the plastic rotation capacities corresponding to the FEMA 350 (FEMA 2000) $L_c/d$ limits of 5 and 8.

The beam flange and web slenderness modifier is based on the same modifications to beam acceptance criteria contained in Tables 5-6 and 5-7. While not an aspect of the connection itself, beam flange and web slenderness affects the behavior of the connection assembly.

Type FR connections designed to promote yielding of the beam in the span, remote from the column face, are discussed in FEMA 350.

### 5.4.2.5 Rehabilitation Measures

FR moment frame components that do not meet the acceptance criteria for the selected Rehabilitation Objective shall be rehabilitated. Rehabilitation measures shall meet the requirements of Section 5.3.3 and other provisions of this standard.

### C5.4.2.5 Rehabilitation Measures

The following measures, which are presented in greater detail in FEMA 351 (FEMA 2000), may be effective in rehabilitating FR moment frames:

1. Add steel braces to one or more bays of each story to form concentric or eccentric braced frames to increase the stiffness of the frames. The attributes and design criteria for braced frames shall be as specified in Section 5.5. The location of added braces should be selected so as to not increase horizontal torsion in the system;
2. Add ductile concrete or masonry shear walls or infill walls to one or more bays of each story to increase the stiffness and strength of the structure. The attributes and design requirements of concrete

and masonry shear walls shall be as specified in Sections 6.8 and 7.4, respectively. The attributes and design requirements of concrete and masonry infills shall be as specified in Sections 6.7 and 7.5, respectively. The location of added walls should be selected so as not to increase horizontal torsion in the system;

3. Attach new steel frames to the exterior of the building. The rehabilitated structure should be checked for the effects of the change in the distribution of stiffness, the seismic load path, and the connections between the new and existing frames. The rehabilitation scheme of attaching new steel frames to the exterior of the building has been used in the past and has been shown to be very effective under certain conditions. This rehabilitation approach may be structurally efficient, but it changes the architectural appearance of the building. The advantage is that the rehabilitation may take place without disrupting the use of the building;
4. Reinforce moment-resisting connections to force plastic hinge locations in the beam material away from the joint region to reduce the stresses in the welded connection, thereby reducing the possibility of brittle fractures. This scheme should not be used if the full-pen connection of the existing structure did not use weld material of sufficient toughness to avoid fracture at stresses lower than yield or where strain-hardening at the new hinge location would produce larger stresses than the existing ones at the weld. The rehabilitation measures to reinforce selected moment-resisting connections shall consist of providing horizontal cover plates, vertical stiffeners, or haunches. Removal of beam material to force the plastic hinge into the beam and away from the joint region shall also be permitted subject to the above restrictions. Guidance on the design of these modifications of FR moment connections is discussed in FEMA 351 (FEMA 2000);
5. Add energy dissipation devices as specified in Chapter 9; and
6. Increase the strength and stiffness of existing frames by welding steel plates or shapes to selected members.

### 5.4.3 Partially Restrained Moment Frames

#### 5.4.3.1 General

PR moment frames shall be defined as those moment frames with connections identified as PR in Table 5-4. Moment frames with connections not included in Table 5-4 shall be defined as PR if the deformations of the beam-to-column joints contribute greater than 10% to the total lateral deflection of the frame or where the strength of the connections is less than the strength of the weaker of the two members being joined. For a PR connection with two or more failure modes, the weakest failure mechanism shall be considered to govern the behavior of the joint. Design provisions for PR frames specified in AISC 341 (AISC 2002) or ASCE 7 (ASCE 2005) shall apply unless superseded by the provisions in this standard. Equations for calculating nominal design strength shall be used for determining the expected strength, except $f = 1$, and either the expected strength or lower-bound strength shall be used in place of $F_y$, as further indicated in this standard.

### C5.4.3.1 General

Table 5-4 includes simple shear or pinned connections classified as PR connections. Although the gravity load-carrying beams and columns are typically neglected in the lateral analysis of steel moment frame structures, SAC research contained in FEMA 355D (FEMA 2000) indicates that these connections are capable of contributing non-negligible stiffness through very large drift demands. Including gravity load-carrying elements (subject to the modeling procedures and acceptance criteria in this section) in the mathematical model could be used by the design engineer to reduce the demands on the moment frame elements.

### 5.4.3.2 Stiffness

#### 5.4.3.2.1 Linear Static and Dynamic Procedures

1. **Beams, columns, and panel zones.** Axial area, shear area, moment of inertia, and panel zone stiffness shall be determined as specified in Section 5.4.2.2 for FR frames.
2. **Connections.** The rotational stiffness $K_\theta$ of each PR connection for use in PR frame analysis shall be determined by the procedure of this section, by experiment, or by an approved rational analysis. The deformation of the connection shall be included where calculating frame displacements.

The rotational spring stiffness, $K_\theta$, shall be calculated in accordance with Eq. 5-15:

$$K_\theta = \frac{M_{CE}}{0.005} \qquad \text{(Eq. 5-15)}$$

where

$M_{CE}$ = expected moment strength of connection for the following PR connections:

1. PR connections encased in concrete, where the nominal resistance, $M_{CE}$, determined for the connection shall include the composite action provided by the concrete encasement;
2. PR connections encased in masonry, where composite action shall not be included in the determination of connection resistance, $M_{CE}$; and
3. Bare steel PR connections.

For PR connections not listed above, the rotational spring stiffness shall be calculated in accordance with Eq. 5-16:

$$K_\theta = \frac{M_{CE}}{0.003} \qquad \text{(Eq. 5-16)}$$

As a simplified alternative, modeling the frame as for FR joints but with the beam stiffness, $EI_b$, adjusted to account for the flexibility of the joints in accordance with Eq. 5-17 shall be permitted:

$$EI_b adjusted = \frac{1}{\dfrac{6h}{L_b^2 K_\theta} + \dfrac{1}{EI_b}} \qquad \text{(Eq. 5-17)}$$

where

$K_\theta$ = equivalent rotational spring stiffness of connection per Eq. 5-15 or 5-16;
$M_{CE}$ = expected moment strength;
$I_b$ = moment of inertia of the beam;
$E$ = modulus of elasticity;
$h$ = average story height of the columns; and
$L_b$ = centerline span of the beam.

Where Eq. 5-17 is used, the adjusted beam stiffness shall be used in standard rigid-connection frame analysis and the rotation of the connection shall be taken as the rotation of the beam at the joint.

#### C5.4.3.2.1 Linear Static and Dynamic Procedures
FEMA 274 (FEMA 1997) is a useful reference for information concerning stiffness properties and modeling guidelines for PR connections.

#### 5.4.3.2.2 Nonlinear Static Procedure
If the Nonlinear Static Procedure (NSP) of Chapter 3 is used, the following criteria shall apply:

1. The elastic component properties shall be modeled as specified in Section 5.4.3.2.1;
2. The nonlinear moment-curvature or load–deformation behavior for beams, beam-columns,

and panel zones shall be modeled as specified in Section 5.4.2.2 for FR frames; and

3. In lieu of relationships derived from experiment or analysis, the generalized load-deformation curve shown in Fig. 5-1 with its parameters *a*, *b*, and *c* as defined in Tables 5-6 and 5-7, shall be used to represent moment-rotation behavior for PR connections in accordance with Section 5.4.2.2.2. The value for $q_y$ shall be 0.005 for connections, for which Eq. 5-15 in Section 5.4.3.2.1 applies, or 0.003 for all other connections.

*C5.4.3.2.2 Nonlinear Static Procedure* FEMA 355D (FEMA 2000) is a useful reference for information concerning nonlinear behavior of various tested connection configurations.

*5.4.3.2.3 Nonlinear Dynamic Procedure* The complete hysteretic behavior of each component shall be modeled as verified by experiment or by other procedures approved by the authority having jurisdiction.

*C5.4.3.2.3 Nonlinear Dynamic Procedure* FEMA 355D (FEMA 2000) is a useful reference for information concerning nonlinear behavior of various tested connection configurations.

### 5.4.3.3 Strength

*5.4.3.3.1 General* Component strengths shall be computed in accordance with the general requirements of Section 5.3.2 and the specific requirements of this section.

*5.4.3.3.2 Linear Static and Dynamic Procedures* The strength of steel beams and columns in PR Frames being analyzed using linear procedures shall be computed in accordance with Section 5.4.2.3.2 for FR Frames.

The expected strength, $Q_{CE}$, for PR connections shall be based on procedures specified in *Load and Resistance Factor Design Specification for Structural Steel Buildings (LRFD)* (AISC 1999) , based on experiment or based on the procedures listed in the subsequent sections.

1. **Top and Bottom Clip Angle Connection.** The moment strength, $M_{CE}$, of the riveted or bolted clip angle connection, as shown in Fig. 5-3, shall be the

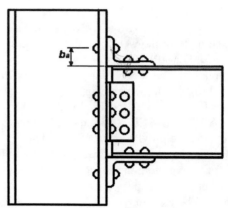

**FIGURE 5-3. Top and Bottom Clip Angle Connection.**

smallest value of $M_{CE}$ computed for the following four limit states:

1.1 **Limit State 1.** If the shear connectors between the beam flange and the flange angle control the capacity of the connection, $Q_{CE}$ shall be computed in accordance with Eq. 5-18:

$$Q_{CE} = M_{CE} = d_b(F_{ve}A_bN_b) \qquad \text{(Eq. 5-18)}$$

where

$A_b$ = gross area of rivet or bolt;
$d_b$ = overall beam depth;
$F_{ve}$ = unfactored nominal shear strength of the bolts or rivets given in *Load and Resistance Factor Design Specification for Structural Steel Buildings (LRFD)* (AISC 1999); and
$N_b$ = least number of bolts or rivets connecting the top or bottom angle to the beam flange.

1.2 **Limit State 2.** If the tensile capacity of the horizontal leg of the connection controls the capacity, $P_{CE}$ shall be taken as the smaller of that computed by Eq. 5-19 or 5-20:

$$P_{CE} \leq F_{ye}A_g \qquad \text{(Eq. 5-19)}$$

$$P_{CE} \leq F_{te}A_e \qquad \text{(Eq. 5-20)}$$

and $Q_{CE}$ shall be calculated in accordance with Eq. 5-21:

$$Q_{CE} = M_{CE} \leq P_{CE}(d_b + t_a) \qquad \text{(Eq. 5-21)}$$

where

$F_{ye}$ = expected yield strength of the angle;
$F_{te}$ = expected tensile strength of the angle;
$A_e$ = effective net area of the horizontal leg;
$A_g$ = gross area of the horizontal leg; and
$t_a$ = thickness of angle.

1.3 **Limit State 3.** If the tensile capacity of the rivets or bolts attaching the vertical outstanding leg to the column flange controls the capacity of the connection, $Q_{CE}$ shall be computed in accordance with Eq. 5-22:

$$Q_{CE} = M_{CE} = (d_b + b_a)(F_{te}A_bN_b) \qquad \text{(Eq. 5-22)}$$

where

$A_b$ = gross area of rivet or bolt;
$b_a$ = dimension in Fig. 5-3;
$F_{te}$ = expected tensile strength of the bolts or rivets; and
$N_b$ = least number of bolts or rivets connecting top or bottom angle to column flange.

1.4 **Limit State 4.** If the flexural yielding of the flange angles controls the capacity of the connection, $Q_{CE}$ shall be given by Eq. 5-23:

$$Q_{CE} = M_{CE} = \frac{wt_a^2F_{ye}}{4\left[b_a - \dfrac{t_a}{2}\right]}(d_b + b_a) \qquad \text{(Eq. 5-23)}$$

where

$b_a$ = dimension shown in Fig. 5-3; and
$w$ = length of the flange angle.

2. **Double Split Tee Connection.** The moment strength, $M_{CE}$, of the double split tee (T-stub) connection, as shown in Fig. 5-4, shall be the smallest value of $M_{CE}$ computed for the following four limit states:

2.1 **Limit State 1.** If the shear connectors between the beam flange and the web of the split tee control the capacity of the connection, $Q_{CE}$ shall be calculated using Eq. 5-18.

2.2 **Limit State 2.** If the tension capacity of the bolts or rivets connecting the flange of the split tee to the column flange control the capacity of the connection, $Q_{CE}$ shall be calculated using Eq. 5-24:

$$Q_{CE} = M_{CE} = (d_b + 2b_t + t_s)(F_{te}A_bN_b) \qquad \text{(Eq. 5-24)}$$

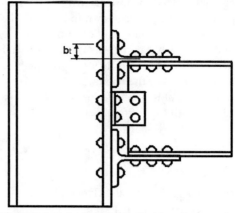

**FIGURE 5-4. Double Split Tee Connection.**

where

$d_b$ = overall beam depth;
$b_t$ = distance between one row of fasteners in the split tee flange and the centerline of the stem as shown in Fig. 5-4;
$t_s$ = thickness of the split tee stem;
$F_{te}$ = expected tensile strength of the bolts or rivets;
$A_b$ = gross area of rivet or bolt; and
$N_b$ = number of fasteners in tension connecting the flanges of one split tee to the column flange.

2.3 **Limit State 3.** If tension in the stem of the split tee controls the capacity of the connection, Eqs. 5-21 and 5-22 shall be used to determine $Q_{CE}$, with $A_g$ and $A_e$ being the gross and net areas of the split tee stem and replacing $t_a$ with $t_s$.

2.4 **Limit State 4.** If flexural yielding of the flanges of the split tee controls the capacity of the connection, $Q_{CE}$ shall be determined in accordance with Eq. 5-25:

$$Q_{CE} = M_{CE} = \frac{(d_b + t_s)wt_f^2F_{ye}}{2(b_t - k_1)} \qquad \text{(Eq. 5-25)}$$

where

$k_1$ = distance from the center of the split tee stem to the edge of the split tee flange fillet;
$b_t$ = distance between one row of fasteners in the split tee flange and the centerline of the stem as shown in Fig. 5-4;
$w$ = length of split tee; and
$t_f$ = thickness of split tee flange.

3. **Bolted Flange Plate Connections.** For bolted flange plate connections, as shown in Fig. 5-5, the flange plate shall be welded to the column and welded or bolted to the beam flange. This connection shall be considered fully restrained if its strength equals or exceeds the strength of the connected beam. The expected strength of the connection shall be calculated in accordance with Eq. 5-26:

$$Q_{CE} = M_{CE} = P_{CE}(d_b + t_p) \qquad \text{(Eq. 5-26)}$$

where

$P_{CE}$ = expected strength of the flange plate connection as governed by the net section of the flange plate, the shear capacity of the bolts, or the strength of the welds to the column flange;

$t_p$ = thickness of flange plate; and

$d_b$ = overall beam depth.

4. **Bolted End Plate Connections.** Bolted end plate connections, as shown in Fig. 5-6, shall be considered FR if their expected and lower-bound strengths equal or exceed the expected strength of the connecting beam. The lower-bound strength, $Q_{CL} = M_{CL}$, shall be the value determined for the limit state of the bolts under combined shear and tension and the expected strength, $Q_{CE} = M_{CE}$, shall be determined for the limit state of bending in the end plate calculated in accordance with the procedures of the *Load and Resistance Factor Design Specification for Structural Steel Buildings (LRFD) (AISC 1999)* or by another procedure approved by the authority having jurisdiction.

5. **Composite Partially Restrained Connections.** Strength and deformation acceptance criteria of composite partially restrained connections shall be based on approved rational analysis procedures and experimental evidence.

*5.4.3.3.3 Nonlinear Static Procedure* The complete load–deformation relationship of each component as depicted by Fig. 5-1 shall be determined in accordance with Section 5.4.2.2.2. The values for expected strength, $Q_{CE}$, of PR connections shall be the same as those used for linear procedures as specified in Section 5.4.3.3.2.

*5.4.3.3.4 Nonlinear Dynamic Procedure* The complete hysteretic behavior of each component shall be determined experimentally or by other procedures approved by the authority having jurisdiction.

*C5.4.3.3.4 Nonlinear Dynamic Procedure* FEMA 355D (FEMA 2000) is a useful reference for information concerning nonlinear behavior of various tested connection configurations.

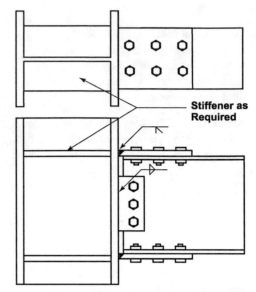

**FIGURE 5-5. Bolted Flange Plate Connection.**

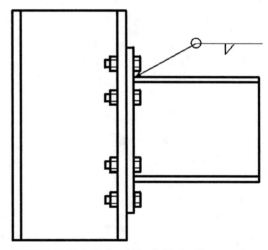

**FIGURE 5-6. Bolted End Plate Connection.**

*5.4.3.4 Acceptance Criteria*

*5.4.3.4.1 General* Component acceptance criteria shall be computed in accordance with the general requirements of Section 5.3.2 and the specific requirements of this section.

*C5.4.3.4.1 General* The strength and behavior of PR steel moment-resisting frames is typically governed by the connections. The design professional is urged to consider the acceptance criteria for the mechanism that controls the system.

*5.4.3.4.2 Linear Static and Dynamic Procedures*
Design actions shall be compared with design strengths in accordance with Section 3.4.2. The *m*-factors for steel components and connections of PR frames shall be selected from Table 5-5. Limit states for which no m-factors are provided in Table 5-5 shall be considered force-controlled.

Acceptance criteria for steel beams and columns in PR frames shall be computed in accordance with Section 5.4.2.4.2.

*5.4.3.4.3 Nonlinear Static and Dynamic Procedures*
Calculated component actions shall satisfy the requirements of Section 3.4.3. Maximum permissible inelastic deformations shall be taken from Tables 5-6 and 5-7.

### 5.4.3.5 Rehabilitation Measures

PR moment frames that do not meet the acceptance criteria for the selected Rehabilitation Objective shall be rehabilitated. Rehabilitation measures shall meet the requirements of Section 5.3.3 and other provisions of this standard.

### C5.4.3.5 Rehabilitation Measures

The rehabilitation measures for FR moment frames described in C5.4.2.5 may be effective for PR moment frames as well. PR moment frames are often too flexible to provide adequate seismic performance. Adding concentric or eccentric bracing, or reinforced concrete or masonry infills, may be a cost-effective rehabilitation measure.

Connections in PR moment frames are usually components that are weak, flexible, or both. Connections may be rehabilitated by replacing rivets with high-strength bolts, adding weldment to supplement rivets or bolts, or welding stiffeners to connection pieces or combinations of these measures. Refer to FEMA 351 (FEMA 2000) for additional information concerning the rehabilitation of PR moment frames.

## 5.5 STEEL BRACED FRAMES

### 5.5.1 General

Steel braced frames shall be defined as those frames that develop seismic resistance primarily through axial forces in the components.

Modeling procedures and rehabilitation measures for concentric braced frames and eccentric braced frames shall be as specified in Sections 5.5.2 and 5.5.3, respectively. Components of concentric and eccentric braced frames shall include columns, beams, braces, and connections. Eccentric braced frames shall also include link beam components.

### C5.5.1 General

Steel braced frames act as vertical trusses where the columns are the chords and the beams and braces are the web members.

Components can be either bare steel, steel with a nonstructural coating for fire protection, or steel with concrete or masonry encasement.

### 5.5.2 Concentric Braced Frames

#### 5.5.2.1 General

Concentric braced frames (CBF) shall be defined as braced frame systems where component worklines intersect at a single point in a joint, or at multiple points such that the distance between points of intersection, or eccentricity, *e*, is less than or equal to the width of the smallest member connected at the joint. Bending due to such eccentricities shall be considered in the design of the components.

#### 5.5.2.2 Stiffness

*5.5.2.2.1 Linear Static and Dynamic Procedures*
Axial area, shear area, and moment of inertia shall be calculated as specified for FR frames in Section 5.4.2.2.1.

FR connections shall be modeled as specified in Section 5.4.2.2.1. PR connections shall be modeled as specified in Section 5.4.3.2.1.

Braces shall be modeled as columns as specified in Section 5.4.2.2.1.

*5.5.2.2.2 Nonlinear Static Procedure* If the NSP of Chapter 3 is used, the following criteria shall apply:

1. The elastic component properties shall be modeled as specified in Section 5.5.2.2.1;
2. The nonlinear moment-curvature or load–deformation behavior to represent yielding and buckling shall be as specified in Section 5.4.2.2.2 for beams and columns and Section 5.4.3.2.2 for PR connections; and
3. In lieu of relationships derived from experiment or analysis, the nonlinear load–deformation behavior of braces shall be modeled as shown in Fig. 5-1 with parameters as defined in Tables 5-6 and 5-7. For braces loaded in compression, the parameter D in Fig. 5-1 shall represent total elastic and plastic axial deformation. The parameter $D_c$ shall represent the axial deformation at the expected buckling load,

which occurs at point B in the curve in Fig. 5-1. The reduction in strength of a brace after buckling shall be included in the model. Modeling of the compression brace behavior using elasto-plastic behavior shall be permitted if the yield force is assumed as the residual strength after buckling, as defined by parameter $c$ in Fig. 5-1 and Tables 5-6 and 5-7. Implications of forces higher than this lower-bound force shall be evaluated relative to other components to which the brace is connected. For braces in tension, the parameter $D_T$ shall be the axial deformation at development of the expected tensile yield load in the brace, which occurs at point B in the curve in Fig. 5-1.

*C5.5.2.2.2 Nonlinear Static Procedure* FEMA 274 (FEMA 1997) is a useful reference for information regarding nonlinear load-deformation behavior of braces.

*5.5.2.2.3 Nonlinear Dynamic Procedure* The complete hysteretic behavior of each component shall be based on experiment or other approved method.

*C5.5.2.2.3 Nonlinear Dynamic Procedure* FEMA 274 (FEMA 1997) is a useful reference for information concerning hysteretic behavior of braced frame components.

### 5.5.2.3 Strength

*5.5.2.3.1 General* Component strengths shall be computed in accordance with the general requirements of Section 5.3.2 and the specific requirements of this section.

*5.5.2.3.2 Linear Static and Dynamic Procedures* The expected strength, $Q_{CE}$, of steel braces under axial compression shall be the lowest value obtained for the limit states of component buckling or local buckling. The effective design strength, $P_{CE}$, shall be calculated in accordance with *Load and Resistance Factor Design Specification for Structural Steel Buildings (LRFD)* (AISC 1999), taking $\phi = 1.0$ and using the expected yield strength, $F_{ye}$, for yield strength.

For common cross-bracing configurations where both braces cross at their midpoints and are attached to a common gusset plate, the effective length of each brace shall be taken as 0.5 times the total length of the brace, including gusset plates for both axes of buckling. For other bracing configurations (chevron, V, single brace), the length shall be taken as the total length

of the brace, including gusset plates, and the effective length shall be taken as 0.8 times the total length for in-plane buckling and 1.0 times the total length for out-of-plane buckling.

The expected strength, $Q_{CE}$, of steel braces in tension shall be calculated as for columns, in accordance with Section 5.4.2.3.2.

Expected, $Q_{CE}$, and lower-bound, $Q_{CL}$, strengths of beams and columns shall be calculated as for FR beams and columns in Section 5.4.2.3. Strength of beams with axial load that exceeds 10% of the axial strength shall be as calculated for FR columns.

The lower-bound strength of brace connections shall be calculated in accordance with the *Load and Resistance Factor Design Specification for Structural Steel Buildings (LRFD)* (AISC 1999), taking $\phi = 1.0$ and using the lower-bound yield strength, $F_{yLB}$, for yield strength.

*5.5.2.3.3 Nonlinear Static Procedure* In lieu of relationships derived by experiment or analysis, the complete load–deformation behavior of each component as depicted by Fig. 5-1 shall be determined in accordance with Section 5.4.2.2.2. The values for expected strength, $Q_{CE}$, shall as specified in Section 5.5.2.3.2 for linear procedures.

*5.5.2.3.4 Nonlinear Dynamic Procedure* The complete hysteretic behavior of each component shall be determined experimentally or by other procedures approved by the authority having jurisdiction.

*C5.5.2.3.4 Nonlinear Dynamic Procedure* FEMA 274 (FEMA 1997) is a useful reference for information concerning hysteretic behavior of braced frame components.

### 5.5.2.4 Acceptance Criteria

*5.5.2.4.1 General* Component acceptance criteria shall be computed in accordance with the general requirements of Section 5.3.2 and the specific requirements of this section.

Axial tension and compression in braces shall be considered deformation-controlled. Actions on beams and columns with non-negligible axial load shall be considered force- or deformation-controlled as determined for FR frame columns in Section 5.4.2.4. Compression, tension, shear, and bending actions on brace connections including gusset plates, bolts, welds, and other connectors shall be considered force-controlled.

## 5.5.2.4.2 Linear Static and Dynamic Procedures

Design actions shall be compared with design strengths in accordance with Section 3.4.2. The *m*-factors for steel components shall be selected from Table 5-5.

Beams in chevron braced frames shall be evaluated as force-controlled actions to resist the unbalanced load effects in combination with gravity loads in accordance with Section 3.2.8. The unbalanced load effects shall be calculated using the expected yield capacity of the brace in tension and 30% of the expected compression capacity of the brace in compression.

## 5.5.2.4.3 Nonlinear Static and Dynamic Procedures

Calculated component actions shall satisfy the requirements of Section 3.4.3. Deformations limits shall be taken from Tables 5-6 and 5-7.

### 5.5.2.5 Rehabilitation Measures

Concentric braced frame components that do not meet the acceptance criteria for the selected Rehabilitation Objective shall be rehabilitated. Rehabilitation measures shall meet the requirements of Section 5.3.3 and other provisions of this standard.

### C5.5.2.5 Rehabilitation Measures

The rehabilitation measures for FR moment frames described in Section C5.4.2.5 may be effective for braced frames. Other modifications, which may be effective, include replacement or modification of connections that are insufficient in strength and/or ductility, and encasement of columns in concrete to improve their performance.

### 5.5.3 Eccentric Braced Frames

### 5.5.3.1 General

Eccentric braced frames (EBF) shall be defined as braced frames where component worklines do not intersect at a single point and the distance between points of intersection, or eccentricity, *e*, exceeds the width of the smallest member connected at the joint. The component segment between these points is defined as the link component with a span equal to the eccentricity.

### 5.5.3.2 Stiffness

## 5.5.3.2.1 Linear Static and Dynamic Procedures

The elastic stiffness of beams, columns, braces, and connections shall be the same as those specified for FR and PR moment frames and concentric braced frames. The load–deformation model for a link beam shall include shear deformation and flexural deformation.

The elastic stiffness of the link beam, $K_e$, shall be computed in accordance with Eq. 5-27:

$$K_e = \frac{K_s K_b}{K_s + K_b} \qquad \text{(Eq. 5-27)}$$

where

$$K_s = \frac{GA_w}{e} \qquad \text{(Eq. 5-28)}$$

$$K_b = \frac{12EI_b}{e^3} \qquad \text{(Eq. 5-29)}$$

$A_w = (d_b - 2t_f)\, t_w$;
$e$ = length of link beam;
$G$ = shear modulus;
$K_e$ = stiffness of the link beam;
$K_b$ = flexural stiffness;
$K_s$ = shear stiffness;
$d_b$ = beam depth;
$t_f$ = thickness of flange; and
$t_w$ = thickness of web.

## 5.5.3.2.2 Nonlinear Static Procedure

In lieu of relationships derived from experiment or analysis, the nonlinear load–deformation behavior of members of EBFs shall be modeled as shown in Fig. 5-1 and in accordance with Section 5.4.2.2.2.

Nonlinear models for beams, columns, and connections for FR and PR moment frames, and for the braces for a CBF, shall be permitted.

The link rotation at yield shall be calculated in accordance with Eq. 5-30:

$$\theta_y = \frac{Q_{CE}}{K_e e} \qquad \text{(Eq. 5-30)}$$

## 5.5.3.2.3 Nonlinear Dynamic Procedure

If the NDP is used, the complete hysteretic behavior of each component shall be modeled and shall be based on experiment or an approved rational analysis procedure.

### C5.5.3.2.3 Nonlinear Dynamic Procedure

FEMA 274 (FEMA 1997) is a useful reference for guidelines on modeling the link beams and information regarding the hysteretic behavior of eccentric braced frame (EBF) components.

### 5.5.3.3 Strength

*5.5.3.3.1 General* Component strengths shall be computed in accordance with the general requirements of Section 5.3.2 and the specific requirements of this section.

*5.5.3.3.2 Linear Static and Dynamic Procedures* Lower-bound compressive strength, $P_{CL}$, of braces in eccentric braced frames shall be calculated as for columns in accordance with Section 5.4.2.3.2 except that lower-bound yield strength, $F_{yLB}$, shall be used for yield strength.

Expected, $Q_{CE}$, and lower-bound, $Q_{CL}$, strengths of beams and columns shall be calculated as for FR beams and columns in Section 5.4.2.3. Strength of beams with non-negligible axial load shall be as calculated for FR columns.

The lower-bound strength of brace connections shall be calculated in accordance with *Load and Resistance Factor Design Specification for Structural Steel Buildings (LRFD)* (AISC 1999), taking $\phi = 1.0$ and using the lower-bound yield strength, $F_{yLB}$, for yield strength.

The strength of the link beam shall be governed by shear, flexure, or the combination of shear and flexure. $M_{CE}$ shall be taken as the expected moment capacity and $V_{CE}$ shall be taken as $0.6 F_{ye} A_w$.

If $e \leq \dfrac{1.6 M_{CE}}{V_{CE}}$, Eq. 5-31 shall be used to compute the expected strength of the link beam:

$$Q_{CE} = V_{CE} = 0.6 F_{ye} A_w \qquad \text{(Eq. 5-31)}$$

If $e > \dfrac{2.6 M_{CE}}{V_{CE}}$, Eq. 5-32 shall be used to compute the expected strength of the link beam:

$$Q_{CE} = 2 \frac{M_{CE}}{e} \qquad \text{(Eq. 5-32)}$$

Linear interpolation between Eqs. 5-31 and 5-32 shall be used for intermediate values of $e$.

*5.5.3.3.3 Nonlinear Static Procedure* Strengths for the components of EBFs shall be the same as those specified in Section 5.5.2.3.3 for the components of CBFs. In lieu of relationships derived from experiment or analysis, the load–deformation behavior of each component, as depicted by Fig. 5-1, shall be determined in accordance with Section 5.5.3.2.2.

*5.5.3.3.4 Nonlinear Dynamic Procedure* The complete hysteretic behavior of each component shall be determined experimentally or by other procedures approved by the authority having jurisdiction.

### 5.5.3.4 Acceptance Criteria

*5.5.3.4.1 General* Component acceptance criteria shall be computed in accordance with the general requirements of Section 5.3.2 and the specific requirements of this section.

Shear and flexure in link beams shall be considered deformation-controlled actions. All other actions, and actions on other EBF components, shall be considered force-controlled. Compression, tension, shear, and bending actions on brace connections including gusset plates, bolts, welds, and other connectors shall be considered force-controlled.

*5.5.3.4.2 Linear Static and Dynamic Procedures* Design actions shall be compared with design strengths in accordance with Section 3.4.2. The *m*-factors for steel components shall be selected from Table 5-5.

Link beams shall conform to the requirements of the AISC 341 (AISC 2002) with regard to detailing. The brace connecting to a link beam, the columns, and the other components in the EBF shall be designed for 1.25 times the lesser of the link beam flexural or shear expected strength to ensure link yielding without brace or column buckling. Where the link beam is attached to the column flange with full-pen welds, the provisions for these connections shall be the same as for FR frame full-pen connections. *m*-factors for flexure and shear in link beams shall be taken from Table 5-5.

*C5.5.3.4.2 Linear Static and Dynamic Procedures* The acceptance criteria for full-penetration, welded beam-to-column connections are based on testing of typical moment frame proportioning and span ratios.

*5.5.3.4.3 Nonlinear Static and Dynamic Procedures* Calculated component actions shall satisfy the requirements of Section 3.4.3. Deformations limits shall be taken from Tables 5-6 and 5-7.

### 5.5.3.5 Rehabilitation Measures

Eccentric braced frame components that do not meet the acceptance criteria for the selected Rehabilitation Objective shall be rehabilitated. Rehabilitation measures shall meet the requirements of Section 5.3.3 and other provisions of this standard.

## C5.5.3.5 Rehabilitation Measures

The rehabilitation measures described in C5.3.2.4 for FR moment frames and in C5.4.2.4 for concentric braced frames (CBFs) may be effective for many of the beams, columns, and braces. Cover plates and/or stiffeners may be effective in rehabilitating these components. The strength of the link may be increased by adding cover plates to the beam flange(s), adding doubler plates or stiffeners to the web, or changing the brace configuration.

## 5.6 STEEL PLATE SHEAR WALLS

### 5.6.1 General

A steel plate shear wall, with or without perforations, shall be provided with boundary members on all four sides and shall be fastened to these boundary elements. The boundary elements shall be evaluated as beams and/or columns.

### C5.6.1 General

A steel plate wall develops its seismic resistance through shear stress in the plate wall. Although steel plate walls are not common, they have been used to rehabilitate a few essential structures where Immediate Occupancy and operation of a facility is mandatory after a large earthquake. Due to their stiffness, the steel plate walls attract much of the seismic shear. It is essential that the new load paths be carefully established.

The provisions for steel plate walls in this standard assume that the plates are sufficiently stiffened to prevent buckling. The design professional is referred to Timler (2000) for additional information regarding the behavior and design of steel plate shear walls.

### 5.6.2 Stiffness

#### 5.6.2.1 Linear Static and Dynamic Procedures

Use of a plane stress finite element with beams and columns as boundary elements to analyze a steel plate shear wall shall be permitted. The global stiffness of the wall, $K_w$, shall be calculated in accordance with Eq. 5-33 unless another method based on principles of mechanics is used.

$$K_w = \frac{Ga\, t_w}{h} \qquad \text{(Eq. 5-33)}$$

where

$G$ = shear modulus of steel;
$a$ = clear width of wall between vertical boundary elements;

$h$ = clear height of wall between beams; and
$t_w$ = thickness of plate wall.

#### 5.6.2.2 Nonlinear Static Procedure

The elastic stiffness of the load–deformation relationship for the wall shall be as specified in Section 5.6.2.1. The complete nonlinear load–deformation relationship shall be based on experiment or approved rational analysis. Alternatively, use of the generalized load–deformation relationship shown in Fig. 5-1, as specified in Section 5.4.2.2.2, shall be permitted using strength and deformation limits based on the requirements of Sections 5.6.3 and 5.6.4.

#### 5.6.2.3 Nonlinear Dynamic Procedure

The complete hysteretic behavior of each component shall be modeled by a rational procedure verified by experiment.

#### C5.6.2.3 Nonlinear Dynamic Procedure

This procedure is not recommended in most cases.

### 5.6.3 Strength

#### 5.6.3.1 General

Component strengths shall be computed in accordance with the general requirements of Section 5.3.2 and the specific requirements of this section.

#### 5.6.3.2 Linear Static and Dynamic Procedures

The expected strength of the steel wall, $Q_{CE}$, shall be determined using the applicable equations in Part 6 of *Load and Resistance Factor Design Specification for Structural Steel Buildings (LRFD)* (AISC 1999), with $\phi = 1.0$ and the expected yield strength, $F_{ye}$, substituted for $F_y$. The wall shall be permitted to be modeled as the web of a plate girder. If stiffeners are provided to prevent buckling, they shall be spaced according to the requirements for plate girders given in *Load and Resistance Factor Design Specification for Structural Steel Buildings (LRFD)* and the expected strength of the wall shall be determined by Eq. 5-34:

$$Q_{CE} = V_{CE} = 0.6F_{ye}at_w \qquad \text{(Eq. 5-34)}$$

where

$F_{ye}$ = expected yield strength;
$a$ = clear width of the wall between vertical boundary elements; and
$t_w$ = thickness of plate wall.

In lieu of providing stiffeners, the steel wall shall be permitted to be encased in concrete. If buckling is not prevented by the use of stiffeners, equations for $V_{CE}$ given in *Load and Resistance Factor Design Specification for Structural Steel Buildings (LRFD)* for plate girders shall be used to calculate the expected strength of the wall.

### 5.6.3.3 Nonlinear Static and Dynamic Procedures

The generalized load–deformation curve shown in Fig. 5-1, as specified in Section 5.4.2.2.2, shall be used to represent the complete load–deformation behavior of the steel shear wall to failure unless another load–deformation relationship based on experiment or approved rational analysis verified by experiment is used. The expected strength, $Q_{CE}$, shall be calculated in accordance with Eq. 5-34. The yield deformation shall be calculated in accordance with Eq. 5-35:

$$\Delta_y = \frac{Q_{CE}}{K_w} \qquad \text{(Eq. 5-35)}$$

### 5.6.4 Acceptance Criteria

### 5.6.4.1 Linear Static and Dynamic Procedures

Design actions shall be compared with design strengths in accordance with Section 3.4.2. The *m*-factors for steel components shall be selected from Table 5-5.

Shear behavior in steel plate shear walls shall be considered a deformation-controlled action, with acceptance criteria as provided in Table 5-5. Design restrictions for plate girder webs given in *Load and Resistance Factor Design Specification for Structural Steel Buildings (LRFD)* (AISC1999), including those related to stiffener spacing, shall be followed.

### 5.6.4.2 Nonlinear Static and Dynamic Procedures

Calculated component actions shall satisfy the requirements of Section 3.4.3. Deformation limits shall be taken from Tables 5-6 and 5-7.

### 5.6.5 Rehabilitation Measures

Steel plate walls that do not meet the acceptance criteria for the selected Rehabilitation Objective shall be rehabilitated. Rehabilitation measures shall meet the requirements of Section 5.3.3 and other provisions of this standard.

### C5.6.5 Rehabilitation Measures

Rehabilitation measures may include the addition of stiffeners, encasement in concrete, or the addition of concrete or steel plate shear walls.

## 5.7 STEEL FRAMES WITH INFILLS

Steel frames with partial or complete infills of reinforced concrete or reinforced or unreinforced masonry shall be evaluated considering the combined stiffness of the steel frame and infill material.

The engineering properties and acceptance criteria for the infill walls shall comply with the requirements in Chapter 6 for concrete and Chapter 7 for masonry. Infill walls and frames shall be considered to carry the seismic force in composite action, considering the relative stiffness of each element, until complete failure of the walls has occurred. The interaction between the steel frame and infill shall be considered using procedures specified in Chapter 6 for concrete frames with infill. The analysis of each component shall be done in stages, considering the effects of interaction between the elements and carried through each performance level. At the point where the infill has been deemed to fail, as determined by the acceptance criteria specified in Chapter 6 or Chapter 7, the wall shall be removed from the analytical model. The analysis shall be resumed on the bare steel frame, taking into consideration any vertical discontinuity created by the degraded wall. At this point, the engineering properties and acceptance criteria for the frame, as specified in Section 5.4, shall apply.

## C5.7 STEEL FRAMES WITH INFILLS

Seismic evaluation of infill walls is required because, in many cases, these walls are unreinforced or lightly reinforced, and their strength and ductility may be inadequate. Before the loss of the wall, the steel frame adds confining pressure to the wall and enhances its resistance. The actual effective forces on the steel frame components, however, are probably minimal. As the frame components attempt to develop force, they deform and the stiffer concrete or masonry components on the far side of the member pick up load. However, beam end connections, column splices, and steel frame connections at the foundation should be investigated for forces due to interaction with the infill similar to procedures specified for concrete frames in Chapter 6.

The stiffness and resistance provided by concrete and/or masonry infills may be much larger than the stiffness of the steel frame acting alone with or without composite actions. Gaps or incomplete contact between the steel frame and the infill may negate some or all of this stiffness. These gaps may be between the wall and columns of the frame or between the wall and the top beam enclosing the frame. Different strength and stiffness conditions must be expected with different discontinuity types and locations. Therefore, the presence of any gaps or discontinuities between the infill walls and the frame must be determined and considered in the design and rehabilitation process. The resistance provided by infill walls may also be included if proper evaluation of the connection and interaction between the wall and the frame is made and if the strength, ductility, and properties of the wall are properly included.

The stiffness provided by infill masonry walls is excluded from the design and rehabilitation process unless integral action between the steel frame and the wall is verified. If complete or partial interaction between the wall and frame is verified, the stiffness is increased accordingly. The seismic performance of unconfined masonry walls is far inferior to that of confined masonry walls; therefore, the resistance of the attached wall can be used only if strong evidence as to its strength, ductility, and interaction with the steel frame is provided.

## 5.8 DIAPHRAGMS

### 5.8.1 Bare Metal Deck Diaphragms

#### 5.8.1.1 General

Metal deck diaphragms shall be composed of metal plate or gage thickness steel sheets formed in a repeating pattern with ridges and valleys. Decking units shall be attached to each other by welds, crimping, or mechanical fasteners and shall be attached to the structural steel supports by welds or by mechanical fasteners. Bare metal deck diaphragms shall be permitted to resist seismic loads acting alone or in conjunction with supplementary diagonal bracing complying with the requirements of Section 5.8.4. Steel frame elements, to which bare metal deck diaphragms are attached at their boundaries, shall be considered to be the chord and collector elements.

Criteria shall apply to existing diaphragms as well as to stiffened, strengthened, or otherwise rehabilitated diaphragms. Interaction of new and existing elements of rehabilitated diaphragms shall be evaluated to ensure strain compatibility. Load transfer mechanisms between new and existing diaphragm elements shall be evaluated.

#### C5.8.1.1 General

Bare metal deck diaphragms are usually used for roofs of buildings where there are very light gravity loads other than support of roofing materials. Load transfer to frame elements that act as chords or collectors in modern frames is through shear connectors, puddle welds, screws, or shot pins.

#### 5.8.1.2 Stiffness

*5.8.1.2.1 Linear Procedures* Metal deck diaphragms shall be classified as flexible, stiff, or rigid in accordance with Section 3.2.4. Flexibility factors for use in the analysis shall be calculated by an approved rational method.

*C5.8.1.2.1 Linear Procedures* Flexibility factors for various types of metal decks are available from manufacturers' catalogs. In systems for which values are not available, values can be established by interpolating between the most representative systems for which values are available. Flexibility factors for use in the analysis can also be calculated using the Steel Deck Institute (SDI) *Diaphragm Design Manual* (SDI 1981).

*5.8.1.2.2 Nonlinear Static Procedure* Inelastic properties of diaphragms shall not be included in inelastic seismic analyses if the weak link of the diaphragm is connection failure. Procedures for developing models for inelastic response of wood diaphragms in unreinforced masonry (URM) buildings shall be permitted for use as the basis of an inelastic model of a flexible metal diaphragm. A strain-hardening modulus of 3% shall be used in the post-elastic region.

#### 5.8.1.3 Strength

The strength of bare metal deck diaphragms shall be determined in accordance with Section 5.3.2 and the requirements of this section.

Expected strength, $Q_{CE}$, for bare metal deck diaphragms shall be taken as two times allowable values specified in approved codes and standards, unless a larger value is justified by test data. Alternatively, lower-bound strength shall be taken as nominal strength published in codes or standards approved by the authority having jurisdiction, except that the strength reduction factor, $\phi$, shall be taken equal to unity.

Lower-bound strengths, $Q_{CL}$, of welded connectors shall be as specified in the *Welding Code for Sheet Steel*, AWS D1.3 (AWS 1998), or other approved standard.

#### C5.8.1.3 Strength

Capacities of steel deck diaphragms are given in International Code Council (ICC-ES) reports, in manufacturers' literature, or in the publications of the SDI. Where allowable stresses are given, these may be multiplied by 2.0 in lieu of information provided by the manufacturer or other knowledgeable sources.

Connections between metal decks and steel framing commonly use puddle welds. Connection capacities are provided in ICC-ES reports, manufacturers' data, the SDI *Diaphragm Design Manual* (SDI 1981), or AWS D1.3 (AWS 1998). Other attachment systems, such as clips, are sometimes used.

### 5.8.1.4 Acceptance Criteria

Connections of bare metal deck diaphragms shall be considered force-controlled. Connection capacity shall be checked for the ability to transfer the total diaphragm reaction into the steel framing. Diaphragms that are governed by the capacity of the connections shall also be considered force-controlled. Bare metal deck diaphragms not governed by the capacity of the connections shall be considered deformation-controlled. The *m*-factors for shear yielding or plate buckling shall be taken from Table 5-5.

For the Life Safety Structural Performance Level, a loss of bearing support or anchorage of the deck shall not be permitted. For higher performance levels, the amount of damage to the connections shall not impair the load transfer between the diaphragm and the steel frame. Deformations shall not exceed the threshold of deflections that cause unacceptable damage to other elements (either structural or nonstructural) at specified performance levels.

### C5.8.1.4 Acceptance Criteria

If bare deck capacity is controlled by connections to frame members or panel buckling, then inelastic action and ductility are limited and the deck should be considered to be a force-controlled member.

### 5.8.1.5 Rehabilitation Measures

Bare metal diaphragms that do not meet the acceptance criteria for the selected Rehabilitation Objective shall be rehabilitated. Rehabilitation measures shall meet the requirements of Section 5.3.3 and other provisions of this standard.

### C5.8.1.5 Rehabilitation Measures

The following measures may be effective in rehabilitating bare metal diaphragms:

1. Adding shear connectors for transfer of stress to chord or collector elements;
2. Strengthening existing chords or collectors by the addition of new steel plates to existing frame components;
3. Adding puddle welds or other shear connectors at panel perimeters;
4. Adding diagonal steel bracing to form a horizontal truss to supplement diaphragm strength;
5. Adding structural concrete; and
6. Adding connections between deck and supporting members.

### 5.8.2 Metal Deck Diaphragms with Structural Concrete Topping

#### 5.8.2.1 General

Metal deck diaphragms with structural concrete topping, consisting of either a composite deck with indentations, or a noncomposite form deck and the concrete topping slab with reinforcement acting together, shall be permitted to resist diaphragm loads. The concrete fill shall be either normal or lightweight structural concrete, with reinforcing composed of wire mesh or reinforcing steel. Decking units shall be attached to each other by welds, crimping, or mechanical fasteners and shall be attached to structural steel supports by welds or by mechanical fasteners. The steel frame elements to which the topped metal deck diaphragm boundaries are attached shall be considered the chord and collector elements.

Criteria shall apply to existing diaphragms as well as new and rehabilitated diaphragms. Interaction of new and existing elements of rehabilitated diaphragms shall be evaluated for strain compatibility. Load transfer mechanisms between new and existing diaphragm components shall be considered in determining the flexibility of the diaphragm.

#### C5.8.2.1 General

Metal deck diaphragms with structural concrete topping are frequently used on floors and roofs of buildings where there are typical floor gravity loads. Concrete has structural properties that significantly add to diaphragm stiffness and strength. Concrete reinforcing ranges from light mesh reinforcement to a regular grid of small reinforcing bars (No. 3 or No. 4). Metal decking is typically composed of corrugated sheet steel from 22 ga. down to 14 ga. Rib depths vary from $1\frac{1}{2}$ to 3 in. in most cases. Attachment of the metal deck to the steel frame is usually accomplished using puddle welds at 1 to 2 ft on center. For composite behavior, shear studs are welded to the frame before the concrete is cast.

Load transfer to frame elements that act as chords or collectors in modern frames is usually through puddle welds or headed studs. In older construction where the frame is encased for fire protection, load transfer is made through bond.

## 5.8.2.2 Stiffness

*5.8.2.2.1 Linear Procedures* For existing topped metal deck diaphragms, a rigid diaphragm assumption shall be permitted if the span-to-depth ratio is not greater than 5:1. For greater span-to-depth ratios, and in cases with plan irregularities, diaphragm flexibility shall be explicitly included in the analysis in accordance with Section 3.2.4. Diaphragm stiffness shall be calculated using an approved method with a representative concrete thickness.

*C5.8.2.2.1 Linear Procedures* Flexibility factors for topped metal decks are available from manufacturers' catalogs. For combinations for which values are not available, values can be established by interpolating between the most representative systems for which values are available. Flexibility factors for use in the analysis can also be calculated using the SDI *Diaphragm Design Manual* (SDI 1981).

*5.8.2.2.2 Nonlinear Procedures* Inelastic properties of diaphragms shall not be included in inelastic seismic analyses if the weak link in the diaphragm is connection failure. Procedures for developing models for inelastic response of wood diaphragms in URM buildings shall be permitted for use as the basis of an inelastic model of a flexible metal deck diaphragm with structural concrete topping.

### 5.8.2.3 Strength

Capacities of metal deck diaphragms with structural concrete topping shall be established by an approved procedure.

Alternatively, the expected strength, $Q_{CE}$, of topped metal deck diaphragms shall be taken as two times allowable values specified in approved codes and standards unless a larger value is justified by test data. Lower-bound strengths, $Q_{CL}$, of welded connectors shall be as specified in AWS D1.3 (AWS 1998) or other approved standards. Lower-bound strengths, $Q_{CL}$, for headed stud connectors shall be as specified in *Load and Resistance Factor Design Specification for Structural Steel Buildings (LRFD)* (AISC 1999), with $\phi = 1.0$.

### C5.8.2.3 Strength

Member capacities of steel deck diaphragms with structural concrete are given in manufacturers' catalogs, ICC-ES reports, or the SDI *Diaphragm Design Manual* (SDI 1981). If composite deck capacity is controlled by shear connectors, inelastic action and ductility are limited. It would be expected that there

would be little or no inelastic action in steel deck/concrete diaphragms, except in long span conditions; however, perimeter transfer mechanisms and collector forces must be considered to be sure this is the case. SDI calculation procedures or ICC-ES values with a multiplier of 2.0 should be used to bring allowable values to a strength level. Connector capacities may also be found in ICC-ES reports, manufacturers' data, or the SDI *Diaphragm Design Manual* (SDI 1981).

### 5.8.2.4 Acceptance Criteria

Connections of metal deck diaphragms with structural concrete topping shall be considered force-controlled. Connection capacity shall be checked for the ability to transfer the total diaphragm reaction into the steel framing. Diaphragms that are governed by the capacity of the connections shall also be considered force-controlled. Topped metal deck diaphragms not governed by the capacity of the connections shall be considered deformation-controlled. The *m*-factors for shear yielding shall be taken from Table 5-5.

For the Life Safety Structural Performance Level, a loss of bearing support or anchorage shall not be permitted. For higher performance levels, the amount of damage to the connections or cracking in concrete-filled slabs shall not impair the load transfer between the diaphragm and the steel frame. Deformations shall be limited to be below the threshold of deflections that cause damage to other elements (either structural or nonstructural) at specified performance levels. Acceptance criteria for collectors shall be as specified in Section 5.8.6.4.

Shear connectors for steel beams designed to act compositely with the slab shall have the capacity to transfer both diaphragm shears and composite beam shears. Where the beams are encased in concrete, use of bond between the steel and the concrete shall be permitted to transfer loads.

### C5.8.2.4 Acceptance Criteria

Shear failure of topped metal deck diaphragms requires cracking of the concrete or tearing of the metal deck, so *m*-factors have been set at conservative levels.

### 5.8.2.5 Rehabilitation Measures

Metal deck diaphragms with structural concrete topping that do not meet the acceptance criteria for the selected Rehabilitation Objective shall be rehabilitated. Rehabilitation measures shall meet the requirements of Section 5.3.3 and other provisions of this standard.

### C5.8.2.5 Rehabilitation Measures

The following measures may be effective in rehabilitating metal deck diaphragms with structural concrete topping:

1. Adding shear connectors to transfer forces to chord or collector elements;
2. Strengthening existing chords or collectors by the addition of new steel plates to existing frame components, or attaching new plates directly to the slab by embedded bolts or epoxy; and
3. Adding diagonal steel bracing to supplement diaphragm strength.

## 5.8.3 Metal Deck Diaphragms with Nonstructural Topping

### 5.8.3.1 General

Metal deck diaphragms with nonstructural topping shall be evaluated as bare metal deck diaphragms, unless the strength and stiffness of the nonstructural topping are substantiated through approved test data.

### C5.8.3.1 General

Metal deck diaphragms with nonstructural fill are typically used on roofs of buildings where there are very small gravity loads. The fill, such as very lightweight insulating concrete (e.g., vermiculite), usually does not have usable structural properties and is most often unreinforced. Consideration of any composite action must be done with caution after extensive investigation of field conditions. Material properties, force transfer mechanisms, and other similar factors must be verified in order to include such composite action. Typically, the decks are composed of corrugated sheet steel from 22 ga. down to 14 ga., and the rib depths vary from 9/16 to 3 in. in most cases.

### 5.8.3.2 Stiffness

*5.8.3.2.1 Linear Procedures* The potential for composite action and modification of load distribution shall be considered if composite action results in higher demands on components of the lateral-force-resisting system. Otherwise, the composite action shall be permitted to be ignored as described in Section 5.8.3.1. Interaction of new and existing elements of strengthened diaphragms shall be evaluated by maintaining strain compatibility between the two, and the load transfer mechanisms between the new and existing diaphragm elements shall be considered in determining the flexibility of the diaphragm. Similarly, the interaction of new diaphragms with existing frames shall be evaluated, as well as the load transfer mechanisms between them.

*C5.8.3.2.1 Linear Procedures* Flexibility of the diaphragm will depend on the strength and thickness of the topping. It may be necessary to bound the solution in some cases using both rigid and flexible diaphragm assumptions.

*5.8.3.2.2 Nonlinear Procedures* Inelastic response of diaphragms shall not be permitted in inelastic seismic analyses if the weak link in the diaphragm is connection failure. Procedures for developing models for inelastic response of wood diaphragms in URM buildings shall be permitted as the basis of an inelastic model of a flexible bare metal deck diaphragm with nonstructural topping.

### 5.8.3.3 Strength

Capacities of metal deck diaphragms with nonstructural topping shall be taken as specified for bare metal deck in Section 5.8.1. Capacities for welded and headed stud connectors shall be taken as specified in Section 5.8.2.3.

### 5.8.3.4 Acceptance Criteria

Connections of metal deck diaphragms with nonstructural topping to steel framing shall be considered force-controlled. Connection capacity shall be checked for the ability to transfer the total diaphragm reaction into the steel framing. Diaphragms that are governed by the capacity of the connections shall also be considered force-controlled. Topped metal deck diaphragms not governed by the capacity of the connections shall be considered deformation-controlled. The *m*-factors for shear yielding or plate buckling shall be taken from Table 5-5.

For the Life Safety Structural Performance Level, a loss of bearing support or anchorage shall not be permitted. For higher performance levels, the amount of damage to the connections or cracking in concrete filled slabs shall not impair the load transfer mechanism between the diaphragm and the steel frame. Deformations shall be limited to be below the threshold of deflections that cause damage to other elements (either structural or nonstructural) at specified performance levels.

### C5.8.3.4 Acceptance Criteria

Generally, there should be little or no inelastic action in the diaphragms, provided the connections to the framing members are adequate. SDI calculation procedures, or International Conference of Building Officials (ICBO) values with a multiplier of 2, should be used to bring capacities from allowable values to strength levels.

### 5.8.3.5 Rehabilitation Measures

Metal deck diaphragms with nonstructural topping that do not meet the acceptance criteria for the selected Rehabilitation Objective shall be rehabilitated. Rehabilitation measures shall meet the requirements of Section 5.3.3 and other provisions of this standard.

### C5.8.3.5 Rehabilitation Measures

The following measures may be effective in rehabilitating metal deck diaphragms with nonstructural topping:

1. Adding shear connectors to transfer forces to chord or collector elements;
2. Strengthening existing chords or collectors by the addition of new steel plates to existing frame components, or attaching new plates directly to the slab by embedded bolts or epoxy;
3. Adding puddle welds at panel perimeters of diaphragms;
4. Adding diagonal steel bracing to supplement diaphragm strength; and
5. Replacing nonstructural fill with structural concrete.

### 5.8.4 Horizontal Steel Bracing (Steel Truss Diaphragms)

#### 5.8.4.1 General

Horizontal steel bracing (steel truss diaphragms) shall be permitted to act as diaphragms independently or in conjunction with bare metal deck roofs. Where structural concrete fill is provided over the metal decking, relative rigidities between the steel truss and concrete systems shall be considered in the analysis.

Criteria shall apply to existing truss diaphragms, strengthened truss diaphragms, and new diaphragms.

Where steel truss diaphragms are added as part of a rehabilitation plan, interaction of new and existing elements of strengthened diaphragm systems (stiffness compatibility) shall be evaluated and the load transfer mechanisms between new and existing diaphragm elements shall be considered in determining the flexibility of the strengthened diaphragm.

Load transfer mechanisms between new diaphragm elements and existing frames shall be considered in determining the flexibility of the diaphragm/frame system.

#### C5.8.4.1 General

Steel truss diaphragm elements are typically found in conjunction with vertical framing systems that are of structural steel framing. Steel trusses are more common in long span situations, such as special roof structures for arenas, exposition halls, auditoriums, and industrial buildings. Diaphragms with large span-to-depth ratios may often be stiffened by the addition of steel trusses. The addition of steel trusses for diaphragms identified to be deficient may provide a proper method of enhancement.

Horizontal steel bracing (steel truss diaphragms) may be made up of any of the various structural shapes. Often, the truss chord elements consist of wide flange shapes that also function as floor beams to support the gravity loads of the floor. For lightly loaded conditions, such as industrial metal deck roofs without concrete fill, the diagonal members may consist of threaded rod elements, which are assumed to act only in tension. For steel truss diaphragms with large loads, diagonal elements may consist of wide flange members, tubes, or other structural elements that will act in both tension and compression. Truss element connections are generally concentric, to provide the maximum lateral stiffness and ensure that the truss members act under pure axial load. These connections are generally similar to those of gravity-load-resisting trusses.

#### 5.8.4.2 Stiffness

*5.8.4.2.1 Linear Procedures* Truss diaphragm systems shall be modeled as horizontal truss elements (similar to braced steel frames) where axial stiffness controls deflections. Joints shall be permitted to be modeled as pinned except where joints provide moment resistance or where eccentricities exist at the connections. In such cases, joint rigidities shall be modeled. Flexibility of truss diaphragms shall be explicitly considered in distribution of lateral loads to vertical elements.

*5.8.4.2.2 Nonlinear Procedures* Inelastic models similar to those of braced steel frames shall be used for truss elements where nonlinear behavior of truss elements will occur. Elastic properties of truss diaphragms shall be permitted in the model for inelastic seismic analyses where nonlinear behavior of truss elements will not occur.

#### 5.8.4.3 Strength

Capacities of truss diaphragm members shall be calculated as specified for steel braced frame members in Section 5.5. Lateral support of truss diaphragm members provided by metal deck, with or without concrete fill, shall be considered in evaluation of truss diaphragm capacities. Gravity force effects shall be included in the calculations for those members that support gravity loads.

### 5.8.4.4 Acceptance Criteria

Force transfer mechanisms between various members of the truss at the connections, and between trusses and frame elements, shall be evaluated to verify the completion of the load path.

For the Life Safety Structural Performance Level, a loss of bearing support or anchorage shall not be permitted. For higher performance levels, the amount of damage to the connections or bracing elements shall not result in the loss of the load transfer between the diaphragm and the steel frame. Deformations shall be limited to be below the threshold of deflections that cause damage to other elements (either structural or nonstructural) at specified performance levels.

*5.8.4.4.1 Linear Procedures* Linear acceptance criteria for horizontal steel truss diaphragm components shall be as specified for concentric braced frames in Section 5.5.2.4 except that beam and column criteria need not be used. Use of *m*-factors specified for diagonal brace components, in lieu of those for beam and column components of braced frames, shall be permitted for strut and chord members in the truss.

*5.8.4.4.2 Nonlinear Procedures* Nonlinear acceptance criteria for horizontal steel truss diaphragm components shall be as specified for concentric braced frames in Section 5.5.2.4 except that beam and column criteria need not be used. Use of plastic deformations specified for diagonal brace components, in lieu of those specified for beam and column components of braced frames, shall be permitted for strut and chord members in the truss.

### 5.8.4.5 Rehabilitation Measures

Steel truss diaphragms that do not meet the acceptance criteria for the selected Rehabilitation Objective shall be rehabilitated. Rehabilitation measures shall meet the requirements of Section 5.3.3 and other provisions of this standard.

### C5.8.4.5 Rehabilitation Measures

The following measures may be effective in rehabilitating steel truss diaphragms:

1. Diagonal components may be added to form additional horizontal trusses as a method of strengthening a weak existing diaphragm;

2. Existing chords components strengthened by the addition of shear connectors to enhance composite action;
3. Existing steel truss components strengthened by methods specified for braced steel frame members;
4. Truss connections strengthened by the addition of welds, new or enhanced plates, and bolts; and
5. Structural concrete fill added to act in combination with steel truss diaphragms after verifying the effects of the added weight of concrete fill.

## 5.8.5 Archaic Diaphragms

### 5.8.5.1 General

Archaic diaphragms in steel buildings are those consisting of shallow brick arches that span between steel floor beams, with the arches packed tightly between the beams to provide the necessary resistance to thrust forces.

### C5.8.5.1 General

Archaic steel diaphragm elements are almost always found in older steel buildings in conjunction with vertical systems of structural steel framing. The brick arches were typically covered with a very low-strength concrete fill, usually unreinforced. In many instances, various archaic diaphragm systems were patented by contractors.

### 5.8.5.2 Stiffness

*5.8.5.2.1 Linear Procedures* Existing archaic diaphragm systems shall be modeled as a horizontal diaphragm with equivalent thickness of brick arches and concrete fill. Modeling of the archaic diaphragm as a truss with steel beams as tension elements and arches as compression elements shall be permitted. The flexibility of archaic diaphragms shall be considered in calculating the distribution of lateral loads to vertical elements. Analysis results shall be evaluated to verify that diaphragm response remains elastic as assumed.

Interaction of new and existing elements of strengthened diaphragms shall be evaluated by checking the strain compatibility of the two in cases where new structural elements are added as part of a seismic rehabilitation. Load transfer mechanisms between new and existing diaphragm elements shall be considered in determining the flexibility of the strengthened diaphragm.

*5.8.5.2.2 Nonlinear Procedures* Archaic diaphragms shall be required to remain in the elastic range unless otherwise approved.

*C5.8.5.2.2 Nonlinear Procedures* Inelastic properties of archaic diaphragms should be chosen with caution for seismic analyses. For the case of archaic diaphragms, inelastic models similar to those of archaic timber diaphragms in unreinforced masonry buildings may be appropriate. Inelastic deformation limits of archaic diaphragms should be lower than those prescribed for a concrete-filled diaphragm.

### 5.8.5.3 Strength

Member capacities of archaic diaphragm components shall be permitted to be calculated, assuming no tension capacity exists for all components except steel beam members. Gravity force effects shall be included for components of these diaphragms. Force transfer mechanisms between various members and between frame elements shall be evaluated to verify the completion of the load path.

### 5.8.5.4 Acceptance Criteria

Archaic diaphragms shall be considered force-controlled. For the Life Safety Structural Performance Level, diaphragm deformations and displacements shall not lead to a loss of bearing support for the elements of the arches. For higher performance levels, the deformation due to diagonal tension shall not result in the loss of the load transfer mechanism. Deformations shall be limited below the threshold of deflections that cause damage to other elements (either structural or nonstructural) at specified performance levels. These values shall be established in conjunction with those for steel frames.

### 5.8.5.5 Rehabilitation Measures

Archaic diaphragms that do not meet the acceptance criteria for the selected Rehabilitation Objective shall be rehabilitated. Rehabilitation measures shall meet the requirements of Section 5.3.3 and other provisions of this standard.

### C5.8.5.5 Rehabilitation Measures

The following measures may be effective in rehabilitating archaic diaphragms:

1. Adding diagonal members to form a horizontal truss as a method of strengthening a weak archaic diaphragm;
2. Strengthening existing steel members by adding shear connectors to enhance composite action; and

3. Removing weak concrete fill and replacing it with a structural concrete topping slab after verifying the effects of the added weight of concrete fill.

## 5.8.6 Chord and Collector Elements

### 5.8.6.1 General

Steel framing that supports the diaphragm shall be permitted as diaphragm chord and collector elements. Where structural concrete is present, additional slab reinforcing shall be permitted to act as the chord or collector for tensile loads, while the slab carries chord or collector compression. Where the steel framing acts as a chord or collector, it shall be attached to the deck with spot welds or by mechanical fasteners.

### C5.8.6.1 General

Where reinforcing acts as the chord or collector, load transfer occurs through bond between the reinforcing bars and the concrete.

### 5.8.6.2 Stiffness

Modeling assumptions specified for equivalent steel frame members in this chapter shall be used for chord and collector elements.

### 5.8.6.3 Strength

Capacities of structural steel chords and collectors shall be as specified for FR beams and columns in Section 5.4.2.3.2. Capacities for reinforcing steel embedded in concrete slabs and acting as chords or collectors shall be determined in accordance with the provisions of Chapter 6.

### 5.8.6.4 Acceptance Criteria

Inelastic action in chords and collectors shall be permitted if it is permitted in the diaphragm. Where such actions are permissible, chords and collectors shall be considered deformation-controlled. The *m*-factors shall be taken from Table 5-5 and inelastic acceptance criteria shall be taken from FR beam and column components in Section 5.4. Where inelastic action is not permitted, chords and collectors shall be considered force-controlled components. Where chord and collector elements are force-controlled, $Q_{UD}$ need not exceed the total force that can be delivered to the component by the expected strength of the diaphragm or the vertical elements of the lateral-force-resisting system. For the Life Safety Structural Performance Level, the deformations and displacements of chord and collector components shall not result in the loss of vertical support. For higher performance levels, chords and collectors shall not impair the load path.

Welds and connectors joining the diaphragms to the chords and collectors shall be considered force-controlled. If all connections meet the acceptance criteria, the diaphragm shall be considered to prevent buckling of the chord member within the plane of the diaphragm. Where chords or collectors carry gravity loads in combination with seismic loads, they shall be checked as members with combined axial load and bending in accordance with Section 5.4.2.4.2.

### 5.8.6.5 Rehabilitation Measures

Chord and collector elements that do not meet the acceptance criteria for the selected Rehabilitation Objective shall be rehabilitated. Rehabilitation measures shall meet the requirements of Section 5.3.3 and other provisions of this standard.

### C5.8.6.5 Rehabilitation Measures

The following measures may be effective in rehabilitating chord and collector elements:

1. Strengthen the connection between diaphragms and chords or collectors;
2. Strengthen steel chords or collectors with steel plates attached directly to the slab with embedded bolts or epoxy, and strengthen slab chord or collectors with added reinforcing bars; and
3. Add chord members.

## 5.9 STEEL PILE FOUNDATIONS

### 5.9.1 General

A pile shall provide strength and stiffness to the foundation either by bearing directly on soil or rock, by friction along the pile length in contact with the soil, or by a combination of these mechanisms. Foundations shall be evaluated as specified in Chapter 4. Concrete components of foundations shall conform with Chapter 6. The design of the steel piles shall comply with the requirements of this section.

### C5.9.1 General

Steel piles of wide flange shape (H-piles) or structural tubes, with and without concrete infills, shall be permitted to be used to support foundation loads. Piles driven in groups should have a pile cap to transfer loads from the superstructure to the piles.

In poor soils or soils subject to liquefaction, bending of the piles may be the only dependable resistance to lateral loads.

### 5.9.2 Stiffness

If the pile cap is below grade, the foundation stiffness from the pile cap bearing against the soil shall be permitted to be represented by equivalent soil springs derived as specified in Chapter 4. Additional stiffness of the piles shall be permitted to be derived through bending and bearing against the soil. For piles in a group, the reduction in each pile's contribution to the total foundation stiffness and strength shall be made to account for group effects. Additional requirements for calculating the stiffness shall be as specified in Chapter 4.

### 5.9.3 Strength

Except in sites subject to liquefaction of soils, it shall be permitted to neglect buckling of portions of piles embedded in the ground. Flexural demands in piles shall be calculated either by nonlinear methods or by elastic methods for which the pile is treated as a cantilever column above a calculated point of fixity.

### 5.9.4 Acceptance Criteria

The acceptance criteria for the axial force and maximum bending moments for the pile strength shall be as specified for a steel column in Section 5.4.2.4.2 for linear methods and in Section 5.4.2.4.3 for nonlinear methods, where the lower-bound axial compression, expected axial tension, and flexural strengths shall be computed for an unbraced length equal to zero for those portions of piles that are embedded in nonliquefiable soils.

Connections between steel piles and pile caps shall be considered force-controlled.

### C5.9.4 Acceptance Criteria

Nonlinear methods require the use of a computer program. FEMA 274 (FEMA 1997) is a useful reference for additional information.

### 5.9.5 Rehabilitation Measures

Steel pile foundations that do not meet the acceptance criteria for the selected Rehabilitation Objective shall be rehabilitated. Rehabilitation measures shall meet the requirements of Section 5.3.3 and other provisions of this standard.

### C5.9.5 Rehabilitation Measures

Rehabilitation of the concrete pile cap is specified in Chapter 6. Criteria for the rehabilitation of the foundation element are specified in Chapter 4. The following measure may be effective in rehabilitating steel pile foundations: driving additional piles near existing groups and then adding a new pile cap to increase stiffness and strength of the pile foundation. Monolithic behavior gained by connecting the new and old pile caps with epoxied dowels may also be effective. In most cases, it is not possible to rehabilitate the existing piles.

## 5.10 CAST AND WROUGHT IRON

### 5.10.1 General

Existing components of cast and wrought iron shall be permitted to participate in resisting seismic forces in combination with concrete or masonry walls. Cast iron frames, in which beams and columns are integrally cast, shall not be permitted to resist seismic forces as primary elements of the lateral-force-resisting system. The ability of cast iron elements to resist the design displacements at the selected earthquake hazard level shall be evaluated.

### 5.10.2 Stiffness

The axial and flexural stiffness of cast iron shall be calculated using elastic section properties and a modulus of elasticity, $E$, of 25,000 kips/in.$^2$ unless a different value is obtained by testing or other methods approved by the authority having jurisdiction.

### 5.10.3 Strength and Acceptance Criteria

Axial and flexural loads on cast iron components shall be considered to be force-controlled behaviors. Lower-bound material properties for cast iron shall be based on Table 5-1.

The lower-bound strength of a cast iron column shall be calculated as:

$$Q_{CL} = P_{CL} = A_g F_{cr} \qquad \text{(Eq. 5-36)}$$

where

$A_g$ = gross area of column;
$F_{cr}$ = 12 ksi for $l_c/r \le 108$; or
$= \dfrac{1.40 \times 10^5}{(l_c/r)^2}$ ksi for $l_c/r > 108$.

Cast iron columns shall only be permitted to carry axial compression.

## 6.0 CONCRETE

## 6.1 SCOPE

This chapter sets forth requirements for the Systematic Rehabilitation of concrete components of the lateral-force-resisting system of an existing building. The requirements of this chapter shall apply to existing concrete components of a building system, rehabilitated concrete components of a building system, and new concrete components that are added to an existing building system.

Section 6.2 specifies data collection procedures for obtaining material properties and performing condition assessments. Section 6.3 specifies general analysis and design requirements for concrete components. Sections 6.4, 6.5, 6.6, 6.7, 6.8, and 6.9 provide modeling procedures, component strengths, acceptance criteria, and rehabilitation measures for concrete and precast concrete moment frames, braced frames, and shear walls. Sections 6.10, 6.11, and 6.12 provide modeling procedures, strengths, acceptance criteria, and rehabilitation measures for concrete diaphragms and concrete foundation systems.

## C6.1 SCOPE

Techniques for repair of earthquake-damaged concrete components are not included in this standard. The design professional is referred to FEMA 306 (FEMA 1998), FEMA 307 (FEMA 1998), and FEMA 308 (FEMA 1998) for information on evaluation and repair of damaged concrete wall components.

## 6.2 MATERIAL PROPERTIES AND CONDITION ASSESSMENT

### 6.2.1 General

Mechanical properties of concrete materials and components shall be obtained from available drawings, specifications, and other documents for the existing construction in accordance with the requirements of Section 2.2. Where such documents fail to provide adequate information to quantify concrete material properties or the condition of concrete components of the structure, such information shall be supplemented by materials tests and assessments of existing conditions in compliance with requirements of this chapter as specified in Section 2.2.6.

Material properties of existing concrete components shall be determined in accordance with Section 6.2.2. A condition assessment shall be conducted in accordance with Section 6.2.3. The extent of materials testing and condition assessment performed shall be used to determine the knowledge factor as specified in Section 6.2.4.

Use of default material properties shall be permitted in accordance with Section 6.2.2.5. Use of material properties based on historical information as default values shall be permitted as specified in Section 6.2.2.5.

**C6.2.1 General**

This section identifies properties requiring consideration and provides guidelines for determining the properties of buildings. Also described is the need for a thorough condition assessment and utilization of knowledge gained in analyzing component and system behavior. Personnel involved in material property quantification and condition assessment should be experienced in the proper implementation of testing practices and the interpretation of results.

The form, function, concrete strength, concrete quality, reinforcing steel strength, quality and detailing, forming techniques, and concrete placement techniques have constantly evolved and have had a significant impact on the seismic resistance of a concrete building. Innovations such as prestressed and precast concrete, post tensioning, and lift slab construction have created a multivariant inventory of existing concrete structures.

It is important to investigate the local practices relative to seismic design where trying to analyze a concrete building. Specific benchmark years can be determined for the implementation of earthquake-resistant design in most locations, but caution should be exercised in assuming optimistic characteristics for any specific building.

Particularly with concrete materials, the date of original building construction significantly influences seismic performance. In the absence of deleterious conditions or materials, concrete gains compressive strength from the time it is originally cast and in-place. Strengths typically exceed specified design values (28-day or similar). Early uses of concrete did not specify any design strength, and low-strength concrete was not uncommon. Also, early use of concrete in buildings often employed reinforcing steel with relatively low strength and ductility, limited continuity, and reduced bond development. Continuity between specific existing components and elements (e.g., beams and columns, diaphragms, and shear walls) is also particularly difficult to assess, given the presence of concrete cover and other barriers to inspection.

Properties of welded wire fabric for various periods of construction can be obtained from the Wire Reinforcement Institute.

Documentation of properties and grades of material used in component and connection construction is invaluable and may be effectively used to reduce the amount of in-place testing required. The design professional is encouraged to research and acquire all available records from original construction.

**6.2.2 Properties of In-Place Materials and Components**

*6.2.2.1 Material Properties*

*6.2.2.1.1 General* The following component and connection material properties shall be obtained for the as-built structure:

Concrete compressive strength; and

Yield and ultimate strength of conventional and prestressing reinforcing steel and metal connection hardware.

Where materials testing is required by Section 2.2.6, the test methods to quantify material properties shall comply with the requirements of Section 6.2.2.3. The frequency of sampling, including the minimum number of tests for property determination, shall comply with the requirements of Section 6.2.2.4.

*C6.2.2.1.1 General* Other material properties that may be of interest for concrete components include:

1. Tensile strength and modulus of elasticity of concrete, which can be derived from the compressive strength, do not warrant the damage associated with the extra coring required;
2. Ductility, toughness, and fatigue properties of concrete;
3. Carbon equivalent present in the reinforcing steel; and
4. Presence of any degradation such as corrosion, bond with concrete, and chemical composition.

The effort required to determine these properties depends on the availability of accurate updated construction documents and drawings, the quality and type of construction (absence of degradation), accessibility, and the condition of materials. The method of analysis selected [e.g., Linear Static Procedure (LSP), Nonlinear Static Procedure (NSP)] may also influence the scope of the testing.

The size of the samples and removal practices to be followed are referenced in FEMA 274 (FEMA 1997). Generally, mechanical properties for both concrete and reinforcing steel can be established from combined core and specimen sampling at similar locations, followed by laboratory testing. Core drilling should minimize damage of the existing reinforcing steel as much as is practicable.

*6.2.2.1.2 Nominal or Specified Properties* Nominal material properties, or properties specified in construction documents, shall be taken as lower-bound material properties. Corresponding expected material properties shall be calculated by multiplying lower-bound values by a factor taken from Table 6-4 to translate from lower-bound to expected values. Alternative factors shall be permitted where justified by test data.

### 6.2.2.2 Component Properties

The following component properties and as-built conditions shall be established:

1. Cross-sectional dimensions of individual components and overall configuration of the structure;
2. Configuration of component connections, size of anchor bolts, thickness of connector material, anchorage and interconnection of embedments, and the presence of bracing or stiffening components;
3. Modifications to components or overall configuration of the structure;
4. Current physical condition of components and connections, and the extent of any deterioration present; and
5. Presence of conditions that influence building performance.

### C6.2.2.2 Component Properties

Component properties may be needed to characterize building performance properly in the seismic analysis. The starting point for assessing component properties and condition should be retrieval of available construction documents. Preliminary review of these documents should be performed to identify primary gravity- and lateral-force-resisting elements, systems, and their critical components and connections. In the absence of a complete set of building drawings, the design professional must perform a thorough investigation of the building to identify these elements, systems and components as indicated in Section 6.2.3.

### 6.2.2.3 Test Methods to Quantify Material Properties

*6.2.2.3.1 General* Destructive and nondestructive test methods used to obtain in-place mechanical properties of materials identified in Section 6.2.2.1, and component properties identified in Section 6.2.2.2 shall comply with the requirements of this section. Samples of concrete and reinforcing and connector steel shall be examined for physical condition as specified in Section 6.2.3.2.

If the determination of material properties is accomplished through removal and testing of samples for laboratory analysis, sampling shall take place in primary gravity- and lateral-force-resisting components in regions with the least stress.

Where Section 6.2.2.4.1 does not apply and the coefficient of variation is greater than 14%, the expected concrete strength shall not exceed the mean minus one standard deviation.

*6.2.2.3.2 Sampling* For testing of concrete material, the sampling program shall consist of the removal of standard cores. Core drilling shall be preceded by nondestructive location of the reinforcing steel, and core holes shall be located to minimize damage to or drilling through the reinforcing steel. Core holes shall be filled with concrete or grout of comparable strength. If conventional reinforcing and bonded prestressing steel are tested, sampling shall consist of the removal of local bar segments and installation of replacement spliced material to maintain continuity of the rebar for transfer of bar force.

Removal of core samples and performance of laboratory destructive testing shall be permitted as a method of determining existing concrete strength properties. Removal of core samples shall employ the procedures contained in ASTM C42/C42M-03 (ASTM 2003). Testing shall follow the procedures contained in ASTM C42/C42M-03, ASTM C39/C39M-01 (ASTM 2001), and ASTM C496-96 (ASTM 1996). Core strength shall be converted to in situ concrete compressive strength ($f_c$) by an approved procedure.

Removal of bar or tendon length samples and performance of laboratory destructive testing shall be permitted as a method of determining existing reinforcing steel strength properties. The tensile yield strength and ultimate strength for reinforcing and prestressing steels shall be obtained using the procedures contained in ASTM A370-03 (ASTM 2003). Prestressing materials also shall meet the supplemental requirements in ASTM A416/A416M-02 (ASTM 2002), ASTM A421/A421M-02 (ASTM 2002), or ASTM A722/A722M-98 (ASTM 2003), depending on material type. Properties of connector steels shall be permitted to be determined by wet and dry chemical composition tests, and by direct tensile and compressive strength tests as specified by ASTM A370-03. Where strengths of embedded connectors are required, in situ testing shall satisfy the provisions of ASTM E488-96 (ASTM 2003).

### C6.2.2.3 Test Methods to Quantify Material Properties

ACI 318 (ACI 2002) and FEMA 274 (FEMA 1997) provide further guidance on correlating core strength to in-place strength and provide references for various test methods that may be used to estimate material properties. The chemical composition may also be determined from the retrieved samples. FEMA 274 provides references for these tests.

Usually, the reinforcing steel system used in the construction of a specific building is of a common grade and strength. Occasionally, one grade of reinforcement is used for small-diameter bars (e.g., those used for stirrups and hoops) and another grade for large-diameter bars (e.g., those used for longitudinal reinforcement). Furthermore, it is possible that a number of different concrete design strengths (or "classes") have been employed. Historical research and industry documents also contain insight on material mechanical properties used in different construction eras.

### 6.2.2.4 Minimum Number of Tests

Materials testing is not required if material properties are available from original construction documents that include material test records or material test reports.

The minimum number of tests necessary to quantify properties by in-place testing for comprehensive data collection shall be as specified in Sections 6.2.2.4.1 through 6.2.2.4.4. The minimum number of tests for usual data collection shall be as specified in Section 6.2.2.4.5. If the existing gravity- or lateral-force-resisting system is being replaced in the rehabilitation process, material testing shall be required only to quantify properties of existing material at new connection points.

### C6.2.2.4 Minimum Number of Tests

In order to quantify in-place properties accurately, it is important that a minimum number of tests be conducted on primary components of the lateral-force-resisting system. The minimum number of tests is dictated by the data available from original construction, the type of structural system employed, the desired accuracy, and the quality and condition of in-place materials. The accessibility of the structural system may also influence the testing program scope. The focus of this testing shall be on primary lateral-force-resisting components and on specific properties needed for analysis. The test quantities provided in this section are minimum numbers; the design professional should determine whether further testing is needed to evaluate as-built conditions.

Testing generally is not required on components other than those of the lateral-force-resisting system.

The design professional (and subcontracted testing agency) should carefully examine test results to verify that suitable sampling and testing procedures were followed and that appropriate values for the analysis were selected from the data.

*6.2.2.4.1 Comprehensive Testing* Unless specified otherwise, a minimum of three tests shall be conducted to determine any property. If the coefficient of variation exceeds 14%, additional tests shall be performed until the coefficient of variation is equal to or less than 14%.

*6.2.2.4.2 Concrete Materials* For each concrete element type (such as a shear wall), a minimum of three core samples shall be taken and subjected to compression tests. A minimum of six total tests shall be performed on a building for concrete strength determination, subject to the limitations of this section. If varying concrete classes/grades were employed in the construction of the building, a minimum of three samples and tests shall be performed for each class. The modulus of elasticity shall be permitted to be estimated from the data of strength testing. Samples shall be taken from randomly selected components critical to structural behavior of the building. Tests also shall be performed on samples from components that are damaged or degraded, if such damage or degradation is identified, to quantify their condition. Test results shall be compared with strength values specified in the construction documents. If test values less than the specified strength in the construction documents are found, further strength testing shall be performed to determine the cause or identify the extent of the condition.

The minimum number of tests to determine compressive and tensile strength shall conform to the following criteria:

For concrete elements for which the specified design strength is known and test results are not available, a minimum of three cores/tests shall be conducted for each floor level, 400 yd³ of concrete, or 10,000 sf of surface area, whichever requires the most frequent testing; and

For concrete elements for which the design strength is unknown and test results are not available, a minimum of six cores/tests shall be conducted for each floor level, 400 yd³ of concrete, or 10,000 sf of surface area, whichever requires the most frequent testing. Where the results indicate that different classes of concrete were employed, the degree of testing shall be increased to confirm class use.

Quantification of concrete strength via ultrasonics or other nondestructive test methods shall not be substituted for core sampling and laboratory testing.

*C6.2.2.4.2 Concrete Materials* Ultrasonics and nondestructive test methods should not be substituted for core sampling and laboratory testing since they do not yield accurate strength values directly.

*6.2.2.4.3 Conventional Reinforcing and Connector Steels* The minimum number of tests required to determine reinforcing and connector steel strength properties shall be as follows. Connector steel shall be defined as additional structural steel or miscellaneous metal used to secure precast and other concrete shapes to the building structure. Tests shall determine both yield and ultimate strengths of reinforcing and connector steel. A minimum of three tensile tests shall be conducted on conventional reinforcing steel samples from a building for strength determination, subject to the following supplemental conditions:

1. If original construction documents defining properties exist, at least three strength coupons shall be randomly removed from each element or component type and tested; and
2. If original construction documents defining properties do not exist but the approximate date of construction is known and a common material grade is confirmed, at least three strength coupons shall be randomly removed from each element or component type for every three floors of the building. If the date of construction is unknown, at least six such samples/tests, for every three floors, shall be performed.

All sampled steel shall be replaced with new fully spliced and connected material unless an analysis confirms that replacement of original components is not required.

*6.2.2.4.4 Prestressing Steels* The sampling of prestressing steel tendons for laboratory testing shall be required only for those prestressed components that are a part of the lateral-force-resisting system. Prestressed components in diaphragms shall be permitted to be excluded from testing.

Tendon or prestress removal shall be avoided if possible by sampling of either the tendon grip or the extension beyond the anchorage.

All sampled prestressed steel shall be replaced with new fully connected and stressed material and anchorage hardware unless an analysis confirms that replacement of original components is not required.

*6.2.2.4.5 Usual Testing* The minimum number of tests to determine concrete and reinforcing steel material properties for usual data collection shall be based on the following criteria:

1. If the specified design strength of the concrete is known, at least one core shall be taken from sam-

ples of each different concrete strength used in the construction of the building, with a minimum of three cores taken for the entire building;
2. If the specified design strength of the concrete is not known, at least one core shall be taken from each type of component, with a minimum of six cores taken for the entire building;
3. If the specified design strength of the reinforcing steel is known, use of nominal or specified material properties shall be permitted without additional testing; and
4. If the specified design strength of the reinforcing steel is not known, at least two strength coupons of reinforcing steel shall be removed from the building for testing.

*C6.2.2.4.5 Usual Testing* For other material properties, such as hardness and ductility, no minimum number of tests is prescribed. Similarly, standard test procedures may not exist. The design professional should examine the particular need for this type of testing and establish an adequate protocol.

### 6.2.2.5 Default Properties

Use of default material properties to determine component strengths shall be permitted in conjunction with the linear analysis procedures of Chapter 3.

Default lower-bound concrete compressive strengths shall be taken from Table 6-3. Default expected concrete compressive strengths shall be determined by multiplying lower-bound values by an appropriate factor selected from Table 6-4 unless another factor is justified by test data. The appropriate default compressive strength—lower-bound or expected strength, as specified in Section 2.4.4—shall be used to establish other strength and performance characteristics for the concrete as needed in the structural analysis.

Default lower-bound values for reinforcing steel shall be taken from Table 6-1 or 6-2.

Default expected strength values for reinforcing steel shall be determined by multiplying lower-bound values by an appropriate factor selected from Table 6-4 unless another factor is justified by test data. Where default values are assumed for existing reinforcing steel, welding or mechanical coupling of new reinforcement to the existing reinforcing steel shall not be used.

The default lower-bound yield strength for steel connector material shall be taken as 27,000 psi. The default expected yield strength for steel connector material shall be determined by multiplying lower-bound values by an appropriate factor selected from Table 6-4 unless another value is justified by test data.

**Table 6-1. Default Lower-Bound Tensile and Yield Properties of Reinforcing for Various Periods[1]**

|  | | Structural[2] | Intermediate[2] | Hard[2] | | | | |
|---|---|---|---|---|---|---|---|---|
|  | Grade | 33 | 40 | 50 | 60 | 65 | 70 | 75 |
|  | Minimum Yield[2] (psi) | 33,000 | 40,000 | 50,000 | 60,000 | 65,000 | 70,000 | 75,000 |
| Year | Minimum Tensile[2] (psi) | 55,000 | 70,000 | 80,000 | 90,000 | 75,000 | 80,000 | 100,000 |
| 1911–1959 | | x | x | x | — | x | — | — |
| 1959–1966 | | x | x | x | x | x | x | x |
| 1966–1972 | | — | x | x | x | x | x | — |
| 1972–1974 | | — | x | x | x | x | x | — |
| 1974–1987 | | — | x | x | x | x | x | — |
| 1987–Present | | — | x | x | x | x | x | x |

[1]An entry of "x" indicates the grade was available in those years.
[2]The terms Structural, Intermediate, and Hard became obsolete in 1968.

**Table 6-2. Default Lower-Bound Tensile and Yield Properties of Reinforcing for Various ASTM Specifications and Periods[1]**

|  |  |  |  | Structural[2] | Intermediate[2] | Hard[2] | | | | |
|---|---|---|---|---|---|---|---|---|---|---|
|  |  |  | ASTM Grade | 33 | 40 | 50 | 60 | 65 | 70 | 75 |
|  |  |  | Minimum Yield (psi) | 33,000 | 40,000 | 50,000 | 60,000 | 65,000 | 70,000 | 75,000 |
| ASTM Designation[3] | Steel Type | Year Range | Minimum Tensile (psi) | 55,000 | 70,000 | 80,000 | 90,000 | 75,000 | 80,000 | 100,000 |
| A15 | Billet | 1911–1966 | | x | x | x | — | — | — | — |
| A16 | Rail[4] | 1913–1966 | | — | — | x | — | — | — | — |
| A61 | Rail[4] | 1963–1966 | | — | — | — | x | — | — | — |
| A160 | Axle | 1936–1964 | | x | x | x | — | — | — | — |
| A160 | Axle | 1965–1966 | | x | x | x | x | — | — | — |
| A185 | WWF | 1936–Present | | — | — | — | — | x | — | — |
| A408 | Billet | 1957–1966 | | x | x | x | — | — | — | — |
| A431 | Billet | 1959–1966 | | — | — | — | — | — | — | x |
| A432 | Billet | 1959–1966 | | — | — | — | x | — | — | — |
| A497 | WWF | 1964–Present | | — | — | — | — | — | x | — |
| A615 | Billet | 1968–1972 | | — | x | — | x | — | — | x |
| A615 | Billet | 1974–1986 | | — | x | — | x | — | — | — |
| A615 | Billet | 1987–Present | | — | x | — | x | — | — | x |
| A616[5] | Rail[4] | 1968–Present | | — | — | — | — | — | — | — |
| A617 | Axle | 1968–Present | | — | x | — | x | — | — | — |
| A706 | Low-Alloy | 1974–Present | | — | — | — | x | — | x | — |
| A955 | Stainless | 1996–Present | | — | x | — | x | — | — | x |

[1]An entry of "x" indicates the grade was available in those years.
[2]The terms Structural, Intermediate, and Hard became obsolete in 1968.
[3]ASTM steel is marked with the letter W."
[4]Rail bars are marked with the letter "R."
[5]Bars marked "s!" (ASTM 616) have supplementary requirements for bend tests.
[6]ASTM A706 has a minimum tensile strength of 80 ksi, but not less than 1.25 times the actual yield strength.

**Table 6-3. Default Lower-Bound Compressive Strength of Structural Concrete (psi)**

| Time Frame | Footings | Beams | Slabs | Columns | Walls |
|---|---|---|---|---|---|
| 1900–1919 | 1,000–2,500 | 2,000–3,000 | 1,500–3,000 | 1,500–3,000 | 1,000–2,500 |
| 1920–1949 | 1,500–3,000 | 2,000–3,000 | 2,000–3,000 | 2,000–4,000 | 2,000–3,000 |
| 1950–1969 | 2,500–3,000 | 3,000–4,000 | 3,000–4,000 | 3,000–6,000 | 2,500–4,000 |
| 1970–Present | 3,000–4,000 | 3,000–5,000 | 3,000–5,000 | 3,000–1,0000 | 3,000–5,000 |

**Table 6-4. Factors to Translate Lower-Bound Material Properties to Expected Strength Material Properties**

| Material Property | Factor |
|---|---|
| Concrete Compressive Strength | 1.50 |
| Reinforcing Steel Tensile and Yield Strength | 1.25 |
| Connector Steel Yield Strength | 1.50 |

Default values for prestressing steel in prestressed concrete construction shall not be used.

### C6.2.2.5 Default Properties

Default values provided in this standard are generally conservative. While the strength of reinforcing steel may be fairly consistent throughout a building, the strength of concrete in a building could be highly variable, given variability in concrete mix designs and sensitivity to water/cement ratio and curing practices. It is recommended to conservatively assume the minimum value of the concrete compressive strength in the given range unless a higher strength is substantiated by construction documents, test reports, or material testing; it would be conservative to assume the maximum value in a given range where determining the force-controlled actions on other components.

Until about 1920, a variety of proprietary reinforcing steels was used. Yield strengths are likely to be in the range of 33,000 to 55,000 psi, but higher values are possible and actual yield and tensile strengths may exceed minimum values. Once commonly used to designate reinforcing steel grade, the terms structural, intermediate, and hard became obsolete in 1968. Plain and twisted square bars were sometimes used between 1900 and 1949.

Factors to convert default reinforcing steel strength to expected strength include consideration of material overstrength and strain-hardening.

### 6.2.3 Condition Assessment

#### 6.2.3.1 General

A condition assessment of the existing building and site conditions shall be performed as specified in this section.

The condition assessment shall include the following:

1. The physical condition of primary and secondary components shall be examined and the presence of any degradation shall be noted;
2. The presence and configuration of components and their connections, and the continuity of load paths between components, elements, and systems shall be verified or established;
3. Other conditions, including neighboring party walls and buildings, presence of nonstructural components, prior remodeling, and limitations for rehabilitation that may influence building performance shall be reviewed and documented;
4. Information needed to select a knowledge factor in accordance with Section 6.2.4 shall be obtained; and
5. Component orientation, plumbness, and physical dimensions shall be confirmed.

#### 6.2.3.2 Scope and Procedures

The scope of the condition assessment shall include all accessible structural components involved in lateral load resistance.

#### C6.2.3.2 Scope and Procedures

The degree to which the condition assessment is performed will affect the knowledge factor ($\kappa$) as specified in Section 6.2.4.

*6.2.3.2.1 Visual Condition Assessment* Direct visual inspection of accessible and representative primary components and connections shall be performed to identify any configurational issues, determine whether degradation is present, establish continuity of load paths, establish the need for other test methods to quantify the presence and degree of degradation, and measure dimensions of existing construction to compare with available design information and reveal any permanent deformations.

Visual inspection of the building shall include visible portions of foundations, lateral-force-resisting members, diaphragms (slabs), and connections. As a minimum, a representative sampling of at least 20% of the components and connections shall be visually inspected at each floor level. If significant damage or degradation is found, the assessment sample of all critical components of similar type in the building shall be increased to 40%.

If coverings or other obstructions exist, partial visual inspection through the obstruction, using drilled holes and a fiberscope, shall be permitted.

*6.2.3.2.2 Comprehensive Co\ndition Assessment* Exposure is defined as local minimized removal of cover concrete and other materials to allow inspection of reinforcing system details. All damaged concrete cover shall be replaced after inspection. The following criteria shall be used for assessing primary connections in the building for comprehensive data collection:

1. If detailed design drawings exist, exposure of at least three different primary connections shall occur, with the connection sample including different types of connections. If no deviations from the drawings exist, it shall be permitted to consider the sample as being representative of installed conditions. If deviations are noted, then at least 25% of the specific connection type shall be inspected to identify the extent of deviation; and
2. In the absence of detailed design drawings, at least three connections of each primary connection type shall be exposed for inspection. If common detailing among the three connections is observed, it shall be permitted to consider this condition as representative of installed conditions. If variations are observed among like connections, additional connections shall be inspected until an accurate understanding of building construction is gained.

*6.2.3.2.3 Additional Testing* If additional destructive and nondestructive testing are required to determine the degree of damage or presence of deterioration or to understand the internal condition and quality of concrete, approved test methods shall be used.

*C6.2.3.2.3 Additional Testing* The physical condition of components and connectors will affect their performance. The need to accurately identify the physical condition may also dictate the need for certain additional destructive and nondestructive test methods. Such methods may be used to determine the degree of damage or presence of deterioration, and to improve understanding of the internal condition and quality of

the concrete. Further guidelines and procedures for destructive and nondestructive tests that may be used in the condition assessment are provided in FEMA 274 (FEMA 1997) and FEMA 306 (FEMA 1998). The following paragraphs identify those nondestructive examination (NDE) methods having the greatest use and applicability to condition assessment.

- Surface NDE methods include infrared thermography, delamination sounding, surface hardness measurement, and crack mapping. These methods may be used to find surface degradation in components such as service-induced cracks, corrosion, and construction defects.
- Volumetric NDE methods, including radiography and ultrasonics, may be used to identify the presence of internal discontinuities, as well as to identify loss of section. Impact-echo ultrasonics is particularly useful because of ease of implementation and proven capability in concrete.
- Structural condition and performance may be assessed through on-line monitoring using acoustic emissions and strain gauges, and in-place static or dynamic load tests. Monitoring is used to determine if active degradation or deformations are occurring, while nondestructive load testing provides direct insight on load-carrying capacity.
- Locating, sizing, and initial assessment of the reinforcing steel may be completed using electromagnetic methods (such as a pachometer) or radiography. Further assessment of suspected corrosion activity should use electrical half-cell potential and resistivity measurements.
- Where it is absolutely essential, the level of prestress remaining in an unbonded prestressed system may be measured using lift-off testing (assuming original design and installation data are available), or another nondestructive method such as "coring stress relief" specified in ASCE 11 (ASCE 1999).

### 6.2.3.3 Basis for the Mathematical Building Model

The results of the condition assessment shall be used to quantify the following items needed to create the mathematical building model:

1. Component section properties and dimensions;
2. Component configuration and the presence of any eccentricities or permanent deformation;
3. Connection configuration and the presence of any eccentricities;
4. Presence and effect of alterations to the structural system since original construction; and
5. Interaction of nonstructural components and their involvement in lateral load resistance.

All deviations between available construction records and as-built conditions obtained from visual inspection shall be accounted for in the structural analysis.

Unless concrete cracking, reinforcing corrosion, or other mechanisms are observed in the condition assessment to be causing damage or reduced capacity, the cross-sectional area and other sectional properties shall be taken as those from the design drawings. If some sectional material loss has occurred, the loss shall be quantified by direct measurement and sectional properties shall be reduced accordingly, using principles of structural mechanics.

### 6.2.4 Knowledge Factor

A knowledge factor, $\kappa$, for computation of concrete component capacities and permissible deformations shall be selected in accordance with Section 2.2.6.4, with the following additional requirements specific to concrete components.

A knowledge factor, $\kappa$, equal to 0.75 shall be used if any of the following criteria are met:

1. Components are found damaged or deteriorated during assessment, and further testing is not performed to quantify their condition or justify the use of $\kappa = 1.0$;
2. Component mechanical properties have a coefficient of variation exceeding 25%; and
3. Components contain archaic or proprietary material and the condition is uncertain.

## 6.3 GENERAL ASSUMPTIONS AND REQUIREMENTS

### 6.3.1 Modeling and Design

#### 6.3.1.1 General Approach

Seismic rehabilitation of concrete structural components of existing buildings shall comply with the requirements of ACI 318 (ACI 2002), except as otherwise indicated in this standard. Seismic evaluation shall identify brittle or low-ductility failure modes of force-controlled actions as defined in Section 2.4.4.

Evaluation of demands and capacities of reinforced concrete components shall include consideration of locations along the length where lateral and gravity loads produce maximum effects, where changes in cross section or reinforcement result in reduced strength, and where abrupt changes in cross section or reinforcement, including splices, may produce stress concentrations, resulting in premature failure.

#### C6.3.1.1 General Approach

Brittle or low-ductility failure modes typically include behavior in direct or nearly-direct compression, shear in slender components and in component connections, torsion in slender components, and reinforcement development, splicing, and anchorage. It is recommended that the stresses, forces, and moments acting to cause these failure modes be determined from a limit-state analysis considering probable resistances at locations of nonlinear action.

#### 6.3.1.2 Stiffness

Component stiffnesses shall be calculated considering shear, flexure, axial behavior, and reinforcement slip deformations. Consideration shall be given to the state of stress on the component due to volumetric changes from temperature and shrinkage, and to deformation levels to which the component will be subjected under gravity and earthquake loading.

#### C6.3.1.2 Stiffness

For columns with low axial loads, deformations due to bar slip can account for as much as 50% of the total deformations at yield. The design professional is referred to Elwood and Eberhard (2006) for further guidance regarding calculation of effective stiffness of reinforced concrete columns to include the effects of flexure, shear, and bar slip.

6.3.1.2.1 Linear Procedures Where design actions are determined using the linear procedures of Chapter 3, component effective stiffnesses shall correspond to the secant value to the yield point of the component. The use of higher stiffnesses shall be permitted where it is demonstrated by analysis to be appropriate for the design loading. Alternatively, the use of effective stiffness values in Table 6-5 shall be permitted.

6.3.1.2.2 Nonlinear Procedures Where design actions are determined using the nonlinear procedures of Chapter 3, component load–deformation response shall be represented by nonlinear load–deformation relations. Linear relations shall be permitted where nonlinear response will not occur in the component. The nonlinear load–deformation relation shall be based on experimental evidence or taken from quantities specified in Sections 6.4 through 6.12. For the Nonlinear Static Procedure (NSP), use of the generalized load–deformation relation shown in Fig. 6-1 or other curves defining behavior under monotonically increasing deformation shall be permitted. For the Nonlinear Dynamic Procedure (NDP), load–deformation relations shall define behavior under monotonically increasing lateral deformation and

**Table 6-5. Effective Stiffness Values[1]**

| Component | Flexural Rigidity | Shear Rigidity | Axial Rigidity |
|---|---|---|---|
| Beams—Non-prestressed | $0.5E_cI_g$ | $0.4E_cA_w$ | — |
| Beams—Prestressed | $E_cI_g$ | $0.4E_cA_w$ | — |
| Columns with Compression Due to Design Gravity Loads $\geq 0.5A_gf'_c$ | $0.7E_cI_g$ | $0.4E_cA_w$ | $E_cA_g$ |
| Columns with Compression Due to Design Gravity Loads $\leq 0.3A_gf'_c$ or with Tension | $0.5E_cI_g$ | $0.4E_cA_w$ | $E_sA_s$ |
| Walls—Uncracked (on inspection) | $0.8E_cI_g$ | $0.4E_cA_w$ | $E_cA_g$ |
| Walls—Cracked | $0.5E_cI_g$ | $0.4E_cA_w$ | $E_cA_g$ |
| Flat Slabs—Non-prestressed | See Section 6.5.4.2 | $0.4E_cA_g$ | — |
| Flat Slabs—Prestressed | See Section 6.5.4.2 | $0.4E_cA_g$ | — |

[1]It shall be permitted to take $I_g$ for T-beams as twice the value of $I_g$ of the web alone. Otherwise, $I_g$ shall be based on the effective width as defined in Section 6.3.1.3. For columns with axial compression falling between the limits provided, linear interpolation shall be permitted. Alternatively, the more conservative effective stiffnesses shall be used.

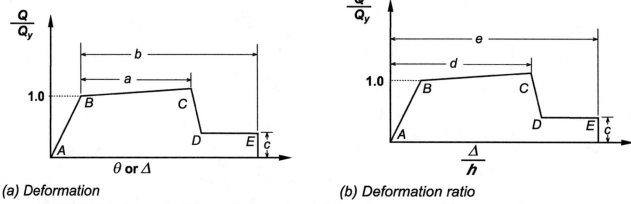

**(a) Deformation**    **(b) Deformation ratio**

**FIGURE 6-1. Generalized Force–Deformation Relations for Concrete Elements or Components.**

under multiple reversed deformation cycles as specified in Section 6.3.2.1.

The generalized load–deformation relation shown in Fig. 6-1 shall be described by linear response from $A$ (unloaded component) to an effective yield B, then a linear response at reduced stiffness from point $B$ to $C$, then sudden reduction in lateral load resistance to point $D$, then response at reduced resistance to $E$, and final loss of resistance thereafter. The slope from point $A$ to $B$ shall be determined according to Section 6.3.1.2.1. The slope from point $B$ to $C$, ignoring effects of gravity loads acting through lateral displacements, shall be taken between zero and 10% of the initial slope unless an alternate slope is justified by experiment or analysis. Point $C$ shall have an ordinate equal to the strength of the component and an abscissa equal to the deformation at which significant strength degradation begins. Representation of the load–

deformation relation by points $A$, $B$, and $C$ only (rather than all points $A$–$E$) shall be permitted if the calculated response does not exceed point $C$. Numerical values for the points identified in Fig. 6-1 shall be as specified in Sections 6.4 through 6.12. Other load–deformation relations shall be permitted if justified by experimental evidence or analysis.

*C6.3.1.2.2 Nonlinear Procedures* Typically, the responses shown in Fig. 6-1 are associated with flexural response or tension response. In this case, the resistance at $Q/Q_y = 1.0$ is the yield value, and subsequent strain-hardening accommodates strain hardening in the load–deformation relation as the member is deformed toward the expected strength. Where the response shown in Fig. 6-1 is associated with compression, the resistance at $Q/Q_y = 1.0$ typically is the value at which concrete begins to spall, and strain-hardening in well-confined sections may be associated with strain-hardening of the longitudinal reinforcement and the confined concrete. Where the response shown in Fig. 6-1 is associated with shear, the resistance at $Q/Q_y = 1.0$ typically is the value at which the design shear strength is reached, and no strain-hardening follows.

The deformations used for the load–deformation relation of Fig. 6-1 shall be defined in one of two ways, as follows:

1. **Deformation, or Type I.** In this curve, deformations are expressed directly using terms such as strain, curvature, rotation, or elongation. The parameters $a$ and $b$ shall refer to those portions of the deformation that occur after yield; that is, the plastic deformation. The parameter $c$ is the reduced resistance after the sudden reduction from $C$ to $D$. Parameters $a$, $b$, and $c$ are defined numerically in various tables in this chapter. Alternatively, it shall be permitted to determine the parameters $a$, $b$, and $c$ directly by analytical procedures justified by experimental evidence.

2. **Deformation Ratio, or Type II.** In this curve, deformations are expressed in terms such as shear angle and tangential drift ratio. The parameters $d$ and $e$ refer to total deformations measured from the origin. Parameters $c$, $d$, and $e$ are defined numerically in various tables in this chapter. Alternatively, it shall be permitted to determine the parameters $c$, $d$, and $e$ directly by analytical procedures justified by experimental evidence.

Provisions for determining alternative modeling parameters and acceptance criteria based on experimental evidence are given in Section 2.8.

### 6.3.1.3 Flanged Construction

In beams consisting of a web and flange that act integrally, the combined stiffness and strength for flexural and axial loading shall be calculated considering a width of effective flange on each side of the web equal to the smaller of: (1) the provided flange width; (2) eight times the flange thickness; (3) half the distance to the next web; or (4) one-fifth of the span for beams. Where the flange is in compression, both the concrete and reinforcement within the effective width shall be considered effective in resisting flexure and axial load. Where the flange is in tension, longitudinal reinforcement within the effective width and what is developed beyond the critical section shall be considered fully effective for resisting flexural and axial loads. The portion of the flange extending beyond the width of the web shall be assumed ineffective in resisting shear.

In walls, effective flange width shall be in accordance with Chapter 21 of ACI 318 (ACI 2002).

### 6.3.2 Strength and Deformability

#### 6.3.2.1 General

Actions in a structure shall be classified as being either deformation-controlled or force-controlled, as defined in Section 2.4.4. Design strengths for deformation-controlled and force-controlled actions shall be calculated in accordance with Sections 6.3.2.2 and 6.3.2.3, respectively.

Components shall be classified as having low, moderate, or high ductility demands according to Section 6.3.2.4.

Where strength and deformation capacities are derived from test data, the tests shall be representative of proportions, details, and stress levels for the component and comply with requirements specified in Section 2.8.1.

The strength and deformation capacities of concrete members shall correspond to values resulting from earthquake loadings involving three fully reversed cycles to the design deformation level unless a larger or smaller number of deformation cycles is determined considering earthquake duration and the dynamic properties of the structure.

#### C6.3.2.1 General

Strengths and deformation capacities given in this chapter are for earthquake loadings involving three fully reversed deformation cycles to the design deformation levels, in addition to similar cycles to lesser deformation levels. In some cases—including some short-period buildings and buildings subjected to a long-duration design earthquake—a building may be expected to be subjected to additional cycles to the design deformation levels. The increased number of cycles may lead to reductions in resistance and deformation capacity. The effects on strength and deformation capacity of additional deformation cycles should be considered in design. Large earthquakes will cause additional cycles.

#### 6.3.2.2 Deformation-Controlled Actions

Strengths used for deformation-controlled actions shall be taken as equal to expected strengths, $Q_{CE}$, obtained experimentally, or calculated using accepted principles of mechanics. Expected strength is defined as the mean maximum resistance expected over the range of deformations to which the concrete component is likely to be subjected. Where calculations are used to define expected strength, expected material properties shall be used. Unless other procedures are specified in this standard, procedures specified in ACI 318 (ACI 2002) to calculate design strengths shall be permitted except that the strength reduction factor, $\phi$, shall be taken equal to unity. Deformation capacities for acceptance of deformation-controlled actions calculated by nonlinear procedures shall be as specified in Sections 6.4 to Section 6.12. For components constructed of lightweight concrete, $Q_{CE}$ shall be modified in accordance with ACI 318 procedures for lightweight concrete.

### C6.3.2.2 Deformation-Controlled Actions

Expected yield strength of reinforcing steel, as specified in this standard, includes consideration of material overstrength and strain-hardening.

### 6.3.2.3 Force-Controlled Actions

Strengths used for force-controlled actions shall be taken as lower-bound strengths, $Q_{CL}$, obtained experimentally, or calculated using established principles of mechanics. Lower-bound strength is defined as the mean minus one standard deviation of resistance expected over the range of deformations and loading cycles to which the concrete component is likely to be subjected. Where calculations are used to define lower-bound strengths, lower-bound estimates of material properties shall be used. Unless other procedures are specified in this standard, procedures specified in ACI 318 (ACI 2002) to calculate design strengths shall be permitted, except that the strength reduction factor, $\phi$, shall be taken equal to unity. For components constructed of lightweight concrete, $Q_{CL}$ shall be modified in accordance with ACI 318 procedures for lightweight concrete.

### 6.3.2.4 Component Ductility Demand Classification

Where procedures in this chapter require classification of component ductility demand, components shall be classified as having low, moderate, or high ductility demands, based on the maximum value of the demand capacity ratio (DCR) defined in Section 2.4.1 for linear procedures, or the calculated displacement ductility for nonlinear procedures in accordance with Table 6-6.

### 6.3.3 Flexure and Axial Loads

Flexural strength and deformation capacity of members with and without axial loads shall be

**Table 6-6. Component Ductility Demand Classification**

| Maximum Value of DCR or Displacement Ductility | Descriptor |
|---|---|
| < 2 | Low Ductility Demand |
| 2 to 4 | Moderate Ductility Demand |
| > 4 | High Ductility Demand |

calculated according to the procedures of ACI 318 (ACI 2002) or by other approved methods. Strengths and deformation capacities of components with monolithic flanges shall be calculated considering concrete and developed longitudinal reinforcement within the effective flange width as defined in Section 6.3.1.3.

Strength and deformation capacities shall be determined considering available development of longitudinal reinforcement. Where longitudinal reinforcement has embedment or development length that is insufficient for development of reinforcement strength, flexural strength shall be calculated based on limiting stress capacity of the embedded bar as defined in Section 6.3.5.

Where flexural deformation capacities are calculated from basic principles of mechanics, reductions in deformation capacity due to applied shear shall be taken into consideration. Where using analytical models for flexural deformability that do not directly consider effect of shear, and where design shear equals or exceeds $6\sqrt{f_c'}\, A_w$, where $f_c'$ is in psi and $A_w$ is gross area of web in in.[2], the design value shall not exceed 80% of the value calculated using the analytical model.

For concrete columns under combined axial load and biaxial bending, the combined strength shall be evaluated considering biaxial bending. Where using linear procedures, the design axial load, $P_{UF}$, shall be calculated as a force-controlled action in accordance with Section 3.4. The design moments, $M_{UD}$, shall be calculated about each principal axis in accordance with Section 3.4. Acceptance shall be based on the following equation:

$$\left(\frac{M_{UDx}}{m_x \kappa M_{CEx}}\right)^2 + \left(\frac{M_{UDy}}{m_y \kappa M_{CEy}}\right)^2 \leq 1 \qquad \text{(Eq. 6-1)}$$

where

$M_{UDx}$ = design bending moment about x-axis for axial load $P_{UF}$, kip-in.;

$M_{UDy}$ = design bending moment about y-axis for axial load $P_{UF}$, kip-in.;

$M_{CEx}$ = expected bending moment strength about x-axis, kip-in.;

$M_{CEy}$ = expected bending moment strength about y-axis, kip-in.;

$m_x$ = m-factor for column for bending about x-axis in accordance with Table 6-12; and

$m_y$ = m-factor for column for bending about y-axis in accordance with Table 6-12.

Alternative approaches based on principles of mechanics shall be permitted.

## C6.3.3 Flexure and Axial Loads

Laboratory tests indicate that flexural deformability may be reduced as coexisting shear forces increase. As flexural ductility demands increase, shear capacity decreases, which may result in a shear failure before theoretical flexural deformation capacities are reached. Caution should be exercised where flexural deformation capacities are determined by calculation. FEMA 306 (FEMA 1998) is a resource for guidance regarding the interaction between shear and flexure.

### 6.3.3.1 Usable Strain Limits

Without confining transverse reinforcement, the maximum usable strain at the extreme concrete compression fiber shall not exceed 0.002 for components in nearly pure compression and 0.005 for other components unless larger strains are substantiated by experimental evidence and approved by the authority having jurisdiction. Maximum usable compressive strains for confined concrete shall be based on experimental evidence and shall consider limitations posed by fracture of transverse reinforcement, buckling of longitudinal reinforcement, and degradation of component resistance at large deformation levels. Maximum compressive strains in longitudinal reinforcement shall not exceed 0.02, and maximum tensile strains in longitudinal reinforcement shall not exceed 0.05.

## 6.3.4 Shear and Torsion

Strengths in shear and torsion shall be calculated according to ACI 318 (ACI 2002) except as modified in this standard.

Within yielding regions of components with moderate or high ductility demands, shear and torsional strength shall be calculated according to procedures for ductile components, such as the provisions in Chapter 21 of ACI 318. Within yielding regions of components with low ductility demands and outside yielding regions for all ductility demands, calculation of design shear strength using procedures for effective

elastic response such as the provisions in Chapter 11 of ACI 318 shall be permitted.

Where the longitudinal spacing of transverse reinforcement exceeds half the component effective depth measured in the direction of shear, transverse reinforcement shall be assumed not more than 50% effective in resisting shear or torsion. Where the longitudinal spacing of transverse reinforcement exceeds the component effective depth measured in the direction of shear, transverse reinforcement shall be assumed ineffective in resisting shear or torsion. For beams and columns in which perimeter hoops are either lap-spliced or have hooks that are not adequately anchored in the concrete core, transverse reinforcement shall be assumed not more than 50% effective in regions of moderate ductility demand and shall be assumed ineffective in regions of high ductility demand.

Shear friction strength shall be calculated according to ACI 318, taking into consideration the expected axial load due to gravity and earthquake effects. Where rehabilitation involves the addition of concrete requiring overhead work with dry-pack, the shear friction coefficient, $\mu$, shall be taken as equal to 70% of the value specified by ACI 318.

## 6.3.5 Development and Splices of Reinforcement

Development of straight bars, hooked bars, and lap-spliced bars shall be calculated according to the provisions of ACI 318 (ACI 2002), with the following modifications:

1. Deformed straight bars, hooked bars, and lap-spliced bars shall meet the development requirements of Chapter 12 of ACI 318 except requirements for lap splices shall be the same as those for straight development of bars in tension without consideration of lap splice classifications;
2. Where existing deformed straight bars, hooked bars, and lap-spliced bars do not meet the development requirements of (1) above, the capacity of existing reinforcement shall be calculated using Eq. 6-2:

$$f_s = \frac{l_b}{l_d} f_y \qquad \text{(Eq. 6-2)}$$

where $f_s$ = maximum stress that can be developed in the bar for the straight development, hook, or lap splice length $l_b$ provided; $f_y$ = yield strength of reinforcement; and $l_d$ = length required by Chapter 12 of ACI 318 for straight development, hook development, or lap splice length, except required splice lengths may be taken as straight bar development lengths in tension. Where transverse reinforcement is distributed along the development length with spacing not exceeding one-third of the effective depth of the component, it shall be permitted to assume the reinforcement retains the calculated maximum stress to high ductility demands. For larger spacings of transverse reinforcement, the developed stress shall be assumed to degrade from $f_s$ at a ductility demand or DCR equal to 1.0 to $0.2f_s$ at a ductility demand or DCR equal to 2.0;

3. Strength of deformed straight, discontinuous bars embedded in concrete sections or beam–column joints, with clear cover over the embedded bar not less than $3d_b$, shall be calculated according to Eq. 6-3:

$$f_s = \frac{2500}{d_b} l_e \le f_y \qquad \text{(Eq. 6-3)}$$

where

$f_s$ = maximum stress (in psi) that can be developed in an embedded bar having embedment length $l_e$ (in in.);

$d_b$ = diameter of embedded bar (in in.); and

$f_y$ = bar yield stress (in psi).

Where $f_s$ is less than $f_y$, and the calculated stress in the bar due to design loads equals or exceeds $f_s$, the maximum developed stress shall be assumed to degrade from $f_s$ to $0.2f_s$ at a ductility demand or DCR equal to 2.0. In beams with short bottom bar embedments into beam–column joints, flexural strength shall be calculated considering the stress limitation of Eq. 6-3;

4. For plain straight bars, hooked bars, and lap-spliced bars, development and splice lengths shall be taken as twice the values determined in accordance with ACI 318 unless other lengths are justified by approved tests or calculations considering only the chemical bond between the bar and the concrete;

5. Doweled bars added in seismic rehabilitation shall be assumed to develop yield stress where all the following conditions are satisfied:
   5.1. Drilled holes for dowel bars are cleaned with a stiff brush that extends the length of the hole; and
   5.2. Embedment length $l_e$ is not less than $10d_b$; and
   5.3. Minimum spacing of dowel bars is not less than $4l_e$ and minimum edge distance is not less than $2l_e$. Design values for dowel bars not satisfying these conditions shall be verified by test data. Field samples shall be obtained to

ensure design strengths are developed in accordance with Section 6.3.

**C6.3.5 Development and Splices of Reinforcement**

Development requirements in accordance with Chapter 12 of ACI 318 (ACI 2002) will be applicable to development of bars in all components. Chapter 21 of ACI 318 provides development requirements that are only intended for use in yielding components of reinforced concrete moment frames that comply with the cover and confinement provisions of Chapter 21. Chapter 12 permits reductions in lengths if minimum cover and confinement exist in an existing component.

Experimental tests by Melek and Wallace (2004) and Lynn (2001) have demonstrated that lap splices can achieve a higher flexural capacity than that calculated using the effective steel stress given in Eq. 6-2. The possibility of a shear failure in lap-spliced columns may go undetected if the flexural capacity is underestimated. Cho and Pincheira (2006) suggest an alternative model for the effective steel stress in lap-splice bars which provides a better estimate of the mean flexural strength observed in experimental tests.

For buildings constructed prior to 1950, the bond strength developed between reinforcing steel and concrete may be less than present-day strength. Current equations for development and splices of reinforcement account for mechanical bond due to deformations present in deformed bars in addition to chemical bond. The length required to develop plain bars will be much greater than that required for deformed bars, and will be more sensitive to cracking in the concrete. Procedures for testing and assessment of tensile lap splices and development length of plain reinforcing steel may be found in *Evaluation of Reinforcing Steel Systems in Old Reinforced Concrete Structures* (CRSI 1981).

**6.3.5.1 Square Reinforcing Bars**

Square reinforcing bars in a building shall be classified as either twisted or straight. The developed strength of twisted square bars shall be as specified for deformed bars in Section 6.3.5, using an effective diameter calculated based on the gross area of the square bar. Straight square bars shall be considered as plain bars, and the developed strength shall be as specified for plain bars in Section 6.3.5.

**6.3.6 Connections to Existing Concrete**

Connections used to connect two or more components shall be classified according to their anchoring systems as cast-in-place or as post-installed.

### 6.3.6.1 Cast-In-Place Systems

Component actions on cast-in-place connection systems, including shear forces, tension forces, bending moments, and prying actions, shall be considered force-controlled. Lower-bound strength of connections shall be ultimate values as specified in an approved building code with $\phi = 1.0$.

The capacity of anchors placed in areas where cracking is expected shall be reduced by a factor of 0.5.

### 6.3.6.2 Drilled-In Anchors

Component actions on drilled-in anchor connection systems shall be considered force-controlled. The lower-bound capacity of drilled-in anchors shall be mean minus one standard deviation of ultimate values published in approved test reports.

### 6.3.6.3 Quality Assurance

Connections between existing concrete components and new components added to rehabilitate the structure shall be subject to the quality assurance provisions specified in Section 2.7. The design professional shall specify the required inspection and testing of cast-in-place and post-installed anchors as part of the Quality Assurance Plan.

## 6.3.7 Rehabilitation

### 6.3.7.1 General Requirements

Upon determining that concrete components in an existing building are deficient for the selected Rehabilitation Objective, these components shall be rehabilitated or replaced or the structure shall be otherwise rehabilitated so that the component is no longer deficient for the selected Rehabilitation Objective. If replacement of the component is selected, the new component shall be designed in accordance with this standard and detailed and constructed in accordance with a building code approved by the authority having jurisdiction.

Rehabilitation measures shall be evaluated in accordance with the requirements of this standard, to assure that the completed rehabilitation achieves the selected Rehabilitation Objective. The effects of rehabilitation on stiffness, strength, and deformability shall be taken into account in an analytical model of the rehabilitated structure. The compatibility of new and existing components shall be checked at displacements consistent with the selected performance level.

Connections required between existing and new components shall satisfy the requirements of Section 6.3.6 and other requirements of this standard.

## 6.4 CONCRETE MOMENT FRAMES

### 6.4.1 Types of Concrete Moment Frames

Concrete moment frames shall be defined as elements comprising primarily horizontal framing components (beams and/or slabs), vertical framing components (columns) and joints connecting horizontal and vertical framing components. These elements resist lateral loads acting alone, or in conjunction with shear walls, braced frames, or other elements.

Frames that are cast monolithically, including monolithic concrete frames created by the addition of new material, shall meet the provisions of this section. Frames covered under this section include reinforced concrete beam–column moment frames, prestressed concrete beam–column moment frames, and slab–column moment frames. Precast concrete frames, concrete frames with infills, and concrete braced frames shall meet the provisions of Sections 6.5, 6.6, and 6.9, respectively.

### 6.4.1.1 Reinforced Concrete Beam-Column Moment Frames

Reinforced concrete beam–column moment frames shall satisfy the following conditions:

1. Framing components shall be beams (with or without slabs), columns, and their connections;
2. Beams and columns shall be of monolithic construction that provides for moment transfer between beams and columns; and
3. Primary reinforcement in components contributing to lateral load resistance shall be nonprestressed.

Special Moment Frames, Intermediate Moment Frames, and Ordinary Moment Frames as defined in ASCE 7 (ASCE 2005) shall be deemed to satisfy the above conditions. This classification shall include existing construction, new construction, and existing construction that has been rehabilitated.

### 6.4.1.2 Post-Tensioned Concrete Beam-Column Moment Frames

Post-tensioned concrete beam–column moment frames shall satisfy the following conditions:

1. Framing components shall be beams (with or without slabs), columns, and their connections;
2. Frames shall be of monolithic construction that provides for moment transfer between beams and columns; and
3. Primary reinforcement in beams contributing to lateral load resistance shall include post-tensioned reinforcement with or without mild reinforcement.

This classification shall include existing construction, new construction, and existing construction that has been rehabilitated.

### 6.4.1.3 Slab–Column Moment Frames

Slab–column moment frames shall satisfy the following conditions:

1. Framing components shall be slabs (with or without beams in the transverse direction), columns, and their connections;
2. Frames shall be of monolithic construction that provides for moment transfer between slabs and columns; and
3. Primary reinforcement in slabs contributing to lateral load resistance shall include nonprestressed reinforcement, prestressed reinforcement, or both.

This classification shall include frames intended as part of the lateral-force-resisting system and frames not intended as part of the lateral-force-resisting system in the original design, including existing construction, new construction, and existing construction that has been rehabilitated.

## 6.4.2 Reinforced Concrete Beam-Column Moment Frames

### 6.4.2.1 General Considerations

The analytical model for a beam-column frame element shall represent strength, stiffness, and deformation capacity of beams, columns, beam–column joints, and other components of the frame, including connections with other elements. Potential failure in flexure, shear, and reinforcement development at any section along the component length shall be considered. Interaction with other elements, including non-structural components, shall be included.

Analytical models representing a beam-column frame using line elements with properties concentrated at component centerlines shall be permitted. Where beam and column centerlines do not intersect, the effects of the eccentricity between centerlines of framing shall be taken into account. Where the centerline of the narrower component falls within the middle third of the adjacent framing component measured transverse to the framing direction; however, this eccentricity need not be considered. Where larger eccentricities occur, the effect shall be represented either by reductions in effective stiffness, strength, and deformation capacity, or by direct modeling of the eccentricity.

The beam–column joint in monolithic construction shall be represented as a stiff or rigid zone having horizontal dimensions equal to the column cross-sectional dimensions and vertical dimension equal to the beam depth, except that a wider joint shall be permitted where the beam is wider than the column and where justified by experimental evidence. The model of the connection between the columns and foundation shall be selected based on the details of the column–foundation connection and rigidity of the foundation–soil system in accordance with Section 6.12.

Action of the slab as a diaphragm interconnecting vertical components shall be represented. Action of the slab as a composite beam flange shall be considered in developing stiffness, strength, and deformation capacities of the beam component model, according to Section 6.3.1.3.

Inelastic action shall be restricted to those components and actions listed in Tables 6-7 through 6-9, except where it is demonstrated by experimental evidence and analysis that other inelastic action is acceptable for the selected performance level. Acceptance criteria shall be as specified in Section 6.4.2.4.

### 6.4.2.2 Stiffness for Analysis

*6.4.2.2.1 Linear Static and Dynamic Procedures* Beams shall be modeled considering flexural and shear stiffnesses, including the effect of the slab acting as a flange in monolithic construction. Columns shall be modeled considering flexural, shear, and axial stiffnesses. Joints shall be modeled as either stiff or rigid components. Effective stiffnesses shall be according to Section 6.3.1.2.

*6.4.2.2.2 Nonlinear Static Procedure* Nonlinear load–deformation relations shall follow the requirements of Section 6.3.1.2.

Beams and columns shall be modeled using concentrated plastic hinge models or distributed plastic hinge models. Other models whose behavior has been demonstrated to represent the behavior of reinforced concrete beam and column components subjected to lateral loading shall be permitted. The beam and column model shall be capable of representing inelastic response along the component length, except where it is shown by equilibrium that yielding is restricted to the component ends. Where nonlinear response is expected in a mode other than flexure, the model shall be established to represent these effects.

Monotonic load–deformation relations shall be according to the generalized load–deformation relation shown in Fig. 6-1, except that different relations shall be

**Table 6-7. Modeling Parameters and Numerical Acceptance Criteria for Nonlinear Procedures—Reinforced Concrete Beams**

| | | | Modeling Parameters[1] | | | Acceptance Criteria[1,2] | | | | |
| | | | | | | Plastic Rotations Angle, radians | | | | |
| | | | | | | | Performance Level | | | |
| | | | | | | | | Component Type | | |
| | | | Plastic Rotations Angle, radians | | Residual Strength Ratio | | Primary | | Secondary | |
| Conditions | | | $a$ | $b$ | $c$ | IO | LS | CP | LS | CP |
|---|---|---|---|---|---|---|---|---|---|---|
| **i. Beams Controlled by Flexure[3]** | | | | | | | | | | |
| $\dfrac{\rho - \rho'}{\rho_{bal}}$ | Transverse Reinforcement[4] | $\dfrac{V}{b_w d\sqrt{f_c'}}$ | | | | | | | | |
| $\leq 0.0$ | C | $\leq 3$ | 0.025 | 0.05 | 0.2 | 0.010 | 0.02 | 0.025 | 0.02 | 0.05 |
| $\leq 0.0$ | C | $\geq 6$ | 0.02 | 0.04 | 0.2 | 0.005 | 0.01 | 0.02 | 0.02 | 0.04 |
| $\geq 0.5$ | C | $\leq 3$ | 0.02 | 0.03 | 0.2 | 0.005 | 0.01 | 0.02 | 0.02 | 0.03 |
| $\geq 0.5$ | C | $\geq 6$ | 0.015 | 0.02 | 0.2 | 0.005 | 0.005 | 0.015 | 0.015 | 0.02 |
| $\leq 0.0$ | NC | $\leq 3$ | 0.02 | 0.03 | 0.2 | 0.005 | 0.01 | 0.02 | 0.02 | 0.03 |
| $\leq 0.0$ | NC | $\geq 6$ | 0.01 | 0.015 | 0.2 | 0.0015 | 0.005 | 0.01 | 0.01 | 0.015 |
| $\geq 0.5$ | NC | $\leq 3$ | 0.01 | 0.015 | 0.2 | 0.005 | 0.01 | 0.01 | 0.01 | 0.015 |
| $\geq 0.5$ | NC | $\geq 6$ | 0.005 | 0.01 | 0.2 | 0.0015 | 0.005 | 0.005 | 0.005 | 0.01 |
| **ii. Beams Controlled by Shear[3]** | | | | | | | | | | |
| Stirrup Spacing $\leq d/2$ | | | 0.0030 | 0.02 | 0.2 | 0.0015 | 0.0020 | 0.0030 | 0.01 | 0.02 |
| Stirrup Spacing $> d/2$ | | | 0.0030 | 0.01 | 0.2 | 0.0015 | 0.0020 | 0.0030 | 0.005 | 0.01 |
| **iii. Beams Controlled by Inadequate Development or Splicing along the Span[3]** | | | | | | | | | | |
| Stirrup Spacing $\leq d/2$ | | | 0.0030 | 0.02 | 0.0 | 0.0015 | 0.0020 | 0.0030 | 0.01 | 0.02 |
| Stirrup Spacing $> d/2$ | | | 0.0030 | 0.01 | 0.0 | 0.0015 | 0.0020 | 0.0030 | 0.005 | 0.01 |
| **iv. Beams Controlled by Inadequate Embedment into Beam–Column Joint[3]** | | | | | | | | | | |
| | | | 0.015 | 0.03 | 0.2 | 0.01 | 0.01 | 0.015 | 0.02 | 0.03 |

[1]Linear interpolation between values listed in the table shall be permitted.

[2]Primary and secondary component demands shall be within secondary component acceptance criteria where the full backbone curve is explicitly modeled including strength degradation and residual strength in accordance with Section 3.4.3.2.

[3]Where more than one of the conditions i, ii, iii, and iv occurs for a given component, use the minimum appropriate numerical value from the table.

[4]"C" and "NC" are abbreviations for conforming and nonconforming transverse reinforcement. A component is conforming if, within the flexural plastic hinge region, hoops are spaced at $\leq d/3$, and if, for components of moderate and high ductility demand, the strength provided by the hoops ($V_s$) is at least three-fourths of the design shear. Otherwise, the component is considered nonconforming.

permitted where verified by experiments. The overall load–deformation relation shall be established so that the maximum resistance is consistent with the design strength specifications of Sections 6.3.2 and 6.4.2.3.

For beams and columns, the generalized deformation in Fig. 6-1 shall be either the chord rotation or the plastic hinge rotation. For beam-column joints, the generalized deformation shall be shear strain. Values of the generalized deformation at points B, C, and D shall be derived from experiments or rational analyses, and shall take into account the interactions between flexure, axial load, and shear.

*C6.4.2.2.2 Nonlinear Static Procedure* Refer to Sections C6.3.1.2 and C6.4.2.3.1 for discussion of alternative modeling parameters for reinforced concrete columns.

**Table 6-8. Modeling Parameters and Numerical Acceptance Criteria for Nonlinear Procedures—Reinforced Concrete Columns**

| | | | Modeling Parameters[1] | | | Acceptance Criteria[1,2] | | | | |
| --- | --- | --- | --- | --- | --- | --- | --- | --- | --- | --- |
| | | | | | | Plastic Rotations Angle, radians | | | | |
| | | | | | | Performance Level | | | | |
| | | | | | Residual Strength Ratio | | Component Type | | | |
| | | | Plastic Rotations Angle, radians | | | | Primary | | Secondary | |
| Conditions | | | $a$ | $b$ | $c$ | IO | LS | CP | LS | CP |
| **i. Columns Controlled by Flexure[3]** | | | | | | | | | | |
| $\dfrac{P}{A_g f_c'}$ | Transverse Reinforcement[6] | $\dfrac{V}{b_w d \sqrt{f_c'}}$ | | | | | | | | |
| $\leq 0.1$ | C | $\leq 3$ | 0.02 | 0.03 | 0.2 | 0.005 | 0.015 | 0.02 | 0.02 | 0.03 |
| $\leq 0.1$ | C | $\geq 6$ | 0.016 | 0.024 | 0.2 | 0.005 | 0.012 | 0.016 | 0.016 | 0.024 |
| $\geq 0.4$ | C | $\leq 3$ | 0.015 | 0.025 | 0.2 | 0.003 | 0.012 | 0.015 | 0.018 | 0.025 |
| $\geq 0.4$ | C | $\geq 6$ | 0.012 | 0.02 | 0.2 | 0.003 | 0.01 | 0.012 | 0.013 | 0.02 |
| $\leq 0.1$ | NC | $\leq 3$ | 0.006 | 0.015 | 0.2 | 0.005 | 0.005 | 0.006 | 0.01 | 0.015 |
| $\leq 0.1$ | NC | $\geq 6$ | 0.005 | 0.012 | 0.2 | 0.005 | 0.005 | 0.005 | 0.008 | 0.012 |
| $\geq 0.4$ | NC | $\leq 3$ | 0.003 | 0.01 | 0.2 | 0.002 | 0.002 | 0.003 | 0.006 | 0.01 |
| $\geq 0.4$ | NC | $\geq 6$ | 0.002 | 0.008 | 0.2 | 0.002 | 0.002 | 0.002 | 0.005 | 0.008 |
| **ii. Columns Controlled by Shear[3,4]** | | | | | | | | | | |
| All cases[5] | | | — | — | — | — | — | — | 0.0030 | 0.0040 |
| **iii. Columns Controlled by Inadequate Development or Splicing along the Clear Height[3,4]** | | | | | | | | | | |
| Hoop Spacing $\leq d/2$ | | | 0.01 | 0.02 | 0.4 | 0.005 | 0.005 | 0.01 | 0.01 | 0.02 |
| Hoop Spacing $\leq d/2$ | | | 0.0 | 0.01 | 0.2 | 0.0 | 0.0 | 0.0 | 0.005 | 0.01 |
| **iv. Columns with Axial Loads Exceeding $0.70 P_o$[3,4]** | | | | | | | | | | |
| Conforming Hoops over the Entire Length | | | 0.015 | 0.025 | 0.02 | 0.0 | 0.005 | 0.01 | 0.01 | 0.02 |
| All Other Cases | | | 0.0 | 0.0 | 0.0 | 0.0 | 0.0 | 0.0 | 0.0 | 0.0 |

[1]Linear interpolation between values listed in the table shall be permitted.

[2]

[3]Where more than one of the conditions i, ii, iii, and iv occurs for a given component, use the minimum appropriate numerical value from the table.

[4]To qualify, columns must have transverse reinforcement consisting of hoops. Otherwise, actions shall be treated as force-controlled.

[5]For columns controlled by shear, see Section 6.4.2.4.2 for primary component acceptance criteria. Primary and secondary component demands shall be within secondary component acceptance criteria where the full backbone curve is explicitly modeled including strength degradation and residual strength in accordance with Section 3.4.3.2.

[6]"C" and "NC" are abbreviations for conforming and nonconforming transverse reinforcement. A component is conforming if, within the flexural plastic hinge region, hoops are spaced at $\leq d/3$, and if, for components of moderate and high ductility demand, the strength provided by the hoops ($V_s$) is at least three-fourths of the design shear. Otherwise, the component is considered nonconforming.

**Table 6-9. Modeling Parameters and Numerical Acceptance Criteria for Nonlinear Procedures—Reinforced Concrete Beam–Column Joints**

| | | | Modeling Parameters[1] | | | Acceptance Criteria[1,2] | | | | |
|---|---|---|---|---|---|---|---|---|---|---|
| | | | | | | Plastic Rotations Angle, radians | | | | |
| | | | | | | | Performance Level | | | |
| | | | | | Residual Strength Ratio | | | Component Type | | |
| | | | Plastic Rotations Angle, radians | | | | Primary | | Secondary | |
| | Conditions | | $a$ | $b$ | $c$ | IO | LS | CP | LS | CP |
| **i. Interior Joints[3,4]** | | | | | | | | | | |
| $\dfrac{P}{A_g f_c'}$ | Transverse Reinforcement[5] | $\dfrac{V}{V_n}$[4] | | | | | | | | |
| ≤ 0.1 | C | ≤ 1.2 | 0.015 | 0.03 | 0.2 | 0.0 | 0.0 | 0.0 | 0.02 | 0.03 |
| ≤ 0.1 | C | ≥ 1.5 | 0.015 | 0.03 | 0.2 | 0.0 | 0.0 | 0.0 | 0.015 | 0.02 |
| ≥ 0.4 | C | ≤ 1.2 | 0.015 | 0.025 | 0.2 | 0.0 | 0.0 | 0.0 | 0.015 | 0.025 |
| ≥ 0.4 | C | ≥ 1.5 | 0.015 | 0.02 | 0.2 | 0.0 | 0.0 | 0.0 | 0.015 | 0.02 |
| ≤ 0.1 | NC | ≤ 1.2 | 0.005 | 0.02 | 0.2 | 0.0 | 0.0 | 0.0 | 0.015 | 0.02 |
| ≤ 0.1 | NC | ≥ 1.5 | 0.005 | 0.015 | 0.2 | 0.0 | 0.0 | 0.0 | 0.01 | 0.015 |
| ≥ 0.4 | NC | ≤ 1.2 | 0.005 | 0.015 | 0.2 | 0.0 | 0.0 | 0.0 | 0.01 | 0.015 |
| ≥ 0.4 | NC | ≥ 1.5 | 0.005 | 0.015 | 0.2 | 0.0 | 0.0 | 0.0 | 0.01 | 0.015 |
| **ii. Other Joints[3,4]** | | | | | | | | | | |
| $\dfrac{P}{A_g f_c'}$ | Transverse Reinforcement[5] | $\dfrac{V}{V_n}$[4] | | | | | | | | |
| ≤ 0.1 | C | ≤ 1.2 | 0.01 | 0.02 | 0.2 | 0.0 | 0.0 | 0.0 | 0.015 | 0.02 |
| ≤ 0.1 | C | ≥ 1.5 | 0.01 | 0.015 | 0.2 | 0.0 | 0.0 | 0.0 | 0.01 | 0.015 |
| ≥ 0.4 | C | ≤ 1.2 | 0.01 | 0.02 | 0.2 | 0.0 | 0.0 | 0.0 | 0.015 | 0.02 |
| ≥ 0.4 | C | ≥ 1.5 | 0.01 | 0.015 | 0.2 | 0.0 | 0.0 | 0.0 | 0.01 | 0.015 |
| ≤ 0.1 | NC | ≥ 1.2 | 0.005 | 0.01 | 0.2 | 0.0 | 0.0 | 0.0 | 0.0075 | 0.01 |
| ≤ 0.1 | NC | > 1.5 | 0.005 | 0.01 | 0.2 | 0.0 | 0.0 | 0.0 | 0.0075 | 0.01 |
| ≥ 0.4 | NC | ≤ 1.2 | 0.0 | 0.0 | — | 0.0 | 0.0 | 0.0 | 0.005 | 0.0075 |
| ≥ 0.4 | NC | ≥ 1.5 | 0.0 | 0.0 | — | 0.0 | 0.0 | 0.0 | 0.005 | 0.0075 |

[1]Linear interpolation between values listed in the table shall be permitted.

[2]Primary and secondary component demands shall be within secondary component acceptance criteria where the full backbone curve is explicitly modeled including strength degradation and residual strength in accordance with Section 3.4.3.2.

[3]$P$ is the design axial force on the column above the joint calculated using limit-state analysis procedures in accordance with Section 6.4.2.4 and $A_g$ is the gross cross-sectional area of the joint.

[4]$V$ is the design shear force and $V_n$ is the shear strength for the joint. The shear strength shall be calculated according to Section 6.4.2.3.

[5]"C" and "NC" are abbreviations for conforming and nonconforming transverse reinforcement. A joint is conforming if hoops are spaced at ≤ $h_c/3$ within the joint. Otherwise, the component is considered nonconforming.

*6.4.2.2.3 Nonlinear Dynamic Procedure* For the NDP, the complete hysteretic behavior of each component shall be modeled using properties verified by experimental evidence. The use of the generalized load–deformation relation described by Fig. 6-1 to represent the envelope relation for the analysis shall be permitted. Unloading and reloading properties shall represent significant stiffness and strength degradation characteristics.

*6.4.2.3 Strength*

Component strengths shall be computed according to the general requirements of Sections 6.3.2 as modified in this section.

The maximum component strength shall be determined considering potential failure in flexure, axial load, shear, torsion, development, and other actions at all points along the length of the component under the actions of design gravity and earthquake load combinations.

*6.4.2.3.1 Columns* For columns, the shear strength, $V_n$, calculated according to Eq. 6-4 shall be permitted.

$$V_n = k \frac{A_v f_y d}{s} + \lambda k \left( \frac{6\sqrt{f_c'}}{M/Vd} \sqrt{1 + \frac{N_u}{6\sqrt{f_c'} A_g}} \right) 0.8 A_g$$

$$\text{(Eq. 6-4)}$$

where

$k = 1.0$ in regions where displacement ductility is less than or equal to 2, 0.7 in regions where displacement ductility is greater than or equal to 6, and varies linearly for displacement ductility between 2 and 6;

$\lambda = 0.75$ for lightweight aggregate concrete and 1.0 for normal weight aggregate concrete;

$N_u =$ axial compression force in pounds ($= 0$ for tension force);

$M/Vd =$ the largest ratio of moment to shear times effective depth under design loadings for the column but shall not be taken greater than 4 or less than 2;

$d =$ the effective depth; and

$A_g =$ the gross cross-sectional area of the column.

It shall be permitted to assume $d = 0.8h$, where $h$ is the dimension of the column in the direction of shear. Where axial force is calculated from the linear procedures of Chapter 3, the maximum compressive axial load for use in Eq. 6-4 shall be taken as equal to the value calculated using Eq. 3-4 considering design gravity load only, and the minimum compression axial load shall be calculated according to Eq. 3-18. Alternatively, limit analysis as specified in Section 3.4.2.1.2 shall be permitted to be used to determine design axial loads for use with the linear analysis procedures of Chapter 3. Alternative formulations for column strength that consider effects of reversed cyclic, inelastic deformations and that are verified by experimental evidence shall be permitted.

For columns satisfying the detailing and proportioning requirements of Chapter 21 of ACI 318 (ACI 2002), the shear strength equations of ACI 318 shall be permitted to be used.

For beam-column joints, the nominal cross-sectional area, $A_j$, shall be defined by a joint depth equal to the column dimension in the direction of framing and a joint width equal to the smallest of (1) the column width, (2) the beam width plus the joint depth, and (3) twice the smaller perpendicular distance from the longitudinal axis of the beam to the column side. Design forces shall be calculated based on development of flexural plastic hinges in adjacent framing members, including effective slab width, but need not exceed values calculated from design gravity and earthquake-load combinations. Nominal joint shear strength, $V_n$, shall be calculated according to the general procedures of ACI 318, as modified by Eq. 6-5:

$$Q_{CL} = V_n = \lambda \gamma \sqrt{f_c'} A_j \text{ psi} \qquad \text{(Eq. 6-5)}$$

in which $\lambda = 0.75$ for lightweight aggregate concrete and 1.0 for normal weight aggregate concrete, $A_j$ is the effective horizontal joint area with dimensions as defined above, and $\gamma$ is as defined in Table 6-10.

*C6.4.2.3.1 Columns* As discussed in C6.3.3, experimental evidence indicates that flexural deformability may be reduced as coexisting shear forces increase. As flexural ductility demands increase, shear capacity decreases, which may result in a shear failure before theoretical flexural deformation capacities are reached. Caution should be exercised when flexural deformation capacities are determined by calculation.

The modeling parameters and acceptance criteria in Table 6-8 are generally conservative, and may be relaxed based on experimental evidence. The design professional is referred to reports by Berry and Eberhard (2005); Elwood and Moehle (2005a; 2005b); Fardis and Biskinis (2003); Biskinis et al. (2004); Panagiotakos and Fardis (2001); Lynn et al. (1996); Sezen (2002); and Elwood and Moehle (2004) for further guidance regarding determination of model-

### Table 6-10. Values of $\gamma$ for Joint Strength Calculation

| | Value of $\gamma$ | | | | |
|---|---|---|---|---|---|
| $\rho''$ [1] | Interior Joint with Transverse Beams | Interior Joint without Transverse Beams | Exterior Joint with Transverse Beams | Exterior Joint without Transverse Beams | Knee Joint with or without Transverse Beams |
| $< 0.003$ | 12 | 10 | 8 | 6 | 4 |
| $\geq 0.003$ | 20 | 15 | 15 | 12 | 8 |

[1]$\rho'' =$ volumetric ratio of horizontal confinement reinforcement in the joint.

ing parameters and acceptance criteria for reinforced concrete columns.

Elwood and Moehle (2005a) have demonstrated based on experimental evidence that Eq. 6-4 does not provide a reliable estimate of the displacement ductility at shear failure.

### 6.4.2.4 Acceptance Criteria

*6.4.2.4.1 Linear Static and Dynamic Procedures* All actions shall be classified as being either deformation-controlled or force-controlled, as defined in Section 2.4.4. In primary components, deformation-controlled actions shall be restricted to flexure in beams (with or without slab) and columns. In secondary components, deformation-controlled actions shall be restricted to flexure in beams (with or without slab), plus restricted actions in shear and reinforcement development, as identified in Tables 6-11 through 6-13. All other actions shall be defined as being force-controlled actions.

Design actions on components shall be determined as prescribed in Chapter 3. Where the calculated DCR values exceed unity, the following design actions shall be determined using limit analysis principles as prescribed in Chapter 3: (1) moments, shears,

### Table 6-11. Numerical Acceptance Criteria for Linear Procedures— Reinforced Concrete Beams

| | | | | $m$-Factors[1] | | | | |
|---|---|---|---|---|---|---|---|---|
| | | | | Performance Level | | | | |
| | | | | | Component Type | | | |
| | | | | | Primary | | Secondary | |
| Conditions | | | IO | LS | CP | LS | CP |
| **i. Beams Controlled by Flexure[2]** | | | | | | | |
| $\dfrac{\rho - \rho'}{\rho_{bal}}$ | Transverse Reinforcement[3] | $\dfrac{V}{b_w d \sqrt{f_c'}}$ [4] | | | | | |
| $\leq 0.0$ | C | $\leq 3$ | 3 | 6 | 7 | 6 | 10 |
| $\leq 0.0$ | C | $\geq 6$ | 2 | 3 | 4 | 3 | 5 |
| $\geq 0.5$ | C | $\leq 3$ | 2 | 3 | 4 | 3 | 5 |
| $\geq 0.5$ | C | $\geq 6$ | 2 | 2 | 3 | 2 | 4 |
| $\leq 0.0$ | NC | $\leq 3$ | 2 | 3 | 4 | 3 | 5 |
| $\leq 0.0$ | NC | $\geq 6$ | 1.25 | 2 | 3 | 2 | 4 |
| $\geq 0.5$ | NC | $\leq 3$ | 2 | 3 | 3 | 3 | 4 |
| $\geq 0.5$ | NC | $\geq 6$ | 1.25 | 2 | 2 | 2 | 3 |
| **ii. Beams Controlled by Shear[2]** | | | | | | | |
| Stirrup Spacing $\leq d/2$ | | | 1.25 | 1.5 | 1.75 | 3 | 4 |
| Stirrup Spacing $> d/2$ | | | 1.25 | 1.5 | 1.75 | 2 | 3 |
| **iii. Beams Controlled by Inadequate Development or Splicing along the Span[2]** | | | | | | | |
| Stirrup Spacing $\leq d/2$ | | | 1.25 | 1.5 | 1.75 | 3 | 4 |
| Stirrup Spacing $> d/2$ | | | 1.25 | 1.5 | 1.75 | 2 | 3 |
| **iv. Beams Controlled by Inadequate Embedment into Beam–Column Joint[2]** | | | | | | | |
| | | | 2 | 2 | 3 | 3 | 4 |

[1]Linear interpolation between values listed in the table shall be permitted.

[2]Where more than one of the conditions i, ii, iii, and iv occurs for a given component, use the minimum appropriate numerical value from the table.

[3]"C" and "NC" are abbreviations for conforming and nonconforming transverse reinforcement. A component is conforming if, within the flexural plastic hinge region, hoops are spaced at $\leq d/3$, and if, for components of moderate and high ductility demand, the strength provided by the hoops ($V_s$) is at least three-fourths of the design shear. Otherwise, the component is considered nonconforming.

[4]$V$ is the design shear force calculated using limit-state analysis procedures in accordance with Section 6.4.2.4.1.

torsions, and development and splice actions corresponding to development of component strength in beams and columns; (2) joint shears corresponding to development of strength in adjacent beams and columns; and (3) axial load in columns and joints, considering likely plastic action in components above the level in question.

Design actions shall be compared with design strengths in accordance with Section 3.4.2.2. $m$-factors shall be selected from Tables 6-11 through 6-13. Those components that satisfy Eq. 3-20 or 3-21, as applicable, shall comply with the performance criteria.

Where the average DCR of columns at a level exceeds the average value of beams at the same level, and

**Table 6-12. Numerical Acceptance Criteria for Linear Procedures—
Reinforced Concrete Columns**

| | | | $m$-Factors[1] | | | | |
| | | | IO | Primary LS | Primary CP | Secondary LS | Secondary CP |
|---|---|---|---|---|---|---|---|
| **Conditions** | | | | | | | |
| **i. Columns Controlled by Flexure[2]** | | | | | | | |
| $\dfrac{P}{A_g f'_c}$ [4] | Transverse Reinforcement[3] | $\dfrac{V}{b_w d \sqrt{f'_c}}$ [5] | | | | | |
| $\le 0.1$ | C | $\le 3$ | 2 | 3 | 4 | 4 | 5 |
| $\le 0.1$ | C | $\ge 6$ | 2 | 2.4 | 3.2 | 3.2 | 4 |
| $\ge 0.4$ | C | $\le 3$ | 1.25 | 2 | 3 | 3 | 4 |
| $\ge 0.4$ | C | $\ge 6$ | 1.25 | 1.6 | 2.4 | 2.4 | 3.2 |
| $\le 0.1$ | NC | $\le 3$ | 2 | 2 | 3 | 2 | 3 |
| $\le 0.1$ | NC | $\ge 6$ | 2 | 2 | 2.4 | 1.6 | 2.4 |
| $\ge 0.4$ | NC | $\le 3$ | 1.25 | 1.5 | 2 | 1.5 | 2 |
| $\ge 0.4$ | NC | $\ge 6$ | 1.25 | 1.5 | 1.75 | 1.5 | 1.75 |
| **ii. Columns Controlled by Shear[2,6]** | | | | | | | |
| Hoop Spacing $\le d/2$, or $\dfrac{P}{A_g f'_c} \le 0.1$ | | | — | — | — | 2 | 3 |
| Other Cases | | | — | — | — | 1.5 | 2 |
| **iii. Columns Controlled by Inadequate Development or Splicing along the Clear Height[2,6]** | | | | | | | |
| Hoop Spacing $\le d/2$ | | | 1.25 | 1.5 | 1.75 | 3 | 4 |
| Hoop Spacing $> d/2$ | | | — | — | — | 2 | 3 |
| **iv. Columns with Axial Loads Exceeding $0.70 P_o$[2,6]** | | | | | | | |
| Conforming Hoops over the Entire Length | | | 1 | 1 | 2 | 2 | 2 |
| All Other Cases | | | — | — | — | 1 | 1 |

[1]Linear interpolation between values listed in the table shall be permitted.

[2]Where more than one of the conditions i, ii, iii, and iv occurs for a given component, use the minimum appropriate numerical value from the table.

[3]"C" and "NC" are abbreviations for conforming and nonconforming transverse reinforcement. A component is conforming if, within the flexural plastic hinge region, hoops are spaced at $\le d/3$, and if, for components of moderate and high ductility demand, the strength provided by the hoops ($Vs$) is at least three-fourths of the design shear. Otherwise, the component is considered nonconforming.

[4]$P$ is the design axial force in the member. Alternatively, use of axial loads determined based on a limit-state analysis shall be permitted.

[5]$V$ is the design shear force calculated using limit-state analysis procedures in accordance with Section 6.4.2.4.1.

[6]To qualify, columns must have transverse reinforcement consisting of hoops. Otherwise, actions shall be treated as force-controlled.

exceeds the greater of 1.0 and $m/2$ for all columns, the level shall be defined as a weak story element. For weak story elements, one of the following shall be satisfied:

1. The check of average DCR values at the level shall be repeated, considering all primary and secondary components at the level with a weak story element. If the average of the DCR values for vertical components exceeds the average value for horizontal components at the level, and exceeds 2.0, the structure shall be reanalyzed using a nonlinear proce-

dure, or the structure shall be rehabilitated to eliminate this deficiency;

2. The structure shall be reanalyzed using either the NSP or the NDP of Chapter 3; and

3. The structure shall be rehabilitated to remove the weak story element.

*6.4.2.4.2. Nonlinear Static and Dynamic Procedures*
Calculated component actions shall satisfy the requirements of Section 3.4.3.2. Where the generalized deformation is taken as rotation in the flexural plastic hinge

## Table 6-13. Numerical Acceptance Criteria for Linear Procedures—Reinforced Concrete Beam–Column Joints

| | | | | | $m$-Factors[1] | | | |
| | | | | | Performance Level | | | |
| | | | | | | Component Type | | |
| | | | | | | Primary[2] | | Secondary |
| | Conditions | | | IO | LS | CP | LS | CP |
|---|---|---|---|---|---|---|---|---|
| **i. Interior Joints[3,4]** | | | | | | | | |
| $\dfrac{P}{A_g f'_c}$ | Transverse Reinforcement[5] | $\dfrac{V}{V_n}$ | | | | | | |
| ≤ 0.1 | C | ≤ 1.2 | | — | — | — | 3 | 4 |
| ≤ 0.1 | C | ≥ 1.5 | | — | — | — | 2 | 3 |
| ≥ 0.4 | C | ≤ 1.2 | | — | — | — | 3 | 4 |
| ≥ 0.4 | C | ≥ 1.5 | | — | — | — | 2 | 3 |
| ≤ 0.1 | NC | ≤ 1.2 | | — | — | — | 2 | 3 |
| ≤ 0.1 | NC | ≥ 1.5 | | — | — | — | 2 | 3 |
| ≥ 0.4 | NC | ≤ 1.2 | | — | — | — | 2 | 3 |
| ≥ 0.4 | NC | ≥ 1.5 | | — | — | — | 2 | 3 |
| **ii. Other Joints[3,4]** | | | | | | | | |
| $\dfrac{P}{A_g f'_c}$ | Transverse Reinforcement[5] | $\dfrac{V}{V_n}$ | | | | | | |
| ≤ 0.1 | C | ≤ 1.2 | | — | — | — | 3 | 4 |
| ≤ 0.1 | C | ≥ 1.5 | | — | — | — | 2 | 3 |
| ≥ 0.4 | C | ≤ 1.2 | | — | — | — | 3 | 4 |
| ≥ 0.4 | C | ≥ 1.5 | | — | — | — | 2 | 3 |
| ≤ 0.1 | NC | ≤ 1.2 | | — | — | — | 2 | 3 |
| ≤ 0.1 | NC | ≥ 1.5 | | — | — | — | 2 | 3 |
| ≥ 0.4 | NC | ≤ 1.2 | | — | — | — | 1.5 | 2.0 |
| ≥ 0.4 | NC | ≥ 1.5 | | — | — | — | 1.5 | 2.0 |

[1]Linear interpolation between values listed in the table shall be permitted.

[2]For linear procedures, all primary joints shall be force-controlled; $m$-factors shall not apply.

[3]$P$ is the design axial force on the column above the joint calculated using limit-state analysis procedures in accordance with Section 6.4.2.4. $A_g$ is the gross cross-sectional area of the joint.

[4]$V$ is the design shear force and $V_n$ is the shear strength for the joint. The design shear force and shear strength shall be calculated according to Section 6.4.2.4.1 and Section 6.4.2.3, respectively.

[5]"C" and "NC" are abbreviations for conforming and nonconforming transverse reinforcements. A joint is conforming if hoops are spaced at ≤ $h_c/3$ within the joint. Otherwise, the component is considered nonconforming.

zone in beams and columns, the plastic hinge rotation capacities shall be as defined by Tables 6-7 and 6-8. Where the generalized deformation is shear distortion of the beam-column joint, shear angle capacities shall be as defined by Table 6-9. For columns designated as primary components and for which calculated design shear exceeds design shear strength, the permissible deformation for the Collapse Prevention Performance Level shall not exceed the deformation at which shear strength is calculated to be reached; the permissible deformation for the Life Safety Performance Level shall not exceed three-quarters of that value. Where inelastic action is indicated for a component or action not listed in these tables, the performance shall be deemed unacceptable. Alternative approaches or values shall be permitted where justified by experimental evidence and analysis.

*C6.4.2.4.2 Nonlinear Static and Dynamic Procedures*
Refer to Section C6.4.2.3.1 for discussion of alternative acceptance criteria for reinforced concrete columns.

### 6.4.2.5 Rehabilitation Measures

Concrete beam–column moment frame components that do not meet the acceptance criteria for the selected rehabilitation objective shall be rehabilitated. Rehabilitation measures shall meet the requirements of Section 6.3.7 and other provisions of this standard.

### C6.4.2.5 Rehabilitation Measures

The following rehabilitation measures may be effective in rehabilitating reinforced concrete beam–column moment frames:

1. **Jacketing existing beams, columns, or joints with new reinforced concrete, steel, or fiber wrap overlays.** The new materials should be designed and constructed to act compositely with the existing concrete. Where reinforced concrete jackets are used, the design should provide detailing to enhance ductility. Component strength should be taken to not exceed any limiting strength of connections with adjacent components. Jackets should be designed to provide increased connection strength and improved continuity between adjacent components;

2. **Post-tensioning existing beams, columns, or joints using external post-tensioned reinforcement.** Post-tensioned reinforcement should be unbonded within a distance equal to twice the effective depth from sections where inelastic action is expected. Anchorages should be located away from regions where inelastic action is anticipated,

and should be designed considering possible force variations due to earthquake loading;

3. **Modification of the element by selective material removal from the existing element.** Examples include: (1) where nonstructural components interfere with the frame, removing or separating the nonstructural components to eliminate the interference; (2) weakening, due to removal of concrete or severing of longitudinal reinforcement, to change response mode from a nonductile mode to a more ductile mode (e.g., weakening of beams to promote formation of a strong-column, weak-beam system); and (3) segmenting walls to change stiffness and strength;

4. **Improvement of deficient existing reinforcement details.** Removal of cover concrete for modification of existing reinforcement details should avoid damage to core concrete and the bond between existing reinforcement and core concrete. New cover concrete should be designed and constructed to achieve fully composite action with the existing materials;

5. **Changing the building system to reduce the demands on the existing element.** Examples include addition of supplementary lateral-force-resisting elements such as walls or buttresses, seismic isolation, and mass reduction; and

6. **Changing the frame element to a shear wall, infilled frame, or braced frame element by addition of new material.** Connections between new and existing materials should be designed to transfer the forces anticipated for the design load combinations. Where the existing concrete frame columns and beams act as boundary components and collectors for the new shear wall or braced frame, these should be checked for adequacy, considering strength, reinforcement development, and deformability. Diaphragms, including ties and collectors, should be evaluated and, if necessary, rehabilitated to ensure a complete load path to the new shear wall or braced frame element.

## 6.4.3 Post-Tensioned Concrete Beam-Column Moment Frames

### 6.4.3.1 General Considerations

The analytical model for a post-tensioned concrete beam–column frame element shall be established following the criteria specified in Section 6.4.2.1 for reinforced concrete beam–column moment frames. In addition to potential failure modes described in Section 6.4.2.1, the analysis model shall consider potential failure of tendon anchorages.

The analysis procedures described in Chapter 3 shall apply to frames with post-tensioned beams satisfying the following conditions:

1. The average prestress, $f_{pc}$, calculated for an area equal to the product of the shortest cross-sectional dimension and the perpendicular cross-sectional dimension of the beam, does not exceed the greater of 750 psi or $f'_c/12$ at locations of nonlinear action;
2. Prestressing tendons do not provide more than one-quarter of the strength for both positive moments and negative moments at the joint face; and
3. Anchorages for tendons are demonstrated to have performed satisfactorily for seismic loadings in compliance with the requirements of ACI 318 (ACI 2002). These anchorages occur outside hinging areas or joints, except in existing components where experimental evidence demonstrates that the connection will meet the performance objectives under design loadings.

Alternative procedures shall be provided where these conditions are not satisfied.

### 6.4.3.2 Stiffness

*6.4.3.2.1 Linear Static and Dynamic Procedures* Beams shall be modeled considering flexural and shear stiffnesses, including the effect of the slab acting as a flange in monolithic and composite construction. Columns shall be modeled considering flexural, shear, and axial stiffnesses. Joints shall be modeled as either stiff or rigid components. Effective stiffnesses shall be according to Section 6.3.1.2.

*6.4.3.2.2 Nonlinear Static Procedure* Nonlinear load–deformation relations shall comply with the requirements of Section 6.3.1.2 and the reinforced concrete frame requirements of Section 6.4.2.2.2.

Values of the generalized deformation at points *B*, *C*, and *D* in Fig. 6-1 shall be either derived from experiments or approved rational analyses, and shall take into account the interactions between flexure, axial load, and shear. Alternatively, where the generalized deformation is taken as rotation in the flexural plastic hinge zone, and where the three conditions of Section 6.4.3.1 are satisfied, beam plastic hinge rotation capacities shall be as defined by Table 6-7. Columns and joints shall be modeled as described in Section 6.4.2.2.

*6.4.3.2.3 Nonlinear Dynamic Procedure* For the NDP, the complete hysteretic behavior of each component

shall be modeled using properties verified by experimental evidence. The relation of Fig. 6-1 shall be taken to represent the envelope relation for the analysis. Unloading and reloading properties shall represent significant stiffness and strength degradation characteristics as influenced by prestressing.

### 6.4.3.3 Strength

Component strengths shall be computed according to the general requirements of Sections 6.3.2 and the additional requirements of Section 6.4.2.3. Effects of prestressing on strength shall be considered.

For deformation-controlled actions, prestress shall be assumed to be effective for the purpose of determining the maximum actions that may be developed associated with nonlinear response of the frame. For force-controlled actions, the effects on strength of prestress loss shall also be considered as a design condition, where these losses are possible under design load combinations including inelastic deformation reversals.

### 6.4.3.4 Acceptance Criteria

Acceptance criteria for post-tensioned concrete beam-column moment frames shall follow the criteria for reinforced concrete beam-column frames specified in Section 6.4.2.4.

Modeling parameters and acceptance criteria shall be based on Tables 6-7 through 6-9 and 6-11 through 6-13.

### 6.4.3.5 Rehabilitation Measures

Post-tensioned concrete beam–column moment frame components that do not meet the acceptance criteria for the selected Rehabilitation Objective shall be rehabilitated. Rehabilitation measures shall meet the requirements of Section 6.3.7 and other provisions of this standard.

### C6.4.3.5 Rehabilitation Measures

The rehabilitation measures described in C6.5.2.5 for reinforced concrete beam–column moment frames may also be effective in rehabilitating post-tensioned concrete beam–column moment frames.

## 6.4.4 Slab-Column Moment Frames

### 6.4.4.1 General Considerations

The analytical model for a slab-column frame element shall represent strength, stiffness, and deformation capacity of slabs, columns, slab–column connections, and other components of the frame. Potential failure in flexure, shear, shear-moment transfer, and reinforcement development at any section along the component length shall be considered. Interaction with other components, including nonstructural components, shall be included.

The analytical model that represents the slab–column frame, using either line elements with properties concentrated at component centerlines or a combination of line elements (to represent columns) and plate-bending elements (to represent the slab), based on any of the following approaches, shall be permitted:

1. An effective beam width model, in which the columns and slabs are represented by line elements that are rigidly interconnected at the slab–column joint. The effective width shall be calculated in accordance with the provisions of ACI 318 (ACI 2002);
2. An equivalent frame model in which the columns and slabs are represented by line elements that are interconnected by connection springs; and
3. A finite element model in which the columns are represented by line elements and the slab is represented by plate-bending elements.

In any model, the effects of changes in cross section, including slab openings, shall be considered.

The connection between the columns and foundation shall be modeled based on the details of the column–foundation connection and rigidity of the foundation–soil system.

Action of the slab as a diaphragm interconnecting vertical elements shall be represented.

In the design model, inelastic deformations in primary components shall be restricted to flexure in slabs and columns, plus nonlinear response in slab–column connections. Other inelastic deformations shall be permitted as part of the design in secondary components. Acceptance criteria shall be as specified in Section 6.4.4.4.

### 6.4.4.2 Stiffness

*6.4.4.2.1 Linear Static and Dynamic Procedures* Slabs shall be modeled considering flexural, shear, and torsional (in the slab adjacent to the column) stiffnesses. Columns shall be modeled considering flexural, shear, and axial stiffnesses. Joints shall be modeled as either stiff or rigid components. The effective stiffnesses of components shall be determined according to the general principles of Section 6.3.1.2, but adjustments on the basis of experimental evidence shall be permitted.

*6.4.4.2.2 Nonlinear Static Procedure* Nonlinear load–deformation relations shall comply with the requirements of Section 6.3.1.2.

Slabs and columns shall be modeled using concentrated plastic hinge models, distributed plastic hinge models, or other models whose behavior has been demonstrated to adequately represent behavior of reinforced concrete slab and column components subjected to lateral loading. The model shall be capable of representing inelastic response along the component length, except where it is shown by equilibrium that yielding is restricted to the component ends. Slab–column connections shall be modeled separately from the slab and column components in order to identify potential failure in shear and moment transfer; alternatively, the potential for connection failure shall be otherwise checked as part of the analysis. Where nonlinear response is expected in a mode other than flexure, the model shall be established to represent these effects.

Monotonic load–deformation relations shall be according to the generalized relation shown in Fig. 6-1, with definitions according to Section 6.4.2.2.2. The overall load–deformation relation shall be established so that the maximum resistance is consistent with the design strength specifications of Sections 6.3.2 and 6.4.4.3. Where the generalized deformation shown in Fig. 6-1 is taken as the flexural plastic hinge rotation for the column, the plastic hinge rotation capacities shall be as defined by Table 6-8. Where the generalized deformation shown in Fig. 6-1 is taken as the rotation of the slab–column connection, the plastic rotation capacities shall be as defined by Table 6-14.

*6.4.4.2.3 Nonlinear Dynamic Procedure* The requirements of Sections 6.3.2 and 6.4.2.2.3 for reinforced concrete beam-column moment frames shall apply to slab-column moment frames.

### 6.4.4.3 Strength

Component strengths shall be computed according to the general requirements of Sections 6.4.2, as modified in this section.

The maximum component strength shall be determined considering potential failure in flexure, axial load, shear, torsion, development, and other actions at all points along the length of the component under the actions of design gravity and earthquake load combinations. The strength of slab–column connections also shall be determined and incorporated in the analytical model.

**Table 6-14. Modeling Parameters and Numerical Acceptance Criteria for Nonlinear Procedures—Two-way Slabs and Slab–Column Connections**

| | | Modeling Parameters[1] | | | Acceptance Criteria[1,2] | | | | |
|---|---|---|---|---|---|---|---|---|---|
| | | | | | Plastic Rotations Angle, radians | | | | |
| | | | | | | Performance Level | | | |
| | | Plastic Rotations Angle, radians | | Residual Strength Ratio | | | Component Type | | |
| | | | | | | Primary | | Secondary | |
| Conditions | | $a$ | $b$ | $c$ | IO | LS | CP | LS | CP |
| i. Slabs Controlled by Flexure, and Slab–Column Connections[3] | | | | | | | | | |
| $\dfrac{V_g}{V_o}$[4] | Continuity Reinforcement[5] | | | | | | | | |
| $\leq 0.2$ | Yes | 0.02 | 0.05 | 0.2 | 0.01 | 0.015 | 0.02 | 0.03 | 0.05 |
| $\geq 0.4$ | Yes | 0.0 | 0.04 | 0.2 | 0.0 | 0.0 | 0.0 | 0.03 | 0.04 |
| $\leq 0.2$ | No | 0.02 | 0.02 | — | 0.01 | 0.015 | 0.02 | 0.015 | 0.02 |
| $\geq 0.4$ | No | 0.0 | 0.0 | — | 0.0 | 0.0 | 0.0 | 0.0 | 0.0 |
| ii. Slabs Controlled by Inadequate Development or Splicing along the Span[3] | | | | | | | | | |
| | | 0.0 | 0.02 | 0.0 | 0.0 | 0.0 | 0.0 | 0.01 | 0.02 |
| iii. Slabs Controlled by Inadequate Embedment into Slab–Column Joint[3] | | | | | | | | | |
| | | 0.015 | 0.03 | 0.2 | 0.01 | 0.01 | 0.015 | 0.02 | 0.03 |

[1]Linear interpolation between values listed in the table shall be permitted.

[2]Primary and secondary component demands shall be within secondary component acceptance criteria where the full backbone curve is explicitly modeled, including strength degradation and residual strength in accordance with Section 3.4.3.2.

[3]Where more than one of the conditions i, ii, and iii occurs for a given component, use the minimum appropriate numerical value from the table.

[4]$V_g$ = the gravity shear acting on the slab critical section as defined by ACI 318 (ACI 2002); $V_o$ = the direct punching shear strength as defined by ACI 318.

[5]Under the heading "Continuity Reinforcement," use "Yes" where at least one of the main bottom bars in each direction is effectively continuous through the column cage. Where the slab is post-tensioned, use "Yes" where at least one of the post-tensioning tendons in each direction passes through the column cage. Otherwise, use "No."

The flexural strength of a slab to resist moment due to lateral deformations shall be calculated as $M_{nCS}\,M_{gCS}$, where $M_{nCS}$ is the design flexural strength of the column strip and $M_{gCS}$ is the column strip moment due to gravity loads. $M_{gCS}$ shall be calculated according to the procedures of ACI 318 (ACI 2002) for the design gravity load specified in Chapter 3.

For columns, the evaluation of shear strength according to Section 6.4.2.3 shall be permitted.

Shear and moment transfer strength of the slab–column connection shall be calculated considering the combined action of flexure, shear, and torsion acting in the slab at the connection with the column. The procedures described below shall be permitted to satisfy this requirement.

For interior connections without transverse beams, and for exterior connections with moment about an axis perpendicular to the slab edge, the shear and moment transfer strength calculated as the minimum of the following strengths shall be permitted:

The strength calculated considering eccentricity of shear on a slab critical section due to combined shear and moment, as prescribed in ACI 318;

The moment transfer strength equal to $\Sigma M_n / \gamma_f$, where $\Sigma M_n$ = the sum of positive and negative flexural strengths of a section of slab between lines that are two and one-half slab or drop panel thicknesses ($2.5h$) outside opposite faces of the column or capital; $\gamma_f$ = the fraction of the moment resisted by flexure per ACI 318; and $h$ = slab thickness.

For moment about an axis parallel to the slab edge at exterior connections without transverse beams, where the shear on the slab critical section due to gravity loads does not exceed $0.75V_c$, or the shear at a corner support does not exceed $0.5V_c$, the moment transfer strength shall be permitted to be taken as equal to the flexural strength of a section of slab between lines that are a distance, $c_1$, outside opposite faces of the column or capital. $V_c$ is the direct punching shear strength defined by ACI 318.

### 6.4.4.4 Acceptance Criteria

*6.4.4.4.1 Linear Static and Dynamic Procedures* All component actions shall be classified as being either deformation-controlled or force-controlled, as defined in Section 2.4.4. In primary components, deformation-controlled actions shall be restricted to flexure in slabs and columns, and shear and moment transfer in slab–column connections. In secondary components, deformation-controlled actions shall also be permitted in shear and reinforcement development, as identified in Table 6-15. All other actions shall be defined as being force-controlled actions.

Design actions on components shall be determined as prescribed in Chapter 3. Where the calculated DCR values exceed unity, the following design actions shall be determined using limit analysis principles as prescribed in Chapter 3: (1) moments, shears, torsions, and development and splice actions corresponding to development of component strength in slabs and columns; and (2) axial load in columns, considering likely plastic action in components above the level in question.

Design actions shall be compared with design strengths in accordance with Section 3.4.2.2 and Tables 6-12 and 6-15. Those components that satisfy Eqs. 3-20 and 3-21 shall satisfy the performance criteria. Components that reach their design strengths shall be further evaluated according to this section to determine performance acceptability.

Where the average of the DCRs for columns at a level exceeds the average value for slabs at the same level, and exceeds the greater of 1.0 and $m/2$, the element shall be defined as a weak story element and shall be evaluated by the proce-dure for weak story elements described in Section 6.4.2.4.1.

### Table 6-15. Numerical Acceptance Criteria for Linear Procedures—Two-way Slabs and Slab–Column Connections

| | | $m$-Factors[1] | | | | |
| | | Performance Level | | | | |
| | | | Component Type | | | |
| | | | Primary | | Secondary | |
| Conditions | | IO | LS | CP | LS | CP |
|---|---|---|---|---|---|---|
| i. Slabs Controlled by Flexure, and Slab–Column Connections[2] | | | | | | |
| $\dfrac{V_g}{V_o}$[3] | Continuity Reinforcement[4] | | | | | |
| ≤ 0.2 | Yes | 2 | 2 | 3 | 3 | 4 |
| ≥ 0.4 | Yes | 1 | 1 | 1 | 2 | 3 |
| ≤ 0.2 | No | 2 | 2 | 3 | 2 | 3 |
| ≥ 0.4 | No | 1 | 1 | 1 | 1 | 1 |
| ii. Slabs Controlled by Inadequate Development or Splicing along the Span[2] | | | | | | |
| | | — | — | — | 3 | 4 |
| iii. Slabs Controlled by Inadequate Embedment into Slab–Column Joint[2] | | | | | | |
| | | 2 | 2 | 3 | 3 | 4 |

[1]Linear interpolation between values listed in the table shall be permitted.
[2]Where more than one of the conditions i, ii, and iii occurs for a given component, use the minimum appropriate numerical value from the table.
[3]$V_g$ = the gravity shear acting on the slab critical section as defined by ACI 318 (ACI 2002); $V_o$ = the direct punching shear strength as defined by ACI 318.
[4]Under the heading "Continuity Reinforcement," use "Yes" where at least one of the main bottom bars in each direction is effectively continuous through the column cage. Where the slab is post-tensioned, use "Yes" where at least one of the post-tensioning tendons in each direction passes through the column cage. Otherwise, use "No."

**6.4.4.4.2 Nonlinear Static and Dynamic Procedures** In the design model, inelastic response shall be restricted to those components and actions listed in Tables 6-8 and 6-14, except where it is demonstrated by experimental evidence and analysis that other inelastic action is acceptable for the selected performance levels.

Calculated component actions shall satisfy the requirements of Chapter 3. Maximum permissible inelastic deformations shall be as listed in Tables 6-8 and 6-14. Where inelastic action is indicated for a component or action not listed in these tables, the performance shall be deemed unacceptable. Alternative approaches or values shall be permitted where justified by experimental evidence and analysis.

### 6.4.4.5 Rehabilitation Measures

Reinforced concrete slab–column moment frame components that do not meet the acceptance criteria for the selected Rehabilitation Objective shall be rehabilitated. Rehabilitation measures shall meet the requirements of Section 6.3.7 and other provisions of this standard.

### C6.4.4.5 Rehabilitation Measures

The rehabilitation measures described in C6.5.2.5 for reinforced concrete beam–column moment frames may also be effective in rehabilitating reinforced concrete slab-column moment frames.

## 6.5 PRECAST CONCRETE FRAMES

### 6.5.1 Types of Precast Concrete Frames

Precast concrete frames shall be defined as those elements that are constructed from individually made beams and columns that are assembled to create gravity-load-carrying systems. These systems shall include those that are considered in design to resist design lateral loads, and those that are considered in design as secondary elements that do not resist design lateral loads but must resist the effects of deformations resulting from design lateral loads.

### 6.5.1.1 Precast Concrete Frames Expected to Resist Lateral Load

Frames of this classification shall be assembled using either reinforcement and wet concrete or dry joints (connections are made by bolting, welding, post-tensioning, or other similar means) in a way that results in significant lateral force resistance in the framing element. Frames of this classification resist lateral loads either acting alone, or acting in conjunc-

tion with shear walls, braced frames, or other lateral-load-resisting elements.

### C6.5.1.1 Precast Concrete Frames Expected to Resist Lateral Load

These systems are recognized and accepted by FEMA 450 (FEMA 2004) and are based on ACI 318 (ACI 2002), which specifies safety and serviceability levels expected from precast concrete frame construction. In the referenced documents, precast frames are classified not by the method of construction (wet or dry joints) but by the expected behavior resulting from the detailing used. In addition to recognizing varying levels of ductile performance as a result of overall frame detailing, ACI 318 (in Section 21.6) acknowledges three types of unit-to-unit connections that can result in the highest level of performance. Such connections are either "strong" or "ductile" as defined in Sections 21.1 and 21.6 of ACI 318, or have demonstrated acceptable performance where tested in accordance with ACI T1.1-01 (ACI 2001).

### 6.5.1.2 Precast Concrete Frames Not Expected to Resist Lateral Load Directly

Frames of this classification shall be assembled using dry joints in a way that does not result in significant lateral force resistance in the framing element. Shear walls, braced frames, or moment frames provide the entire lateral load resistance, with the precast concrete frame system deforming in a manner that is compatible with the structure as a whole.

### 6.5.2 Precast Concrete Frames Expected to Resist Lateral Load

### 6.5.2.1 General Considerations

The analytical model for a frame element of this classification shall represent strength, stiffness, and deformation capacity of beams, columns, beam–column joints, and other components of the frame. Potential failure in flexure, shear, and reinforcement development at any section along the component length shall be considered. Interaction with other components, including nonstructural components, shall be included. All other considerations of Section 6.4.2.1 shall be taken into account. In addition, the effects of shortening due to creep, and other effects of prestressing and post-tensioning on member behavior, shall be evaluated. Where dry joints are used in assembling the precast system, consideration shall be given to the effect of those joints on overall behavior. Where connections yield under design lateral loads, the analysis model shall take this into account.

### 6.5.2.2 Stiffness

Stiffness for analysis shall be as defined in Section 6.4.2.2. The effects of prestressing shall be considered where computing the effective stiffness values using Table 6-5. Flexibilities associated with connections shall be included in the analytical model.

### 6.5.2.3 Strength

Component strength shall be computed according to the requirements of Section 6.4.2.3, with the additional requirement that the following effects be included in the analysis:

1. Effects of prestressing that are present, including, but not limited to, reduction in rotation capacity, secondary stresses induced, and amount of effective prestress force remaining;
2. Effects of construction sequence, including the possibility of construction of the moment connections occurring after portions of the structure are subjected to dead loads; and
3. Effects of restraint due to interaction with interconnected wall or brace components.

Effects of connection strength shall be considered in accordance with Section 6.3.6.

### 6.5.2.4 Acceptance Criteria

Acceptance criteria for precast concrete frames expected to resist lateral load shall be as specified in Section 6.4.2.4, except that the factors defined in Section 6.4.2.3 shall also be considered. Connections shall comply with the requirements of Section 6.3.6.

### 6.5.2.5 Rehabilitation Measures

Precast concrete frame components that do not meet the acceptance criteria for the selected Rehabilitation Objective shall be rehabilitated. Rehabilitation measures shall meet the requirements of Section 6.3.7 and other provisions of this standard.

### C6.5.2.5 Rehabilitation Measures

The rehabilitation measures described in C6.5.2.5 for reinforced concrete beam–column moment frames may also be effective in rehabilitating precast concrete moment frames. When installing new components or materials to the existing system, existing prestressing strands should be protected.

### 6.5.3 Precast Concrete Frames Not Expected to Resist Lateral Loads Directly

### 6.5.3.1 General Considerations

The analytical model for precast concrete frames that are not expected to resist lateral loads directly shall comply with the requirements of Section 6.5.2.1 and shall include the effects of deformations that are calculated to occur under the design earthquake loadings.

### 6.5.3.2 Stiffness

The analytical model shall include either realistic lateral stiffness of these frames to evaluate the effects of deformations under lateral loads or, if the lateral stiffness is ignored in the analytical model, the effects of calculated building drift on these frames shall be evaluated separately. The analytical model shall consider the negative effects of connection stiffness on component response where that stiffness results in actions that may cause component failure.

### C6.5.3.2 Stiffness

The stiffness used in the analysis should consider possible resistance that may develop under lateral deformation. In some cases, it may be appropriate to assume zero lateral stiffness. However, the Northridge earthquake graphically demonstrated that there are few instances where the precast column can be considered to be completely pinned top and bottom and, as a consequence, not resist any shear from building drift. Several parking structures collapsed as a result of this defect. Conservative assumptions should be made.

### 6.5.3.3 Strength

Component strength shall be computed according to the requirements of Section 6.5.2.3. All components shall have sufficient strength and ductility to transmit induced forces from one member to another and to the designated lateral-force-resisting system.

### 6.5.3.4 Acceptance Criteria

Acceptance criteria for components in precast concrete frames not expected to resist lateral loads directly shall be as specified in Section 6.5.3.4. All moments, shear forces, and axial loads induced through the deformation of the structural system shall be checked using appropriate criteria in the referenced section.

### 6.5.3.5 Rehabilitation Measures

Precast concrete frame components that do not meet the acceptance criteria for the selected Rehabilitation Objective shall be rehabilitated. Rehabilitation measures shall meet the requirements of Section 6.3.7 and other provisions of this standard.

### C6.5.3.5 Rehabilitation Measures

The rehabilitation measures described in C6.4.2.5 for reinforced concrete beam–column moment frames may also be effective in rehabilitating precast concrete frames not expected to resist lateral loads directly. When installing new components or materials to the existing system, existing prestressing strands should be protected.

## 6.6 CONCRETE FRAMES WITH INFILLS

### 6.6.1 Types of Concrete Frames with Infills

Concrete frames with infills are elements with complete gravity-load-carrying concrete frames infilled with masonry or concrete, constructed in such a way that the infill and the concrete frame interact when subjected to vertical and lateral loads.

Isolated infills are infills isolated from the surrounding frame complying with the minimum gap requirements specified in Section 7.5.1. If all infills in a frame are isolated infills, the frame shall be analyzed as an isolated frame according to provisions given elsewhere in this chapter, and the isolated infill panels shall be analyzed according to the requirements of Chapter 7.

The provisions of Section 6.6 shall apply to concrete frames with existing infills, frames that are rehabilitated by addition or removal of material, and concrete frames that are rehabilitated by the addition of new infills.

### 6.6.1.1 Types of Frames

The provisions of Section 6.6 shall apply to concrete frames as defined in Sections 6.4, 6.5, and 6.9, where those frames interact with infills.

### 6.6.1.2 Masonry Infills

The provisions of Section 6.6 shall apply to masonry infills as defined in Chapter 7, where those infills interact with concrete frames.

### 6.6.1.3 Concrete Infills

The provisions of Section 6.6 shall apply to concrete infills that interact with concrete frames, where

the infills were constructed to fill the space within the bay of a complete gravity frame without special provision for continuity from story to story. The concrete of the infill shall be evaluated separately from the concrete of the frame.

### C6.6.1.3 Concrete Infills

The construction of concrete-infilled frames is very similar to that of masonry-infilled frames, except that the infill is of concrete instead of masonry units. In older existing buildings, the concrete infill commonly contains nominal reinforcement, which is unlikely to extend into the surrounding frame. The concrete is likely to be of lower quality than that used in the frame, and should be investigated separately from investigations of the frame concrete.

### 6.6.2 Concrete Frames with Masonry Infills

### 6.6.2.1 General Considerations

The analytical model for a concrete frame with masonry infills shall represent strength, stiffness, and deformation capacity of beams, slabs, columns, beam–column joints, masonry infills, and all connections and components of the element. Potential failure in flexure, shear, anchorage, reinforcement development, or crushing at any section shall be considered. Interaction with nonstructural components shall be included.

For a concrete frame with masonry infill resisting lateral forces within its plane, modeling of the response using a linear elastic model shall be permitted provided that the infill will not crack when subjected to design lateral forces. If the infill will not crack when subjected to design lateral forces, modeling the assemblage of frame and infill as a homogeneous medium shall be permitted.

For a concrete frame with masonry infills that will crack when subjected to design lateral forces, modeling of the response using a diagonally braced frame model, in which the columns act as vertical chords, the beams act as horizontal ties, and the infill acts as an equivalent compression strut, shall be permitted. Requirements for the equivalent compression strut analogy shall be as specified in Chapter 7.

Frame components shall be evaluated for forces imparted to them through interaction of the frame with the infill, as specified in Chapter 7. In frames with full-height masonry infills, the evaluation shall include the effect of strut compression forces applied to the column and beam, eccentric from the beam–column joint. In frames with partial-height masonry infills, the evaluation shall include the reduced effective length of the columns above the infilled portion of the bay.

### C6.6.2.1 General Considerations

The design professional is referred to FEMA 274 (FEMA 1997) and FEMA 306 (FEMA 1998) for additional information regarding the behavior of masonry infills.

### 6.6.2.2 Stiffness

*6.6.2.2.1 Linear Static and Dynamic Procedures* In frames having infills in some bays and no infill in other bays, the restraint of the infill shall be represented as described in Section 6.6.2.1, and the non-infilled bays shall be modeled as frames as specified in appropriate portions of Sections 6.4, 6.5, and 6.9. Where infills create a discontinuous wall, the effects of the discontinuity on overall building performance shall be evaluated. Effective stiffnesses shall be in accordance with Section 6.3.1.2.

*6.6.2.2.2 Nonlinear Static Procedure* Nonlinear load–deformation relations for use in analysis by NSP shall follow the requirements of Section 6.3.1.2.2.

Modeling beams and columns using nonlinear truss elements shall be permitted in infilled portions of the frame. Beams and columns in non-infilled portions of the frame shall be modeled using the relevant specifications of Sections 6.4, 6.5, and 6.9. The model shall be capable of representing inelastic response along the component lengths.

Monotonic load–deformation relations shall be according to the generalized relation shown in Fig. 6-1, except different relations shall be permitted where verified by tests. Numerical quantities in Fig. 6-1 shall be derived from tests or by analyses procedures as specified in Chapter 2, and shall take into account the interactions between frame and infill components. Alternatively, the following procedure shall be permitted for monolithic reinforced concrete frames:

1. For beams and columns in non-infilled portions of frames, where the generalized deformation is taken as rotation in the flexural plastic hinge zone, the

### Table 6-16. Modeling Parameters and Numerical Acceptance Criteria for Nonlinear Procedures— Reinforced Concrete Infilled Frames

| | Modeling Parameters[1] | | | Acceptance Criteria[1,2] | | | | |
| | | | | Total Strain | | | | |
| | | | | | Performance Level | | | |
| | Plastic Rotations Angle, radians | | Residual Strength Ratio | | | Component Type | | |
| | | | | | Primary | | Secondary | |
| Conditions | $d$ | $e$ | $c$ | IO | LS | CP | LS | CP |
|---|---|---|---|---|---|---|---|---|
| i. Columns Modeled as Compression Chords[3] | | | | | | | | |
| Columns Confined along Entire Length[4] | 0.02 | 0.04 | 0.4 | 0.003 | 0.015 | 0.020 | 0.03 | 0.04 |
| All Other Cases | 0.003 | 0.01 | 0.2 | 0.002 | 0.002 | 0.003 | 0.01 | 0.01 |
| ii. Columns Modeled as Tension Chords[3] | | | | | | | | |
| Columns with Well-Confined Splices, or No Splices | 0.05 | 0.05 | 0.0 | 0.01 | 0.03 | 0.04 | 0.04 | 0.05 |
| All Other Cases | See Note 5 | 0.03 | 0.2 | See Note 5 | | | 0.02 | 0.03 |

[1]Interpolation shall not be permitted.

[2]Primary and secondary component demands shall be within secondary component acceptance criteria where the full backbone curve is explicitly modeled including strength degradation and residual strength in accordance with Section 3.4.3.2.

[3]If load reversals will result in both conditions i and ii applying to a single column, both conditions shall be checked.

[4]A column shall be permitted to be considered to be confined along its entire length where the quantity of hoops along the entire story height including the joint is equal to three-quarters of that required by ACI 318 (ACI 2002) for boundary components of concrete shear walls. The maximum longitudinal spacing of sets of hoops shall not exceed either $h/3$ or $8d_b$.

[5]Potential for splice failure shall be evaluated directly to determine the modeling and acceptance criteria. For these cases, refer to the generalized procedure of Sections 6.3.2. For primary components, Collapse Prevention Performance Level shall be defined as the deformation at which strength degradation begins. Life Safety Performance Level shall be taken as three-quarters of that value.

plastic hinge rotation capacities shall be as defined by Table 6-18;

2. For masonry infills, the generalized deformations and control points shall be as defined in Chapter 7; and

3. For beams and columns in infilled portions of frames, where the generalized deformation is taken as elongation or compression displacement of the beams or columns, the tension and compression strain capacities shall be as specified in Table 6-16.

*6.6.2.2.3 Nonlinear Dynamic Procedure* Nonlinear load–deformation relations for use in analysis by NDP shall model the complete hysteretic behavior of each component using properties verified by tests. Unloading and reloading properties shall represent stiffness and strength degradation characteristics.

### 6.6.2.3 Strength

Strengths of reinforced concrete components shall be calculated according to the general requirements of Sections 6.3.2, as modified by other specifications of this chapter. Strengths of masonry infills shall be calculated according to the requirements of Chapter 7. Strength calculations shall consider:

1. Limitations imposed by beams, columns, and joints in non-infilled portions of frames;
2. Tensile and compressive capacity of columns acting as boundary components of infilled frames;
3. Local forces applied from the infill to the frame;
4. Strength of the infill; and
5. Connections with adjacent components.

### 6.6.2.4 Acceptance Criteria

*6.6.2.4.1 Linear Static and Dynamic Procedures* All component actions shall be classified as either deformation-controlled or force-controlled, as defined in Section 2.4.4. In primary components, deformation-controlled actions shall be restricted to flexure and axial actions in beams, slabs, and columns, and lateral deformations in masonry infill panels. In secondary components, deformation-controlled actions shall be restricted to those actions identified for the isolated frame in Sections 6.4, 6.5, and 6.9, as appropriate, and for the masonry infill in Section 7.4.

Design actions shall be determined as prescribed in Chapter 3. Where calculated DCR values exceed unity, the following design actions shall be determined using limit analysis principles as prescribed in Chapter 3: (1) moments, shears, torsions, and development and splice actions corresponding to development of

component strength in beams, columns, or masonry infills; and (2) column axial load corresponding to development of the flexural capacity of the infilled frame acting as a cantilever wall.

Design actions shall be compared with design strengths in accordance with Section 3.4.2.2.

Values of *m*-factors shall be as specified in Section 7.4.2.3 for masonry infills; applicable portions of Sections 6.4, 6.5, and 6.9 for concrete frames; and Table 6-17 for columns modeled as tension and compression chords. Those components that have design actions less than design strengths shall be assumed to satisfy the performance criteria for those components.

*6.6.2.4.2 Nonlinear Static and Dynamic Procedures* In the design model, inelastic response shall be restricted to those components and actions that are permitted for isolated frames as specified in Sections 6.4, 6.5, and 6.9, as well as for masonry infills as specified in Section 7.4.

Calculated component actions shall satisfy the requirements of Section 3.4.3.2, and shall not exceed the numerical values listed in Table 6-16, the relevant tables for isolated frames given in Sections 6.4, 6.5, and 6.9, and the relevant tables for masonry infills given in Chapter 7. Component actions not listed in Tables 6-7 through 6-9 shall be treated as force-controlled. Alternative approaches or values shall be permitted where justified by experimental evidence and analysis.

### 6.6.2.5 Rehabilitation Measures

Concrete frames with masonry infill that do not meet the acceptance criteria for the selected Rehabilitation Objective shall be rehabilitated. Rehabilitation measures shall meet the requirements of Section 6.3.7 and other provisions of this standard.

### C6.6.2.5 Rehabilitation Measures

The rehabilitation measures described in relevant commentary of Sections 6.4, 6.5, and 6.9 for isolated frames, and rehabilitation measures described in relevant commentary or Section 7.4 for masonry infills, may also be effective in rehabilitating concrete frames with masonry infills. The design professional is referred to FEMA 308 (FEMA 1998) for further information in this regard. In addition, the following rehabilitation measures may be effective in rehabilitating concrete frames with infills:

1. **Post-tensioning existing beams, columns, or joints using external post-tensioned reinforcement.** Vertical post-tensioning may be effective in increasing tensile capacity of columns acting as

**Table 6-17. Numerical Acceptance Criteria for Linear Procedures—Reinforced Concrete Infilled Frames**

| Conditions | m-Factors[1] | | | | |
| | | Performance Level | | | |
| | | | Component Type | | |
| | | Primary | | Secondary | |
| | IO | LS | CP | LS | CP |
|---|---|---|---|---|---|
| i. Columns Modeled as Compression Chords[2] | | | | | |
| Columns Confined along Entire Length[3] | 1 | 3 | 4 | 4 | 5 |
| All Other Cases | 1 | 1 | 1 | 1 | 1 |
| ii. Columns Modeled as Tension Chords[2] | | | | | |
| Columns with Well-Confined Splices, or No Splices | 3 | 4 | 5 | 5 | 6 |
| All Other Cases | 1 | 2 | 2 | 3 | 4 |

[1]Interpolation shall not be permitted.

[2]If load reversals will result in both conditions i and ii applying to a single column, both conditions shall be checked.

[3]A column may be considered to be confined along its entire length where the quantity of hoops along the entire story height including the joint is equal to three-quarters of that required by ACI 318 (ACI 2002) for boundary components of concrete shear walls. The maximum longitudinal spacing of sets of hoops shall not exceed either $h/3$ or $8d_b$.

boundary zones. Anchorages should be located away from regions where inelastic action is anticipated, and should be designed considering possible force variations due to earthquake loading;

2. **Modification of the element by selective material removal from the existing element.** Either the infill should be completely removed from the frame, or gaps should be provided between the frame and the infill. In the latter case, the gap requirements of Chapter 7 should be satisfied; and

3. **Changing the building system to reduce the demands on the existing element.** Examples include the addition of supplementary lateral-force-resisting elements such as walls, steel braces, or buttresses; seismic isolation; and mass reduction.

### 6.6.3 Concrete Frames with Concrete Infills

#### 6.6.3.1 General Considerations

The analytical model for a concrete frame with concrete infills shall represent the strength, stiffness, and deformation capacity of beams, slabs, columns, beam-column joints, concrete infills, and all connections and components of the elements. Potential failure in flexure, shear, anchorage, reinforcement development, or crushing at any section shall be considered. Interaction with nonstructural components shall be included.

The analytical model shall be established considering the relative stiffness and strength of the frame and the infill, as well as the level of deformations and associated damage. For low deformation levels, and for cases where the frame is relatively flexible, the infilled frame shall be permitted to be modeled as a shear wall, with openings modeled where they occur. In other cases, the frame–infill system shall be permitted to be modeled using a braced-frame analogy such as that described for concrete frames with masonry infills in Section 6.6.2.

Frame components shall be evaluated for forces imparted to them through interaction of the frame with the infill as specified in Chapter 7. In frames with full-height infills, the evaluation shall include the effect of strut compression forces applied to the column and beam eccentric from the beam–column joint. In frames with partial-height infills, the evaluation shall include the reduced effective length of the columns above the infilled portion of the bay.

In frames having infills in some bays and no infills in other bays, the restraint of the infill shall be represented as described in this section, and the non-infilled bays shall be modeled as frames as specified in appropriate portions of Sections 6.4, 6.5, and 6.9. Where infills create a discontinuous wall, the effects of the discontinuity on overall building performance shall be evaluated.

**Table 6-18. Modeling Parameters and Numerical Acceptance Criteria for Nonlinear Procedures— Reinforced Concrete Shear Walls and Associated Components Controlled by Flexure**

| Conditions | | | Plastic Hinge Rotation (radians) | | Residual Strength Ratio | Acceptable Plastic Hinge Rotation[1,2] (radians) | | | | |
|---|---|---|---|---|---|---|---|---|---|---|
| | | | | | | Performance Level | | | | |
| | | | | | | Component Type | | | | |
| | | | | | | Primary | | | Secondary | |
| | | | $a$ | $b$ | $c$ | IO | LS | CP | LS | CP |
| **i. Shear Walls and Wall Segments** | | | | | | | | | | |
| $\dfrac{(A_s - A_s')f_y + P}{t_w l_w f_c'}$ | $\dfrac{V}{t_w l_w \sqrt{f_c'}}$ | Confined Boundary[3] | | | | | | | | |
| ≤ 0.1 | ≤ 3 | Yes | 0.015 | 0.020 | 0.75 | 0.005 | 0.010 | 0.015 | 0.015 | 0.020 |
| ≤ 0.1 | ≥ 6 | Yes | 0.010 | 0.015 | 0.40 | 0.004 | 0.008 | 0.010 | 0.010 | 0.015 |
| ≥ 0.25 | ≤ 3 | Yes | 0.009 | 0.012 | 0.60 | 0.003 | 0.006 | 0.009 | 0.009 | 0.012 |
| ≥ 0.25 | ≥ 6 | Yes | 0.005 | 0.010 | 0.30 | 0.0015 | 0.003 | 0.005 | 0.005 | 0.010 |
| ≤ 0.1 | ≤ 3 | No | 0.008 | 0.015 | 0.60 | 0.002 | 0.004 | 0.008 | 0.008 | 0.015 |
| ≤ 0.1 | ≥ 6 | No | 0.006 | 0.010 | 0.30 | 0.002 | 0.004 | 0.006 | 0.006 | 0.010 |
| ≥ 0.25 | ≤ 3 | No | 0.003 | 0.005 | 0.25 | 0.001 | 0.002 | 0.003 | 0.003 | 0.00 |
| ≥ 0.25 | ≥ 6 | No | 0.002 | 0.004 | 0.20 | 0.001 | 0.001 | 0.002 | 0.002 | 0.004 |
| **ii. Columns Supporting Discontinuous Shear Walls** Transverse Reinforcement[4] | | | | | | | | | | |
| Conforming | | | 0.010 | 0.015 | 0.20 | 0.003 | 0.007 | 0.010 | n.a. | n.a. |
| Nonconforming | | | 0.0 | 0.0 | 0.0 | 0.0 | 0.0 | 0.0 | n.a. | n.a. |
| **iii. Shear Wall Coupling Beams[5]** | | | | | | | | | | |
| Longitudinal Reinforcement and Transverse Reinforcement[6] | $\dfrac{V}{t_w l_w \sqrt{f_c'}}$ | | | | | | | | | |
| Conventional Longitudinal Reinforcement with Conforming Transverse Reinforcement | ≤ 3 | | 0.025 | 0.050 | 0.75 | 0.010 | 0.02 | 0.025 | 0.025 | 0.050 |
| | ≥ 6 | | 0.020 | 0.040 | 0.50 | 0.005 | 0.010 | 0.020 | 0.020 | 0.040 |
| Conventional Longitudinal Reinforcement with Nonconforming Transverse Reinforcement | ≤ 3 | | 0.020 | 0.035 | 0.50 | 0.006 | 0.012 | 0.020 | 0.020 | 0.035 |
| | ≥ 6 | | 0.010 | 0.025 | 0.25 | 0.005 | 0.008 | 0.010 | 0.010 | 0.025 |
| Diagonal Reinforcement | n.a. | | 0.030 | 0.050 | 0.80 | 0.006 | 0.018 | 0.030 | 0.030 | 0.050 |

[1]Primary and secondary component demands shall be within secondary component acceptance criteria where the full backbone curve is explicitly modeled including strength degradation and residual strength in accordance with Section 3.4.3.2.

[2]Linear interpolation between values listed in the table shall be permitted.

[3]Requirements for a confined boundary are the same as those given in ACI 318 (ACI 2002).

[4]Requirements for conforming transverse reinforcement in columns are: (a) hoops over the entire length of the column at a spacing ≤ $d/2$, and (b) strength of hoops $V_s$ ≥ required shear strength of column.

[5]For secondary coupling beams spanning <8 ft, 0 in., with bottom reinforcement continuous into the supporting walls, secondary values shall be permitted to be doubled.

[6]Conventional longitudinal reinforcement consists of top and bottom steel parallel to the longitudinal axis of the coupling beam. Conforming transverse reinforcement consists of: (a) closed stirrups over the entire length of the coupling beam at a spacing ≤ $d/3$, and (b) strength of closed stirrups $V_s$ ≥ three-fourths of required shear strength of the coupling beam.

### 6.6.3.2 Stiffness

#### 6.6.3.2.1 Linear Static and Dynamic Procedures
Effective stiffnesses shall be calculated according to the principles of Section 6.3.1.2 and the procedure of Section 6.6.2.2.1.

#### 6.6.3.2.2 Nonlinear Static Procedure
Nonlinear load–deformation relations for use in analysis by NSP shall follow the requirements of Section 6.3.1.2.2.

Monotonic load–deformation relations shall be according to the generalized relation shown in Fig. 6-1, except different relations shall be permitted where verified by tests. Numerical quantities in Fig. 6-1 shall be derived from tests or by analysis procedures specified in Section 2.8, and shall take into account the interactions between frame and infill components. Alternatively, the procedure of Section 6.6.2.2.2 shall be permitted for the development of nonlinear modeling parameters for concrete frames with concrete infills.

*6.6.3.2.3 Nonlinear Dynamic Procedure* Nonlinear load–deformation relations for use in analysis by NDP shall model the complete hysteretic behavior of each component using properties verified by tests. Unloading and reloading properties shall represent stiffness and strength degradation characteristics.

### 6.6.3.3 Strength

Strengths of reinforced concrete components shall be calculated according to the general requirements of Sections 6.4.2, as modified by other specifications of this chapter. Strength calculations shall consider:

1. Limitations imposed by beams, columns, and joints in unfilled portions of frames;
2. Tensile and compressive capacity of columns acting as boundary components of infilled frames;
3. Local forces applied from the infill to the frame;
4. Strength of the infill; and
5. Connections with adjacent components.

Strengths of existing concrete infills shall be determined considering shear strength of the infill panel. For this calculation, procedures specified in Section 6.7.2.3 shall be used for calculation of the shear strength of a wall segment.

Where the frame and concrete infill are assumed to act as a monolithic wall, flexural strength shall be based on continuity of vertical reinforcement in both (1) the columns acting as boundary components, and (2) the infill wall, including anchorage of the infill reinforcement in the boundary frame.

### 6.6.3.4 Acceptance Criteria

The acceptance criteria for concrete frames with concrete infills shall comply with relevant acceptance criteria of Sections 6.6.2.4, 6.7, and 6.8.

### 6.6.3.5 Rehabilitation Measures

Concrete frames with concrete infills that do not meet the acceptance criteria for the selected Rehabilitation Objective shall be rehabilitated. Rehabilitation measures shall meet the requirements of Section 6.3.7 and other provisions of this standard.

### C6.6.3.5 Rehabilitation Measures

Rehabilitation measures described in C6.6.2.5 for concrete frames with masonry infills may also be effective in rehabilitating concrete frames with concrete infills. In addition, application of shotcrete to the face of an existing wall to increase the thickness and shear strength may be effective. For this purpose, the face of the existing wall should be roughened, a mat of reinforcing steel should be doweled into the existing structure, and shotcrete should be applied to the desired thickness. The design professional is referred to FEMA 308 (FEMA 1998) for further information regarding rehabilitation of concrete frames with concrete infill.

## 6.7 CONCRETE SHEAR WALLS

### 6.7.1 Types of Concrete Shear Walls and Associated Components

The provisions of Section 6.7 shall apply to all shear walls in all types of structural systems that incorporate shear walls. This includes isolated shear walls, shear walls used in wall–frame systems, coupled shear walls, and discontinuous shear walls. Shear walls shall be permitted to be considered as solid walls if they have openings that do not significantly influence the strength or inelastic behavior of the wall. Perforated shear walls shall be defined as walls having a regular pattern of openings in both horizontal and vertical directions that creates a series of wall pier and deep beam components referred to as wall segments.

Coupling beams and columns that support discontinuous shear walls shall comply with provisions of Section 6.7.2. These special frame components associated with shear walls shall be exempted from the provisions for beams and columns of frame components covered in Section 6.4.

### C6.7.1 Types of Concrete Shear Walls and Associated Components

Concrete shear walls are planar vertical elements or combinations of interconnected planar elements that serve as lateral-load-resisting elements in concrete structures. Shear walls (or wall segments) shall be considered slender if their aspect ratio (height/length) is > 3.0, and shall be considered short or squat if their aspect ratio is < 1.5. Slender shear walls are normally controlled by flexural behavior; short walls are normally controlled by shear behavior. The response of walls with intermediate aspect ratios is influenced by both flexure and shear.

Identification of component types in concrete shear wall elements depends, to some degree, on the relative strengths of the wall segments. Vertical segments are often termed wall piers, while horizontal segments may be called coupling beams or spandrels. The design professional is referred to FEMA 306 (FEMA 1998) for additional information regarding the behavior of concrete wall components. Selected information from FEMA 306 has been reproduced in the commentary of this standard, in Table C6-1 and Fig. C6-1 to clarify wall component identification.

### 6.7.1.1 Monolithic Reinforced Concrete Shear Walls and Wall Segments

Monolithic reinforced concrete shear walls shall consist of vertical cast-in-place elements, either uncoupled or coupled, in open or closed shapes. These walls shall have relatively continuous cross sections and reinforcement and shall provide both vertical and lateral force resistance, in contrast with infilled walls defined in Section 6.6.1.3.

Shear walls or wall segments with axial loads greater than $0.35P_o$ shall not be considered effective in resisting seismic forces. For the purpose of determining effectiveness of shear walls or wall segments, the use of axial loads based on a limit state analysis shall be permitted. The maximum spacing of horizontal and vertical reinforcement shall not exceed 18 in. Walls with horizontal and vertical reinforcement ratios less than 0.0025, but with reinforcement spacings less than 18 in., shall be permitted where the shear force demand does not exceed the reduced nominal shear strength of the wall calculated in accordance with Section 6.7.2.3.

### C6.7.1.1 Monolithic Reinforced Concrete Shear Walls and Wall Segments

The wall reinforcement is normally continuous in both the horizontal and vertical directions, and bars are typically lap spliced for tension continuity. The reinforcement mesh may also contain horizontal ties around vertical bars that are concentrated either near the vertical edges of a wall with constant thickness, or in boundary members formed at the wall edges. The amount and spacing of these ties is important for determining how well the concrete at the wall edge is confined, and thus for determining the lateral deformation capacity of the wall.

In general, slender reinforced concrete shear walls will be governed by flexure and will tend to form a plastic flexural hinge near the base of the wall under severe lateral loading. The ductility of the wall will be a function of the percentage of longitudinal reinforcement concentrated near the boundaries of the wall, the level of axial load, the amount of lateral shear required

#### Table C6-1. Reinforced Concrete Shear Wall Component Types[1]

| Component Type per FEMA 306 | | Description | ASCE 41 Designation |
|---|---|---|---|
| RC1 | Isolated Wall or Stronger Wall Pier | Stronger than beam or spandrel components that may frame into it so that nonlinear behavior (and damage) is generally concentrated at the base, with a flexural plastic hinge, shear failure, etc. Includes isolated (cantilever) walls. If the component has a major setback or cutoff of reinforcement above the base, this section should be also checked for nonlinear behavior. | Monolithic reinforced concrete wall or vertical wall segment |
| RC2 | Weaker Wall Pier | Weaker than the spandrels to which it connects; characterized by flexural hinging top and bottom, or shear failure, etc. | |
| RC3 | Weaker Spandrel or Coupling Beam | Weaker than the wall piers to which it connects; characterized by hinging at each end, shear failure, sliding shear failure, etc. | Horizontal wall segment or coupling beam |
| RC4 | Stronger Spandrel | Should not suffer damage because it is stronger than attached wall piers. If this component is damaged, it should probably be reclassified as RC3. | |
| RC5 | Pier–Spandrel Panel Zone | Typically not a critical area in RC walls. | Wall segment |

[1]Source: FEMA. (1998). "Evaluation of earthquake-damaged concrete and masonry wall buildings—Basic procedures manual." FEMA 306, prepared by the Applied Technology Council (ATC-43 Project), for the Federal Emergency Management Agency, Washington, D.C.

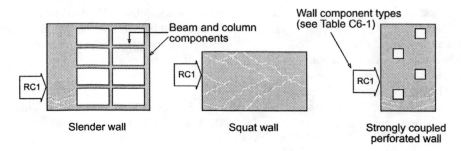

*(a) Cantilever Wall Mechanisms*

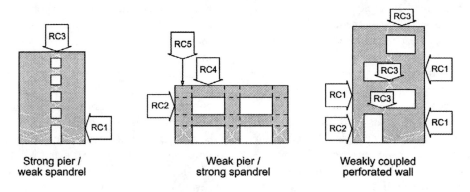

*(b) Pier / Spandrel Mechanisms*

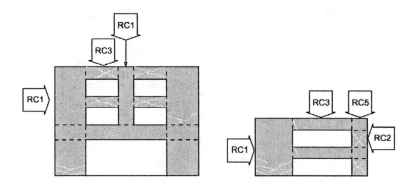

*(c) Mixed Mechanisms*

**FIGURE C6-1. Identification of Component Types in Concrete Shear Wall Elements.**

From: FEMA. (1998). "Evaluation of earthquake-damaged concrete and masonry wall buildings—Basic procedures manual." *FEMA 306*, prepared by the Applied Technology Council (ATC-43 Project), for the Federal Emergency Management Agency, Washington, D.C.

to cause flexural yielding, and the thickness and reinforcement used in the web portion of the shear wall. In general, higher axial load stresses and higher shear stresses will reduce the flexural ductility and energy-absorbing capability of the shear wall. Short or squat shear walls will normally be governed by shear. These walls will normally have a limited ability to deform beyond the elastic range and continue to carry lateral loads. Thus, these walls are typically

designed either as displacement-controlled components with low ductility capacities or as force-controlled components.

### 6.7.1.2 Reinforced Concrete Columns Supporting Discontinuous Shear Walls

Reinforced concrete columns supporting discontinuous shear walls shall be evaluated and rehabilitated to comply with the requirements of Section 6.7.2.

## C6.7.1.2 Reinforced Concrete Columns Supporting Discontinuous Shear Walls

In shear wall buildings it is not uncommon to find that some walls are terminated either to create commercial space in the first story or to create parking spaces in the basement. In such cases, the walls are commonly supported by columns. Such designs are not recommended in seismic zones because very large demands may be placed on these columns during earthquake loading. In older buildings such columns will often have "standard" longitudinal and transverse reinforcement; the behavior of such columns during past earthquakes indicates that tightly spaced closed ties with well-anchored 135-degree hooks will be required for the building to survive severe earthquake loading.

### 6.7.1.3 Reinforced Concrete Coupling Beams

Reinforced concrete coupling beams used to link two shear walls together shall be evaluated and rehabilitated to comply with the requirements of Section 6.7.2.

### C6.7.1.3 Reinforced Concrete Coupling Beams

The coupled walls are generally much stiffer and stronger than they would be if they acted independently. Coupling beams typically have a small span-to-depth ratio, and their inelastic behavior is normally affected by the high shear forces acting in these components. Coupling beams in most older reinforced concrete buildings will commonly have "conventional" reinforcement that consists of longitudinal flexural steel and transverse steel for shear. In some, more modern buildings, or in buildings where coupled shear walls are used for seismic rehabilitation, the coupling beams may use diagonal reinforcement as the primary reinforcement for both flexure and shear. The inelastic behavior of coupling beams that use diagonal reinforcement has been shown experimentally to be much better with respect to retention of strength, stiffness, and energy dissipation capacity than the observed behavior of coupling beams with conventional reinforcement.

### 6.7.2 Reinforced Concrete Shear Walls, Wall Segments, Coupling Beams, and Reinforced Concrete Columns Supporting Discontinuous Shear Walls

### 6.7.2.1 General Considerations

The analytical model for a shear wall element shall represent the stiffness, strength, and deformation capacity of the shear wall. Potential failure in flexure, shear, and reinforcement development at any point in the shear wall shall be considered. Interaction with other structural and nonstructural components shall be included.

Slender shear walls and wall segments shall be permitted to be modeled as equivalent beam-column elements that include both flexural and shear deformations. The flexural strength of beam-column elements shall include the interaction of axial load and bending. The rigid-connection zone at beam connections to this equivalent beam-column element shall represent the distance from the wall centroid to the edge of the wall. Unsymmetrical wall sections shall model the different bending capacities for the two loading directions.

A beam element that incorporates both bending and shear deformations shall be used to model coupling beams. The element inelastic response shall account for the loss of shear strength and stiffness during reversed cyclic loading to large deformations. For coupling beams that have diagonal reinforcement satisfying ACI 318 (ACI 2002), a beam element representing flexure only shall be permitted.

For columns supporting discontinuous shear walls, the model shall account for axial compression, axial tension, flexure, and shear response, including rapid loss of resistance where this behavior is likely under design loadings. The diaphragm action of concrete slabs that interconnect shear walls and frame columns shall be represented in the model.

### C6.7.2.1 General Considerations

For rectangular shear walls and wall segments with $h/l_w \leq 2.5$, and flanged wall sections with $h/l_w \leq 3.5$, either a modified beam–column analogy or a multiple-node, multiple-spring approach should be used. Because shear walls usually respond in single curvature over a story height, the use of one multiple-spring element per story should be permitted for modeling shear walls. Wall segments should be modeled with either the beam–column element or with a multiple-spring model with two elements over the length of the wall segment.

Coupling beams that have diagonal reinforcement satisfying FEMA 450 (FEMA 2004) will commonly have a stable hysteretic response under large load reversals. Therefore, these members could adequately be modeled with beam elements used for typical frame analyses.

### 6.7.2.2 Stiffness

The effective stiffness of all the elements discussed in Section 6.7 shall be defined based on the material properties, component dimensions, reinforcement quantities, boundary conditions, and current state of the member with respect to cracking and stress levels. Alternatively, use of values for effective stiffness given in Table 6-5 shall be permitted. To obtain a proper distribution of lateral forces in bearing wall buildings, all of the walls shall be assumed to be either cracked or uncracked. In buildings where lateral load resistance is provided by either structural walls only, or a combination of walls and frame members, all shear walls and wall segments discussed in this section shall be considered to be cracked.

For coupling beams, the effective stiffness values given in Table 6-5 for non-prestressed beams shall be used unless alternative stiffnesses are determined by more detailed analysis. The effective stiffness of columns supporting discontinuous shear walls shall change between the values given for columns in tension and compression, depending on the direction of the lateral load being resisted by the shear wall.

*6.7.2.2.1 Linear Static and Dynamic Procedures* Shear walls and associated components shall be modeled considering axial, flexural, and shear stiffness. For closed and open wall shapes, such as box, T, L, I, and C sections, the effective tension or compression flange widths shall be as specified in Section 6.3.1.3. The calculated stiffnesses to be used in analysis shall be in accordance with the requirements of Section 6.3.1.2.

Joints between shear walls and frame elements shall be modeled as stiff components or rigid components, as appropriate.

*6.7.2.2.2 Nonlinear Static Procedure* Nonlinear load–deformation relations for use in analysis by nonlinear static and dynamic procedures shall comply with the requirements of Section 6.3.1.2.

Monotonic load–deformation relationships for analytical models that represent shear walls, wall elements, coupling beams, and columns that support discontinuous shear walls shall be in accordance with the generalized relation shown in Fig. 6-1.

For shear walls and wall segments having inelastic behavior under lateral loading that is governed by flexure, as well as columns supporting discontinuous shear walls, the following approach shall be permitted. The load–deformation relationship in Fig. 6-1 shall be used with the x-axis of Fig. 6-1 taken as the rotation over the plastic hinging region at the end of the member shown in Fig. 6-2. The hinge rotation at point B in Fig. 6-1 corresponds to the yield point, $\theta_y$, and shall be calculated in accordance with Eq. 6-6:

$$\theta_y = \left(\frac{M_y}{E_c I}\right) l_p \qquad \text{(Eq. 6-6)}$$

where

$M_y$ = yield moment capacity of the shear wall or wall segment;

$E_c$ = concrete modulus;

$I$ = member moment of inertia; and

$l_p$ = assumed plastic hinge length.

For analytical models of shear walls and wall segments, the value of $l_p$ shall be set equal to 0.5 times the flexural depth of the element, but less than one story height for shear walls and less than 50% of the element length for wall segments. For columns supporting discontinuous shear walls, $l_p$ shall be set equal to 0.5 times the flexural depth of the component.

Values for the variables *a, b,* and *c* required to define the location of points *C, D,* and *E* in Fig. 6-1(a), shall be as specified in Table 6-18.

For shear walls and wall segments whose inelastic response is controlled by shear, the following approach shall be permitted. The load–deformation relationship in Fig. 6-1(b) shall be used, with the x-axis of Fig. 6-1(b) taken as lateral drift. For shear walls, this drift shall be the story drift as shown in Fig. 6-3. For wall segments, Fig. 6-3 shall represent the member drift.

For coupling beams, the following approach shall be permitted. The load–deformation relationship in Fig. 6-1(b) shall be used, with the x-axis of Fig. 6-1(b) taken as the chord rotation as defined in Fig. 6-4.

Values for the variables *d, e,* and *c* required to find the points *C, D,* and *E* in Fig. 6-1(b), shall be as specified in Table 6-19 for the appropriate members. Linear interpolation between tabulated values shall be used if the member under analysis has conditions that are between the limits given in the tables.

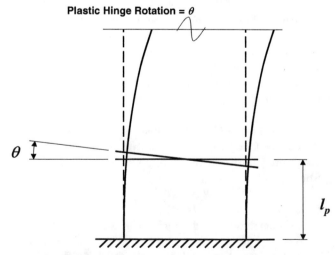

**FIGURE 6-2. Plastic Hinge Rotation in Shear Wall where Flexure Dominates Inelastic Response.**

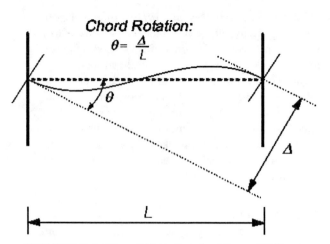

**FIGURE 6-4. Chord Rotation for Shear Wall Coupling Beams.**

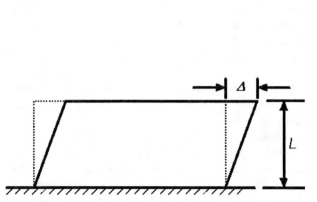

**FIGURE 6-3. Story Drift in Shear Wall where Shear Dominates Inelastic Response.**

**Table 6-19. Modeling Parameters and Numerical Acceptance Criteria for Nonlinear Procedures— Reinforced Concrete Shear Walls and Associated Components Controlled by Shear**

| | | Total Drift Ratio (%), or Chord Rotation (radians)[2] | | Residual Strength Ratio | Acceptable Total Drift (%) or Chord[1] Rotation (radians)[2] | | | | |
|---|---|---|---|---|---|---|---|---|---|
| | | | | | Performance Level | | | | |
| | | | | | | | Component Type | | |
| | | | | | IO | Primary | | Secondary | |
| Conditions | | $d$ | $e$ | $c$ | IO | LS | CP | LS | CP |
| i. Shear Walls and Wall Segments | | | | | | | | | |
| All Shear Walls and Wall Segments[3] | | 0.75 | 2.0 | 0.40 | 0.40 | 0.60 | 0.75 | 0.75 | 1.5 |
| ii. Shear wall coupling beams[4] | | | | | | | | | |
| Longitudinal Reinforcement and Transverse Reinforcement[5] | $\dfrac{V}{t_w l_w \sqrt{f'_c}}$ | | | | | | | | |
| Conventional Longitudinal Reinforcement with Conforming Transverse Reinforcement | $\leq 3$ | 0.02 | 0.030 | 0.60 | 0.006 | 0.015 | 0.020 | 0.020 | 0.030 |
| | $\geq 6$ | 0.016 | 0.024 | 0.30 | 0.005 | 0.012 | 0.016 | 0.016 | 0.024 |
| Conventional Longitudinal Reinforcement with Nonconforming Transverse Reinforcement | $\leq 3$ | 0.012 | 0.025 | 0.40 | 0.006 | 0.008 | 0.010 | 0.010 | 0.020 |
| | $\geq 6$ | 0.008 | 0.014 | 0.20 | 0.004 | 0.006 | 0.007 | 0.007 | 0.012 |

[1]Primary and secondary component demands shall be within secondary component acceptance criteria where the full backbone curve is explicitly modeled including strength degradation and residual strength in accordance with Section 3.4.3.2.

[2]For shear walls and wall segments, use drift; for coupling beams, use chord rotation; refer to Figures 6-3 and 6-4.

[3]For shear walls and wall segments where inelastic behavior is governed by shear, the axial load on the member must be $\leq 0.15\, A_g f'_c$; otherwise, the member must be treated as a force-controlled component.

[4]For secondary coupling beams spanning <8 ft, 0 in., with bottom reinforcement continuous into the supporting walls, secondary values shall be permitted to be doubled.

[5]Conventional longitudinal reinforcement consists of top and bottom steel parallel to the longitudinal axis of the coupling beam. Conforming transverse reinforcement consists of: (1) closed stirrups over the entire length of the coupling beam at a spacing $\leq d/3$, and (2) strength of closed stirrups $V_s \geq$ three-fourths of required shear strength of the coupling beam.

*6.7.2.2.3 Nonlinear Dynamic Procedure* For the NDP, the complete hysteretic behavior of each component shall be modeled using properties verified by experimental evidence. Use of the generalized load–deformation relation shown in Fig. 6-1 to represent the envelope relation for the analysis shall be permitted. The unloading and reloading stiffnesses and strengths, and any pinching of the load-versus-rotation hysteresis loops, shall reflect the behavior experimentally observed for wall elements similar to the one under investigation.

### 6.7.2.3 Strength

Component strengths shall be computed according to the general requirements of Sections 6.3.2, with the additional requirements of this section. Strength shall be determined considering the potential for failure in flexure, shear, or development under combined gravity and lateral load.

Nominal flexural strength of shear walls or wall segments, $M_n$, shall be determined using the fundamental principles given in Chapter 10 of ACI 318 (ACI 2002). For calculation of nominal flexural strength, the effective compression and tension flange widths defined in Section 6.7.2.2 shall be used, except that the first limit shall be changed to one-tenth of the wall height. Where determining the flexural yield strength of a shear wall, as represented by point *B* in Fig. 6-1(a), only the longitudinal steel in the boundary of the wall shall be included. If the wall does not have a boundary member, then only the longitudinal steel in the outer 25% of the wall section shall be included in the calculation of the yield strength. Where calculating the nominal flexural strength of the wall, as represented by point *C* in Fig. 6-1(a), all longitudinal steel (including web reinforcement) shall be included in the calculation. For all moment strength calculations, the strength of the longitudinal reinforcement shall be taken as the expected yield strength to account for material overstrength and strain-hardening, and the axial load acting on the wall shall include gravity loads as defined in Chapter 3.

The nominal shear strength of a shear wall or wall segment, $V_n$, shall be determined based on the principles and equations given in Chapter 21 of ACI 318. The nominal shear strength of columns supporting discontinuous shear walls shall be determined based on the principles and equations given in Chapter 21 of ACI 318. For all shear strength calculations, 1.0 times the specified reinforcement yield strength shall be used. There shall be no difference between the yield and nominal shear strengths, as represented by points *B* and *C* in Fig. 6-1.

Where a shear wall or wall segment has a transverse reinforcement percentage, $\rho_n$, less than the minimum value of 0.0025 but greater than 0.0015 and reinforcement is spaced no greater than 18 in., the shear strength of the wall shall be analyzed using the ACI 318 equations noted above. For transverse reinforcement percentages less than 0.0015, the contribution from the wall reinforcement to the shear strength of the wall shall be held constant at the value obtained using $\rho_n = 0.0015$.

Splice lengths for primary longitudinal reinforcement shall be evaluated using the procedures given in Section 6.3.5. Reduced flexural strengths shall be evaluated at locations where splices govern the usable stress in the reinforcement. The need for confinement reinforcement in shear wall boundary members shall be evaluated by the procedure in ACI 318 or other approved procedure.

The nominal flexural and shear strengths of coupling beams shall be evaluated using the principles and equations contained in Chapter 21 of ACI 318. The expected strength of longitudinal or diagonal reinforcement shall be used.

The nominal shear and flexural strengths of columns supporting discontinuous shear walls shall be evaluated as defined in Section 6.4.2.3.

### C6.7.2.3 Strength

Data presented by Wood (1990) indicate that wall strength is insensitive to the quantity of transverse reinforcement where it drops below a steel ratio of 0.0015.

The need for confinement reinforcement in shear wall boundary members may be evaluated by the method recommended by Wallace and Thomsen (1995) for determining maximum lateral deformations in the wall and the resulting maximum compression strains in the wall boundary.

Strength calculations based on ACI 318 (ACI 2002), excluding Chapter 22, assume a maximum spacing of wall reinforcement. No data are available to justify performance for walls that do not meet the maximum spacing requirements. If plain concrete is encountered in an existing building, Chapter 22 of ACI 318 can be used to derive capacities, while Section 2.8 of this standard can be used to develop acceptance criteria.

### 6.7.2.4 Acceptance Criteria

*6.7.2.4.1 Linear Static and Dynamic Procedures* Shear walls, wall segments, coupling beams, and columns supporting discontinuous shear walls shall be classified as either deformation- or force-controlled, as defined in Section 2.4.4. For columns supporting discontinuous shear walls, deformation-controlled actions shall be restricted to flexure. In other components, deformation-controlled actions shall be restricted to flexure or shear. All other actions shall be defined as being force-controlled actions.

The nominal flexukral strength of a shear wall or wall segment shall be used to determine the maximum shear force in shear walls, wall segments, and columns supporting discontinuous shear walls. For cantilever shear walls and columns supporting discontinuous shear walls, the design shear force shall be equal to the magnitude of the lateral force required to develop the nominal flexural strength at the base of the wall, assuming the lateral force is distributed uniformly over the height of the wall. For wall segments, the design force shall be equal to the shear corresponding to the development of the positive and negative nominal moment strengths at opposite ends of the wall segment.

Design actions (flexure, shear, axial, or force transfer at rebar anchorages and splices) on components shall be determined as prescribed in Chapter 3. Where determining the appropriate value for the design actions, proper consideration shall be given to gravity loads and to the maximum forces that can be transmitted considering nonlinear action in adjacent components. Design actions shall be compared with design strengths in accordance with Section 3.4.2.2. Tables 6-20 and 6-21 specify $m$ values for use in Eq. 3-20. Alternate $m$ values shall be permitted where justified by experimental evidence and analysis.

*C6.7.2.4.1 Linear Static and Dynamic Procedures* For shear-controlled coupling beams, ductility is a function of the shear in the member as determined by the expected shear capacity of the member. In accordance with Section 6.3.2, expected strengths are calculated using the procedures specified in ACI 318 (ACI 2002). For coupling beams, $V_c$ is nearly always zero.

*6.7.2.4.2 Nonlinear Static and Dynamic Procedures* In the design model, inelastic response shall be restricted to those components and actions listed in Tables 6-18 and 6-19, except where it is demonstrated that other inelastic actions are justified for the selected performance levels. For members experiencing inelastic behavior, the magnitude of other actions (forces, moments, or torque) in the member shall correspond to the magnitude of the action causing inelastic behavior. The magnitude of these other actions shall be shown to be below their nominal capacities.

Components experiencing inelastic response shall satisfy the requirements of Section 3.4.3.2, and the maximum plastic hinge rotations, drifts, or chord rotation angles shall not exceed the values given in Tables 6-18 and 6-19, for the selected performance level. Linear interpolation between tabulated values shall be used if the member under analysis has conditions that are between the limits given in the tables.

### 6.7.2.5 Rehabilitation Measures

Reinforced shear walls, wall segments, coupling beams, and columns supporting discontinuous shear walls that do not meet the acceptance criteria for the selected Rehabilitation Objective shall be rehabilitated. Rehabilitation measures shall meet the requirements of Section 6.3.7 and other provisions of this standard.

### C6.7.2.5 Rehabilitation Measures

The following measures may be effective in rehabilitating reinforced shear walls, wall segments, coupling beams, and reinforced concrete columns supporting discontinuous shear walls:

1. **Addition of wall boundary components.** Addition of boundary components may be an effective measure in strengthening shear walls or wall segments that have insufficient flexural strength. These members may be either cast-in-place reinforced concrete components or steel sections. In both cases, proper connections should be made between the existing wall and the added components. The shear capacity of the rehabilitated wall should be reevaluated;
2. **Addition of confinement jackets at wall boundaries.** Increasing the confinement at the wall boundaries by the addition of a steel or reinforced concrete jacket may be an effective measure in improving the flexural deformation capacity of a shear wall. For both types of jackets, the longitudinal steel should not be continuous from story to story unless the jacket is also being used to increase the flexural capacity. The minimum thickness for a concrete jacket should be 3 in. Carbon fiber wrap should be permitted for improving the confinement of concrete in compression;
3. **Reduction of flexural strength.** Reduction in the flexural capacity of a shear wall to change the governing failure mode from shear to flexure may be an effective rehabilitation measure. It may be accomplished by saw-cutting a specified number of longitudinal bars near the edges of the shear wall;
4. **Increased shear strength of wall.** Increasing the shear strength of the web of a shear wall by casting additional reinforced concrete adjacent to the wall

web may be an effective rehabilitation measure. The new concrete should be at least 4 in. thick and should contain horizontal and vertical reinforcement. The new concrete should be properly bonded to the existing web of the shear wall. The use of carbon fiber sheets, epoxied to the concrete surface, should also be permitted to increase the shear capacity of a shear wall;

**Table 6-20. Numerical Acceptance Criteria for Linear Procedures— Reinforced Concrete Shear Walls and Associated Components Controlled by Flexure**

| | | | m-Factors[1] | | | | |
|---|---|---|---|---|---|---|---|
| | | | | Performance Level | | | |
| | | | | | Component Type | | |
| | | | | Primary | | Secondary | |
| Conditions | | | IO | LS | CP | LS | CP |
| **i. Shear Walls and Wall Segments** | | | | | | | |
| $\dfrac{(A_s - A'_s)f_y + P}{t_w l_w f'_c}$ [2] | | $\dfrac{V}{t_w l_w \sqrt{f'_c}}$ [3] | Confined Boundary[4] | | | | |
| $\leq 0.1$ | $\leq 3$ | Yes | 2 | 4 | 6 | 6 | 8 |
| $\leq 0.1$ | $\geq 6$ | Yes | 2 | 3 | 4 | 4 | 6 |
| $\geq 0.25$ | $\leq 3$ | Yes | 1.5 | 3 | 4 | 4 | 6 |
| $\geq 0.25$ | $\geq 6$ | Yes | 1.25 | 2 | 2.5 | 2.5 | 4 |
| $\leq 0.1$ | $\leq 3$ | No | 2 | 2.5 | 4 | 4 | 6 |
| $\leq 0.1$ | $\geq 6$ | No | 1.5 | 2 | 2.5 | 2.5 | 4 |
| $\geq 0.25$ | $\leq 3$ | No | 1.25 | 1.5 | 2 | 2 | 3 |
| $\geq 0.25$ | $\geq 6$ | No | 1.25 | 1.5 | 1.75 | 1.75 | 2 |
| **ii. Columns supporting Discontinuous Shear Walls** | | | | | | | |
| Transverse Reinforcement[5] | | | | | | | |
| Conforming | | | 1 | 1.5 | 2 | n.a. | n.a. |
| Nonconforming | | | 1 | 1 | 1 | n.a. | n.a. |
| **iii. Shear Wall Coupling Beams[6]** | | | | | | | |
| Longitudinal Reinforcement and Transverse Reinforcement[7] | $\dfrac{V}{t_w l_w \sqrt{f'_c}}$ [3] | | | | | | |
| Conventional Longitudinal Reinforcement with Conforming Transverse Reinforcement | $\leq 3$ | | 2 | 4 | 6 | 6 | 9 |
| | $\geq 6$ | | 1.5 | 3 | 4 | 4 | 7 |
| Conventional Longitudinal Reinforcement with Nonconforming Transverse Reinforcement | $\leq 3$ | | 1.5 | 3.5 | 5 | 5 | 8 |
| | $\geq 6$ | | 1.2 | 1.8 | 2.5 | 2.5 | 4 |
| Diagonal Reinforcement | n.a. | | 2 | 5 | 7 | 7 | 10 |

[1]Linear interpolation between values listed in the table shall be permitted.

[2]$P$ is the design axial force in the member. Alternatively, use of axial loads determined based on a limit-state analysis shall be permitted.

[3]$V$ is the design shear force calculated using limit-state analysis procedures in accordance with Section 6.7.2.4.

[4]Requirements for a confined boundary are the same as those given in ACI 318 (ACI 2002).

[5]Requirements for conforming transverse reinforcement in columns are: (1) hoops over the entire length of the column at a spacing $\leq d/2$, and (2) strength of hoops $V_s \geq$ required shear strength of column.

[6]For secondary coupling beams spanning $<8$ ft, 0 in., with bottom reinforcement continuous into the supporting walls, secondary values shall be permitted to be doubled.

[7]Conventional longitudinal reinforcement consists of top and bottom steel parallel to the longitudinal axis of the coupling beam. Conforming transverse reinforcement consists of: (1) closed stirrups over the entire length of the coupling beam at a spacing $\leq d/3$, and (2) strength of closed stirrups $V_s \geq$ three-fourths of required shear strength of the coupling beam.

5. **Confinement jackets to improve deformation capacity of coupling beams and columns supporting discontinuous shear walls.** The use of confinement jackets specified earlier as a rehabilitation measure for wall boundaries, and in Section 6.4 for frame elements, may also be effective in increasing both the shear capacity and the deformation capacity of coupling beams and columns supporting discontinuous shear walls; and

6. **Infilling between columns supporting discontinuous shear walls.** Where a discontinuous shear wall is supported on columns that lack either sufficient strength or deformation capacity to satisfy design criteria, making the wall continuous by infilling the opening between these columns may be an effective rehabilitation measure. The infill and existing columns should be designed to satisfy all the requirements for new wall construction, including any strengthening of the existing columns required by adding a concrete or steel jacket for strength and

increased confinement. The opening below a discontinuous shear wall should also be permitted to be "infilled" with steel bracing. The bracing members should be sized to satisfy all design requirements and the columns should be strengthened with a steel or a reinforced concrete jacket.

All of the above rehabilitation measures require an evaluation of the wall foundation, diaphragms, and connections between existing structural elements and any elements added for rehabilitation purposes.

## 6.8 PRECAST CONCRETE SHEAR WALLS

### 6.8.1 Types of Precast Shear Walls

Precast concrete shear walls shall consist of story-high or half-story-high precast wall segments that are made continuous through the use of either mechanical connectors or reinforcement splicing techniques with

**Table 6-21. Numerical Acceptance Criteria for Linear Procedures— Reinforced Concrete Shear Walls and Associated Components Controlled by Shear**

| | | *m*-Factors | | | | |
|---|---|---|---|---|---|---|
| | | Performance Level | | | | |
| | | | Component Type | | | |
| | | | Primary | | Secondary | |
| Conditions | | IO | LS | CP | LS | CP |
| i. Shear Walls and Wall Segments<br>All Shear Walls and Wall Segments[1] | | 2 | 2 | 3 | 2 | 3 |
| ii. Shear Wall Coupling Beams[2]<br>Longitudinal Reinforcement<br>and Transverse Reinforcement[3] | $\dfrac{V}{t_w l_w \sqrt{f_c'}}$ [4] | | | | | |
| Conventional Longitudinal<br>Reinforcement with Conforming<br>Transverse Reinforcement | $\leq 3$<br>$\geq 6$ | 1.5<br>1.2 | 3<br>2 | 4<br>2.5 | 4<br>2.5 | 6<br>3.5 |
| Conventional Longitudinal<br>Reinforcement with Nonconforming Transverse Reinforcement | $\leq 3$<br>$\geq 6$ | 1.5<br>1.2 | 2.5<br>1.2 | 3<br>1.5 | 3<br>1.5 | 4<br>2.5 |

[1]The shear shall be considered to be a force-controlled action for shear walls and wall segments where inelastic behavior is governed by shear and the design axial load is greater than $0.15 A_g f_c'$. It shall be permitted to calculate the axial load based on a limit-state analysis.

[2]For secondary coupling beams spanning <8 ft, 0 in., with bottom reinforcement continuous into the supporting walls, secondary values shall be permitted to be doubled.

[3]Conventional longitudinal reinforcement consists of top and bottom steel parallel to the longitudinal axis of the coupling beam. Conforming transverse reinforcement consists of: (1) closed stirrups over the entire length of the coupling beam at a spacing $\leq d/3$, and (2) strength of closed stirrups $V_s \geq$ three-fourths of required shear strength of the coupling beam.

[4] For the purpose of determining m, $V$ is the coupling beam expected shear strength.

or without a cast-in-place connection strip. Connections between precast segments shall be permitted along both the horizontal and vertical edges of a wall segment.

The design of the following types of precast shear walls shall meet the requirements of Section 6.8:

1. Effectively monolithic construction, defined as that construction in which the reinforcement connections are made to be stronger than the adjacent precast panels so that the lateral load response of the precast wall system will be comparable to that for monolithic shear walls;
2. Jointed construction, defined as construction in which inelastic action is permitted to occur at the connections between precast panels; and
3. Tilt-up construction, defined as a special technique for precast wall construction where there are vertical joints between adjacent panels and horizontal joints at the foundation level, and where the roof or floor diaphragm connects with the tilt-up panel.

### 6.8.1.1 Effectively Monolithic Construction

For this type of precast wall, the connections between precast wall elements shall be designed and detailed to be stronger than the panels they connect. Precast shear walls and wall segments of effectively monolithic construction shall be evaluated by the criteria defined in Section 6.7.

### C6.8.1.1 Effectively Monolithic Construction

When the precast shear wall is subjected to lateral loading, any yielding and inelastic behavior should take place in the panel elements away from the connections. If the reinforcement detailing in the panel is similar to that for cast-in-place shear walls, then the inelastic response of a precast shear wall should be very similar to that for a cast-in-place wall.

Modern building codes permit the use of precast shear wall construction in high-seismic zones if it satisfies the criteria for cast-in-place shear wall construction.

### 6.8.1.2 Jointed Construction

Precast shear walls and wall segments of jointed construction shall be evaluated by the criteria defined in Section 6.8.2.

### C6.8.1.2 Jointed Construction

For most older structures that contain precast shear walls, and for some modern construction, inelastic activity can be expected in the connections between precast wall panels during severe lateral loading. Because joints between precast shear walls in older buildings have often exhibited brittle behavior during inelastic load reversals, jointed construction had not been permitted in high-seismic zones. Therefore, where evaluating older buildings that contain precast shear walls that are likely to respond as jointed construction, the permissible ductilities and rotation capacities given in Section 6.7 should be reduced.

For some modern structures, precast shear walls have been constructed with special connectors that are detailed to exhibit ductile response and energy absorption characteristics. Many of these connectors are proprietary and only limited experimental evidence concerning their inelastic behavior is available. Although this type of construction is clearly safer than jointed construction in older buildings, the experimental evidence is not sufficient to permit the use of the same ductility and rotation capacities given for cast-in-place construction. Thus, the permissible values given in Section 6.7 should be reduced.

Section 9.6 of FEMA 450 (FEMA 2004) provides testing criteria that may be used to validate design values consistent with the highest performance of monolithic shear wall construction.

### 6.8.1.3 Tilt-Up Construction

Shear walls and wall segments of tilt-up type of precast walls shall be evaluated by the criteria defined in Section 6.8.2.

### C6.8.1.3 Tilt-Up Construction

Tilt-up construction should be considered to be a special case of jointed construction. The walls for most buildings constructed by the tilt-up method are longer than their height. Shear would usually govern their in-plane design. The major concern for most tilt-up construction is the connection between the tilt-up wall and the roof diaphragm. That connection should be analyzed carefully to be sure the diaphragm forces can be transmitted safely to the precast wall system.

### 6.8.2 Precast Concrete Shear Walls and Wall Segments

### 6.8.2.1 General Considerations

The analytical model for a precast concrete shear wall or wall segment shall represent the stiffness, strength, and deformation capacity of the overall member, as well as the connections and joints between any precast panel components that comprise the wall. Potential failure in flexure, shear, and reinforcement development at any point in the shear wall panels or connections shall be considered. Interaction with other structural and nonstructural components shall be included.

Modeling of precast concrete shear walls and wall segments within the precast panels as equivalent beam-columns that include both flexural and shear deformations shall be permitted. The rigid-connection zone at beam connections to these equivalent beam-columns shall represent the distance from the wall centroid to the edge of the wall or wall segment. The different bending capacities for the two loading directions of unsymmetrical precast wall sections shall be modeled.

For precast shear walls and wall segments where shear deformations have a more significant effect on behavior than flextural deformation, a multiple spring model shall be used.

The diaphragm action of concrete slabs interconnecting precast shear walls and frame columns shall be represented in the model.

### 6.8.2.2 Stiffness

The modeling assumptions defined in Section 6.7.2.2 for monolithic concrete shear walls and wall segments shall also be used for precast concrete walls. In addition, the analytical model shall model the axial, shear, and rotational deformations of the connections between the precast components that comprise the wall by either softening the model used to represent the precast panels or by adding spring elements between panels.

*6.8.2.2.1 Linear Static and Dynamic Procedures* The modeling procedures given in Section 6.7.2.2.1, combined with a procedure for including connection deformations as noted above, shall be used.

*6.8.2.2.2 Nonlinear Static Procedure* Nonlinear load–deformation relations shall comply with the requirements of Section 6.3.1.2. The monotonic load–deformation relationships for analytical models that represent precast shear walls and wall elements within precast panels shall be in accordance with the generalized relation shown in Fig. 6-1, except that alternative approaches shall be permitted where verified by experiments. Where the relations are according to Fig. 6-1, the following approach shall be permitted.

Values for plastic hinge rotations or drifts at points *B, C,* and *E* for the two general shapes shall be as defined below. The strength levels at points *B* and *C* shall correspond to the yield strength and nominal strength, as defined in Section 6.7.2.3. The residual strength for the line segment *D–E* shall be as defined below.

For precast shear walls and wall segments whose inelastic behavior under lateral loading is governed by flexure, the general load–deformation relationship shall be defined as in Fig. 6-1(a). For these members, the *x*-axis of Fig. 6-1(a) shall be taken as the rotation over the plastic hinging region at the end of the member as shown in Fig. 6-2. If the requirements for effectively monolithic construction are satisfied, the value of the hinge rotation at point *B* shall correspond to the yield rotation, $\theta_y$, and shall be calculated by Eq. 6-6. The same expression shall also be used for wall segments within a precast panel if flexure controls the inelastic response of the segment. If the precast wall is of jointed construction and flexure governs the inelastic response of the member, then the value of $\theta_y$ shall be increased to account for rotation in the joints between panels or between the panel and the foundation.

For precast shear walls and wall segments whose inelastic behavior under lateral loading is governed by shear, the general load–deformation relationship shall be defined as in Fig. 6-1(b). For these members, the *x*-axis of Fig. 6-1(b) shall be taken as the story drift for shear walls, and as the element drift for wall segments as shown in Fig. 6-3.

For effectively monolithic construction, the values for the variables *a, b,* and *c* required to define the location of points *C, D,* and *E* in Fig. 6-1(a), shall be as specified in Table 6-18. For construction classified as jointed construction, the values of *a, b,* and *c* specified in Table 6-18 shall be reduced to 50% of the given values, unless experimental evidence available to justify higher values is approved by the authority having jurisdiction. In no case, however, shall values larger than those specified in Table 6-18 be used.

For effectively monolithic construction, values for the variables *d, e,* and *c* required to find the points *C, D,* and *E* in Fig. 6-1(b), shall be as specified in Table 6-19 for the appropriate member conditions. For construction classified as jointed construction, the values of *d, e,* and *c* specified in Table 6-19 shall be reduced to 50% of the specified values unless experimental evidence available to justify higher values is approved by the authority having jurisdiction. In no case, however, shall values larger than those specified in Table 6-19 be used.

For Tables 6-18 and 6-19, linear interpolation between tabulated values shall be permitted if the member under analysis has conditions that are between the limits given in the tables.

*6.8.2.2.3 Nonlinear Dynamic Procedure* Nonlinear load–deformation relations for use in analysis by NDP shall model the complete hysteretic behavior of each component using properties verified by experimental evidence. The generalized relation shown in Fig. 6-1 shall be taken to represent the envelope for the analysis. The unloading and reloading stiffnesses and strengths, and any pinching of the load versus rotation hysteresis loops, shall reflect the behavior experimentally observed for wall elements similar to the one under investigation.

### 6.8.2.3 Strength

The strength of precast concrete shear walls and wall segments within the panels shall be computed according to the general requirement of Section 6.3.2, except as modified here. For effectively monolithic construction, the strength calculation procedures given in Section 6.7.2.3 shall be followed.

For jointed construction, calculations of axial, shear, and flexural strength of the connections between panels shall be based on fundamental principles of structural mechanics. Expected yield strength for steel reinforcement of connection hardware used in the connections shall be used where calculating the axial and flexural strength of the connection region. The unmodified specified yield strength of the reinforcement and connection hardware shall be used where calculating the shear strength of the connection region.

For all precast concrete shear walls of jointed construction, no difference shall be taken between the computed yield and nominal strengths in flexure and shear. The values for strength represented by the points *B* and *C* in Fig. 6-1 shall be computed following the procedures given in Section 6.7.2.3.

### C6.8.2.3 Strength

In older construction, particular attention must be given to the technique used for splicing reinforcement extending from adjacent panels into the connection. These connections may be insufficient and often can govern the strength of the precast shear wall system.

### 6.8.2.4 Acceptance Criteria
**Linear Static and Dynamic Procedures**

For precast shear wall construction that is effectively monolithic and for wall segments within a precast panel, the acceptance criteria defined in Section 6.7.2.4.1 shall be followed. For precast shear wall construction defined as jointed construction, the acceptance criteria procedure given in Section 6.7.2.4.1 shall be followed; however, the *m* values specified in Tables 6-20 and 6-21 shall be reduced by 50%, unless experimental evidence justifies the use of a larger value. An *m* value need not be taken as less than 1.0,

and in no case shall be taken larger than the values specified in these tables.

### 6.8.2.4.2 Nonlinear Static and Dynamic Procedures

Inelastic response shall be restricted to those shear walls (and wall segments) and actions listed in Tables 6-18 and 6-19, except where it is demonstrated by experimental evidence and analysis that other inelastic action is acceptable for the selected performance levels. For components experiencing inelastic behavior, the magnitude of the other actions (forces, moments, or torques) in the component shall correspond to the magnitude of the action causing the inelastic behavior. The magnitude of these other actions shall be shown to be below their nominal capacities.

For precast shear walls that are effectively monolithic and wall segments within a precast panel, the maximum plastic hinge rotation angles or drifts during inelastic response shall not exceed the values specified in Tables 6-18 and 6-19. For precast shear walls of jointed construction, the maximum plastic hinge rotation angles or drifts during inelastic response shall not exceed one-half of the values specified in Tables 6-18 and 6-19 unless experimental evidence justifies a higher value. However, in no case shall deformation values larger than those specified in these tables be used for jointed type construction.

If the maximum deformation value exceeds the corresponding tabular value, the element shall be considered to be deficient, and either the element or structure shall be rehabilitated.

Alternative approaches or values shall be permitted where justified by experimental evidence and analysis.

### C6.8.2.4 Acceptance Criteria

The procedures outlined in Section 9.6 of FEMA 450 (FEMA 2004) may be used to establish acceptance criteria for precast shear walls.

### 6.8.2.5 Rehabilitation Measures

Precast concrete shear walls or wall segments that do not meet the acceptance criteria for the selected Rehabilitation Objective shall be rehabilitated. Rehabilitation measures shall meet the requirements of Section 6.3.7 and other provisions of this standard.

### C6.8.2.5 Rehabilitation Measures

Precast concrete shear wall systems may suffer from some of the same deficiencies as cast-in-place walls. These may include inadequate flexural capacity, inadequate shear capacity with respect to flexural capacity, lack of confinement at wall boundaries, and inadequate splice lengths for longitudinal reinforcement in wall boundaries. A few deficiencies unique to precast wall construction are inadequate connections between panels, to the foundation, and to floor or roof diaphragms.

The rehabilitation measures described in Section 6.7.2.5 may be effective in rehabilitating precast concrete shear walls. In addition, the following rehabilitation measures may be effective:

1. **Enhancement of connections between adjacent or intersecting precast wall panels.** Mechanical connectors such as steel shapes and various types of drilled-in anchors, or cast-in-place strengthening methods, or a combination of the two, may be effective in strengthening connections between precast panels. Cast-in-place strengthening methods may include exposing the reinforcing steel at the edges of adjacent panels, adding vertical and transverse (tie) reinforcement, and placing new concrete;

2. **Enhancement of connections between precast wall panels and foundations.** Increasing the shear capacity of the wall panel-to-foundation connection by using supplemental mechanical connectors or by using a cast-in-place overlay with new dowels into the foundation may be an effective rehabilitation measure. Increasing the overturning moment capacity of the panel-to-foundation connection by using drilled-in dowels within a new cast-in-place connection at the edges of the panel may also be an effective rehabilitation measure. Adding connections to adjacent panels may also be an effective rehabilitation measure in eliminating some of the forces transmitted through the panel-to-foundation connection; and

3. **Enhancement of connections between precast wall panels and floor or roof diaphragms.** Strengthening these connections by using either supplemental mechanical devices or cast-in-place connectors may be an effective rehabilitation measure. Both in-plane shear and out-of-plane forces should be considered where strengthening these connections.

## 6.9 CONCRETE-BRACED FRAMES

### 6.9.1 Types of Concrete-Braced Frames

Reinforced concrete-braced frames shall be defined as those frames with monolithic, non-prestressed, reinforced concrete beams, columns, and diagonal braces that are coincident at beam–column joints and that resist lateral loads primarily through truss action.

Where masonry infills are present in concrete-braced frames, requirements for masonry infilled frames as specified in Section 6.6 shall also apply.

The provisions of Section 6.9 shall apply to existing reinforced concrete-braced frames and existing reinforced concrete-braced frames rehabilitated by addition or removal of material.

### 6.9.2 General Considerations

The analytical model for a reinforced concrete-braced frame shall represent the strength, stiffness, and deformation capacity of beams, columns, braces, and all connections and components of the element. Potential failure in tension, compression (including instability), flexure, shear, anchorage, and reinforcement development at any section along the component length shall be considered. Interaction with other structural and nonstructural components shall be included.

The analytical model that represents the framing, using line elements with properties concentrated at component centerlines, shall be permitted. The analytical model also shall comply with the requirements specified in Section 6.4.2.1.

In frames having braces in some bays and no braces in other bays, the restraint of the brace shall be represented in the analytical model as specified above, and the nonbraced bays shall be modeled as frames in compliance with the applicable provisions in other sections of this chapter. Where braces create a vertically discontinuous frame, the effects of the discontinuity on overall building performance shall be considered.

Inelastic deformations in primary components shall be restricted to flexure and axial load in beams, columns, and braces. Other inelastic deformations shall be permitted in secondary components. Acceptance criteria for design actions shall be as specified in Section 6.9.5.

### 6.9.3 Stiffness

#### 6.9.3.1 Linear Static and Dynamic Procedures

Modeling of beams, columns, and braces in braced portions of the frame considering only axial tension and compression flexibilities shall be permitted. Nonbraced portions of frames shall be modeled according to procedures described elsewhere for frames. Effective stiffnesses shall be according to Section 6.3.1.2.

### 6.9.3.2 Nonlinear Static Procedure

Nonlinear load–deformation relations shall comply with the requirements of Section 6.3.1.2.

Beams, columns, and braces in braced portions shall be modeled using nonlinear truss components or other models whose behavior has been demonstrated to adequately represent behavior of concrete components dominated by axial tension and compression loading. Models for beams and columns in nonbraced portions shall comply with requirements for frames specified in Section 6.4.2.2.2. The model shall be capable of representing inelastic response along the component lengths, as well as within connections.

Monotonic load–deformation relations shall be according to the generalized load–deformation relation shown in Fig. 6-1, except that different relations are permitted where verified by experiments. The overall load–deformation relation shall be established so that the maximum resistance is consistent with the design strength specifications of Sections 6.3.2 and 6.4.2.3. Numerical quantities in Fig. 6-1 shall be derived from tests, rational analyses, or criteria of Section 6.6.2.2.2, with braces modeled as columns in accordance with Table 6-16.

### 6.9.3.3 Nonlinear Dynamic Procedure

Nonlinear load–deformation relations for use in analysis by NDP shall model the complete hysteretic behavior of each component using properties verified by experimental evidence. Unloading and reloading properties shall represent stiffness and strength degradation characteristics.

### 6.9.4 Strength

Component strengths shall be computed according to the general requirements of Sections 6.3.2 and the additional requirements of Section 6.4.2.3. The possibility of instability of braces in compression shall be considered.

### 6.9.5 Acceptance Criteria

### 6.9.5.1 Linear Static and Dynamic Procedure

All actions shall be classified as being either deformation-controlled or force-controlled, as defined in Section 2.4.4. In primary components, deformation-controlled actions shall be restricted to flexure and axial actions in beams and columns, and axial actions in braces. In secondary components, deformation-controlled actions shall be restricted to those actions identified for the braced or isolated frame in this chapter.

Calculated component actions shall satisfy the requirements of Section 3.4.2.2. The *m*-factors for concrete frames shall be as specified in other applicable sections of this chapter, and *m*-factors for beams, columns, and braces modeled as tension and compression components shall be as specified for columns in Table 6-17. The *m*-factors shall be reduced to one-half the values in that table, but need not be less than 1.0, where component buckling is a consideration. Alternate approaches or values shall be permitted where justified by experimental evidence and analysis.

### 6.9.5.2 Nonlinear Static and Dynamic Procedures

Calculated component actions shall satisfy the requirements of Section 3.4.2.2 and shall not exceed the numerical values listed in Table 6-16 or the relevant tables for isolated frames specified in other sections of this chapter. Where inelastic action is indicated for a component or action not listed in these tables, the performance shall be deemed unacceptable. Alternate approaches or values shall be permitted where justified by experimental evidence and analysis.

### 6.9.6 Rehabilitation Measures

Concrete-braced frame components that do not meet the acceptance criteria for the selected Rehabilitation Objective shall be rehabilitated. Rehabilitation measures shall meet the requirements of Section 6.3.7 and other provisions of this standard.

### C6.9.6 Rehabilitation Measures

Rehabilitation measures that may be effective in rehabilitating concrete-braced frames include the general approaches listed for other concrete elements in this chapter, plus other approaches based on rational principles.

## 6.10 CAST-IN-PLACE CONCRETE DIAPHRAGMS

### 6.10.1 Components of Concrete Diaphragms

Cast-in-place concrete diaphragms transmit inertial forces within a structure to vertical lateral-force-resisting elements.

Concrete diaphragms shall be made up of slabs, struts, collectors, and chords. Alternatively, diaphragm action may be provided by a structural truss in the horizontal plane. Diaphragms consisting of structural concrete topping on metal deck shall comply with the requirements of Section 5.9.2.

### 6.10.1.1 Slabs

Slabs shall consist of cast-in-place concrete systems that, in addition to supporting gravity loads, transmit inertial loads developed within the structure from one vertical lateral-force-resisting element to another, and provide out-of-plane bracing to other portions of the building.

### 6.10.1.2 Struts and Collectors

Collectors are components that serve to transmit the inertial forces within the diaphragm to elements of the lateral-force-resisting system. Struts are components of a structural diaphragm used to provide continuity around an opening in the diaphragm. Struts and collectors shall be monolithic with the slab, occurring either within the slab thickness or being thicker than the slab.

### 6.10.1.3 Diaphragm Chords

Diaphragm chords are components along diaphragm edges with increased longitudinal and transverse reinforcement, acting primarily to resist tension and compression forces generated by bending in the diaphragm. Exterior walls shall be permitted to serve as chords provided there is adequate strength to transfer shear between the slab and wall.

### C6.10.1.3 Diaphragm Chords

When evaluating an existing building, special care should be taken to evaluate the condition of the lap splices. Where the splices are not confined by closely spaced transverse reinforcement, splice failure is possible if stress levels reach critical values. In rehabilitation construction, new laps should be confined by closely spaced transverse reinforcement.

## 6.10.2 Analysis, Modeling, and Acceptance Criteria

### 6.10.2.1 General Considerations

The analytical model for a diaphragm shall represent the strength, stiffness, and deformation capacity of each component and the diaphragm as a whole. Potential failure in flexure, shear, buckling, and reinforcement development shall be considered.

Modeling of the diaphragm as a continuous or simple span horizontal beam supported by elements of varying stiffness shall be permitted. The beam shall be modeled as rigid, stiff, or flexible considering the deformation characteristics of the actual system.

### C6.10.2.1 General Considerations

Some computer models assume a rigid diaphragm. Few cast-in-place diaphragms would be considered flexible; however, a thin concrete slab on a metal deck might be stiff depending on the length-to-width ratio of the diaphragm.

### 6.10.2.2 Stiffness

Diaphragm stiffness shall be modeled according to Section 6.10.2.1 and shall be determined using a linear elastic model and gross section properties. The modulus of elasticity used shall be that of the concrete as specified in ACI 318 (ACI 2002). Where the length-to-width ratio of the diaphragm exceeds 2.0 (where the length is the distance between vertical elements), the effects of diaphragm flexibility shall be considered where assigning lateral forces to the resisting vertical elements.

### C6.10.2.2 Stiffness

The concern is for relatively flexible vertical members that may be displaced by the diaphragm, and for relatively stiff vertical members that may be overloaded by the same diaphragm displacement.

### 6.10.2.3 Strength

Strength of cast-in-place concrete diaphragm components shall comply with the requirements of Sections 6.3.2 as modified in this section.

The maximum component strength shall be determined considering potential failure in flexure, axial load, shear, torsion, development, and other actions at all points in the component under the actions of design gravity and lateral load combinations. The shear strength shall be as specified in Chapter 21 of ACI 318 (ACI 2002). Strut, collector, and chord strengths shall be as determined for frame components in Section 6.4.2.3.

### 6.10.2.4 Acceptance Criteria

Diaphragm shear and flexure shall be considered deformation-controlled. Acceptance criteria for slab component actions shall be as specified for shear walls in Section 6.7.2.4, with m-values taken according to similar components in Tables 6-20 and 6-21 for use in Eq. 3-20. Acceptance criteria for struts, chords, and collectors shall be as specified for frame components in Section 6.4.2.4. Connections shall be considered force-controlled.

## 6.10.3 Rehabilitation Measures

Concrete diaphragms that do not meet the acceptance criteria for the selected Rehabilitation Objective shall be rehabilitated. Rehabilitation measures shall meet the requirements of Section 6.3.7 and other provisions of this standard.

## C6.10.3 Rehabilitation Measures

Cast-in-place concrete diaphragms can have a wide variety of deficiencies; see Chapter 10 and ASCE 31 (ASCE 2002).

Two general alternatives may be effective in correcting deficiencies: either improve the strength and ductility, or reduce the demand in accordance with FEMA 172 (FEMA 1992). Providing additional reinforcement and encasement may be an effective measure to strengthen or improve individual components. Increasing the diaphragm thickness may also be effective, but the added weight may overload the footings and increase the seismic loads. Lowering seismic demand by providing additional lateral-force-resisting elements, introducing additional damping, or base-isolating the structure may also be effective rehabilitation measures.

## 6.11 PRECAST CONCRETE DIAPHRAGMS

### 6.11.1 Components of Precast Concrete Diaphragms

Precast concrete diaphragms are elements comprising primarily precast components with or without topping, that transmit shear forces from within a structure to vertical lateral-force-resisting elements.

Precast concrete diaphragms shall be classified as topped or untopped. A topped diaphragm shall be defined as one that includes a reinforced structural concrete topping slab poured over the completed precast horizontal system. An untopped diaphragm shall be defined as one constructed of precast components without a structural cast-in-place topping.

### C6.11.1 Components of Precast Concrete Diaphragms

Section 6.10 provided a general overview of concrete diaphragms. Components of precast concrete diaphragms are similar in nature and function to those of cast-in-place diaphragms, with a few critical differences. One is that precast diaphragms do not possess the inherent unity of cast-in-place monolithic construction. Additionally, precast components may be highly stressed due to prestressed forces. These forces cause long-term shrinkage and creep, which shorten the component over time. This shortening tends to fracture connections that restrain the component.

Most floor systems have a topping system, but some hollow-core floor systems do not. The topping slab generally bonds to the top of the precast components, but may have an inadequate thickness at the center of the span, or may be inadequately reinforced. Also, extensive cracking of joints may be present along the panel joints. Shear transfer at the edges of precast concrete diaphragms is especially critical.

Some precast roof systems are constructed as untopped systems. Untopped precast concrete diaphragms have been limited to lower seismic zones by recent versions of the *Uniform Building Code* (ICBO 1997). This limitation has been imposed because of the brittleness of connections and lack of test data concerning the various precast systems. Special consideration shall be given to diaphragm chords in precast construction.

### 6.11.2 Analysis, Modeling, and Acceptance Criteria

Analysis and modeling of precast concrete diaphragms shall conform to Section 6.10.2.2, with the added requirement that the analysis and modeling shall account for the segmental nature of the individual components.

Component strengths shall be determined in accordance with Section 6.10.2.3. Welded connection strength shall be based on rational procedures, and connections shall be assumed to have little ductility capacity unless test data verify higher ductility values. Precast concrete diaphragms with reinforced concrete topping slabs shall be considered deformation-controlled in shear and flexure. *m*-factors shall be taken as 1.0, 1.25, and 1.5 for Immediate Occupancy, Life Safety, and Collapse Prevention Performance Levels, respectively.

Untopped precast concrete diaphragms shall be considered force-controlled.

### C6.11.2 Analysis, Modeling, and Acceptance Criteria

Welded connection strength can be determined using the latest version of the *Precast Concrete Institute (PCI) Handbook* (PCI 1999). A discussion of design provisions for untopped precast diaphragms can be found in the Appendix to Chapter 9 of FEMA 368 (FEMA 2001).

The appendix to Chapter 9 of FEMA 450 (FEMA 2004) provides discussion of the behavior of untopped precast diaphragms and outlines a design approach that may be used for such diaphragms to satisfy the requirements of this standard.

### 6.11.3 Rehabilitation Measures

Precast concrete diaphragms that do not meet the acceptance criteria for the selected Rehabilitation Objective shall be rehabilitated. Rehabilitation measures shall meet the requirements of Section 6.3.7 and other provisions of this standard.

### C6.11.3 Rehabilitation Measures

Section 6.10.3 provides guidance for rehabilitation measures for concrete diaphragms in general.

Special care should be taken to overcome the segmental nature of precast concrete diaphragms, and to avoid damaging prestressing strands when adding connections.

## 6.12 CONCRETE FOUNDATION COMPONENTS

### 6.12.1 Types of Concrete Foundations

Foundations shall be defined as those components that serve to transmit loads from the vertical structural subsystems (columns and walls) of a building to the supporting soil or rock. Concrete foundations for buildings shall be classified as either shallow or deep foundations as defined in Chapter 4. Requirements of Section 6.12 shall apply to shallow foundations that include spread or isolated footing, strip or line footing, combination footing, and concrete mat footing; and to deep foundations that include pile foundations and cast-in-place piers. Concrete grade beams shall be permitted in both shallow and deep foundation systems and shall comply with the requirements of Section 6.12

The provisions of Section 6.12 shall apply to existing foundation components and to new materials or components that are required to rehabilitate an existing building.

#### 6.12.1.1 Shallow Foundations

Existing spread footings, strip footings, and combination footings are reinforced or unreinforced. Vertical loads are transmitted by these footings to the soil by direct bearing; lateral loads are transmitted by a combination of friction between the bottom of the footing and the soil, and passive pressure of the soil on the vertical face of the footing.

Concrete mat footings shall be reinforced to resist the flexural and shear stresses resulting from the superimposed concentrated and line structural loads and the distributed resisting soil pressure under the footing. Lateral loads shall be resisted by friction between the soil and the bottom of the footing, and by passive pressure developed against foundation walls that are part of the system.

#### 6.12.1.2 Deep Foundations

6.12.1.2.1 Driven Pile Foundations Concrete pile foundations shall be composed of a reinforced concrete pile cap supported on driven piles. The piles shall be concrete (with or without prestressing), steel shapes, steel pipes, or composite (concrete in a driven steel shell). Vertical loads shall be transmitted to the

piles by the pile cap. Pile foundation resistance to vertical loads shall be calculated based on the direct bearing of the pile tip in the soil, the skin friction or cohesion of the soil on the surface area of the pile, or based on a combination of these mechanisms. Lateral loads resistance shall be calculated based on passive pressure of the soil on the vertical face of the pile cap, in combination with interaction of the piles in bending and passive soil pressure on the pile surface.

6.12.1.2.2 Cast-in-Place Pile Foundations Cast-in-place concrete pile foundations shall consist of reinforced concrete placed in a drilled or excavated shaft. Cast-in-place pile or pier foundations resistance to vertical and lateral loads shall be calculated in the same manner as that of driven pile foundations specified in Section 6.12.1.2.1.

### C6.12.1.2 Deep Foundations

C6.12.1.2.1 Driven Pile Foundations In poor soils, or soils subject to liquefaction, bending of the piles may be the only dependable resistance to lateral loads.

C6.12.1.2.2 Cast-in-Place Pile Foundations Segmented steel cylindrical liners are available to form the shaft in weak soils and allow the liner to be removed as the concrete is placed. Various slurry mixes are often used to protect the drilled shaft from caving soils. The slurry is then displaced as the concrete is placed by the tremie method.

### 6.12.2 Analysis of Existing Foundations

For concrete buildings, components shall be considered fixed against rotation at the top of the foundation if the connections between components and foundations, the foundations and supporting soil are shown to be capable of resisting the induced moments. Where components are not designed to resist flexural moments, or the connections between components and foundations are not capable of resisting the induced moments, they shall be modeled with pinned ends. In such cases, the column base shall be evaluated for the resulting axial and shear forces as well as the ability to accommodate the necessary end rotation of the columns. The effects of base fixity of columns shall be taken into account at the point of maximum displacement of the superstructure.

If a more rigorous analysis procedure is used, appropriate vertical, lateral, and rotational soil springs shall be incorporated in the analytical model as described in Section 4.4.2. The spring characteristics shall be as specified in Chapter 4. Rigorous analysis of structures with deep foundations in soft soils shall be based on special soil–pile interaction studies to determine the probable location of the point of fixity in the foundation and the resulting distribution of forces and displacements in the superstructure. In these analyses, the appropriate representation of the connection of the pile to the pile cap shall be included in the model. Piles with less than 6 in. of embedment without any dowels into the pile cap shall be modeled as being "pinned" to the cap. Unless the pile and pile cap connection detail is identified as otherwise from the available construction documents, the pinned connection shall be used in the analytical model.

Where the foundations are included in the analytical model, the responses of the foundation components shall be considered. The reactions of structural components attached at the foundation (axial loads, shears, and moments) shall be used to evaluate the individual components of the foundation system.

### C6.12.2 Analysis of Existing Foundations

Overturning moments and economics may dictate the use of more rigorous analysis procedures.

### 6.12.3 Evaluation of Existing Condition

Allowable soil capacities (subgrade modulus, bearing pressure, passive pressure) and foundation displacements for the selected performance level shall be as prescribed in Chapter 4 or as established with project-specific data. All components of existing foundation systems and all new material, components, or components required for rehabilitation shall be evaluated as force-controlled actions. However, the capacity of the foundation components need not exceed 1.25 times the capacity of the supported vertical structural component or element (column or wall).

### 6.12.4 Rehabilitation Measures

Existing foundations that do not meet the acceptance criteria for the selected Rehabilitation Objective shall be rehabilitated. Rehabilitation measures shall meet the requirements of Section 6.3.7 and other provisions of this standard.

### C6.12.4 Rehabilitation Measures

Rehabilitation measures described in Section C6.12.4.1 for shallow foundations and in Section C6.12.4.2 for deep foundations may be effective in rehabilitating existing foundations.

### C6.12.4.1 Rehabilitation Measures for Shallow Foundations

1. **Enlarging the existing footing by lateral additions.** Enlarging the existing footing may be an effective rehabilitation measure. The enlarged footing may be considered to resist subsequent actions produced by the design loads, provided that adequate shear and moment transfer capacity are provided across the joint between the existing footing and the additions;
2. **Underpinning the footing.** Underpinning an existing footing involves the removal of unsuitable soil underneath, coupled with replacement using concrete, soil cement, suitable soil, or other material. Underpinning should be staged in small increments to prevent endangering the stability of the structure. This technique may be used to enlarge an existing footing or to extend it to a more competent soil stratum;
3. **Providing tension tie-downs.** Tension ties (soil and rock anchors—prestressed and unstressed) may be drilled and grouted into competent soils and anchored in the existing footing to resist uplift. Increased soil bearing pressures produced by the ties should be checked against the acceptance criteria for the selected Performance Level specified in Chapter 4. Piles or drilled piers may also be effective in providing tension tie-downs of existing footings;
4. **Increasing effective depth of footing.** This method involves pouring new concrete to increase shear and moment capacity of the existing footing. The new concrete must be adequately doweled or otherwise connected so that it is integral with the existing footing. New horizontal reinforcement should be provided, if required, to resist increased moments;
5. **Increasing the effective depth of a concrete mat foundation with a reinforced concrete overlay.** This method involves pouring an integral topping slab over the existing mat to increase shear and moment capacity;
6. **Providing pile supports for concrete footings or mat foundations.** Adding new piles may be effective in providing support for existing concrete footing or mat foundations, provided the pile locations and spacing are designed to avoid overstressing the existing foundations;
7. **Changing the building structure to reduce the demand on the existing elements.** This method involves removing mass or height of the building or adding other materials or components (such as

energy dissipation devices) to reduce the load transfer at the base level. New shear walls or braces may be provided to reduce the demand on existing foundations;

8. **Adding new grade beams.** This approach involves the addition of grade beams to tie existing footings together where poor soil exists, to provide fixity to column bases, and to distribute lateral loads between individual footings, pile caps, or foundation walls; and

9. **Improving existing soil.** This approach involves grouting techniques to improve existing soil.

### C6.12.4.2 Rehabilitation Measures for Deep Foundations

1. **Providing additional piles or piers.** Providing additional piles or piers may be effective, provided extension and additional reinforcement of existing pile caps comply with the requirements for extending existing footings in Section C6.12.4.1;

2. **Increasing the effective depth of the pile cap.** New concrete and reinforcement to the top of the pile cap may be effective in increasing its shear and moment capacity, provided the interface is designed to transfer actions between the existing and new materials;

3. **Improving soil adjacent to existing pile cap.** Soil improvement adjacent to existing pile caps may be effective if undertaken in accordance with guidance provided in Section 4.3;

4. **Increasing passive pressure bearing area of pile cap.** Addition of new reinforced concrete extensions to the existing pile cap may be effective in increasing the vertical foundation bearing area and load resistance;

5. **Changing the building system to reduce the demands on the existing elements.** New lateral-load-resisting elements may be effective in reducing demand;

6. **Adding batter piles or piers.** Adding batter piles or piers to existing pile or pier foundation may be effective in resisting lateral loads. It should be noted that batter piles have performed poorly in recent earthquakes where liquefiable soils were present. This is especially important to consider around wharf structures and in areas having a high water table. Addition of batter piles to foundations in areas of such seismic hazards should be in accordance with requirements in Section 4.4; and

7. **Increasing tension tie capacity from pile or pier to superstructure.** Added reinforcement should satisfy the requirements of Section 6.3.

## 7.0 MASONRY

## 7.1 SCOPE

This chapter sets forth requirements for the Systematic Rehabilitation of concrete- or clay-unit masonry components of the lateral-force-resisting system of an existing building. The requirements of this chapter shall apply to existing masonry components of a building system, rehabilitated masonry components of a building system, and new masonry components that are added to an existing building system.

Section 7.2 specifies data collection procedures for obtaining material properties and performing condition assessments. Sections 7.3 and 7.4 provide modeling procedures, component strengths, acceptance criteria, and rehabilitation measures for masonry walls and masonry infills. Section 7.5 specifies requirements for anchorage to masonry walls. Section 7.6 specifies requirements for masonry foundation elements.

## C7.1 SCOPE

The provisions of this chapter should be applied to solid or hollow clay-unit masonry, solid or hollow concrete-unit masonry, and hollow clay tile. Stone or glass block masonry is not covered in this chapter.

Portions of masonry buildings that are not subject to systematic rehabilitation provisions include parapets, cladding, and partition walls.

If the Simplified Rehabilitation Method of Chapter 10 is followed, unreinforced masonry buildings with flexible floor and roof diaphragms may be evaluated using the procedures given in ASCE 31 (ASCE 2002).

Techniques for repair of earthquake-damaged masonry components are not included in this standard. The design professional is referred to FEMA 306 (FEMA 1998), FEMA 307 (FEMA 1998), and FEMA 308 (FEMA 1998) for information on evaluation and repair of masonry wall components.

## 7.2 MATERIAL PROPERTIES AND CONDITION ASSESSMENT

### 7.2.1 General

Mechanical properties for masonry materials and components shall be based on available drawings, specifications, and other documents for the existing construction in accordance with requirements of Section 2.2. Where such documents fail to provide adequate information to quantify masonry material properties or the condition of masonry components of the structure, such information shall be supplemented by materials tests and assessments of existing conditions as required in Section 2.2.6.

Material properties of existing masonry components shall be determined in accordance with Section 7.2.2. A condition assessment shall be conducted in accordance with Section 7.2.3. The extent of materials testing and condition assessment performed shall be used to determine the knowledge factor as specified in Section 7.2.4.

Use of default material properties shall be permitted in accordance with Section 7.2.2.10.

Use of material properties based on historical information as default values shall be as specified in Section 7.2.2.10. Other values of material properties shall be permitted if rationally justified, based on available historical information for a particular type of masonry construction, prevailing codes, and assessment of existing conditions.

Procedures for defining masonry structural systems and assessing masonry condition shall be conducted in accordance with the provisions stated in Section 7.2.3.

## C7.2.1 General

Construction of existing masonry buildings in the United States dates back to the 1500s in the southeastern and southwestern regions, to the 1770s in the central and eastern regions, and to the 1850s in the western half of the nation. The stock of existing masonry buildings in the United States is composed largely of structures constructed in the last 150 years. Since the types of units, mortars, and construction methods changed during this time, knowing the age of a masonry building may be useful in identifying the characteristics of its construction. Although structural properties cannot be inferred solely from age, some background on typical materials and methods for a given period can help to improve engineering judgment and provide some direction in the assessment of an existing building. The design professional should be aware that values given in some existing documents are working stress values rather than the expected or lower-bound strengths used in this standard.

As indicated in Chapter 1, great care should be exercised in selecting the appropriate rehabilitation approaches and techniques for application to historic buildings to preserve their unique characteristics.

## 7.2.2 Properties of In-Place Materials

### 7.2.2.1 General

The following component and connection material properties shall be obtained for the as-built structure in accordance with Sections 7.2.2.1 through 7.2.2.10:

1. Masonry compressive strength;
2. Masonry tensile strength;
3. Masonry shear strength;
4. Masonry elastic modulus;
5. Masonry shear modulus; and
6. Strength and modulus of elasticity of reinforcing steel.

Where material testing is required by Section 2.2.6, test methods to quantify masonry material properties shall comply with Sections 7.2.2.2 through 7.2.2.8. The minimum number of tests shall comply with the requirements of Section 7.2.2.9.

Expected material properties shall be based on mean values from test data unless specified otherwise. Lower-bound material properties shall be based on mean values from test data minus one standard deviation unless specified otherwise.

The condition of existing masonry shall be classified as good, fair, or poor as defined as follows, or based on other approved procedures that consider the nature and extent of damage or deterioration present.

**Good Condition:** Masonry found during condition assessment to have mortar and units intact with no visible cracking.

**Fair Condition:** Masonry found during condition assessment to have mortar and units intact but with minor cracking.

**Poor condition:** Masonry found during condition assessment to have degraded mortar, degraded masonry units, or significant cracking.

### C7.2.2.1 General

The design professional is referred to FEMA 306 (FEMA 1998), FEMA 307 (FEMA 1998), and FEMA 308 (FEMA 1998) for additional information regarding the condition of masonry. The classification of the condition of masonry requires consideration of the type of component, the anticipated mode of inelastic behavior, and the nature and extent of damage or deterioration. These documents also contain extensive information regarding the effects of damage on strength, stiffness, and displacement limits for masonry components. Included are damage classification guides with visual representations of typical earthquake-related damage of masonry components, which may be useful in classifying the condition of masonry for this standard. The severity of damage described in FEMA 306, FEMA 307, and FEMA 308 is categorized as Insignificant, Slight, Moderate, Heavy, and Extreme. Masonry in good condition has severity of damage not exceeding Insignificant or Slight, as defined by FEMA 306. Masonry in fair condition has severity of damage not exceeding Moderate. Masonry with Heavy or Extreme damage is classified as Poor.

### 7.2.2.2 Nominal or Specified Properties

Nominal material properties, or properties specified in construction documents, shall be taken as lower-bound material properties. Corresponding expected material properties shall be calculated by multiplying lower-bound values by a factor as specified in Table 7-2 to translate from lower-bound to expected values.

### 7.2.2.3 Masonry Compressive Strength

Expected masonry compressive strength, $f_{me}$, shall be measured using one of the following three methods:

1. Test prisms shall be extracted from an existing wall and tested in accordance with Section 1.4.B.3 of ACI 530.1/ASCE 6/TMS 602, *Specifications for Masonry Structures* (ACI 2002);
2. Prisms shall be fabricated from actual extracted masonry units, and a surrogate mortar designed on the basis of a chemical analysis of actual mortar samples. The test prisms shall be tested in accordance with Section 1.4.B.3 of ACI 530.1/ASCE 6/TMS 602; or
3. For solid unreinforced masonry, the strength of the masonry can be estimated using a flatjack test in accordance with ASTM C1196-03 (ASTM 2003).

For each of the three methods enumerated in this section, the expected compressive strength shall be based on the net mortared area.

### C7.2.2.3 Masonry Compressive Strength

The three test methods are further described in Section C7.2.2.1 of FEMA 274 (FEMA 1997). As an alternative to the test methods given in this section of this standard, the expected masonry compressive strength may be deduced from a nominal value prescribed in ACI 530.1/ASCE 6/TMS 602 (ACI 2002).

### 7.2.2.4 Masonry Elastic Modulus in Compression

Expected values of elastic modulus for masonry in compression, $E_{me}$, shall be measured using one of the following two methods:

1. Test prisms shall be extracted from an existing wall and tested in compression. Stresses and deforma-

tions shall be measured to determine modulus values; or
2. For solid unreinforced masonry, the modulus can be measured using a flatjack test in accordance with ASTM C1197-03 (ASTM 2003).

### C7.2.2.4 Masonry Elastic Modulus in Compression

Both methods measure vertical strain between two gauge points to infer strain, and thus elastic modulus. They are further described in FEMA 274 (FEMA 1997), Section C7.2.2.2.

### 7.2.2.5 Masonry Flexural Tensile Strength

Expected flexural tensile strength, $f_{te}$, for out-of-plane bending shall be measured using one of the following three methods:

1. Test samples shall be extracted from an existing wall and subjected to minor-axis bending using the bond-wrench method of ASTM C1072-00 (ASTM 2000);
2. Test samples shall be tested in situ using the bond-wrench method; or
3. Sample wall panels shall be extracted and subjected to minor-axis bending in accordance with ASTM E518-02 (ASTM 2002).

Flexural tensile strength for unreinforced masonry (URM) walls subjected to in-plane lateral forces shall be assumed to be equal to that for out-of-plane bending, unless testing is done to define the expected tensile strength for in-plane bending.

### C7.2.2.5 Masonry Flexural Tensile Strength

The flexural tensile strength of older brick masonry walls constructed with lime mortars may often be neglected. The tensile strength of newer concrete- and clay-unit masonry walls can result in appreciable flexural strengths.

The three test methods for out-of-plane bending are further described in Section C7.2.2.3 of FEMA 274 (FEMA 1997). For in-plane bending, flexural stress gradients across the section width are much lower than for out-of-plane bending. Thus, data from tests described in this section will be very conservative and should be used only in lieu of data on in-plane tensile strength.

### 7.2.2.6 Masonry Shear Strength

For URM components, lower-bound shear strength shall be measured using an approved in-place shear test. Lower-bound masonry shear strength, $v_{mL}$, shall be determined in accordance with Eq. 7-1:

$$v_{mL} = \frac{0.75\left(0.75v_{tL} + \frac{P_D}{A_n}\right)}{1.5} \qquad \text{(Eq. 7-1)}$$

where

$P_D$ = superimposed dead load at the top of the wall or wall pier under consideration;

$A_n$ = area of net mortared/grouted section of a wall or wall pier; and

$v_{tL}$ = lower-bound bed-joint shear strength defined as lower 20th percentile of $v_{to}$, given in Eq. 7-2.

Values for the lower-bound mortar shear strength, $v_{tL}$, shall not exceed 100 psi for the determination of $v_{me}$ in Eq. 7-1.

Individual bed joint shear strength test values, $v_{to}$, shall be determined in accordance with Eq. 7-2:

$$v_{to} = \frac{V_{test}}{A_b} - p_{D+L} \qquad \text{(Eq. 7-2)}$$

where

$V_{test}$ = test load at first movement of a masonry unit;

$A_b$ = sum of net mortared area of bed joints above and below the test unit; and

$p_{D+L}$ = gravity compressive stress at the test location considering actual dead plus live loads in place at the time of testing.

The in-place shear test shall not be used to estimate shear strength of reinforced masonry components. The expected shear strength of reinforced masonry components shall be determined in accordance with Section 7.3.4.2.

### C7.2.2.6 Masonry Shear Strength

The available standard for masonry shear strength test is UBC 21-6 (ICBO 1997a). Section C7.3.2.4 of FEMA 274 (FEMA 1997) further describes this test and also an alternate procedure.

### 7.2.2.7 Masonry Shear Modulus

The expected shear modulus of masonry (unreinforced or reinforced), $G_{me}$, shall be permitted to be taken as 0.4 times the elastic modulus in compression.

### C7.2.2.7 Masonry Shear Modulus

Shear stiffness of post-cracked masonry should be taken as a fraction of the initial uncracked masonry shear stiffness value. The design professional is referred to FEMA 274 (FEMA 1997),

Section C7.3.2.5 for additional information regarding masonry shear modulus.

### 7.2.2.8 Strength and Modulus of Reinforcing Steel

The expected yield strength of reinforcing bars, $f_{ye}$, shall be based on mill test data, or tension tests of actual reinforcing bars taken from the subject building. Tension tests shall be performed in accordance with ASTM A615/A615M-03 (ASTM 2003).

The modulus of elasticity of steel reinforcement, $E_{se}$, shall be assumed to be 29,000,000 psi.

### 7.2.2.9 Minimum Number of Tests

Materials testing is not required if material properties are available from original construction documents that include material test records or material test reports. Otherwise, minimum number of tests shall be performed as specified in 7.2.2.9.1 or 7.2.2.9.2, as applicable.

*7.2.2.9.1 Usual Testing* The minimum number of tests to determine masonry and reinforcing steel material properties for usual data collection shall be based on the following criteria:

1. If the specified design strength of the masonry is known, and the masonry is in good or fair condition, at least one test shall be performed on samples of each different masonry strength used in the construction of the building, with a minimum of three tests performed for the entire building. If the masonry is in poor condition, additional tests shall be performed to determine the extent of the reduced material properties;
2. If the specified design strength of the masonry is not known, at least one test shall be performed on each type of component, with a minimum of six tests performed on the entire building;
3. If the specified design strength of the reinforcing steel is known, use of nominal or specified material properties shall be permitted without additional testing; and
4. If the specified design strength of the reinforcing steel is not known, at least two strength coupons of reinforcing steel shall be removed from a building for testing.

*7.2.2.9.2 Comprehensive Testing* The minimum number of tests necessary to quantify properties by in-place testing for comprehensive data collection shall be based on the following criteria:

1. For masonry in good or fair condition as defined in this standard, a minimum of three tests shall be performed for each masonry type, and for each three

floors of construction or 3,000 sf of wall surface, if original construction records are available that specify material properties; six tests shall be performed if original construction records are not available. At least two tests shall be performed for each wall or line of wall elements providing a common resistance to lateral forces. A minimum of eight tests shall be performed for each building; and

2. For masonry in poor condition as defined in this standard, additional tests shall be done to estimate material strengths in regions where properties differ, or nondestructive condition assessment tests in accordance with Section 7.2.3.2 shall be used to quantify variations in material strengths.

Samples for tests shall be taken at locations representative of the material conditions throughout the entire building, taking into account variations in workmanship at different story levels, variations in weathering of the exterior surfaces, and variations in the condition of the interior surfaces due to deterioration caused by leaks and condensation of water and/or the deleterious effects of other substances contained within the building.

An increased sample size shall be permitted to improve the confidence level. The relation between sample size and confidence shall be as defined in ASTM E139-00 (ASTM 2000).

If the coefficient of variation in test measurements exceeds 25%, the number of tests performed shall be doubled.

If mean values from in situ material tests are less than the default values prescribed in Section 7.2.2.10, the number of tests performed shall be doubled.

### C7.2.2.9 Minimum Number of Tests

The number and location of material tests should be selected to provide sufficient information to adequately define the existing condition of materials in the building. Test locations should be identified in those masonry components that are determined to be critical to the primary path of lateral-force resistance.

### 7.2.2.10 Default Properties

Use of default material properties to determine component strengths shall be permitted with the linear analysis procedures in Chapter 3.

Default lower-bound values for masonry compressive strength, elastic modulus in compression, flexural tensile strength, and masonry shear strength shall be based on Table 7-1. Default lower-bound masonry properties for fair condition shall be equal to two-thirds of the values for masonry in good condition. Default lower-bound masonry properties for poor condition shall be equal to one-third of the values for good condition. Default expected strength values for masonry compressive strength, elastic modulus in compression, flexural tensile strength, and masonry shear strength shall be determined by multiplying lower-bound values by an appropriate factor taken from Table 7-2.

Default lower-bound and expected strength yield stress values for reinforcing bars shall be determined in accordance with Section 6.2.2.5.

### Table 7-1. Default Lower-Bound Masonry Properties

| Property | Masonry Condition[1] | | |
| --- | --- | --- | --- |
| | Good | Fair | Poor |
| Compressive Strength $(f'_m)$[2] | 900 psi | 600 psi | 300 psi |
| Elastic Modulus in Compression | $550f'_m$ | $550f'_m$ | $550f'_m$ |
| Flexural Tensile Strength[3] | 20 psi | 10 psi | 0 |
| Shear Strength[4] | | | |
| Masonry with a Running Bond Lay-Up | 27 psi | 20 psi | 13 psi |
| Fully Grouted Masonry with a Lay-Up Other Than Running Bond | 27 psi | 20 psi | 13 psi |
| Partially Grouted or Ungrouted Masonry with a Lay-Up Other Than Running Bond | 11 psi | 8 psi | 5 psi |

[1]Masonry condition shall be classified as good, fair, or poor as defined in Section 7.2.2.1
[2]It shall be permitted to take default lower-bound values for masonry compressive strength in good condition from Table 1 and 2 from Section 1.4 of ACI 530.1/ASCE 6/TMS 602 (ACI 2002).
[3]It shall be permitted to take default lower-bound values for masonry flexural tensile strength in good condition from Table 3.1.7.2.1 of ACI 530/ASCE 5/TMS 402 (ACI 2002).
[4]It shall be permitted to take default lower-bound shear strength of unreinforced masonry in good condition from Section 3.3.4 of ACI 530/ASCE 5/TMS 402 (ACI 2002).

**Table 7-2. Factors to Translate Lower-Bound Masonry Properties to Expected Strength Masonry Properties[1]**

| Property | Factor |
|---|---|
| Compressive Strength ($f_{me}$) | 1.3 |
| Elastic Modulus in Compression[2] | — |
| Flexural Tensile Strength | 1.3 |
| Shear Strength | 1.3 |

[1]See Chapter 6 for properties of reinforcing steel
[2]The expected elastic modulus in compression shall be taken as $550f_{me}$ where $f_{me}$ is the expected masonry compressive strength.

### C7.2.2.10 Default Properties

Default properties for masonry based on the tables in current code provisions are applicable to buildings built with materials similar to those specified in current codes. Where materials are different (i.e., type of mortar, unit strength, air-entrainment), default properties should be based on Table 7-1.

Default values of compressive strength are set at very low stresses to reflect an absolute lower bound. Masonry in poor condition is given a strength equal to one-third of that for masonry in good condition, to reflect the influence of mortar deterioration and unit cracking on compressive strength. The coefficient of 550 for default values of elastic modulus in compression in Table 7-1 is set lower than values given in the *International Building Code* (ICC 2003) to compensate for larger values of expected strength. Default values for flexural tensile strength are set low even for masonry in good condition because of its dependence on the unit-mortar bonding, which can be highly variable due to the variability of the condition of the mortar. Comparison of default masonry shear values with values that may be obtained from Eq. 7-1 shows that if in-place shear tests are done, a significant increase in strength over default values is possible.

### 7.2.3 Condition Assessment

A condition assessment of the existing building and site conditions shall be performed as specified in Sections 7.2.3.1 through 7.2.3.3.

A condition assessment shall include the following:

1. The physical condition of primary and secondary components shall be examined and the presence of any degradation shall be noted;
2. The presence and configuration of components and their connections, and the continuity of load paths between components, elements, and systems shall be verified or established; and

3. Other conditions, including the presence and attachment of veneer, neighboring party walls and buildings, presence of nonstructural components, prior remodeling, and limitations for rehabilitation that may influence building performance, shall be identified and documented.

The condition of the masonry shall be classified as good, fair, or poor as defined in this standard, based on the results of visual examination conducted in accordance with Section 7.2.3.1.

### C7.2.3 Condition Assessment

Buildings are often constructed with veneer as an architectural finish, which may make the wall appear thicker than the actual structural thickness. In many areas of the country, the veneer wythe is separated from the structural wall by an air space to provide ventilation and moisture control. This is called cavity wall construction. In this case, the veneer may be anchored but does not add any strength to the assembly.

In areas of the southwest United States and along the California coast (as well as other regions), the veneer is placed directly against the building wall. It will be in a running bond pattern without a header course. Other patterns are also seen. If the veneer is not anchored or has a layer of building paper between it and the inner wythe, it cannot be considered as part of the structural wall.

Veneer on modern buildings may be adhered or anchored. In either case, the veneer is a weight to be considered but does not contribute to a wall's strength. In all cases, the veneer must be anchored to prevent its detaching during an earthquake. Requirements for veneer are specified in Chapter 11.

Face brick bonded to the inner wythes with a regular pattern of header courses is not veneer. In this case, the outer wythes are part of the structural wall and can be used in evaluating the height-to-thickness ratio of the wall.

See Section C7.2.2.1 regarding the use of FEMA 306 (FEMA 1998), FEMA 307 (FEMA 1998), and FEMA 308 (FEMA 1998) for additional information in classifying the condition of masonry.

### 7.2.3.1 Visual Condition Assessment

The size and location of all masonry shear and bearing walls shall be determined by visual examination. The orientation and placement of the walls shall be noted. Overall dimensions of masonry components shall be measured or determined from plans, including wall heights, lengths, and thicknesses. Locations and sizes of window and door openings shall be measured or determined from plans. The distribution of gravity loads to bearing walls shall be estimated.

Walls shall be classified as reinforced or unreinforced, composite or noncomposite, and grouted, partially grouted, or ungrouted. For reinforced masonry (RM) construction, the size and spacing of horizontal and vertical reinforcement shall be estimated. For multi-wythe construction, the number of wythes shall be noted, as well as the distance between wythes, and the placement of inter-wythe ties. The condition and attachment of veneer wythes shall be noted. For grouted construction, the quality of grout placement shall be assessed. For partially grouted walls, the locations of grout placement shall be identified.

The type and condition of the mortar and mortar joints shall be determined. Mortar shall be examined for weathering, erosion, and hardness, and to identify the condition of any repointing, including cracks, internal voids, weak components, and/or deteriorated or eroded mortar. Horizontal cracks in bed joints, vertical cracks in head joints and masonry units, and diagonal cracks near openings shall be noted.

Vertical components that are not straight shall be identified. Bulging or undulations in walls shall be observed, as well as separation of exterior wythes, out-of-plumb walls, and leaning parapets or chimneys.

Connections between masonry walls and floors or roofs shall be examined to identify details and condition. If construction drawings are available, a minimum of three connections shall be inspected for each connection type. If no deviations from the drawings are found, the sample shall be considered representative. If drawings are unavailable, or if deviations are noted between the drawings and constructed work, then a random sample of connections shall be inspected until a representative pattern of connections is identified.

### 7.2.3.2 Comprehensive Condition Assessment

The following nondestructive tests shall be permitted to quantify and confirm the uniformity of construction quality and the presence and degree of deterioration for comprehensive data collection:

1. Ultrasonic or mechanical pulse velocity to detect variations in the density and modulus of masonry materials and to detect the presence of cracks and discontinuities;
2. Impact-echo test to confirm whether reinforced walls are grouted; and
3. Radiography to confirm location of reinforcing steel.

The location and number of nondestructive tests shall be determined in accordance with the requirements of Section 7.2.2.9.2.

### C7.2.3.2 Comprehensive Condition Assessment

Nondestructive tests may be used to supplement the visual observations required by Section 7.2.3.1.

#### C7.2.3.2.1 Ultrasonic Pulse Velocity
Measurement of the velocity of ultrasonic pulses through a wall can detect variations in the density and modulus of masonry materials as well as the presence of cracks and discontinuities. Transmission times for pulses traveling through a wall (direct method) or between two points on the same side of a wall (indirect method) are measured and used to infer wave velocity.

Test equipment with wave frequencies in the range of 50 kHz has been shown to be appropriate for masonry walls. Use of equipment with higher frequency waves is not recommended because the short wave length and high attenuation are not consistent with typical dimensions of masonry units. Test locations should be sufficiently close to identify zones with different properties. Contour maps of direct transmission wave velocities can be constructed to assess the overall homogeneity of a wall elevation. For indirect test data, vertical or horizontal distance can be plotted versus travel time to identify changes in wave velocity (slope of the curve). Abrupt changes in slope will identify locations of cracks or flaws.

Ultrasonic methods are not applicable for masonry of poor quality or low modulus, or with many flaws and cracks. The method is sensitive to surface condition, the coupling material used between the transducer or receiver and the brick, and the pressure applied to the transducer.

The use of ultrasonic pulse velocity methods with masonry walls has been researched extensively by Calvi (1988), Epperson and Abrams (1989), and Kingsley et al. (1987). A standard for the use of ultrasonic methods for masonry is currently under development in Europe with RILEM Committee 76LUM.

#### C7.2.3.2.2 Mechanical Pulse Velocity
The mechanical pulse velocity test consists of impacting a wall with a hammer blow and measuring the travel time of a sonic wave across a specified gauge distance. An impact hammer is equipped with a load cell or accelerometer to detect the time of impact. A distant accelerometer is fixed to a wall to detect the arrival time of the pulse. Wave velocity is determined by dividing the gauge length by the travel time. The form and duration of the generated wave can be varied by changing the material on the hammer cap.

The generated pulse has a lower frequency and higher energy content than an ultrasonic pulse, resulting in longer travel distances and less sensitivity to small variations in masonry properties and minor cracking. The mechanical pulse method should be used in lieu of the ultrasonic pulse method where overall mean properties of a large portion of masonry are of interest.

The use of mechanical pulse velocity measurements for masonry condition assessments has been confirmed through research by Epperson and Abrams (1989) and Kingsley et al. (1987). Although no standard exists for mechanical pulse velocity tests with masonry, a standard for concrete materials [ASTM C597-02 (ASTM 2002)] does exist.

*C7.2.3.2.3 Impact Echo* The impact-echo technique can be useful for nondestructive determination of the location of void areas within grouted reinforced walls, as reported by Sansalone and Carino (1988). Commercial devices are available or systems can be assembled using available electronic components. Since this technique cannot distinguish between a shrinkage crack at the grout-unit interface and a complete void in the grout, drilling of small holes in the bed joint or examination using an optical borescope should be performed to verify the exact condition.

*C7.2.3.2.4 Radiography* A number of commercial radiographic (x-ray) devices exist that can be used to identify the location of reinforcing steel in masonry walls. They are also useful for locating bed-joint reinforcing steel, masonry ties and anchors, and conduits and pipes. The better devices can locate a No. 6 bar at depths up to approximately 6 in.; however, this means that for a 12-in.-thick concrete masonry wall, a bar located off-center cannot be found where access is limited to only one side of the wall. These devices are not able to locate or determine the length of reinforcing bar splices in walls in most cases. They work best for identifying the location of single isolated bars and become less useful where the congestion of reinforcing bars increases.

**7.2.3.3 Supplemental Tests**

Supplemental tests shall be permitted to enhance the level of confidence in masonry material properties, or the assessment of masonry condition for justifying the use of a higher knowledge factor, as specified in Section 7.2.4.

**C7.2.3.3 Supplemental Tests**

Ancillary tests are recommended, but not required, to enhance the level of confidence in

masonry material properties or to assess condition. Possible supplemental tests are described as follows.

*C7.2.3.3.1 Surface Hardness* The surface hardness of exterior wythe masonry can be evaluated using the Schmidt rebound hammer. Research has shown that the technique is sensitive to differences in masonry strength, but cannot by itself be used to determine absolute strength. A Type N hammer (5,000 lb.) is recommended for normal-strength masonry, while a Type L hammer (1,600 lb.) is recommended for lower-strength masonry. Impacts at the same test location should be continued until consistent readings are obtained, because surface roughness can affect initial readings.

The method is limited to tests of only the surface wythe. Tuckpointing may influence readings and the method is not sensitive to cracks.

Measurement of surface hardness for masonry walls has been studied by Noland et al. (1987).

*C7.2.3.3.2 Vertical Compressive Stress* In situ vertical compressive stress resisted by the masonry can be measured using a thin hydraulic flatjack that is inserted into a removed mortar bed joint. Pressure in the flatjack is increased until distortions in the brickwork are reduced to the precut condition. Existing vertical compressive stress is inferred from the jack hydraulic pressure, using correction factors for the shape and stiffness of the flatjack.

The method is useful for measurement of gravity load distribution, flexural stresses in out-of-plane walls, and stresses in masonry veneer walls that are compressed by a surrounding concrete frame. The test is limited to only the face wythe of masonry.

Not less than three tests should be done for each section of the building for which it is desired to measure in situ vertical stress. The number and location of tests should be determined based on the building configuration and the likelihood of overstress conditions.

*C7.2.3.3.3 Diagonal Compression Test* A square panel of masonry is subjected to a compressive force applied at two opposite corners along a diagonal until the panel cracks. Shear strength is inferred from the measured diagonal compressive force based on a theoretical distribution of shear and normal stress for a homogeneous and elastic continuum. Using the same theory, shear modulus is inferred from measured diagonal compressive stress and strain.

Extrapolation of the test data to actual masonry walls is difficult because the ratio of shear to normal stress is fixed at a constant ratio of 1.0 for the test specimens. Also, the distribution of shear and normal stresses across a bed joint may not be as uniform for a test specimen as for an actual wall. Lastly, any redistribution of stresses after the first cracking will not be represented with the theoretical stress distributions. Thus, the test data cannot be useful to predict nonlinear behavior.

If the size of the masonry units relative to the panel dimension is large, masonry properties will be not continuous, but discrete. Test panels should be a minimum of four ft square. The high cost and disruption of extracting a number of panels this size may be impractical. The standard test method specified in ASTM E519-02 (ASTM 2002) may be used.

*C7.2.3.3.4 Large-Scale Load Tests* Large-scale destructive tests may be done on portions of a masonry component or element to (1) increase the confidence level on overall structural properties; (2) obtain performance data on archaic building materials and construction materials; (3) quantify effects of complex edge and boundary conditions around openings and two-way spanning; and (4) verify or calibrate analytical models. Large-scale load tests do not necessarily have to be run to the ultimate limit state. They may have value for simply demonstrating structural integrity up to some specific performance level.

Out-of-plane strength and behavior of masonry walls can be determined with air-bag tests. Behavior of test panels incorporating connections and edge details can be determined from such a test, in addition to flexural and arching properties of a solid or perforated wall. Strength and deformation capacity under in-plane lateral forces can be determined by loading an individual portion of wall that is cut free of the surrounding masonry. Loading actuators are reacted against adjacent and stronger portions of masonry. Such testing is particularly useful where the wall is composed of different materials that cannot be evaluated by testing an individual unit of an individual wythe.

Visual and nondestructive surveys should be used to identify locations for test samples.

Standards for laboratory test methods are published by ASTM. Procedures for removal and transportation of masonry samples are given in *Building Science Series 62* (NBS 1977).

Large-scale tests are expensive and limited to a single or few samples. They may result in considerable local damage and may require substantial reconstruction near the sample location. Test data must be extrapolated to the remainder of the system, based on a low confidence level.

### 7.2.4 Knowledge Factor

A knowledge factor for computation of masonry component capacities and permissible deformations shall be selected in accordance with Section 2.2.6.4.

## 7.3 MASONRY WALLS

The procedures set forth in this section for determination of stiffness, strength, and deformation of masonry walls shall be applied to building systems comprising any combination of existing masonry walls, masonry walls enhanced for seismic rehabilitation, and new walls added to an existing building for seismic rehabilitation.

Actions in a structure shall be classified as being either deformation-controlled or force-controlled as defined in Section 2.4.4.3. Design strengths for deformation-controlled and force-controlled actions shall be calculated in accordance with this section.

Strengths used for deformation-controlled actions are denoted $Q_{CE}$ and shall be taken as equal to expected strengths obtained experimentally, calculated using accepted mechanics principles, or based on default values listed in Section 7.2.2.10. Expected strength is defined as the mean maximum resistance expected over the range of deformations to which the component is likely to be subjected. Where calculations are used to define expected strength, expected material properties shall be used. Unless otherwise specified in this standard, use of strength design procedures specified in ACI 530/ASCE 5/TMS 402 (ACI 2002) to calculate design strengths shall be permitted except that the strength reduction factor, $\phi$, shall be taken equal to unity.

Force-controlled actions shall be as defined in Section 2.4.4. Strengths used in design for force-controlled actions are denoted $Q_{CL}$ and shall be taken as equal to lower-bound strengths obtained experimentally, calculated using established mechanics principles, or based on default values listed in Section 7.2.2.10. Lower-bound strength is defined as the mean minus one standard deviation of resistance over the range of deformations and loading cycles to which the component is subjected. Where calculations are used to define lower-bound strengths, lower-bound material properties shall be used. It shall be permitted to calculate lower-bound properties from expected properties using the conversion factors in Table 7-2. Unless otherwise specified in this standard, use of strength design procedures specified in ACI 530/ASCE 5/TMS 402 to calculate design strengths shall be permitted except that the strength reduction factor, $\phi$, shall be taken equal to unity. Where alternative definitions of design strength are used, they shall be justified by experimental evidence.

Where design actions are determined using the nonlinear procedures of Chapter 3, component force–deformation response shall be represented by nonlinear force–deformation relations. Force–deformation relations shall be based on experimental evidence or the generalized force–deformation relation shown in Fig. 7-1, with parameters $c$, $d$, and $e$ as defined in Tables 7-4 and 7-7.

## C7.3 MASONRY WALLS

Expected yield strength of reinforcing steel, as specified in this standard, includes consideration of material overstrength and strain-hardening.

Component drift ratios are the ratio of differential displacement, $\Delta_{eff}$, between each end of the component over the effective height, $h_{eff}$, of the component. Depending on the geometry of the wall or wall pier configuration, the elevations at which these parameters are determined may vary within the same wall element, as shown in Fig. C7-1.

Materials having brittle behavior as shown in Fig. 7-1(b) should be considered force-controlled actions. Rocking of unreinforced masonry walls and wall piers have a limited s emiductile behavior similar to that shown in Fig. 7-1(a) if all components in a line of resistance have an in-plane shear capacity greater than their rocking capacity.

### 7.3.1 Types of Masonry Walls

Masonry walls shall be categorized as unreinforced or reinforced; ungrouted, partially grouted, or fully grouted; and composite or noncomposite. Masonry walls shall be capable of resisting forces applied parallel to their plane and normal to their plane, as described in Sections 7.3.2 through 7.3.5.

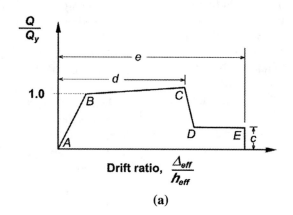

(a)

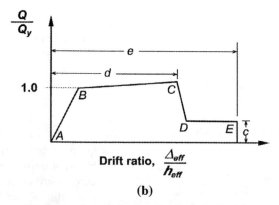

(b)

**FIGURE 7-1. (a) Generalized Force–Deformation Relation for Reinforced Masonry Elements or Components; (b) Generalized Force–Deformation Relation for Unreinforced Masonry Elements or Components.**

## C7.3.1 Types of Masonry Walls

Any of these categories of masonry elements can be used in combination with existing, rehabilitated, or new lateral-force-resisting elements of other materials such as steel, concrete, or timber.

### 7.3.1.1 Existing Masonry Walls

Existing masonry walls shall include all structural walls of a building system that are in place prior to seismic rehabilitation.

Existing masonry walls shall be assumed to behave in the same manner as new masonry walls, provided that the masonry is in fair or good condition as defined in this standard.

### 7.3.1.2 New Masonry Walls

New masonry walls shall include all new wall elements added to an existing lateral-force-resisting system. New walls shall be designed in accordance with the requirements set forth in this standard and detailed and constructed in accordance with a building code approved by the authority having jurisdiction.

### C7.3.1.2 New Masonry Walls

Codes for new buildings include the *International Building Code* (ICC 2003), *National Building Code* (BOCAI 1999), *Standard Building Code* (SBCCI 1999), and the *Uniform Building Code* (ICC 2003). Guidelines for seismic design of new buildings are found in FEMA 302 (FEMA 1997).

### 7.3.1.3 Enhanced Masonry Walls

Enhanced masonry walls shall include existing walls that are rehabilitated by an approved method.

### C7.3.1.3 Enhanced Masonry Walls

Methods of enhancing masonry walls are intended to improve performance of masonry walls subjected to both in-plane and out-of-plane lateral forces.

Possible rehabilitation methods are described in Sections C7.3.1.3.1 through C7.3.1.3.10.

*C7.3.1.3.1 Infilled Openings* An infilled opening may be considered to act compositely with the surrounding masonry if new and old masonry units are interlaced at the boundary with full toothing, or attached with anchorage that provides compatible shear strength at the interface of new and old units.

Stiffness assumptions, strength criteria, and acceptable deformations for masonry walls with infilled openings should be the same as given for non-rehabilitated solid masonry walls; differences in elastic moduli and strengths for the new and old masonry walls should be considered for the composite section.

*C7.3.1.3.2 Enlarged Openings* Openings in unreinforced masonry (URM) shear walls may be enlarged by removing portions of masonry above or below windows or doors.

Openings are enlarged to increase the height-to-length aspect ratio of wall piers so that the limit state may be altered from shear to flexure. This method is only applicable to URM walls.

Stiffness assumptions, strength criteria, and acceptable deformations for URM walls with enlarged openings shall be reassessed to reflect the final condition of the wall.

*C7.3.1.3.3 Shotcrete* An existing masonry wall with an application of shotcrete may be considered to behave as a composite section if anchorage is provided at the shotcrete–masonry interface to transfer the shear forces calculated in accordance with Chapter 3. Stresses in the masonry and shotcrete should be determined considering the difference in elastic moduli for each material, or the existing masonry wall should be neglected and the new shotcrete layer should be designed to resist all of the force.

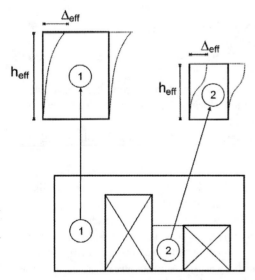

$h_{eff}$ = The effective height of the component under consideration

$\Delta_{eff}$ = The differential displacement between the top and bottom of the component

Depending on the wall and pier geometry, the elevations at which these parameters are defined may vary in the same wall assembly.

**FIGURE C7-1. Effective Height and Differential Displacement of Wall Components.**

Stiffness assumptions, strength criteria, and acceptable deformations for masonry components with shotcrete should be the same as that for new reinforced concrete components. Variations in boundary conditions should be considered.

*C7.3.1.3.4 Coatings for Unreinforced Masonry Walls*
A coated masonry wall may be considered a composite section as long as anchorage is provided at the interface between the coating and the masonry wall to transfer shear forces. Stresses in the masonry and coating should be determined considering the difference in elastic moduli for each material. If stresses exceed expected strengths of the coating material, then the coating should be considered ineffective.

Stiffness assumptions, strength criteria, and acceptable deformations for coated masonry walls should be the same as that for existing URM walls.

*C7.3.1.3.5 Reinforced Cores for Unreinforced Masonry Walls* A reinforced-cored masonry wall should be considered to behave as a reinforced masonry (RM) wall, provided that the bond between the new reinforcement and the grout and between the grout and the cored surface are capable of transferring seismic forces computed in accordance with Chapter 3. Vertical reinforcement should be anchored at the base of the wall to resist the full tensile strength of the wall.

Grout in new reinforced cores should consist of cementitious materials whose hardened properties are compatible with those of the surrounding masonry.

Adequate shear strength must exist, or should be provided, so that the strength of the new vertical reinforcement can be developed.

Stiffness assumptions, strength criteria, and acceptable deformations for URM walls with reinforced cores should be the same as that for existing reinforced walls.

*C7.3.1.3.6 Prestressed Cores for Unreinforced Masonry Walls* A prestressed-cored masonry wall with unbonded tendons should be considered to behave as a URM wall with increased vertical compressive stress.

Losses in prestressing force due to creep and shrinkage of the masonry should be accounted for in analyses conducted in accordance with Chapter 3.

Stiffness assumptions, strength criteria, and acceptable deformations for URM walls with unbonded prestressing tendons should be the same as for existing URM walls subjected to vertical compressive stress.

*C7.3.1.3.7 Grout Injections* Grout used for filling voids and cracks should have strength, modulus, and thermal properties compatible with the existing masonry.

Inspections should be conducted in accordance with Chapter 2 during the grouting process to ensure that voids are completely filled with grout.

Stiffness assumptions, strength criteria, and acceptable deformations for masonry walls with grout injections should be the same as that for existing URM or RM walls.

*C7.3.1.3.8 Repointing* Bond strength of new mortar should be equal to or greater than that of the original mortar. Compressive strength of new mortar should be equal to or less than that of the original mortar.

Stiffness assumptions, strength criteria, and acceptable deformations for repointed masonry walls should be the same as that for existing masonry walls.

*C7.3.1.3.9 Braced Masonry Walls* Masonry walls with height-to-thickness ratios in excess of those permitted by Table 7-5, or out-of-plane bending stresses in excess of those permitted by Section 7.3.3.2, may be braced with external structural elements. Adequate strength should be provided in the bracing element and connections to resist the transfer of forces from the masonry wall to the bracing element. Out-of-plane deflections of braced walls resulting from the transfer of vertical floor or roof loadings should be considered.

Stiffness assumptions, strength criteria, and acceptable deformations for braced masonry walls should be the same as that for existing masonry walls. The reduced span of the masonry wall should be considered.

*C7.3.1.3.10 Stiffening Elements* Masonry walls with inadequate out-of-plane stiffness or strength may be stiffened with external structural members. The stiffening members should be proportioned to resist a tributary portion of lateral load applied normal to the plane of a masonry wall. Connections at the ends of the stiffening element should be provided to transfer the reaction force. Flexibility of the stiffening element should be considered where estimating lateral drift of a masonry wall panel.

Stiffness assumptions, strength criteria, and acceptable deformations for stiffened masonry walls should be the same as that for existing masonry walls. The stiffening action that the new element provides shall be considered.

*C7.3.1.3.11 Veneer Attachment* Veneer not bonded to the structural core of a masonry wall may be rehabilitated by the use of pins inserted through the joints and into the brick substrate. Spacing of pins should match current code requirements given the seismicity of the region.

## 7.3.2 Unreinforced Masonry Walls and Wall Piers In-Plane

Engineering properties of URM walls subjected to lateral forces applied parallel to their plane shall be determined in accordance with this section. Requirements of this section shall apply to cantilevered shear walls that are fixed against rotation at their base, and to wall piers between window or door openings that are fixed against rotation top and bottom.

Stiffness and strength criteria presented in this section shall apply to both the Linear Static Procedures (LSP) and Nonlinear Static Procedures (NSP) prescribed in Chapter 3.

### 7.3.2.1 Stiffness

The lateral stiffness of masonry walls subjected to lateral in-plane forces shall be determined considering both flexural and shear deformations.

The masonry assemblage of units, mortar, and grout shall be considered to be a homogeneous medium for stiffness computations with an expected elastic modulus in compression, $E_{me}$, as specified in Section 7.2.2.4.

For linear procedures, the stiffness of a URM wall or wall pier resisting lateral forces parallel to its plane shall be considered to be linear and proportional with the geometrical properties of the uncracked section, excluding veneer wythes.

Story shears in perforated shear walls shall be distributed to wall piers in proportion to the relative lateral uncracked stiffness of each wall pier.

Stiffnesses for existing and enhanced walls shall be determined using principles of mechanics used for new walls.

### C7.3.2.1 Stiffness

Laboratory tests of solid shear walls have shown that behavior can be depicted at low force levels using conventional principles of mechanics for homogeneous materials. In such cases, the lateral in-plane stiffness of a solid cantilevered shear wall, $k$, can be calculated using Eq. C7-1:

$$k = \frac{1}{\dfrac{h_{eff}^3}{3E_m I_g} + \dfrac{h_{eff}}{A_v G_m}} \qquad \text{(Eq. C7-1)}$$

where

$h_{eff}$ = wall height;
$A_v$ = shear area;
$I_g$ = moment of inertia for the gross section representing uncracked behavior;
$E_m$ = masonry elastic modulus; and
$G_m$ = masonry shear modulus.

Correspondingly, the lateral in-plane stiffness of a wall pier between openings with full restraint against rotation at its top and bottom can be calculated using Eq. C7-2:

$$k = \frac{1}{\dfrac{h_{eff}^3}{12E_m I_g} + \dfrac{h_{eff}}{A_v G_m}} \qquad \text{(Eq. C7-2)}$$

The design professional should be aware that a completely fixed condition is often not present in actual buildings.

The exterior wythe of brick in a URM wall is commonly a veneer that is not bonded to the wall. This veneer should not be used where computing the lateral resistance of the wall.

### 7.3.2.2 Strength

*7.3.2.2.1 Expected Lateral Strength of Unreinforced Masonry Walls and Wall Piers* Expected lateral strength, $Q_{CE}$, of existing and enhanced URM walls or wall pier components shall be the expected rocking strength, calculated in accordance with Eq. 7-3:

$$Q_{CE} = V_r = 0.9\alpha P_D \left(\frac{L}{h_{eff}}\right) \qquad \text{(Eq. 7-3)}$$

where

$h_{eff}$ = height to resultant of lateral force;
$L$ = length of wall or wall pier;
$P_D$ = superimposed dead load at the top of the wall or wall pier under consideration;
$V_r$ = strength of wall or wall pier based on rocking; and
$\alpha$ = factor equal to 0.5 for fixed-free cantilever wall, or equal to 1.0 for fixed-fixed wall pier.

*7.3.2.2.2 Lower-Bound Lateral Strength of Unreinforced Masonry Walls and Wall Piers* Lower-bound lateral strength, $Q_{CL}$, of existing and enhanced URM walls or wall pier components shall be taken as the lesser of the lateral strength values based on lower-bound shear strength or toe compressive stress calculated in accordance with Eqs. 7-4 and 7-5, respectively. $L/h_{eff}$ shall not be taken less than 0.67 for use in Eq. 7-5.

$$Q_{CL} = V_{CL} = v_{mL} A_n \qquad \text{(Eq. 7-4)}$$

$$Q_{CL} = V_{tc} = \alpha Q_G \left( \frac{L}{h_{eff}} \right) \left( 1 - \frac{f_a}{0.7 f'_m} \right) \qquad \text{(Eq. 7-5)}$$

where $h_{eff}$, $L$, and $\alpha$ are the same as given for Eq. 7-3 and:

$v_{mL}$ = lower-bound masonry shear strength, Eq. 7-1;

$A_n$ = area of net mortared/grouted section;

$f_a$ = axial compressive stress due to gravity loads specified in Eq. 3-2;

$f'_m$ = lower-bound masonry compressive strength determined in accordance with Section 7.2.2.3;

$Q_G$ = lower-bound axial compressive force due to gravity loads specified in Eq. 3-3;

$V_{CL}$ = lower-bound shear strength for wall or wall pier; and

$V_{tc}$ = lower-bound shear strength based on toe compressive stress for wall or wall pier.

### 7.3.2.2.3 Lower-Bound Vertical Compressive Strength of Unreinforced Masonry Walls and Wall Piers
Lower-bound vertical compressive strength of existing URM walls or wall pier components shall be limited by lower-bound masonry compressive stress in accordance with Eq. 7-6.

$$Q_{CL} = P_{CL} = 0.80(0.85 f'_m A_n) \qquad \text{(Eq. 7-6)}$$

where:

$f'_m$ = lower-bound compressive strength determined in accordance with Section 7.2.2;

$P_{CL}$ = lower-bound masonry compressive stress; and

$A_n$ = area of net mortared/grouted section.

### 7.3.2.3 Acceptance Criteria
In-plane lateral shear of unreinforced masonry walls and wall piers in a single line of resistance shall be considered a deformation-controlled action if the expected lateral rocking strength of each wall or wall pier in the line of resistance, as specified in Section 7.3.2.2.1, is less than the lower-bound lateral strength of each wall or wall pier limited by shear or toe compressive stress, as specified in Section 7.3.2.2.2. Unreinforced masonry walls not meeting the criteria for deformation-controlled components shall be considered force-controlled components. Expected rocking strength, $V_r$, as specified in Section 7.3.2.2.1 shall be neglected in lines of resistance not considered deformation-controlled. Axial compression on URM wall components shall be considered a force-controlled action.

#### 7.3.2.3.1 Linear Procedures
For the linear procedures in Sections 3.3.1 and 3.3.2, component actions shall be compared with capacities in accordance with Section 3.4.2.2. The $m$-factors for use with corresponding expected strength shall be obtained from Table 7-3.

#### 7.3.2.3.2 Nonlinear Procedures
For the NSP given in Section 3.3.3, wall and wall pier components shall meet the requirements of Section 3.4.3.2. For deformation-controlled components, nonlinear deformations shall not exceed the values given in Table 7-4. Variables $d$ and $e$, representing nonlinear deformation capacities for primary and secondary components, shall be expressed in terms of drift ratio percentages as defined in Fig. 7-1(a).

For the Nonlinear Dynamic Procedure (NDP) given in Section 3.3.4, wall and wall pier components shall meet the requirements of Section 3.4.3.2. Nonlinear force–deflection relations for deformation-controlled wall and wall pier components shall be established based on the information given in

**Table 7-3. Linear Static Procedure—$m$-factors for URM In-Plane Walls and Wall Piers**

| Limiting Behavioral Mode | $m$-factors | | | | |
|---|---|---|---|---|---|
| | | Performance Level | | | |
| | | Primary | | Secondary | |
| | IO | LS | CP | LS | CP |
| Rocking | $1.5h_{eff}/L$ (not less than 1) | $3h_{eff}/L$ (not less than 1.5) | $4h_{eff}/L$ (not less than 2) | $6h_{eff}/L$ (not less than 3) | $8h_{eff}/L$ (not less than 4) |

**Table 7-4. Nonlinear Static Procedure—Simplified Force–Deflection Relations for URM In-Plane Walls and Wall Piers[1]**

| Limiting Behavioral Mode | | | | Acceptance Criteria[2] | | | | |
| --- | --- | --- | --- | --- | --- | --- | --- | --- |
| | | | | | Performance Level | | | |
| | | | | | Primary | | Secondary | |
| | $c$ (%) | $d$ (%) | $e$ (%) | IO % | LS % | CP % | LS % | CP % |
| Rocking | 0.6 | $0.4h_{eff}/L$ | $0.8h_{eff}/L$ | 0.1 | $0.3h_{eff}/L$ | $0.4h_{eff}/L$ | $0.6h_{eff}/L$ | $0.8h_{eff}/L$ |

[1]Interpolation shall be used between table values.
[2]Primary and secondary component demands shall be within secondary component acceptance criteria where the full backbone curve is explicitly modeled including strength degradation and residual strength in accordance with Section 3.4.3.2.

Table 7-4, or an approved procedure based on a comprehensive evaluation of the hysteretic characteristics of those components.

### 7.3.3 Unreinforced Masonry Walls Out-of-Plane

As required by Section 2.6.7, URM walls shall be evaluated for out-of-plane inertial forces as isolated components spanning between floor levels, and/or spanning horizontally between columns or pilasters. URM walls shall not be analyzed out-of-plane with the LSP or NSP prescribed in Chapter 3.

#### 7.3.3.1 Stiffness

The out-of-plane stiffness of walls shall be neglected in analytical models of the global structural system in the orthogonal direction.

#### 7.3.3.2 Strength

Unless arching action is considered, flexural cracking shall be limited by the expected tensile stress values given in Section 7.2.2.5.

Arching action shall be considered only if surrounding floor, roof, column, or pilaster elements have sufficient stiffness and strength to resist thrusts from arching of a wall panel, and a condition assessment has been performed to ensure that there are no gaps between a wall panel and the adjacent structure.

The condition of the collar joint shall be considered where estimating the effective thickness of a wall for out-of-plane behavior. The effective void ratio shall be taken as the ratio of the collar joint area without mortar to the total area of the collar joint. Wythes separated by collar joints that are not bonded,

or have an effective void ratio greater than 50% shall not be considered part of the effective thickness of the wall.

#### C7.3.3.2 Strength

This section applies to treatment of veneer for out-of-plane behavior of walls only. For in-plane resistance, effective thickness is the sum of all wythes without consideration of the condition of the collar joints.

#### 7.3.3.3 Acceptance Criteria

For the Immediate Occupancy Structural Performance Level, flexural cracking in URM walls due to out-of-plane inertial loading shall not be permitted as limited by the tensile stress requirements of Section 7.3.3.2. For the Life Safety and Collapse Prevention Structural Performance Levels, flexural cracking in URM walls due to out-of-plane inertial loading shall be permitted provided that cracked wall segments will remain stable during dynamic excitation. Stability shall be checked using analytical time-step integration models considering acceleration time histories at the top and base of a wall panel. For the Life Safety and Collapse Prevention Structural Performance Levels, stability need not be checked for walls spanning vertically with a height-to-thickness ($h/t$) ratio less than that given in Table 7-5.

#### C7.3.3.3 Acceptance Criteria

For further information on evaluating the stability of unreinforced masonry walls out-of-plane, refer to *Methodology for Mitigation of Seismic Hazards in Existing Unreinforced Masonry Buildings* (ABK 1984).

**Table 7-5. Permissible *h/t* Ratios for URM Out-of-Plane Walls**

| Wall Types | $S_{X1} \leq 0.24$ g | $0.24$ g $< S_{X1} \leq 0.37$ g | $S_{X1} > 0.37$ g |
|---|---|---|---|
| Walls of One-Story Buildings | 20 | 16 | 13 |
| First-Story Wall of Multistory Building | 20 | 18 | 15 |
| Walls in Top Story of Multistory Building | 14 | 14 | 9 |
| All Other Walls | 20 | 16 | 13 |

### 7.3.4 Reinforced Masonry Walls and Wall Piers In-Plane

#### 7.3.4.1 Stiffness
The stiffness of an RM wall or wall pier component in-plane shall be determined as follows:

1. The shear stiffness of RM wall components shall be based on uncracked section properties; and
2. The flexural stiffness of RM wall components shall be based on cracked section properties. Use of a cracked moment of inertia equal to 50% of $I_g$ shall be permitted.

In either case, veneer wythes shall not be considered in the calculation of wall component properties. Stiffnesses for existing and new walls shall be assumed to be the same.

#### 7.3.4.2 Strength
The strength of existing, enhanced, and new RM wall or wall pier components in flexure, shear, and axial compression shall be determined in accordance with Section 7.3.4.2. The strength of flanged RM walls shall also be in accordance with Section 7.3.4.2.1 and 7.3.4.2.2.

*7.3.4.2.1 Flexural Strength of Walls and Wall Piers*
Expected flexural strength of an RM wall or wall pier shall be determined based on strength design procedures specified in ACI 530/ASCE 5/TMS 402 (ACI 2002).

*7.3.4.2.2 Shear Strength of Walls and Wall Piers* The lower-bound shear strength of RM wall or wall pier components, $V_{CL}$, shall be determined based on strength design procedures specified in ACI 530/ASCE 5/TMS 402 (ACI 2002). Design actions (axial, flexure, and shear) on components shall be determined in accordance with Chapter 3 of this standard consid-

ering gravity loads and the maximum forces that can be transmitted based on a limit-state analysis.

*7.3.4.2.3 Vertical Compressive Strength of Walls and Wall Piers* Lower-bound vertical compressive strength of existing RM wall or wall pier components shall be determined based on strength design procedures specified in ACI 530/ASCE 5/TMS 402 (ACI 2002).

*7.3.4.2.4 Strength Considerations for Flanged Walls* Wall intersections shall be considered effective in transferring shear where either condition (1) or (2) and condition (3) are met:

1. The face shells of hollow masonry units are removed and the intersection is fully grouted;
2. Solid units are laid in running bond, and 50% of the masonry units at the intersection are interlocked;
3. Reinforcement from one intersecting wall continues past the intersection a distance not less than 40 bar diameters or 24 in.

The width of flange considered effective in compression on each side of the web shall be taken as the lesser of six times the thickness of the web, half the distance to the next web, or the actual flange on either side of the web wall.

The width of flange considered effective in tension on each side of the web shall be taken as the lesser of three-fourths of the wall height, half the distance to an adjacent web, or the actual flange on either side of the web wall.

#### 7.3.4.3 Acceptance Criteria
The shear required to develop the expected strength of reinforced masonry walls and wall piers in flexure shall be compared to the lower-bound shear strength. For reinforced masonry wall components

214

**Table 7-6. Acceptance Criteria for Linear Procedures—Reinforced Masonry In-Plane Walls**

| | | | $m$-factors[1] | | | | |
|---|---|---|---|---|---|---|---|
| | | | | Performance Level | | | |
| | | | | | Component Type | | |
| | | | | Primary | | Secondary | |
| $f_{ae}/f_{me}$ | $L/h_{eff}$ | $\rho_g f_{ye}/f_{me}$[3] | IO | LS | CP | LS | CP |
| Wall Components Controlled by Flexure | | | | | | | |
| 0.00 | ≤ 0.5 | ≤ 0.01 | 4.0 | 7.0 | 8.0 | 8.0 | 10.0 |
| | | 0.05 | 2.5 | 5.0 | 6.5 | 8.0 | 10.0 |
| | | ≥ 0.20 | 1.5 | 2.0 | 2.5 | 4.0 | 5.0 |
| | 1.0 | ≤ 0.01 | 4.0 | 7.0 | 8.0 | 8.0 | 10.0 |
| | | 0.05 | 3.5 | 6.5 | 7.5 | 8.0 | 10.0 |
| | | ≥ 0.20 | 1.5 | 3.0 | 4.0 | 6.0 | 8.0 |
| | ≥ 2.0 | ≤ 0.01 | 4.0 | 7.0 | 8.0 | 8.0 | 10.0 |
| | | 0.05 | 3.5 | 6.5 | 7.5 | 8.0 | 10.0 |
| | | ≥ 0.20 | 2.0 | 3.5 | 4.5 | 7.0 | 9.0 |
| 0.038 | ≤ 0.5 | ≤ 0.01 | 3.0 | 6.0 | 7.5 | 8.0 | 10.0 |
| | | 0.05 | 2.0 | 3.5 | 4.5 | 7.0 | 9.0 |
| | | ≥ 0.20 | 1.5 | 2.0 | 2.5 | 4.0 | 5.0 |
| | 1.0 | ≤ 0.01 | 4.0 | 7.0 | 8.0 | 8.0 | 10.0 |
| | | 0.05 | 2.5 | 5.0 | 6.5 | 8.0 | 10.0 |
| | | ≥ 0.20 | 1.5 | 2.5 | 3.5 | 5.0 | 7.0 |
| | ≥ 2.0 | ≤ 0.01 | 4.0 | 7.0 | 8.0 | 8.0 | 10.0 |
| | | 0.05 | 3.5 | 6.5 | 7.5 | 8.0 | 10.0 |
| | | ≥ 0.20 | 1.5 | 3.0 | 4.0 | 6.0 | 8.0 |
| 0.075 | ≤ 0.5 | ≤ 0.01 | 2.0 | 3.5 | 4.5 | 7.0 | 9.0 |
| | | 0.05 | 1.5 | 3.0 | 4.0 | 6.0 | 8.0 |
| | | ≥ 0.20 | 1.0 | 2.0 | 2.5 | 4.0 | 5.0 |
| | 1.0 | ≤ 0.01 | 2.5 | 5.0 | 6.5 | 8.0 | 10.0 |
| | | 0.05 | 2.0 | 3.5 | 4.5 | 7.0 | 9.0 |
| | | ≥ 0.20 | 1.5 | 2.5 | 3.5 | 5.0 | 7.0 |
| | ≥ 2.0 | ≤ 0.01 | 4.0 | 7.0 | 8.0 | 8.0 | 10.0 |
| | | 0.05 | 2.5 | 5.0 | 6.5 | 8.0 | 10.0 |
| | | ≥ 0.20 | 1.5 | 3.0 | 4.0 | 4.0 | 8.0 |
| Wall Components Controlled by Shear | | | | | | | |
| All Cases[2] | All Cases[2] | All Cases[2] | 2.0 | 2.0 | 3.0 | 2.0 | 3.0 |

[1]Interpolation shall be used between table values.
[2]For wall components governed by shear, the axial load on the member must be less than or equal to $0.15 A_g f'_m$, otherwise the component shall be treated as force-controlled.
[3]$\rho_g = \rho_v + \rho_h$.

governed by flexure, flexural actions shall be considered deformation-controlled. For reinforced masonry components governed by shear, shear actions shall be considered deformation-controlled. Axial compression on reinforced masonry wall or wall pier components shall be considered a force-controlled action.

*7.3.4.3.1 Linear Procedures* For the linear procedures of Section 3.3.2, component actions shall be compared with capacities in accordance with Section 3.4.2.2. The $m$-factor for use in Eq. 3-20 for those components classified as deformation-controlled shall be as specified in Table 7-6.

For determination of $m$-factors from Table 7-6, the ratio of vertical compressive stress to expected compressive strength, $f_{ae}/f_{me}$, shall be based on gravity compressive force determined in accordance with the load combinations given in Eqs. 3-2 and 3-3.

*7.3.4.3.2 Nonlinear Procedures* For the NSP of Section 3.3.3, wall and wall pier components shall meet the requirements of Section 3.4.3.2. Nonlinear deformations on deformation-controlled components shall not exceed the values given in Table 7-7. Variables *d* and *e*, representing nonlinear deformation capacities for primary and secondary components, shall be expressed in terms of story drift ratio percentages as defined in Fig. 7-1.

For determination of the *c*, *d*, and *e* values and the acceptable drift levels using Table 7-7, the vertical compressive stress, $f_{ae}$, shall be based on gravity compressive force determined in accordance with the load combinations given in Eqs. 3-2 and 3-3.

For the NDP of Section 3.3.4, wall and wall pier components shall meet the requirements of Section 3.4.3.2. Nonlinear force–deflection relations for deformation-controlled wall and wall pier components shall be established based on the information given in Table 7-7, or an approved procedure based on comprehensive evaluation of the hysteretic characteristics of those components.

Acceptable deformations for existing and new walls shall be assumed to be the same.

*C7.3.4.3.2 Nonlinear Procedures* For primary components, collapse is considered at lateral drift percentages exceeding values of *d* in Table 7-7, and the Life Safety Structural Performance Level is considered at approximately 75% of *d*. For secondary components, collapse is considered at lateral drift percentages exceeding the values of *e* in the table, and the Life Safety (LS) Structural Performance Level is considered at approximately 75% of *e*. Story drift ratio percentages based on these criteria are given in Table 7-7.

### 7.3.5 Reinforced Masonry Walls Out-of-Plane

RM walls shall be capable of resisting out-of-plane inertial forces as isolated components spanning between floor levels, and/or spanning horizontally between columns or pilasters. Walls shall not be analyzed out-of-plane with the LSP or NSP prescribed in Chapter 3, but shall be capable of resisting out-of-plane inertial forces as given in Section 2.6.7, or be capable of responding to earthquake motions as determined using the NDP, while satisfying the deflection criteria given in Section 7.3.5.3.

#### 7.3.5.1 Stiffness

RM walls shall be considered local elements spanning out-of-plane between individual story levels.

The out-of-plane stiffness of walls shall be neglected in analytical models of the global structural system.

Stiffness shall be based on the net mortared/grouted area of the uncracked section, provided that net flexural tensile stress does not exceed the expected tensile strength, $f_{te}$, in accordance with Section 7.2.2.5.

Stiffness shall be based on the cracked section for a wall where the net flexural tensile stress exceeds the expected tensile strength.

Stiffnesses for existing and new reinforced out-of-plane walls shall be assumed to be the same.

#### 7.3.5.2 Strength

Expected flexural strength shall be based on Section 7.3.4.2.1. For walls with an *h*/*t* ratio exceeding 20, second-order moment effects due to out-of-plane deflections shall be considered.

The strength of new and existing walls shall be assumed to be the same.

#### 7.3.5.3 Acceptance Criteria

Out-of plane forces on reinforced masonry walls shall be considered force-controlled actions. Out-of-plane RM walls shall be sufficiently strong in flexure to resist the out-of-plane loads prescribed in Section 2.6.7.

If the NDP is used, the following performance criteria shall be based on the maximum out-of-plane deflection normal to the plane of a wall:

1. For the Immediate Occupancy Structural Performance Level, the out-of-plane story drift ratio shall be equal to or less than 2%;
2. For the Life Safety Structural Performance Level, the out-of-plane story drift ratio shall be equal to or less than 3%; and
3. For the Collapse Prevention Structural Performance Level, the out-of-plane story drift ratio shall be equal to or less than 5%.

Acceptable deformations for existing and new walls shall be assumed to be the same.

#### C7.3.5.3 Deformation Acceptance Criteria

The limit states specified in this section are based on the masonry units having significant cracking for Immediate Occupancy (IO), masonry units at a point of being dislodged and falling out of the wall for LS, and masonry units on the verge of collapse for Collapse Prevention (CP).

**Table 7-7. Modeling Parameters and Acceptance Criteria for Nonlinear Procedures—Reinforced Masonry In-Plane Walls**

| $f_{ae}/f_{me}$ | $L/h_{eff}$ | $\rho_g f_{ye}/f_{me}$ | $c$ (%) | $d$ (%) | $e$ (%) | IO (%) | Primary LS (%) | Primary CP (%) | Secondary LS (%) | Secondary CP (%) |
|---|---|---|---|---|---|---|---|---|---|---|
| \multicolumn{11}{l}{Wall Components Controlled by Flexure} |
| 0.00 | ≤ 0.5 | 0.01 | 0.5 | 2.6 | 5.3 | 1.0 | 2.0 | 2.6 | 3.9 | 5.3 |
| | | 0.05 | 0.6 | 1.1 | 2.2 | 0.4 | 0.8 | 1.1 | 1.6 | 2.2 |
| | | 0.20 | 0.7 | 0.5 | 1.0 | 0.2 | 0.4 | 0.5 | 0.7 | 1.0 |
| | 1.0 | 0.01 | 0.5 | 2.1 | 4.1 | 0.8 | 1.6 | 2.1 | 3.1 | 4.1 |
| | | 0.05 | 0.6 | 0.8 | 1.6 | 0.3 | 0.6 | 0.8 | 1.2 | 1.6 |
| | | 0.20 | 0.7 | 0.3 | 0.6 | 0.1 | 0.2 | 0.3 | 0.5 | 0.6 |
| | ≥ 2.0 | 0.01 | 0.5 | 1.6 | 3.3 | 0.6 | 1.2 | 1.6 | 2.5 | 3.3 |
| | | 0.05 | 0.6 | 0.6 | 1.3 | 0.2 | 0.5 | 0.6 | 0.9 | 1.3 |
| | | 0.20 | 0.7 | 0.2 | 0.4 | 0.1 | 0.2 | 0.2 | 0.3 | 0.4 |
| 0.038 | ≤ 0.5 | 0.01 | 0.4 | 1.0 | 2.0 | 0.4 | 0.8 | 1.0 | 1.5 | 2.0 |
| | | 0.05 | 0.5 | 0.7 | 1.4 | 0.3 | 0.5 | 0.7 | 1.0 | 1.4 |
| | | 0.20 | 0.6 | 0.4 | 0.9 | 0.2 | 0.3 | 0.4 | 0.7 | 0.9 |
| | 1.0 | 0.01 | 0.4 | 0.8 | 1.5 | 0.3 | 0.6 | 0.8 | 1.1 | 1.5 |
| | | 0.05 | 0.5 | 0.5 | 1.0 | 0.2 | 0.4 | 0.5 | 0.7 | 1.0 |
| | | 0.20 | 0.6 | 0.3 | 0.6 | 0.1 | 0.2 | 0.3 | 0.4 | 0.6 |
| | ≥ 2.0 | 0.01 | 0.4 | 0.6 | 1.2 | 0.2 | 0.4 | 0.6 | 0.9 | 1.2 |
| | | 0.05 | 0.5 | 0.4 | 0.7 | 0.1 | 0.3 | 0.4 | 0.5 | 0.7 |
| | | 0.20 | 0.6 | 0.2 | 0.4 | 0.1 | 0.1 | 0.2 | 0.3 | 0.4 |
| 0.075 | ≤ 0.5 | 0.01 | 0.3 | 0.6 | 1.2 | 0.2 | 0.5 | 0.6 | 0.9 | 1.2 |
| | | 0.05 | 0.4 | 0.5 | 1.0 | 0.2 | 0.4 | 0.5 | 0.8 | 1.0 |
| | | 0.20 | 0.5 | 0.4 | 0.8 | 0.1 | 0.3 | 0.4 | 0.6 | 0.8 |
| | 1.0 | 0.01 | 0.3 | 0.4 | 0.9 | 0.2 | 0.3 | 0.4 | 0.7 | 0.9 |
| | | 0.05 | 0.4 | 0.4 | 0.7 | 0.1 | 0.3 | 0.4 | 0.5 | 0.7 |
| | | 0.20 | 0.5 | 0.2 | 0.5 | 0.1 | 0.2 | 0.2 | 0.4 | 0.5 |
| | ≥ 2.0 | 0.01 | 0.3 | 0.3 | 0.7 | 0.1 | 0.2 | 0.3 | 0.5 | 0.7 |
| | | 0.05 | 0.4 | 0.3 | 0.5 | 0.1 | 0.2 | 0.3 | 0.4 | 0.5 |
| | | 0.20 | 0.5 | 0.2 | 0.3 | 0.1 | 0.1 | 0.2 | 0.2 | 0.3 |
| \multicolumn{11}{l}{Wall Components Controlled by Shear} |
| All Cases[2] | All Cases[2] | All Cases[2] | 0.4 | 0.75 | 2.0 | 0.4 | 0.6 | 0.75 | 0.75 | 1.5 |

[1]Interpolation shall be used between table values.
[2]For wall components governed by shear, the axial load on the member must be less than or equal to $0.15 A_g f'm$, otherwise the component shall be treated as force-controlled.
[3]Primary and secondary component demands shall be within secondary component acceptance criteria where the full backbone curve is explicitly modeled, including strength degradation and residual strength in accordance with Section 3.4.3.2.

217

## 7.4 MASONRY INFILLS

The requirements of this section shall apply to masonry infill panels composed of any combination of existing panels, panels enhanced for seismic rehabilitation, and new panels added to an existing building for seismic rehabilitation. The procedures for determination of stiffness, strength, and deformation of masonry infills shall be based on this section and used with the analytical methods and acceptance criteria prescribed in Chapter 3, unless noted otherwise.

Masonry infill panels shall be considered as primary elements of a lateral-force-resisting system. For the Collapse Prevention Structural Performance Level, if the analysis shows that the surrounding frame will remain stable following the loss of an infill panel, such infill panels not meeting the acceptance criteria of this section shall be permitted.

## C7.4 ENGINEERING PROPERTIES OF MASONRY INFILLS

The design professional is referred to FEMA 306 (FEMA 1998), FEMA 307 (FEMA 1998), and FEMA 308 (FEMA 1998) for additional information regarding the engineering properties of masonry infills.

### 7.4.1 Types of Masonry Infills

Infills shall include panels built partially or fully within the plane of steel or concrete frames, and bounded by beams and columns.

Infill panel types considered in this standard include unreinforced clay-unit masonry, concrete masonry, and hollow-clay tile masonry. Infills made of stone or glass block are not addressed in this standard.

Infill panels considered isolated from the surrounding frame shall have gaps at top and sides to accommodate maximum expected lateral frame deflections. Isolated panels shall be restrained in the transverse direction to ensure stability under normal forces. Panels in full contact with the frame elements on all four sides are termed "shear infill panels."

Frame members and connections surrounding infill panels shall be evaluated for frame–infill interaction effects. These effects shall include forces transferred from an infill panel to beams, columns, and connections, and bracing of frame members across a partial length.

### 7.4.1.1 Existing Masonry Infills

Existing masonry infills considered in this section shall include all structural infills of a building system that are in place prior to seismic rehabilitation. Infill types included in this section consist of unreinforced and ungrouted panels, and composite or noncomposite panels. Existing infill panels subjected to lateral forces applied parallel with their plane shall be considered separately from infills subjected to forces normal to their plane, as described in Sections 7.4.2 and 7.4.3.

Existing masonry infills shall be assumed to behave the same as new masonry infills, provided that the masonry is in good or fair condition as defined in this standard.

### 7.4.1.2 New Masonry Infills

New masonry infills shall include all new panels added to an existing lateral-force-resisting system for structural rehabilitation. Infill types shall include unreinforced or reinforced, grouted, ungrouted, or partially grouted, and composite or noncomposite. New elements shall be designed in accordance with this standard and detailed and constructed in accordance with a building code approved by the authority having jurisdiction.

### 7.4.1.3 Enhanced Masonry Infills

Enhanced masonry infill panels shall include existing infills that are rehabilitated by an approved method.

### C7.4.1.3 Enhanced Masonry Infills

Masonry infills may be rehabilitated using the methods described in this section. Masonry infills enhanced in accordance with this section should be analyzed using the same procedures and performance criteria used for new infills.

Unless stated otherwise, methods are applicable to unreinforced infills and are intended to improve performance of masonry infills subjected to both in-plane and out-of-plane lateral forces.

Guidelines from the following sections pertaining to enhancement methods for reinforced masonry walls listed in Section C7.3.1.3 may also apply to URM infill panels: (1) Infilled Openings, (2) Shotcrete, (3) Coatings for URM Walls, (4) Grout Injections, (5) Repointing, and (6) Stiffening Elements. In addition, the following two enhancement methods may apply to masonry infill panels.

*C7.4.1.3.1 Boundary Restraints for Infill Panels* Infill panels not in tight contact with perimeter frame members should be restrained for out-of-plane forces. This may be accomplished by installing steel angles or plates on each side of the infills, and welding or bolting the angles or plates to the perimeter frame members.

*C7.4.1.3.2 Joints Around Infill Panels* Gaps between an infill panel and the surrounding frame may be filled if integral infill-frame action is assumed for in-plane response.

## 7.4.2 Masonry Infills In-Plane

The calculation of masonry infill in-plane stiffness and strength based on nonlinear finite element analysis of a composite frame substructure with infill panels that account for the presence of openings and post-yield cracking of masonry shall be permitted. Alternatively, the methods of Sections 7.4.2.1 and 7.4.2.2 shall be used.

## C7.4.2 Masonry Infills In-Plane

Finite element programs such as *FEM/I* may be useful in analyzing masonry infills with openings.

### 7.4.2.1 Stiffness

The elastic in-plane stiffness of a solid unreinforced masonry infill panel prior to cracking shall be represented with an equivalent diagonal compression strut of width, $a$, given by Eq. 7-7. The equivalent strut shall have the same thickness and modulus of elasticity as the infill panel it represents.

$$a = 0.175(\lambda_1 h_{col})^{-0.4} r_{inf} \qquad \text{(Eq. 7-7)}$$

where

$$\lambda_1 = \left[ \frac{E_{me} t_{inf} \sin 2\theta}{4 E_{fe} I_{col} h_{inf}} \right]^{\frac{1}{4}}$$

and

$h_{col}$ = column height between centerlines of beams (in.);

$h_{inf}$ = height of infill panel (in.);

$E_{fe}$ = expected modulus of elasticity of frame material (ksi);

$E_{me}$ = expected modulus of elasticity of infill material (ksi);

$I_{col}$ = moment of inertia of column, (in.$^4$);

$L_{inf}$ = length of infill panel (in.);

$r_{inf}$ = diagonal length of infill panel (in.);

$t_{inf}$ = thickness of infill panel and equivalent strut (in.);

$\theta$ = angle whose tangent is the infill height-to-length aspect ratio (radians); and

$\lambda_1$ = coefficient used to determine equivalent width of infill strut.

For noncomposite infill panels, only the wythes in full contact with the frame elements shall be considered where computing in-plane stiffness unless positive anchorage capable of transmitting in-plane forces from frame members to all masonry wythes is provided on all sides of the walls.

Stiffness of cracked unreinforced masonry infill panels shall be represented with equivalent struts; the strut properties shall be determined from analyses that consider the nonlinear behavior of the infilled frame system after the masonry is cracked.

The equivalent compression strut analogy shall be used to represent the elastic stiffness of a perforated unreinforced masonry infill panel; the equivalent strut properties shall be determined from stress analyses of infill walls with representative opening patterns.

Stiffnesses for existing and new infills shall be assumed to be the same.

### C7.4.2.1 Stiffness

In-plane lateral stiffness of an infilled frame system is not the same as the sum of the frame and infill stiffnesses because of the interaction of the infill with the surrounding frame. Experiments have shown that, under lateral forces, the frame tends to separate from the infill near windward lower and leeward upper corners of the infill panels, causing compressive contact stresses to develop between the frame and the infill at the other diagonally opposite corners. Recognizing this behavior, the stiffness contribution of the infill is represented with an equivalent compression strut connecting windward upper and leeward lower corners of the infilled frame. In such an analytical model, if the thickness and modulus of elasticity of the strut are assumed to be the same as those of the infill, the problem is reduced to determining the effective width of the compression strut. Solidly infilled frames may be modeled with a single compression strut in this fashion.

For global building analysis purposes, the compression struts representing infill stiffness of solid infill panels may be placed concentrically across the diagonals of the frame, effectively forming a concentrically braced frame system (Fig. C7-2). In this

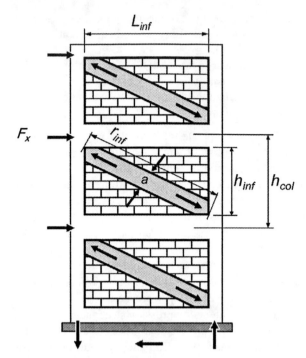

FIGURE C7-2. Compression Strut Analogy—
Concentric Struts.

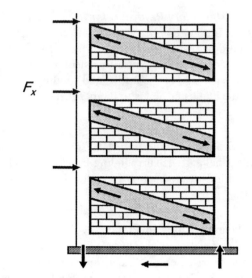

FIGURE C7-3. Compression Strut Analogy—
Eccentric Struts.

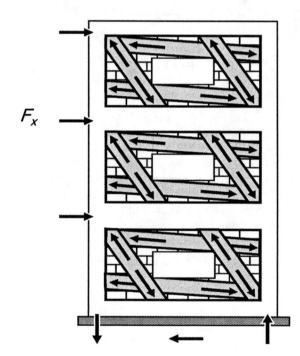

FIGURE C7-4. Compression Strut Analogy—
Perforated Infills.

configuration, however, the forces imposed on columns (and beams) of the frame by the infill are not represented. To account for these effects, compression struts may be placed eccentrically within the frames as shown in Fig. C7-3. If the analytical models incorporate eccentrically located compression struts, the results should yield infill effects on columns directly.

Alternatively, global analyses may be performed using concentric-braced frame models, and the infill effects on columns (or beams) may be evaluated at a local level by applying the strut loads onto the columns (or beams).

Diagonally concentric equivalent struts may also be used to incorporate infill panel stiffnesses into analytical models for perforated infill panels (e.g., infills with window openings), provided that the equivalent stiffness of the infill is determined using appropriate analysis methods (e.g., finite element analysis) in a consistent fashion with the global analytical model. Analysis of local effects, however, must consider various possible stress fields that can potentially develop within the infill. A possible representation of these stress fields with multiple compression struts, as shown in Fig. C7-4, have been proposed by Hamburger (1993). Theoretical work and experimental data for determining multiple strut placement and strut properties, however, are not sufficient to establish reli-

able guidelines; the use of this approach requires judgment on a case-by-case basis.

### 7.4.2.2 Strength

The transfer of story shear across a masonry infill panel confined within a concrete or steel frame shall be considered a deformation-controlled action.

Expected in-plane panel shear strength shall be determined in accordance with the requirements of this section.

Expected infill shear strength, $V_{ine}$, shall be calculated in accordance with Eq. 7-8:

$$Q_{CE} = V_{ine} = A_{ni} f_{vie} \qquad \text{(Eq. 7-8)}$$

where

$A_{ni}$ = area of net mortared/grouted section across infill panel; and

$f_{vie}$ = expected shear strength of masonry infill.

Expected shear strength of existing infills, $f_{vie}$, shall not exceed the expected masonry bed-joint shear strength, $v_{me}$, as determined in accordance with Section 7.2.2.6.

Shear strength of new infill panels, $f_{vie}$, shall not exceed values specified in an approved building code for zero vertical compressive stress.

For noncomposite infill panels, only the wythes in full contact with the frame elements shall be considered where computing in-plane strength, unless positive anchorage capable of transmitting in-plane forces from frame members to all masonry wythes is provided on all sides of the walls.

### 7.4.2.3 Acceptance Criteria

#### 7.4.2.3.1 Required Strength of Column Members Adjacent to Infill Panels
The expected flexural and shear strengths of column members adjacent to an infill panel shall exceed the forces resulting from one of the following conditions:

1.  The application of the horizontal component of the expected infill strut force at a distance $l_{ceff}$ from the top or bottom of the infill panel, where $l_{ceff}$ shall be as defined by Eq. 7-9:

$$l_{ceff} = \frac{a}{\cos \theta_c} \qquad \text{(Eq. 7-9)}$$

where $\tan \theta_c$ shall be as defined by Eq. 7-10:

$$\tan \theta_c = \frac{h_{inf} - \dfrac{a}{\cos \theta_c}}{L_{inf}} \qquad \text{(Eq. 7-10)}$$

2.  The shear force resulting from development of expected column flexural strengths at the top and bottom of a column with a reduced height equal to $l_{ceff}$.

The reduced column length, $l_{ceff}$, in Eq. 7-9 shall be equal to the clear height of opening for a captive column braced laterally with a partial height infill.

The requirements of this section shall be waived if the lower-bound masonry shear strength, $V_{mL}$, as measured in accordance with test procedures of Section 7.2.2.6, is less than 20 psi.

#### 7.4.2.3.2 Required Strength of Beam Members Adjacent to Infill Panels
The expected flexural and shear strengths of beam members adjacent to an infill panel shall exceed forces resulting from one of the following conditions:

1.  The application of the vertical component of the expected infill strut force at a distance, $l_{beff}$, from the top or bottom of the infill panel, where $l_{beff}$ shall be as defined by Eq. 7-11:

$$l_{beff} = \frac{a}{\sin \theta_b} \qquad \text{(Eq. 7-11)}$$

where $\tan \theta_b$ shall be as defined by Eq. 7-12:

$$\tan \theta_b = \frac{h_{inf}}{L_{inf} - \dfrac{a}{\sin \theta_b}} \qquad \text{(Eq. 7-12)}$$

2.  The shear force resulting from development of expected beam flexural strengths at the ends of a beam member with a reduced length equal to $l_{beff}$.

The requirements of this section shall be waived if the expected masonry shear strength, $v_{me}$, as measured using the test procedures of Section 7.2.2.6, is less than 50 psi.

#### 7.4.2.3.3 Linear Procedures
Actions on masonry infills shall be considered deformation-controlled. For the linear procedures of Section 3.3.1, component actions shall be compared with capacities in accordance with Section 3.4.2.2. $m$-factors for use in Eq. 3-20 shall be as specified in Table 7-8. For an infill panel, $Q_E$ shall be the horizontal component of the unreduced axial force in the equivalent strut member

For determination of $m$-factors in accordance with Table 7-8, the ratio of frame to infill strengths, $\beta$, shall be determined considering the expected lateral strength of each component.

#### 7.4.2.3.4 Nonlinear Procedures
For the NSP given in Section 3.3.3, infill panels shall meet the requirements of Section 3.4.3.2. Nonlinear lateral drifts shall not

**Table 7-8. Linear Static Procedure—*m*-Factors for Masonry Infill Panels[1]**

| $\beta = \dfrac{V_{fre}}{V_{ine}}$ | $\dfrac{L_{inf}}{h_{inf}}$ | *m*-Factors IO | *m*-Factors LS | *m*-Factors CP |
|---|---|---|---|---|
| $\beta < 0.7$ | 0.5 | 1.0 | 4.0 | n.a. |
| | 1.0 | 1.0 | 3.5 | n.a. |
| | 2.0 | 1.0 | 3.0 | n.a. |
| $0.7 \leq \beta < 1.3$ | 0.5 | 1.5 | 6.0 | n.a. |
| | 1.0 | 1.2 | 5.2 | n.a. |
| | 2.0 | 1.0 | 4.5 | n.a. |
| $\beta \geq 1.3$ | 0.5 | 1.5 | 8.0 | n.a. |
| | 1.0 | 1.2 | 7.0 | n.a. |
| | 2.0 | 1.0 | 6.0 | n.a. |

[1]Interpolation shall be used between table values.

exceed the values given in Table 7-9. The variable *d*, representing nonlinear deformation capacities, shall be expressed in terms of story drift ratio in percent as defined in Fig. 7-1.

For determination of acceptable drift levels using Table 7-9, the ratio of frame to infill strengths, *b*, shall be determined considering the expected lateral strength of each component.

For the NDP given in Section 3.3.4, infill panels shall meet the requirements of Section 3.4.3.2. Nonlinear force–deflection relations for infill panels shall be established based on the information given in Table 7-9 or an approved procedure based on a comprehensive evaluation of the hysteretic characteristics of those components.

Acceptable deformations for existing and new infills shall be assumed to be the same.

### C7.4.2.3 Acceptance Criteria

Figure C7-5 and Figure C7-6 illustrate how the components of the infill strut force should be applied to columns and beams, respectively.

*C7.4.2.3.4 Nonlinear Procedures* The Immediate Occupancy Structural Performance Level is assumed to be reached when significant visual cracking of an unreinforced masonry infill occurs. The Life Safety Structural Performance Level is assumed to be reached when substantial cracking of the masonry infill occurs and the potential is high for the panel, or some portion of it, to drop out of the frame.

### 7.4.3 Masonry Infills Out-of-Plane

Unreinforced infill panels with $h_{inf}/t_{inf}$ ratios less than those given in Table 7-10, and meeting the requirements for arching action given in the following section, need not be analyzed for out-of-plane seismic forces.

### 7.4.3.1 Stiffness

Infill panels shall be considered local elements spanning out-of-plane vertically between floor levels or horizontally across bays of frames.

The out-of-plane stiffness of infill panels shall be neglected in analytical models of the global structural system in the orthogonal direction.

Flexural stiffness for uncracked masonry infills subjected to transverse forces shall be based on the

**Table 7-9. Nonlinear Static Procedure—Simplified Force–Deflection Relations for Masonry Infill Panels[1]**

| $\beta = \dfrac{V_{fre}}{V_{ine}}$ | $\dfrac{L_{inf}}{h_{inf}}$ | *c* (%) | *d* (%) | *e* (%) | Acceptance Criteria[2] LS (%) | Acceptance Criteria[2] CP (%) |
|---|---|---|---|---|---|---|
| $\beta < 0.7$ | 0.5 | n.a. | 0.5 | n.a. | 0.4 | n.a. |
| | 1.0 | n.a. | 0.4 | n.a. | 0.3 | n.a. |
| | 2.0 | n.a. | 0.3 | n.a. | 0.2 | n.a. |
| $0.7 \leq \beta < 1.3$ | 0.5 | n.a. | 1.0 | n.a. | 0.8 | n.a. |
| | 1.0 | n.a. | 0.8 | n.a. | 0.6 | n.a. |
| | 2.0 | n.a. | 0.6 | n.a. | 0.4 | n.a. |
| $\beta \geq 1.3$ | 0.5 | n.a. | 1.5 | n.a. | 1.1 | n.a. |
| | 1.0 | n.a. | 1.2 | n.a. | 0.9 | n.a. |
| | 2.0 | n.a. | 0.9 | n.a. | 0.7 | n.a. |

[1]Interpolation shall be used between table values.
[2]Primary and secondary component demands shall be within secondary component acceptance criteria where the full backbone curve is explicitly modeled, including strength degradation and residual strength in accordance with Section 3.4.3.2.

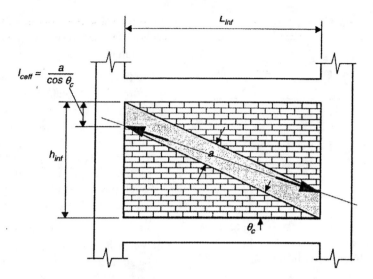

$$l_{ceff} = \frac{a}{\cos \theta_c}$$

**FIGURE C7-5. Estimating Forces Applied to Columns.**

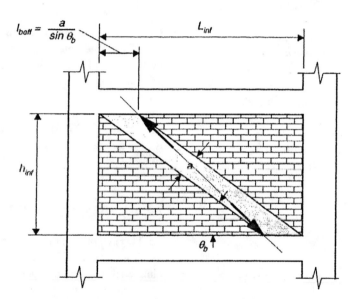

$$l_{beff} = \frac{a}{\sin \theta_b}$$

**FIGURE C7-6. Estimating Forces Applied to Beams.**

minimum net sections of mortared and grouted masonry. Flexural stiffness for unreinforced, cracked infills subjected to transverse forces shall be assumed to be equal to zero unless arching action is considered.

Arching action shall be considered only if all of the following conditions exist.

1. The panel is in full contact with the surrounding frame components;
2. The product of the elastic modulus, $E_{fe}$, times the moment of inertia, $I_f$, of the most flexible frame component exceeds a value of $3.6 \times 10^9$ lb-in.[2];

3. The frame components have sufficient strength to resist thrusts from arching of an infill panel; and
4. The $h_{inf}/t_{inf}$ ratio is less than or equal to 25.

If arching action is considered, mid-height deflection normal to the plane of an infill panel, $\Delta_{inf}$, divided by the infill height, $h_{inf}$, shall be determined in accordance with Eq. 7-13:

$$\frac{\Delta_{inf}}{h_{inf}} = \frac{0.002\left(\frac{h_{inf}}{t_{inf}}\right)}{1 + \sqrt{1 - 0.002\left(\frac{h_{inf}}{t_{inf}}\right)^2}} \quad \text{(Eq. 7-13)}$$

223

**Table 7-10. Maximum $h_{inf}/t_{inf}$ Ratios[1]**

|  | Low Seismic Zone | Moderate Seismic Zone | High Seismic Zone |
|---|---|---|---|
| IO | 14 | 13 | 8 |
| LS | 15 | 14 | 9 |
| CP | 16 | 15 | 10 |

[1]Out-of-plane analysis shall not be required for infills with $h_{inf}/t_{inf}$ ratios less than the values listed herein.

For infill panels not meeting the requirements for arching action, deflections shall be determined in accordance with the procedures given in Sections 7.3.3 or 7.3.5.

Stiffnesses for existing and new infills shall be assumed to be the same.

### 7.4.3.2 Strength

Where arching action is not considered, the lower-bound strength of a URM infill panels shall be limited by the lower-bound masonry flexural tension strength, $f'_{tr}$, which shall be taken as 0.7 times the expected tensile strength, $f_{te}$, as determined in accordance with Section 7.2.2.5.

If arching action is considered, the lower-bound out-of-plane strength of an infill panel in lb/ft$^2$, $q_{in}$, shall be determined using Eq. 7-14:

$$Q_{CL} = q_{in} = \frac{0.7 f'_m \lambda_2}{\left(\dfrac{h_{inf}}{t_{inf}}\right)} \times 144 \qquad \text{(Eq. 7-14)}$$

where

$f'_m$ = lower-bound of masonry compressive strength determined in accordance with Section 7.2.2.3; and

$\lambda_2$ = slenderness parameter as defined in Table 7-11.

### 7.4.3.3 Acceptance Criteria

Infill panels loaded out-of-plane shall not be analyzed with the LSP or NSP prescribed in Chapter 3.

The lower-bound transverse strength of URM infill panels shall exceed normal pressures as prescribed in Section 2.6.7.

**Table 7-11. Values of $l_2$ for Use in Eq. 7-21[1]**

| $h_{inf}/t_{inf}$ | 5 | 10 | 15 | 25 |
|---|---|---|---|---|
| $l_2$ | 0.129 | 0.060 | 0.034 | 0.013 |

[1]Interpolation shall be used.

If the NDP is used, the following performance criteria shall be based on the maximum out-of-plane deflection normal to the plane of the wall:

1. For the Immediate Occupancy Structural Performance Level, the out-of-plane story drift ratio of a panel shall be equal to or less than 2%;
2. For the Life Safety Structural Performance Level, the out-of-plane story drift ratio of a panel shall be equal to or less than 3%; and
3. For the Collapse Prevention Structural Performance Level, the out-of-plane story drift ratio of a panel shall be equal to or less than 5%.

If the surrounding frame is shown to remain stable following the loss of an infill panel, infill panels shall not be subject to limits for the Collapse Prevention Structural Performance Level.

Acceptable deformations of existing and new walls shall be assumed to be the same.

### C7.4.3.3 Acceptance Criteria

The Immediate Occupancy Structural Performance Level is assumed to be reached when significant visual cracking of an unreinforced masonry infill occurs. The Life Safety Structural Performance Level is assumed to be reached when substantial damage of the URM infill occurs and the potential is high for the panel, or some portion of it, to drop out of the frame.

## 7.5 ANCHORAGE TO MASONRY WALLS

### 7.5.1 Types of Anchors

Anchors considered in Section 7.5.2 shall include plate anchors, headed anchor bolts, and bent bar anchor bolts embedded into clay-unit and concrete masonry. Anchors in hollow-unit masonry shall be embedded in grout.

Pullout and shear strength of expansion anchors shall be verified by approved test procedures.

### 7.5.2 Analysis of Anchors

Anchors embedded into existing or new masonry walls shall be considered force-controlled components. Lower-bound values for strengths of embedded anchors with respect to pullout, shear, and combinations of pullout and shear, shall be as specified in an approved building code using load and resistance factor design (LRFD) design procedures taking $\phi = 1.0$.

The minimum effective embedment length or edge distance for considerations of pullout and shear strength of embedded anchors shall be as specified in the building code. Shear strength of anchors with edge distances equal to or less than 1 in. shall be taken as zero.

## C7.5.2 Analysis of Anchors

Anchors in masonry may be analyzed in accordance with FEMA 450 (FEMA 2004).

## 7.6 MASONRY FOUNDATION ELEMENTS

### 7.6.1 Types of Masonry Foundations

Masonry foundations shall be rehabilitated in accordance with this section.

### C7.6.1 Types of Masonry Foundations

Masonry foundations are common in older buildings and are still used for some modern construction. Such foundations may include footings and foundation walls constructed of stone, clay brick, or concrete block. Generally, masonry footings are unreinforced; foundation walls may or may not be reinforced.

Spread footings transmit vertical column and wall loads to the soil by direct bearing. Lateral forces are transferred through friction between the soil and the masonry, as well as by passive pressure of the soil acting on the vertical face of the footing.

### 7.6.2 Analysis of Existing Foundations

The deformability of the masonry footings and the flexibility of the soil under them shall be considered in the lateral force analysis of the building system. The strength and stiffness of the soil shall be determined in accordance with the requirements of Section 4.4.

Masonry footings shall be considered force-controlled components. Masonry footings shall be modeled as elastic components with no inelastic deformation capacity, unless verification tests are done in accordance with Section 2.8 to prove otherwise.

Masonry retaining walls shall be evaluated to resist static and seismic soil pressures in accordance with Section 4.5. Stiffness, strength, and acceptability criteria for masonry retaining walls shall be the same as that for other masonry walls subjected to out-of-plane loadings, as specified in Sections 7.3.3 and 7.3.5.

### 7.6.3 Rehabilitation Measures

Masonry foundation elements shall be rehabilitated in accordance with Section 6.13.4 or by another approved method. New elements shall be designed in accordance with this standard and detailed and constructed in accordance with a building code approved by the authority having jurisdiction.

### C7.6.3 Rehabilitation Measures

Possible rehabilitation methods include:

1. Injection grouting of stone foundations;
2. Reinforcing of URM foundations;
3. Prestressing of masonry foundations;
4. Enlargement of footings by placement of reinforced shotcrete; and
5. Enlargement of footings with additional reinforced concrete sections.

Procedures for rehabilitation should follow provisions for enhancement of masonry walls where applicable, according to Section 7.3.1.3.

## 8.0 WOOD AND LIGHT METAL FRAMING

### 8.1 SCOPE

This chapter sets forth requirements for the Systematic Rehabilitation of wood and light metal frame components of the lateral-force-resisting system of an existing building. The requirements of this chapter shall apply to existing wood and light metal frame components of a building system, rehabilitated wood and light metal frame components of a building system, and new wood and light metal frame components that are added to an existing building system.

Section 8.2 specifies data collection procedures for obtaining material properties and performing condition assessments. Section 8.3 specifies general assumptions and requirements. Sections 8.4 and 8.5 provide modeling procedures, component strengths, acceptance criteria, and rehabilitation measures for wood and light metal frame shear walls and wood diaphragms. Section 8.6 specifies requirements for wood foundations. Section 8.7 specifies requirements for other wood components including, but not limited to, knee-braced frames, rod-braced frames, and braced horizontal diaphragms.

### C8.1 SCOPE

The Linear Static Procedure (LSP) presented in Chapter 3 is most often used for the systematic analysis of wood frame buildings; however, properties of the idealized inelastic performance of various components and connections are included so that nonlinear procedures can be used if desired.

The evaluation and assessment of various structural components of wood frame buildings is found in Section 8.2. For a description and discussion of connections between the various components and elements, see Section 8.2.2.2.2. Properties of shear walls are described in Section 8.4, along with various rehabilitation or strengthening methods. Horizontal floor and roof diaphragms are discussed in Section 8.5, which also covers engineering properties and methods of upgrading or strengthening the elements. Wood foundations and pole structures are addressed in Section 8.6. For additional information regarding foundations, see Chapter 4.

## 8.2 MATERIAL PROPERTIES AND CONDITION ASSESSMENT

### 8.2.1 General

Mechanical properties for wood and light metal framing materials, components, and assemblies shall be based on available construction documents and as-built conditions for the particular structure. Where such information fails to provide adequate information to quantify material properties, capacities of assemblies, or condition of the structure, such information shall be supplemented by materials tests, mock-up tests of assemblies, and assessments of existing conditions as required in Section 2.2.6.

Material properties of existing wood and light metal framing components and assemblies shall be determined in accordance with Section 8.2.2. A condition assessment shall be conducted in accordance with Section 8.2.3. The extent of materials testing and condition assessment performed shall be used to determine the knowledge factor, $\kappa$, as specified in Section 8.2.4.

Use of default material properties shall be permitted in accordance with Section 8.2.2.5. Use of material properties based on historical information for use as default values shall be as specified in Section 8.2.2.5. Other approved values of material properties shall be permitted if based on available historical information for a particular type of wood frame construction, prevailing codes, and assessment of existing condition.

### C8.2.1 General

Various grades and species of wood have been used in a cut dimension form, combined with other structural materials (e.g., steel/wood components), or in multiple layers of construction (e.g., glue-laminated wood components). Wood materials have also been manufactured into hardboard, plywood, and particleboard products, which may have structural or non-structural functions in construction. The condition of the in-place wood materials will greatly influence the future behavior of wood components in the building system.

Quantification of in-place material properties and verification of existing system configuration and condition are necessary to properly analyze the building. The focus of this effort shall be given to the primary components of vertical- and lateral-force-resisting systems. These primary components may be identified through initial analysis and application of loads to the building model.

The extent of in-place materials testing and condition assessment that must be accomplished is related to availability and accuracy of construction documents and as-built records, the quality of materials used and construction performed, and physical condition. A specific problem with wood construction is that structural wood components are often covered with other components, materials, or finishes; in addition, their behavior is influenced by past loading history. Knowledge of the properties and grades of material used in original component/connection fabrication is invaluable, and may be effectively used to reduce the amount of in-place testing required. The design professional is encouraged to research and acquire all available records from the original construction, including design calculations.

Connection configuration also has a very important influence on response to applied loads and motions. A large number of connector types exist, the most prevalent being nails and through bolts. However, more recent construction has included metal straps and hangers, clip angles, and truss plates. An understanding of connector configuration and mechanical properties must be gained to properly analyze the anticipated performance of the building.

Wood frame construction has evolved over the years; wood is the primary building material of most residential and small commercial structures in the United States. It has often been used for the framing of roofs and floors, and in combination with other materials.

Establishing the age and recognizing the location of a building can be helpful in determining what types of lateral-force-resisting systems may be present.

As indicated in Chapter 1, great care should be exercised in selecting the appropriate rehabilitation approaches and techniques for application to historic buildings in order to preserve their unique characteristics.

Based on the approximate age of a building, various assumptions can be made about the design and features of construction. Older wood frame structures that predate building codes and standards usually do not have the types of elements considered essential for predictable seismic performance. These elements will generally have to be added, or the existing elements upgraded by the addition of lateral-load-resisting components to the existing structure in order to obtain predictable performance.

If the age of a building is known, the code in effect at the time of construction and the general quality of the construction usual for the time can be helpful in evaluating an existing building. The level of maintenance of a building may be a useful guide in determining the structure's capacity to resist loads.

Users should be aware that material strengths presented in historical information are typically in allowable stress format. Users should convert allowable stress values to expected strength values in accordance with ASTM D-5457 (ASTM 1998).

The earliest wood frame buildings in the United States were built with post and beam or frame construction adopted from Europe and the British Isles. This was followed by the development of balloon framing in about 1830 in the Midwest, which spread to the East Coast by the 1860s. This, in turn, was followed by the development of western or platform framing shortly after the turn of the century. Platform framing is the system currently in use for multistory construction.

Drywall or wallboard was first introduced in about 1920; however, its use was not widespread until after World War II, when gypsum lath (button board) also came into extensive use as a replacement for wood lath.

With the exception of public schools in high seismic areas, modern wood frame structures detailed to resist seismic loads were generally not built prior to 1934. For most wood frame structures, either general seismic provisions were not provided or the codes that included them were not enforced until the mid-1950s or later, even in the most active seismic areas. This time frame varies somewhat depending on local conditions and practice.

Buildings constructed after 1970 in high seismic areas usually included a well-defined lateral-force-resisting system as a part of the design. However, site inspections and code enforcement varied greatly. Thus, the inclusion of various features and details on the plans does not necessarily mean that they are in place or fully effective. Verification is needed to ensure that good construction practices were followed.

Until about 1950, wood residential buildings were frequently constructed on raised foundations and in some cases included a short stud wall, called a "cripple wall," between the foundation and the first floor

framing. This occurs on both balloon-framed and platform-framed buildings. There may be an extra demand on these cripple walls because most interior partition walls do not continue to the foundation. Special attention is required in these situations. Adequate bracing must be provided for cripple walls as well as the attachment of the sill plate to the foundation.

In more recent times, light gage metal studs and joists have been used in lieu of wood framing for some structures. Lateral-load resistance is either provided by metal straps attached to the studs and top and bottom tracks, or by structural panels attached with sheet metal screws to the studs and the top and bottom track in a manner similar to that of wood construction. The metal studs and joists vary in size, gage, and configuration, depending on the manufacturer and the loading conditions.

For systems using structural panels for bracing, see Section 8.4 for analysis and acceptance criteria. For the all-metal systems using steel strap braces, see Chapter 5 for guidance.

### 8.2.2 Properties of In-Place Materials and Components

#### 8.2.2.1 Material Properties

*8.2.2.1.1 General* The species and grade of wood shall be established by one of the following methods:

1. Construction documents shall be reviewed;
2. An inspection shall be conducted to identify grade by viewing grade stamps or comparing grading rules; or
3. Samples shall be examined by an experienced wood pathologist to establish the species.

Where materials testing is required by Section 2.2.6, grading shall be performed using the ASTM D245-00 (ASTM 2000) grading methodology or an approved grading handbook for the assumed wood species and application. Samples shall be taken from regions where the calculated stress due to applied loads are less than the capacity of the member with the sample removed and tested in accordance with Section 8.2.2.3.

Use of default properties for wood and light metal frame shear walls, wood diaphragms, components, and connectors shall be permitted in accordance with Section 8.2.2.5. For materials comprising individual components, the use of default properties shall be permitted where the species and grade of wood have been determined. Use of default properties for connectors shall be permitted where the species of the connected members has been determined.

227

*8.2.2.1.2 Nominal or Specified Properties* Use of nominal material properties or properties specified in construction documents to compute expected and lower-bound material properties shall be permitted in accordance with Section 8.2.2.5.

*C8.2.2.1.2 Nominal or Specified Properties* Actions associated with wood and light metal framing components generally are deformation-controlled; thus, expected strength material properties will be used most often. Lower-bound values will be used with components supporting discontinuous shear walls, bodies of connections, and axial compression of individual timber frame components, which are force-controlled. Material properties listed in this chapter are expected strength values. If lower-bound material properties are needed, they should be taken as mean minus one standard deviation values, or adjusted from expected strength values in accordance with Section 8.2.2.5.

### 8.2.2.2 Component Properties

*8.2.2.2.1 Elements* The following component properties shall be determined in accordance with Section 8.2.3:

1. Cross-sectional shape and physical dimensions of the primary components and overall configuration of the structure, including any modifications subsequent to original construction;
2. Configuration of elements, size and thickness of connected materials, lumber grade, nail size and spacing, connections, and continuity of load path;
3. Location and dimension of seismic-force-resisting elements, type, materials, and spacing of tie-downs and boundary components; and
4. Current physical condition of components and extent of any deterioration present.

*C8.2.2.2.1 Elements* Structural elements of the lateral-force-resisting system are composed of primary and secondary components, which collectively define element strength and resistance to deformation. Behavior of the components—including shear walls, beams, diaphragms, columns, and braces—is dictated by physical properties such as area; material grade; thickness, depth, and slenderness ratios; lateral torsional buckling resistance; and connection details.

The actual physical dimensions should be measured; for example, 2-in. × 4-in. stud dimensions are generally $1\frac{1}{2}$ in. × $3\frac{1}{2}$ in. Connected members include plywood, bracing, stiffeners, chords, sills, struts, and tie-down posts. Modifications to members include notching, holes, splits, and cracks. The presence of decay or deformation should be noted.

These primary component properties are needed to properly characterize building performance in the seismic analysis. The starting point for establishing component properties should be the available construction documents. Preliminary review of these documents shall be performed to identify vertical- (gravity-) and lateral-force-resisting elements and systems, and their critical components and connections. Site inspections should be conducted to verify conditions and to assure that remodeling has not changed the original design concept. In the absence of a complete set of building drawings, the design professional must thoroughly inspect the building to identify these elements, systems, and components as indicated in Section 8.2.3. Where reliable record drawings do not exist, an as-built set of plans for the building must be created.

*8.2.2.2.2 Connections* Details of the following connections shall be determined or verified in accordance with Section 8.2.3:

1. Connections between horizontal diaphragms and vertical elements of the seismic-force-resisting system;
2. Size and character of all diaphragm ties, including splice connections;
3. Connections at splices in chord members of horizontal diaphragms;
4. Connections of horizontal diaphragms to exterior or interior concrete or masonry walls for both in-plane and out-of-plane loads;
5. Connections of cross-tie members for concrete or masonry buildings;
6. Connections of shear walls to foundations for transfer of shear and overturning forces; and
7. Method of through-floor transfer of wall shear and overturning forces in multistory buildings.

*C8.2.2.2.2 Connections* The method of connecting the various components of the structural system is critical to its performance. The type and character of the connections must be determined by a review of the plans and a field verification of the conditions.

### 8.2.2.3 Test Methods to Quantify Material Properties

The stiffness and strength of wood and light metal framing components and assemblies shall be established through in situ testing or mock-up testing of assemblies in accordance with Section 2.8, unless default values are used in accordance with Section 8.2.2.5. The number of tests required shall be based on Section 8.2.2.4. Expected material properties shall be based on mean values of tests. Lower-bound material properties shall be based on mean values of tests minus one standard deviation.

### C8.2.2.3 Test Methods to Quantify Material Properties

To obtain the desired in-place mechanical properties of materials and components, including expected strength, it is often necessary to use proven destructive and nondestructive testing methods.

Of greatest interest to wood building system performance are the expected orthotropic strengths of the installed materials for anticipated actions (e.g., flexure). Past research and accumulation of data by industry groups have led to published mechanical properties for most wood types and sizes (e.g., dimensional solid-sawn lumber, and glue-laminated or "glulam" beams). Section 8.2.2.5 addresses these established default strengths and distortion properties. This information may be used, together with tests from recovered samples or observation, to establish the expected properties for use in component strength and deformation analyses. Where possible, the load history for the building shall be assessed for possible influence on component strength and deformation properties.

To quantify material properties and analyze the performance of archaic wood construction, shear walls, and diaphragm action, more extensive sampling and testing may be necessary. This testing should include further evaluation of load history and moisture effects on properties, and an examination of wall and diaphragm continuity, and the suitability of in-place connectors.

Where it is desired to use an existing assembly and little or no information about its performance is available, a cyclic load test of a mock-up of the existing structural elements can be used to determine the performance of various assemblies, connections, and load transfer conditions. See Section 2.8 for an explanation of the backbone curve and the establishment of alternative modeling parameters.

### 8.2.2.4 Minimum Number of Tests

*8.2.2.4.1 Usual Testing* The minimum number of tests to quantify expected strength material properties for usual data collection shall be based on the following criteria:

1. If design drawings containing material property and detailing information for the seismic-force-resisting system are available, at least one element of the seismic-force-resisting system for each story, or for

every 100,000 sf of floor area, shall be randomly verified by observation for compliance with the design drawings; and
2. If design drawings are incomplete or not available, at least two locations for each story, or 100,000 sf of floor area, shall be randomly verified by observation or otherwise documented.

*8.2.2.4.2 Comprehensive Testing* The minimum number of tests necessary to quantify expected strength properties for comprehensive data collection shall be defined in accordance with the following requirements:

1. If original construction documents exist that define the grade of wood and mechanical properties, at least one location for each story shall be randomly verified by observing grade stamps, or by compliance with grading rules for each component type identified as having a different material grade;
2. If original construction documents defining properties are not complete or do not exist but the date of construction is known and single material use is confirmed, at least three locations shall be randomly verified—by sampling and testing or by observing grade stamps and conditions—for each component type, for every two floors in the building;
3. If no knowledge of the structural system and materials used exists, at least six locations shall be randomly verified—by sampling and testing or by observing grade stamps and conditions—for each element and component type, for every two floors or 200,000 sf of floor area of construction. If it is determined from testing or observation that more than one material grade exists, additional observations and testing shall be conducted until the extent of use for each grade in component fabrication has been established;
4. In the absence of construction records defining connector features present, the configuration of at least three connectors shall be documented for every floor or 100,000 sf of floor area in the building; and
5. A full-scale mock-up test shall be conducted for archaic assemblies; at least two cyclic tests of each assembly shall be conducted. A third test shall be conducted if the results of the two tests vary by more than 20%.

### C8.2.2.4 Minimum Number of Tests

In order to quantify expected strength and other in-place properties accurately, a minimum number of tests must be conducted on representative components. The minimum number of tests is dictated by available data from original construction, the type of structural system employed, desired accuracy, and quality/condition of in-place materials. Visual access to the structural system also influences testing program definition. As an alternative, the design professional may elect to use the default strength properties in accordance with Section 8.2.2.5. However, using default values without testing is only permitted with the linear analysis procedures. It is strongly encouraged that the expected strengths be derived through testing of assemblies in order to model behavior accurately.

Removal of coverings, including stucco, fireproofing, and partition materials, is generally required to facilitate sampling and observations.

Component types include solid-sawn lumber, glulam beam, and plywood diaphragm. Element types include those that are part of gravity- and lateral-load-resisting systems. The observations shall consist of each connector type present in the building (e.g., nails, bolts, straps), such that the composite strength of the connection can be estimated.

### 8.2.2.5 Default Properties

Use of default properties to determine component strengths shall be permitted in conjunction with the linear analysis procedures of Chapter 3.

Default expected strength and stiffness values for wood and light metal frame shear wall assemblies shall be taken from Table 8-1. Default expected strength and stiffness values for wood diaphragm assemblies shall be taken from Table 8-2

Default expected strength values for wood materials comprising individual components shall be based on design resistance values associated with the AF&PA/ASCE 16 *Standard for Load and Resistance Factor Design (LRFD) for Engineered Wood Construction* (ASCE 1996) as determined in accordance with ASTM D5457-93 (ASTM 1998). All adjustment factors, including the time-effect factor, that are applicable in accordance with ASCE 16 shall be considered. The resistance factor, $\phi$, shall be taken as unity. If components are damaged, reductions in capacity and stiffness shall be applied, considering the position and size of the ineffective cross section.

Default expected strength values for connectors shall be based on design resistance values associated with ASCE 16 as determined in accordance with ASTM D5457-93. All adjustment factors, including the time-effect factor, that are applicable in accordance with ASCE 16 shall be considered. The resistance factor, $\phi$, shall be taken as unity.

Alternatively, expected strength values shall be permitted to be directly computed from allowable stress values listed in an approved code using the method contained in ASTM D5457-93.

Default deformations at yield of connectors shall be taken as:

1. 0.03 in. for wood-to-wood and 0.02 in. for wood-to-metal nailed connections;
2. 0.04 in. for wood-to-wood and 0.03 in. for wood-to-steel screw connections;

### Table 8-1. Default Expected Strength Values for Wood and Light Frame Shear Walls

| Shear Wall Type[1] | Property | |
| --- | --- | --- |
| | Shear Stiffness ($G_d$) (lb/in.) | Expected Strength ($Q_{CE}$) (plf) |
| Single Layer Horizontal Lumber Sheathing or Siding | 2,000 | 80 |
| Single Layer Diagonal Lumber Sheathing | 8,000 | 700 |
| Double Layer Diagonal Lumber Sheathing | 18,000 | 1,300 |
| Vertical Wood Siding | 1,000 | 70 |
| Wood Siding over Horizontal Sheathing | 4,000 | 500 |
| Wood Siding over Diagonal Sheathing | 11,000 | 1,100 |
| Wood Structural Panel Sheathing[2] | — | — |
| Stucco on Studs, Sheathing, or Fiberboard | 14,000 | 350 |
| Gypsum Plaster on Wood Lath | 8,000 | 400 |
| Gypsum Plaster on Gypsum Lath | 10,000 | 80 |
| Gypsum Wallboard | 8,000 | 100 |
| Gypsum Sheathing | 8,000 | 100 |
| Plaster on Metal Lath | 12,000 | 150 |
| Horizontal Lumber Sheathing with Cut-in Braces or Diagonal Blocking | 2,000 | 80 |
| Fiberboard or Particleboard Sheathing | 6,000 | 100 |

[1]As defined in Section 8.4.
[2]See Section 8.4.9 for shear stiffness and expected strength of wood structural panel walls.

**Table 8-2. Default Expected Strength Values for Wood Diaphragms**

| Diaphragm Type[1] | | Property | |
|---|---|---|---|
| | | Shear Stiffness ($G_d$) (lb/in.) | Expected Strength ($Q_{CE}$) (plf) |
| Single Straight Sheathing[2] | | 2,000 | 120 |
| Double Straight Sheathing | Chorded | 15,000 | 600 |
| | Unchorded | 7,000 | 400 |
| Single Diagonally Sheathing | Chorded | 8,000 | 600 |
| | Unchorded | 4,000 | 420 |
| Diagonal Sheathing with Straight Sheathing or Flooring Above | Chorded | 18,000 | 900 |
| | Unchorded | 9,000 | 625 |
| Double Diagonal Sheathing | Chorded | 18,000 | 900 |
| | Unchorded | 9,000 | 625 |
| Wood Structural Panel Sheathing[3] | Unblocked, Chorded | 8,000 | — |
| | Unblocked, Unchorded | 4,000 | — |
| Wood Structural Panel Overlays on: | Unblocked, Chorded | 9,000 | 450 |
| a. Straight or Diagonal Sheathing[4] or | Unblocked, Unchorded | 5,000 | 300 |
| b. Existing Wood Structural Panel Sheathing[5] | Blocked, Chorded | 18,000 | — |
| | Blocked, Unchorded | 7,000 | — |

[1]As defined in Section 8.5.
[2]For single straight sheathing, expected strength shall be multiplied by 1.5 where built-up roofing is present. The value for stiffness shall not be changed.
[3]See Section 8.5.8 for shear stiffness and expected strength of wood structural panel diaphragms.
[4]See Section 8.5.9 for expected strength of wood structural panel overlays on straight or diagonal sheathing.
[5]See Section 8.5.10 for expected strength of wood structural panel overlays on existing wood structural panel sheathing.

3. 0.04 in. for wood-to-wood and 0.027 in. for wood-to-steel lag bolt connections; and
4. 0.045 in. for wood-to-wood and 0.03 in. for wood-to-steel bolted connections.

The estimated deformation of any hardware, including allowance for poor fit or oversized holes, shall be summed to obtain the total deformation of the connection.

Default expected strength values for connection hardware shall be taken as the average ultimate test values from published reports.

Default lower-bound strength values, where required in this chapter, shall be taken as expected strength values multiplied by 0.85.

### C8.2.2.5 Default Properties

The results of any material testing performed should be compared to the default values for the particular era of building construction. If significantly reduced properties from testing are discovered, further evaluation should be undertaken.

Tables 8-1 and 8-2 contain default values for strength and stiffness of shear wall and diaphragm assemblies. The shear stiffness, $G_d$, for the assemblies should not be confused with the modulus of rigidity, $G$, for wood structural panels.

The LRFD methodology of ASCE 16 (ASCE 1996) is based on the concepts of limit state design, similar to the provisions for strength design in steel or concrete. The reference resistance values for wood elements and connections associated with this standard are contained in the *LRFD Manual for Engineered Wood Construction*, including supplements and guidelines (AF&PA LRFD 1996). The resistance values in these documents were developed using ASTM D5457-93 (ASTM 1998), which provides methodologies for calculation directly from data or by format conversion from approved allowable stress values. Use of a format conversion (i.e., the LRFD equivalent of allowable stresses) for computing expected strengths of wood materials comprising individual wood components and for wood connectors (nails, screws, lags, bolts, split rings, and so forth) is permitted. This methodology is not applicable for wood shear wall and diaphragm assemblies covered in Tables 8-1 and 8-2. For use with this chapter, capacities for shear wall and diaphragm assemblies are to be taken directly from the tables or as indicated by the table footnotes.

The LRFD reference resistance is computed as the allowable stress value multiplied by a format conversion factor. The format conversion factor is defined as $K_E = 2.16/\phi$, where $\phi$ is the specified LRFD resistance factor: 0.90 for compression, 0.85 for flexure, 0.80 for tension, 0.75 for shear/torsion, and 0.65 for connections. The allowable stress value shall include all applicable adjustment factors, except for the load duration factor. If allowable values already include consideration of duration effects, the load duration adjustment factor must be divided out prior to format conversion. Note that the time-effect factor specified for LRFD is 1.0 for load combinations that include earthquake loads.

The *NEHRP Recommended Provisions for Seismic Regulations for New Buildings and Other Structures* (BSSC 2000) contain strength-based resistance values for wood structural panel shear walls and diaphragms. Allowable stress values for wood components and connections can be found in the *National Design Specification for Wood Construction* (AF&PA NDS 1997) and the *ASD Manual for Engineered Wood Construction*, including supplements and guidelines (AF&PA ASD 2001).

AF&PA LRFD contains a guideline for calculating resistance values for connection hardware for which published report values are in allowable stress format. Where computing the expected strength of connections, all limit states, including that of the connection hardware, must be considered (e.g., in addition to the published strength of a tie-down device, consider the limit states for the stud bolts, the anchor bolts in the foundation, and so forth).

The connector deformation at yield may be calculated by dividing the load by the load/slip modulus. The load/slip modulus for dowel type connections (bolts, lag screws, screws, and nails) is calculated as $(180)(D)^{1.5}$ kip/in. for wood-to-wood connections and $(270)(D)^{1.5}$ kip/in. for wood-to-steel side plate connections.

Actions associated with wood and light metal framing components generally are deformation-controlled, and expected strength material properties will be used most often. Lower-bound values are needed for actions that are force-controlled. The 0.85 factor included in this standard to convert expected strength to lower-bound values is on the results of shear wall testing. If more precise lower-bound material properties are desired, they should be taken as mean minus one standard deviation from test data for the components in question.

### 8.2.3 Condition Assessment

#### 8.2.3.1 General

A condition assessment of the existing building and site shall be performed as specified in this section.

A condition assessment shall include the following:

1. The physical condition of primary and secondary components shall be examined and the presence of degradation shall be noted.
2. The presence and configuration of components and their connections, and the continuity of load paths between components, elements, and systems shall be verified or established.
3. Other conditions, including neighboring party walls and buildings, presence of nonstructural components, prior remodeling, and limitations for rehabilitation that may influence building performance, shall be reviewed and documented.

#### C8.2.3.1 General

The physical condition of existing components and elements and their connections must be examined for degradation. Degradation may include environmental effects (e.g., decay, splitting, fire damage, and biological, termite, and chemical attack) or past/current loading effects (e.g., overload, damage from past earthquakes, crushing, and twisting). Natural wood also has inherent discontinuities such as knots, checks, and splits that must be noted. Configuration problems observed in recent earthquakes, including effects of discontinuous components, improper nailing or bolting, poor fit-up, and connection problems at the foundation level, should also be evaluated. Often, unfinished areas such as attic spaces, basements, and crawl spaces provide suitable access to wood components and can give a general indication of the condition of the rest of the structure. Invasive inspection of critical components and connections is typically required. Neighboring party walls and buildings, the presence of nonstructural components, prior remodeling, and limitations for rehabilitation should also be noted.

Connections require special consideration and evaluation. The load path for the system must be determined and each connection in the load path(s) must be evaluated. This includes diaphragm-to-component and component-to-component connections. The strength and deformation capacity of connections must be checked where the connection is attached to one or more components that are expected to experience significant inelastic response. Anchorage of exterior walls to roof and floors in concrete and masonry buildings, for which wood diaphragms are used for out-of-plane loading, requires detailed inspection. Bolt holes in relatively narrow straps sometimes preclude the ductile behavior of the steel strap. Twists and kinks in the strap can also have a serious impact on its anticipated behavior. Cross ties, which are part of the wall anchorage system, need to be inspected to confirm their presence, along with the connection of each piece, to ensure that a positive load path exists to tie the building walls together.

The condition assessment also affords an opportunity to review other conditions that may influence wood elements and systems and overall building performance. Of particular importance is the identification of other elements and components that may contribute to or impair the performance of the wood system in question, including infills, neighboring buildings, and equipment attachments. Limitations posed by existing coverings, wall and ceiling space insulation, and other material shall also be defined such that prudent rehabilitation measures can be planned.

### 8.2.3.2 Scope and Procedures

All primary structural components of the gravity- and lateral-load-resistance system shall be included in the condition assessment.

*8.2.3.2.1 Visual Condition Assessment* The dimensions and features of all accessible components shall be measured and compared to available design information. Similarly, the configuration and condition of all accessible connections shall be visually verified, with any deformations or anomalies noted.

*8.2.3.2.2 Comprehensive Condition Assessment* If coverings or other obstructions exist, either partial visual inspection through the use of drilled holes and a fiberscope shall be used, or visual inspection shall be performed by local removal of covering materials based on the following requirements:

1. If detailed design drawings exist, at least three different primary connections shall be exposed for each connection type. If no capacity-reducing deviations from the drawings exist, the sample shall be considered representative. If deviations are noted, then all coverings from primary connections of that type shall be removed unless the connection strength is ignored in the seismic evaluation; and
2. In the absence of accurate drawings, at least 50% of the top and base connections for each type of vertical element in the seismic-force-resisting system as wall as collectors, boundary components, and tie-downs, shall be exposed and inspected or

inspected fiberscopically. If common detailing is observed, this sample shall be considered representative. If any details or conditions are observed that result in a discontinuous load path, all primary connections shall be exposed.

### C8.2.3.2 Scope and Procedures

Accessibility constraints may necessitate the use of instruments such as a fiberscope or video probe to reduce the amount of damage to covering materials and fabrics. The knowledge and insight gained from the condition assessment is invaluable to understanding load paths and the ability of components to resist and transfer loads. The degree of assessment performed also affects the knowledge factor discussed in Section 8.2.4.

Direct visual inspection provides the most valuable information, as it can be used to identify any configuration issues, allows measurement of component dimensions, and identifies the presence of degradation. The continuity of load paths may be established by viewing components and connection condition. From visual inspection, the need for other test methods to quantify the presence and degree of degradation may be established.

The scope of the removal effort is dictated by the component and element design. For example, in a braced frame, exposure of several key connections may suffice if the physical condition is acceptable and the configuration matches the design drawings. However, for shear walls and diaphragms, it may be necessary to expose more connection points because of varying designs and the critical nature of the connections. For encased walls and frames for which no drawings exist, it is necessary to indirectly view or expose all primary end connections for verification.

The physical condition of components and connectors may also support the need to use certain destructive and nondestructive test methods. Devices normally used for the detection of reinforcing steel in concrete or masonry may be used to verify the metal straps and hardware located beneath finish surfaces.

### 8.2.3.3 Basis for the Mathematical Building Model

The results of the condition assessment shall be used to quantify the following items needed to create the mathematical building model:

1. Component section properties and dimensions;
2. Component configuration and eccentricities;
3. Interaction of nonstructural components and their involvement in lateral-load resistance; and
4. Presence and effects of alterations to the structural system.

All deviations noted between available construction records and as-built conditions shall be accounted for in the structural analysis.

### C8.2.3.3 Basis for the Mathematical Building Model

The acceptance criteria for existing components depend on the design professional's knowledge of the condition of the structural system and material properties, as previously noted. Certain damage—such as water staining, evidence of prior leakage, splitting, cracking, checking, warping, and twisting—may be acceptable. The design professional must establish a case-by-case acceptance for such damage on the basis of capacity loss or deformation constraints. Degradation at connection points should be carefully examined; significant capacity reductions may be involved, as well as a loss of ductility.

### 8.2.4 Knowledge Factor

A knowledge factor, $\kappa$, for computation of wood and light metal framing component capacities and permissible deformations shall be selected in accordance with Section 2.2.6.4, with the following additional requirements specific to wood components and assemblies.

If a comprehensive condition assessment is performed in accordance with Section 8.2.3.2.2, a knowledge factor $\kappa = 1.0$ shall be permitted in conjunction with default properties of Section 8.2.2.5, and testing in accordance with Section 8.2.2.4 is not required.

## 8.3 GENERAL ASSUMPTIONS AND REQUIREMENTS

### 8.3.1 Stiffness

Component stiffnesses shall be calculated in accordance with Sections 8.4 through 8.7.

Where design actions are determined using the linear procedures of Chapter 3, stiffnesses for wood materials comprising individual components shall be based on material properties determined in accordance with Section 8.2.2.

Where design actions are determined using the nonlinear procedures of Chapter 3, component force–deformation response shall be represented by nonlinear force–deformation relations. Linear relations shall be permitted where nonlinear response will not occur in the component. The nonlinear force–deformation relation shall be either based on experimental evidence or the generalized force–deformation relation shown in Fig. 8-1, with parameters $c$, $d$, and $e$ as defined in Table 8-4 for wood components and

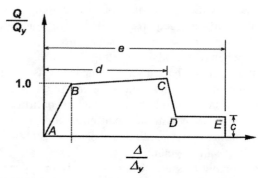

**FIGURE 8-1. Generalized Force–Deformation Relation for Wood Elements or Components.**

assemblies. Distance $d$ is considered the maximum deflection at the point of first loss of strength. Distance $e$ is the maximum deflection at a strength or capacity equal to value $c$. Where the yield strength is not determined by testing in accordance with Section 2.8, the yield strength at point $B$ shall be taken as the expected strength at point $C$ divided by 1.5.

### 8.3.2 Strength and Acceptance Criteria

### 8.3.2.1 General

Actions in a structure shall be classified as being either deformation-controlled or force-controlled, as defined in Section 2.4.4. Design strengths for deformation-controlled and force-controlled actions shall be calculated in accordance with Sections 8.3.2.2 and Sections 8.3.2.3, respectively.

### 8.3.2.2 Deformation-Controlled Actions

Expected strengths for deformation-controlled actions, $Q_{CE}$, shall be taken as the mean maximum strengths obtained experimentally or calculated using accepted principles of mechanics. Unless other procedures are specified in this chapter, expected strengths shall be permitted to be based on 1.5 times the yield strengths. Yield strengths shall be determined using LRFD procedures contained in AF&PA/ASCE 16 *Standard for Load and Resistance Factor Design (LRFD) for Engineered Wood Construction* (ASCE 1996), except that the resistance factor, $\phi$, shall be taken as unity and expected material properties shall be determined in accordance with Section 8.2.2. Acceptance criteria for deformation-controlled actions shall be as specified in Sections 8.4 through 8.7.

### C8.3.2.2 Deformation-Controlled Actions

The relative magnitude of the $m$-factors alone should not be interpreted as a direct indicator of performance. The stiffness of a component and its expected strength, $Q_{CE}$, must be considered where evaluating expected performance. For example, while the *m*-factors for gypsum plaster are higher than those for wood structural panels, the stiffness assigned to gypsum plaster is relatively high and the expected strength values are much lower than those for wood structural panels. As a result, worse performance for a given displacement is predicted.

### 8.3.2.3 Force-Controlled Actions

Where determined by testing, lower-bound strengths for force-controlled actions, $Q_{CL}$, shall be taken as mean minus one standard deviation of the maximum strengths obtained experimentally. Where calculated using established principles of mechanics or based on LRFD procedures contained in AF&PA/ASCE 16 *Standard for Load and Resistance Factor Design (LRFD) for Engineered Wood Construction* (ASCE 1996), the resistance factor, $\phi$, shall be taken as unity, and default lower-bound material properties determined in accordance with Section 8.2.2.5 shall be used.

Where the force-controlled design actions, $Q_{UF}$, calculated in accordance with Section 3.4.2.1.2 are based on a limit-state analysis, the expected strength of the components delivering load to the component under consideration shall be taken as not less than 1.5 times the yield strength.

### C8.3.2.3 Force-Controlled Actions

The maximum forces developed in yielding shear walls and diaphragms are consistently 1.5 to 2 times the yield force. Other wood components and connectors exhibit similar overstrength.

### 8.3.3 Connection Requirements

Unless otherwise specified in this standard, connections between wood components of a lateral-force-resisting system shall be considered in accordance with this section. Demands on connectors, including nails, screws, lags, bolts, split rings, and shear plates used to link wood components to other wood or metal components shall be considered deformation-controlled actions. Demands on bodies of connections, and bodies of connection hardware, shall be considered force-controlled actions.

### C8.3.3 Connection Requirements

In considering connections between wood components in this standard, connectors are distinguished from bodies of connections and bodies of connection hardware. Connectors, which consist of the nails, screws, lags, bolts, split rings, and shear plates used to link pieces of a connection assembly together, are considered to have the ability to deform in a ductile manner, provided the bodies of the connections or bodies of connection hardware do not prematurely fracture. Much of the ductility in a wood shear wall or diaphragm assembly comes from the connectors, such as bending in the nails prior to point where nails pull through the sheathing material. In bolted connections, the connectors, including bolt bending or crushing of the wood around the bolt hole are ductile sources of deformation in an assembly. Brittle failure can occur in the bodies of connections, such as net section fracture or splitting in an end post, or in the bodies of connection hardware such as tie-downs. For this reason, connectors are considered deformation-controlled and bodies of connections and bodies of connection hardware are considered force-controlled. Where determining the demand on force-controlled portions of the connection assembly, use of a limit-state analysis to determine the maximum force that can be delivered to the connection is recommended.

Where computing the strength of connections, all potential limit states should be considered, including those associated with the bodies of connections, the bodies of connection hardware, and connectors with which the assembly may be composed. For example, in addition to the strength of a tie-down device itself, limit states for the stud bolts, foundation bolts, and net section of the end post should be considered. The controlling condition will determine the expected or lower-bound strength of the connection.

### 8.3.4 Rehabilitation Measures

If portions of a wood building structure are deficient for the selected Rehabilitation Objective, the structure shall be rehabilitated, reinforced, or replaced. If replacement of the element is selected or if new elements are added, the new elements shall satisfy the acceptance criteria of this standard and shall be detailed and constructed in accordance with a building code approved by the authority having jurisdiction. If reinforcement of the existing framing system is selected, the following factors shall be considered:

1. Degree of degradation in the component from such mechanisms as biological attack, creep, high static or dynamic loading, moisture, or other effects;
2. Level of steady-state stress in the components to be reinforced and the potential to temporarily remove this stress, if appropriate;
3. Elastic and inelastic properties of existing components; strain compatibility with any new reinforcement materials shall be provided;

4. Ductility, durability, and suitability of existing connectors between components, and access for reinforcement or modification;
5. Efforts necessary to achieve appropriate fit-up for reinforcing components and connections;
6. Load path and deformation of the components at end connections; and
7. Presence of components manufactured with archaic materials, which may contain material discontinuities, shall be examined during the rehabilitation design to ensure that the selected reinforcement is feasible.

### C8.3.4 Rehabilitation Measures

Special attention is required where connections such as bolts and nails are encountered.

Wood structural panels are used to provide lateral strength and stiffness to most modern wood frame buildings and are generally recommended for the rehabilitation of horizontal diaphragms and shear walls of existing buildings. The system relies on the in-plane strength and stiffness of the panels and their connection to the framing. Panels are connected together by nailing into the same structural member to create, in effect, one continuous panel. The various panels are described in Sections 8.4 and 8.5. The performance of the structural panels is dependent to a great degree on the nailing or attachment to the framing. The nail spacing and effectiveness of the attachment should be investigated if the existing panels are expected to withstand significant loads. If nails are to be added to existing panels, they should be the same size as the existing nails.

### 8.3.5 Components Supporting Discontinuous Shear Walls

Axial compression on wood posts and flexure and shear on wood beams that support discontinuous shear walls shall be considered force-controlled actions. Lower-bound strengths shall be determined in accordance with Section 8.3.2.3.

## 8.4 WOOD AND LIGHT FRAME SHEAR WALLS

### 8.4.1 General

Wood and light frame shear walls shall be categorized as primary or secondary components in accordance with Section 2.4.4.2.

Dissimilar wall sheathing materials on opposite sides of a wall shall be permitted to be combined where there are test data to substantiate the stiffness and strength properties of the combined systems. Otherwise, walls sheathed with dissimilar materials shall be analyzed based on only the wall sheathing with the greatest capacity.

For overturning calculations on shear wall elements, stability shall be evaluated in accordance with Section 3.2.10. Net tension due to overturning shall be resisted by uplift connections.

The effects of openings in wood shear walls shall be considered. Where required, reinforcement consisting of chords and collectors shall be added to provide sufficient load capacity around openings to meet the strength requirements for shear walls.

Connections between shear walls and other components, including diaphragm ties, collectors, diaphragms, posts, and foundations, shall be considered in accordance with Section 8.3.3, and designed for forces calculated in accordance with Chapter 3. Components supporting discontinuous shear walls shall be considered in accordance with Section 8.3.5.

The expected strength, $Q_{CE}$, of wood and light frame shear wall assemblies shall be determined in accordance with Sections 8.4.4 through 8.4.18.

### C8.4.1 General

The behavior of wood and light frame shear walls is complex and influenced by many factors, the primary factor being the wall sheathing. Wall sheathings can be divided into many categories (e.g., brittle, elastic, strong, weak, good at dissipating energy, poor at dissipating energy). In many existing buildings, the walls were not expected to act as shear walls (e.g., a wall sheathed with wood lath and plaster). Most shear walls are designed based on values from monotonic load tests and historically accepted values. The allowable shear per unit length used for design was assumed to be the same for long walls, narrow walls, walls with stiff tie-downs, and walls with flexible tie-downs. Only recently have shear wall assemblies—framing, covering, and anchorage—been tested using cyclic loading.

Another major factor influencing the behavior of shear walls is the aspect ratio of the wall. The *NEHRP Recommended Provisions for Seismic Regulations for New Buildings and Other Structures* (BSSC 2000) limit the aspect ratio (height-to-width) for structural panel shear walls to 2:1 for full design shear capacity and permit reduced design shear capacities for walls with aspect ratios up to 3.5:1. The interaction of the floor and roof with the wall, the end conditions of the wall, and the redundancy or number of walls along any wall line would affect the wall behavior for walls with the same aspect ratio. In addition, the rigidity of the tie-downs at the wall ends has an important effect in the behavior of narrow walls.

The presence of any but small openings in wood shear walls will cause a reduction in the stiffness and strength due to a reduced length of wall available to resist lateral forces. Special analysis techniques and detailing are required at the openings. The presence or addition of chord members around the openings will reduce the loss in overall stiffness and limit damage in the area of openings. See the *NEHRP Recommended Provisions for Seismic Regulations for New Buildings and Other Structures* for reinforcement requirements around openings in wood shear walls.

For wood and light frame shear walls, the important limit states are sheathing failure, connection failure, tie-down failure, and excessive deflection. Limit states define the point of life safety and, often, of structural stability. To reduce damage or retain usability immediately after an earthquake, deflection must be limited (see Section 2.5). The ultimate capacity is the maximum capacity of the assembly, regardless of the deflection.

## 8.4.2 Types of Wood Frame Shear Walls

### 8.4.2.1 Existing Wood Frame Shear Walls

*8.4.2.1.1 Single-Layer Horizontal Lumber Sheathing or Siding* Single-layer horizontal lumber sheathing or siding shall include horizontal sheathing or siding applied directly to studs or horizontal boards nailed to studs 2 in. or greater in width.

*C8.4.2.1.1 Single Layer Horizontal Lumber Sheathing or Siding* Typically, 1-in. × horizontal sheathing or siding is applied directly to studs. Forces are resisted by nail couples. Horizontal boards, from 1-in. × 4-in. to 1-in. × 12-in., typically are nailed to 2-in. × or greater width studs with two or more nails (typically 8d or 10d) per stud.

*8.4.2.1.2 Diagonal Lumber Sheathing* Diagonal lumber sheathing shall include sheathing applied at approximately a 45-degree angle to the studs in a single or double layer with three or more nails per stud, sill, and top plates.

*C8.4.2.1.2 Diagonal Lumber Sheathing* Typically, 1-in. × 6-in. to 1-in. × 8-in. diagonal sheathing, applied directly to the studs, resists lateral forces primarily by triangulation (i.e., direct tension and compression). Sheathing boards are installed at a 45-degree angle to studs, with three or more nails (typically 8d or 10d) per stud, and to sill and top plates. A second layer of diagonal sheathing is sometimes added on top of the first layer, at 90 degrees to the first layer (called Double Diagonal Sheathing), for increased load capacity and stiffness.

*8.4.2.1.3 Vertical Wood Siding Only* Vertical wood siding shall include vertical boards nailed directly to studs and blocking 2 in. or greater in width.

*C8.4.2.1.3 Vertical Wood Siding Only* Typically, 1-in. × 8-in., 1-in. × 10-in., or 1-in. × 12-in. vertical boards are nailed directly to 2-in. × or greater width studs and blocking with 8d or 10d galvanized nails. The lateral forces are resisted by nail couples, similarly to horizontal siding.

*8.4.2.1.4 Wood Siding over Horizontal Sheathing* Wood siding over horizontal sheathing shall include siding connected to horizontal sheathing with nails that go through the sheathing to the studs.

*C8.4.2.1.4 Wood Siding over Horizontal Sheathing* Typically, siding is nailed with 8d or 10d galvanized nails through the sheathing to the studs. Lateral forces are resisted by nail couples for both layers.

*8.4.2.1.5 Wood Siding over Diagonal Sheathing* Wood siding over diagonal sheathing shall include siding connected to diagonal sheathing with nails that go through the sheathing to the studs.

*C8.4.2.1.5 Wood Siding over Diagonal Sheathing* Typically, siding is nailed with 8d or 10d galvanized nails to and through the sheathing into the studs. Diagonal sheathing provides most of the lateral resistance by triangulation (see Section 8.4.2.1.2).

*8.4.2.1.6 Wood Structural Panel Sheathing or Siding* Wood structural panel sheathing or siding shall include wood structural panels, as defined in this standard, oriented vertically or horizontally and nailed to studs 2 in. or greater in width.

*C8.4.2.1.6 Wood Structural Panel Sheathing or Siding* Typically, 4-ft × 8-ft panels are applied vertically or horizontally to 2-in. × or greater studs and nailed with 6d to 10d nails. These panels resist lateral forces by panel diaphragm action.

*8.4.2.1.7 Stucco on Studs* Stucco on studs (over sheathing or wire-backed building paper) shall include Portland cement plaster applied to wire lath or expanded metal lath. Wire lath or expanded metal lath shall be nailed to the studs.

*C8.4.2.1.7 Stucco on Studs* Typically, $\frac{7}{8}$-inch Portland cement plaster is applied to wire lath or expanded metal lath. Wire lath or expanded metal lath is nailed to the studs with 11-gage nails or 16-gage staples at

6 in. on center. This assembly resists lateral forces by panel diaphragm action.

*8.4.2.1.8 Gypsum Plaster on Wood Lath* Gypsum plaster on wood lath shall include gypsum plaster keyed onto spaced wood lath that is nailed to the studs.

*C8.4.2.1.8 Gypsum Plaster on Wood Lath* Typically, 1-in. gypsum plaster is keyed onto spaced $1\frac{1}{4}$-in. wood lath that is nailed to studs with 13-gage nails. Gypsum plaster on wood lath resists lateral forces by panel diaphragm-shear action.

*8.4.2.1.9 Gypsum Plaster on Gypsum Lath* Gypsum plaster on gypsum lath shall include plaster that is glued or keyed to gypsum lath nailed to studs.

*C8.4.2.1.9 Gypsum Plaster on Gypsum Lath* Typically, $\frac{1}{2}$-in. plaster is glued or keyed to $16'' \times 48''$ gypsum lath, which is nailed to studs with 13-gage nails. Gypsum plaster on gypsum lath resists lateral loads by panel diaphragm action.

*8.4.2.1.10 Gypsum Wallboard or Drywall* Gypsum wallboard or drywall shall include manufactured panels with a paper facing and gypsum core that are oriented horizontally or vertically and nailed to studs or blocking in a single layer or multiple layers.

*C8.4.2.1.10 Gypsum Wallboard or Drywall* Typically, 4-ft × 8-ft to 4-ft × 12-ft panels are laid-up horizontally or vertically and nailed to studs or blocking with 5d to 8d cooler nails at 4 to 7 in. on center. Multiple layers are used in some situations. The assembly resists lateral forces by panel diaphragm action.

*8.4.2.1.11 Gypsum Sheathing* Gypsum sheathing shall include manufactured gypsum panels that are oriented horizontally or vertically and nailed to studs or blocking.

*C8.4.2.1.11 Gypsum Sheathing* Typically, 4-ft × 8-ft to 4-ft × 12-ft panels are laid-up horizontally or vertically and nailed to studs or blocking with galvanized 11-gage $\frac{7}{16}$-in. diameter head nails at 4 to 7 in. on center. Gypsum sheathing is usually installed on the exterior of structures with siding over it in order to improve fire resistance. Lateral forces are resisted by panel diaphragm action.

*8.4.2.1.12 Plaster on Metal Lath* Plaster on metal lath shall include gypsum plaster applied to expanded wire lath that is nailed to the studs.

*C8.4.2.1.12 Plaster on Metal Lath* Typically, 1-in. gypsum plaster is applied on expanded wire lath that is nailed to the studs. Lateral forces are resisted by panel diaphragm action.

*8.4.2.1.13 Horizontal Lumber Sheathing with Cut-In Braces or Diagonal Blocking* Horizontal lumber sheathing with cut-in braces or diagonal blocking shall include 1-in. × horizontal sheathing or siding applied directly to studs or 1-in. × 4-in. to 1-in. × 12-in. horizontal boards nailed to studs 2 in. or greater in width. The wall shall be braced with diagonal cut-in braces or blocking extending from corner to corner.

*C8.4.2.1.13 Horizontal Lumber Sheathing with Cut-In Braces or Diagonal Blocking* Horizontal sheathing with cut-in braces or diagonal blocking is installed in the same manner as horizontal sheathing, except the wall is braced with cut-in (or let-in) braces or blocking. The bracing is usually installed at a 45-degree angle and nailed with 8d or 10d nails at each stud, and at the top and bottom plates. Bracing provides only nominal increase in resistance.

*8.4.2.1.14 Fiberboard or Particleboard Sheathing* Fiberboard or particleboard sheathing walls shall include fiberboard or particleboard panels that are applied directly to the studs with nails.

*C8.4.2.1.14 Fiberboard or Particleboard Sheathing* Typically, 4-ft × 8-ft panels are applied directly to the studs with nails. Fiberboard requires nails (typically 8d) with large heads such as roofing nails. Lateral loads are resisted by panel diaphragm action.

### 8.4.2.2 Enhanced Wood Frame Shear Walls

Enhanced wood frame shear walls shall include existing shear walls rehabilitated in accordance with an approved method. Enhanced wood shear walls consisting of wood structural panel sheathing added to unfinished stud walls or wood structural panel sheathing overlay on existing shear walls shall be evaluated in accordance with Section 8.4.9. Where wood structural panel sheathing is applied over existing sheathing, the expected strength shall be based on the expected strength of the overlaid material only and reduced by 20% unless a different value is substantiated by testing.

### C8.4.2.2 Enhanced Wood Frame Shear Walls

Possible rehabilitation methods for wood shear walls are described in Sections C8.4.2.2.1 through C8.4.2.2.5.

*C8.4.2.2.1 Wood Structural Panel Sheathing Added to Unfinished Stud Walls* Wood structural panel sheathing may be added to one side of unfinished stud walls to increase the wall shear capacity and stiffness.

Examples of unfinished stud walls are cripple walls and attic end walls.

*C8.4.2.2.2 Wood Structural Panel Sheathing Overlay of Existing Shear Walls* The following types of existing shear walls may be overlaid with wood structural panel sheathing:

1. Single layer horizontal lumber sheathing or siding;
2. Single layer diagonal lumber sheathing;
3. Vertical wood siding only;
4. Gypsum plaster or wallboard on studs (also on gypsum lath and gypsum wallboard);
5. Gypsum sheathing;
6. Horizontal lumber sheathing with cut-in braces or diagonal blocking; and
7. Fiberboard or particleboard sheathing.

The original sheathing should not be included in the evaluation conducted in accordance with Section 8.4.9 and the expected capacity of the overlay material should be reduced by 20%.

This method results in a moderate increase in shear capacity and stiffness and can be applied in most places in most structures. For example, plywood sheathing can be applied over an interior wall finish. For exterior applications, the wood structural panel can be nailed directly through the exterior finish to the studs.

Where existing shear walls are overlaid with wood structural panels, the connections of the overlay to the existing framing must be considered. Splitting can occur in both the wood sheathing and the framing. The length of nails needed to achieve full capacity attachment in the existing framing must be determined. This length will vary with the thickness of the existing wall covering. Sometimes staples are used instead of nails to prevent splitting. The overlay is stapled to the wood sheathing instead of the framing. Nails are recommended for overlay attachment to the underlying framing. In some cases, new blocking at wood structural panel joints may also be needed.

*C8.4.2.2.3 Wood Structural Panel Sheathing Added under Existing Wall Covering* The existing wall covering may be removed; wood structural panel sheathing, connections, and tie-downs may be added and the wall covering may be replaced.

This method will result in a significant increase in shear capacity. In some cases, where earthquake loads are large, this may be the best method of rehabilita-

tion. This rehabilitation procedure can be used on any of the existing shear wall assemblies. Additional framing members can be added if necessary, and the wood structural panels can be cut to fit existing stud spacings.

*C8.4.2.2.4 Increased Attachment* Additional nailing, collector straps, splice straps, tie-downs, or other collectors may be added to existing wood structural panel-sheathed walls to increase their rigidity and capacity.

For existing structural panel-sheathed walls, additional nailing will result in higher capacity and increased stiffness. Other connectors—collector straps, splice straps, or tie-downs—are often necessary to increase the rigidity and capacity of existing structural panel shear walls. Increased ductility will not necessarily result from the additional nailing. Access to these shear walls will often require the removal and replacement of existing finishes.

*C8.4.2.2.5 Connections* Where absent, new connections between shear walls and diaphragms and foundations may be added. Where needed, blocking between floor and roof joists at shear walls may be added. Blocking should be connected to the shear wall and the diaphragm to provide a load path for lateral loads. Wood for framing members or blocking should be kiln-dried or well-seasoned to prevent it from shrinking away from the existing framing, or splitting.

Most shear wall rehabilitation procedures require a check of all existing connections, especially to diaphragms and foundations. Sheet metal framing clips can be used to provide a verifiable connection between the wall framing, the blocking, and the diaphragm. Framing clips are also often used for connecting blocking or rim joists to sill plates.

Frequently, bolting between sill plates and foundations must be added.

The framing in existing buildings is usually very dry, hard, and easily split. Care must be taken not to split the existing framing when adding connectors. Predrilling holes for nails will reduce splitting, and framing clips that use small nails are less likely to split the existing framing.

### 8.4.2.3 New Wood Frame Shear Walls

New wood frame shear walls shall include all new wood structural panel shear walls added to an existing lateral-force-resisting system. Design of new walls shall satisfy the acceptance criteria of this standard. Details of construction for new shear walls, including sill plate anchorage details, tie-down anchor details, nailing details for sheathing, and dimensional limitations for studs and sill plates, shall be in accordance with the requirements of an approved building code.

### C8.4.2.3 New Wood Frame Shear Walls

New shear walls using the existing framing or new framing generally are sheathed with wood structural panels (i.e., plywood or oriented strand board). According to the *NEHRP Recommended Provisions for Seismic Regulations for New Buildings and Other Structures* (BSSC 2000), only wood structural panel sheathing is permitted for use in wood frame shear walls in engineered construction. The thickness and grade of these panels can vary. In most cases, the panels are placed vertically and fastened directly to the studs and plates. This reduces the need for blocking at the joints. All edges of panels must be blocked to obtain full capacity. The thickness, size, and number of fasteners, and aspect ratio and connections will determine the capacity of the new walls. Additional information on the various panels available and their application for shear walls can be found in documents from the American Plywood Association (APA) such as *Design Capacities of APA Performance Rated Structural-Use Panels* (APA 1995) and *Plywood Design Specification* (APA 1997) and Tissell (1993).

### 8.4.3 Types of Light Gage Metal Frame Shear Walls

### 8.4.3.1 Existing Light Gage Metal Frame Shear Walls

*8.4.3.1.1 Plaster on Metal Lath* Plaster on metal lath shall include gypsum plaster applied to metal lath or expanded metal lath that is connected to the metal framing with wire ties.

*C8.4.3.1.1 Plaster on Metal Lath* Typically, 1 in. of gypsum plaster is applied to metal lath or expanded metal that is connected to the metal framing with wire ties.

*8.4.3.1.2 Gypsum Wallboard* Gypsum wallboard shear walls shall include gypsum wallboard panels that are attached to the studs.

*C8.4.3.1.2 Gypsum Wallboard* Typically, 4-ft $\times$ 8-ft to 4-ft $\times$ 12-ft panels are laid-up horizontally and screwed with No. 6 $\times$ 1-in.-long self-tapping screws to studs at 4 to 7 in. on center.

*8.4.3.1.3 Wood Structural Panel Sheathing or Siding* Wood structural panel shear walls shall include structural panels that are attached to the studs and tracks.

*C8.4.3.1.3 Wood Structural Panel Sheathing or Siding* Typically, the wood structural panels are applied vertically and screwed to the studs and tracks with No. 8 to No. 12 self-tapping screws.

### 8.4.3.2 Enhanced Light Gage Metal Frame Shear Walls

Enhanced light gage metal frame shear walls shall include existing shear walls rehabilitated in accordance with an approved method.

### C8.4.3.2 Enhanced Light-Gage Metal Frame Shear Walls

Possible rehabilitation methods for light gage metal frame shear walls are described in Sections C8.4.3.2.1 and C8.4.3.2.2. See Section 8.4.2.2 for additional information concerning enhancement of existing shear walls.

*C8.4.3.2.1 Wood Structural Panel Sheathing Added to Existing Metal Stud Walls* Any existing covering other than wood structural panels shall be removed and replaced with wood structural panels. Connections to the diaphragm(s) and the foundation shall be checked and strengthened where not adequate to resist enhanced wall capacity.

*C8.4.3.2.2 Increased Attachment* Screws and connections shall be added to connect existing wood structural panels to framing.

### 8.4.3.3 New Light Gage Metal Frame Shear Walls

New light gage metal frame shear walls shall include all new wood structural panel elements added to an existing lateral-force-resisting system. Design of new walls shall satisfy the acceptance criteria of this standard. Details of construction for new shear walls, including track anchorage details, tie-down anchor details, fastening details for sheathing, and dimensional limitations for studs and tracks, shall be in accordance with the requirements of an approved building code.

### 8.4.4 Single-Layer Horizontal Lumber Sheathing or Siding Shear Walls

### 8.4.4.1 Stiffness

The deflection of single-layer horizontal lumber sheathing or siding shear walls shall be calculated in accordance with Eq. 8-1:

$$\Delta_y = v_y h / G_d + (h/b) d_a \qquad \text{(Eq. 8-1)}$$

where

$b$ = shear wall width (in.);

$h$ = shear wall height (in.);

$v_y$ = shear at yield in the direction under consideration (lb/in.);

$G_d$ = shear stiffness from Table 8-1 (lb/in.);

$\Delta_y$ = calculated deflection of shear wall at yield (in.); and

$d_a$ = elongation of anchorage at end of wall determined by anchorage details and load magnitude (in.).

Properties used to compute shear wall deflection and stiffness shall be based on Section 8.2.2.

### C8.4.4.1 Stiffness

Horizontal lumber sheathed shear walls are weak and very flexible and have long periods of vibration. The strength and stiffness degrade with cyclic loading. These shear walls are suitable only where earthquake shear loads are low and deflection control is not required.

### 8.4.4.2 Strength

The expected strength of horizontal sheathing or siding shall be determined in accordance with Section 8.2.2.

### C8.4.4.2 Strength

This capacity is dependent on the width of the boards, spacing of the studs, and the size, number, and spacing of the nails. Allowable capacities are listed for various configurations, together with a description of the nail couple method, in the *Western Woods Use Book* (WWPA 1996). See also *Guidelines for the Design of Horizontal Wood Diaphragms*, ATC-7 (ATC 1981) for a discussion of the nail couple method.

### 8.4.4.3 Acceptance Criteria

For linear procedures, *m*-factors for use with deformation-controlled actions shall be taken from Table 8-3. For nonlinear procedures, the coordinates of the generalized force-deformation relations, described by Fig. 8-1, and deformation acceptance criteria for primary and secondary components shall be taken from Table 8-4.

### C8.4.4.3 Acceptance Criteria

Deformation acceptance criteria are determined by the capacity of lateral- and gravity-load-resisting components and elements to deform with limited damage or without failure. Excessive deflection could result in major damage to the structure and/or its contents.

### 8.4.4.4 Connections

The connections between parts of the shear wall assembly and other elements of the lateral-force-resisting system shall be considered in accordance with Section 8.4.1.

### C8.4.4.4 Connections

The capacity and ductility of these connections will often determine the failure mode as well as the capacity of the assembly. Ductile connections with sufficient capacity will give acceptable and expected performance (see Section 8.2.2.2.2).

## 8.4.5 Diagonal Lumber Sheathing Shear Walls

### 8.4.5.1 Stiffness

The deflection of diagonal lumber sheathed shear walls shall be determined using Eq. 8-1. Properties used to compute shear wall deflection and stiffness shall be based on Section 8.2.2.

### C8.4.5.1 Stiffness

Diagonal lumber sheathed shear walls are stiffer and stronger than horizontal sheathed shear walls. They also provide greater stiffness for deflection control, and thereby greater damage control.

### 8.4.5.2 Strength

The expected strength of diagonal sheathing shall be determined in accordance with Section 8.2.2.

### C8.4.5.2 Strength

The strength of diagonal sheathing is dependent on the width of the boards, the spacing of the studs, the size of nails, the number of nails per board, and the boundary conditions. Allowable capacities are listed for various configurations in the *Western Woods Use Book* (WWPA 1996).

### 8.4.5.3 Acceptance Criteria

For linear procedures, *m*-factors for use with deformation-controlled actions shall be taken from Table 8-3. For nonlinear procedures, the coordinates of the generalized force–deformation relation, described by Fig. 8-1, and deformation acceptance criteria for primary and secondary components shall be taken from Table 8-4.

### 8.4.5.4 Connections

The connections between parts of the shear wall assembly and other elements of the lateral-force-resisting system shall be considered in accordance with Section 8.4.1.

**Table 8-3. Numerical Acceptance Factors for Linear Procedures—Wood Components**

| Wood and Light Frame Shear Walls[1,3] | Height/Width Ratio ($h/b$) | $m$-Factors | | | | |
|---|---|---|---|---|---|---|
| | | | Primary | | Secondary | |
| | | IO | LS | CP | LS | CP |
| Horizontal 1-in. × 6-in. Sheathing | $h/b \leq 1.0$ | 1.8 | 4.2 | 5.0 | 5.0 | 5.5 |
| Horizontal 1-in. × 8-in. or 1-in. × 10-in. Sheathing | $h/b \leq 1.0$ | 1.6 | 3.4 | 4.0 | 4.0 | 5.0 |
| Horizontal Wood Siding over Horizontal 1-in. × 6-in. Sheathing | $h/b \leq 1.0$ | 1.4 | 2.6 | 3.0 | 3.1 | 4.0 |
| Horizontal Wood Siding over Horizontal 1-in. × 8-in. or 1-in. × 10-in. Sheathing | $h/b \leq 1.5$ | 1.3 | 2.3 | 2.6 | 2.8 | 3.0 |
| Diagonal 1-in. × 6-in. Sheathing | $h/b \leq 1.5$ | 1.5 | 2.9 | 3.3 | 3.4 | 3.8 |
| Diagonal 1-in. × 8-in. Sheathing | $h/b \leq 1.5$ | 1.4 | 2.7 | 3.1 | 3.1 | 3.6 |
| Horizontal Wood Siding over Diagonal 1-in. × 6-in. Sheathing | $h/b \leq 2.0$ | 1.3 | 2.2 | 2.5 | 2.5 | 3.0 |
| Horizontal Wood Siding over Diagonal 1-in. × 8-in. Sheathing | $h/b \leq 2.0$ | 1.3 | 2.0 | 2.3 | 2.5 | 2.8 |
| Double Diagonal 1-in. × 6-in. Sheathing | $h/b \leq 2.0$ | 1.2 | 1.8 | 2.0 | 2.3 | 2.5 |
| Double Diagonal 1-in. × 8-in. Sheathing | $h/b \leq 2.0$ | 1.2 | 1.7 | 1.9 | 2.0 | 2.5 |
| Vertical 1-in. × 10-in. Sheathing | $h/b \leq 1.0$ | 1.5 | 3.1 | 3.6 | 3.6 | 4.1 |
| Wood Structural Panel Sheathing or Siding[2] | $h/b \leq 2.0$ | 1.7 | 3.8 | 4.5 | 4.5 | 5.5 |
| | $h/b = 3.5$ | 1.4 | 2.6 | 3.0 | 6.0 | 7.0 |
| Stucco on Studs[2] | $h/b \leq 1.0$ | 1.5 | 3.1 | 3.6 | 3.6 | 4.0 |
| | $h/b = 2.0$ | 1.3 | 2.2 | 2.5 | 5.0 | 6.0 |
| Stucco over 1-in. × Horizontal Sheathing | $h/b \leq 2.0$ | 1.5 | 3.0 | 3.5 | 3.5 | 4.0 |
| Gypsum Plaster on Wood Lath | $h/b \leq 2.0$ | 1.7 | 3.9 | 4.6 | 4.6 | 5.1 |
| Gypsum Plaster on Gypsum Lath | $h/b \leq 2.0$ | 1.8 | 4.2 | 5.0 | 4.2 | 5.5 |
| Gypsum Plaster on Metal Lath | $h/b \leq 2.0$ | 1.7 | 3.7 | 4.4 | 3.7 | 5.0 |
| Gypsum Sheathing | $h/b \leq 2.0$ | 1.9 | 4.7 | 5.7 | 4.7 | 6.0 |
| Gypsum Wallboard[2] | $h/b \leq 1.0$ | 1.9 | 4.7 | 5.7 | 4.7 | 6.0 |
| | $h/b = 2.0$ | 1.6 | 3.4 | 4.0 | 3.8 | 4.5 |
| Horizontal 1-in. × 6-in. Sheathing with Cut-In Braces or Diagonal Blocking | $h/b \leq 1.0$ | 1.7 | 3.7 | 4.4 | 4.2 | 4.8 |
| Fiberboard or Particleboard Sheathing | $h/b \leq 1.5$ | 1.6 | 3.2 | 3.8 | 3.8 | 5.0 |

| Diaphragms[5] | Length/Width Ratio ($L/b$) | | | | | |
|---|---|---|---|---|---|---|
| Single Straight Sheathing, Chorded | $L/b \leq 3.0$ | 1 | 2.0 | 2.5 | 2.4 | 3.1 |
| Single Straight Sheathing, Unchorded | $L/b \leq 3.0$ | 1 | 1.5 | 2.0 | 1.8 | 2.5 |
| Double Straight Sheathing, Chorded | $L/b \leq 3.0$ | 1.25 | 2.0 | 2.5 | 2.3 | 2.8 |
| Double Straight Sheathing, Unchorded | $L/b \leq 3.0$ | 1 | 1.5 | 2.0 | 1.8 | 2.3 |
| Single Diagonal Sheathing, Chorded | $L/b \leq 3.0$ | 1.25 | 2.0 | 2.5 | 2.3 | 2.9 |
| Single Diagonal Sheathing, Unchorded | $L/b \leq 3.0$ | 1 | 1.5 | 2.0 | 1.8 | 2.5 |
| Straight Sheathing over Diagonal Sheathing, Chorded | $L/b \leq 3.0$ | 1.5 | 2.5 | 3.0 | 2.8 | 3.5 |
| Straight Sheathing over Diagonal Sheathing, Unchorded | $L/b \leq 3.0$ | 1.25 | 2.0 | 2.5 | 2.3 | 3.0 |
| Double Diagonal Sheathing, Chorded | $L/b \leq 3.5$ | 1.5 | 2.5 | 3.0 | 2.9 | 3.5 |
| Double Diagonal Sheathing, Unchorded | $L/b \leq 3.5$ | 125 | 2.0 | 2.5 | 2.4 | 3.1 |
| Wood Structural Panel, Blocked, Chorded[2] | $L/b \leq 3.0$ | 1.5 | 3.0 | 4.0 | 3.0 | 4.5 |
| | $L/b = 4$ | 1.5 | 2.5 | 3.0 | 2.8 | 3.5 |
| Wood Structural Panel, Unblocked, Chorded[2] | $L/b \leq 3$ | 1.5 | 2.5 | 3.0 | 2.9 | 4.0 |
| | $L/b = 4$ | 1.5 | 2.0 | 2.5 | 2.6 | 3.2 |
| Wood Structural Panel, Blocked, Unchorded[2] | $L/b \leq 2.5$ | 1.25 | 2.5 | 3.0 | 2.9 | 4.0 |
| | $L/b = 3.5$ | 1.25 | 2.0 | 2.5 | 2.6 | 3.2 |

| Diaphragms[5] | Length/Width Ratio ($L/b$) | IO | Primary | | Secondary | |
|---|---|---|---|---|---|---|
| | | | LS | CP | LS | CP |
| | | | *m*-Factors | | | |
| Wood Structural Panel, Unblocked, Unchorded[2] | $L/b \leq 2.5$ | 1.25 | 2.0 | 2.5 | 2.4 | 3.0 |
| | $L/b = 3.5$ | 1.0 | 1.5 | 2.0 | 2.0 | 2.6 |
| Wood Structural Panel Overlay on Sheathing, Chorded[2] | $L/b \leq 3$ | 1.5 | 2.5 | 3.0 | 2.9 | 4.0 |
| | $L/b = 4$ | 1.5 | 2.0 | 2.5 | 2.6 | 3.2 |
| Wood Structural Panel Overlay on Sheathing, Unchorded[2] | $L/b \leq 2.5$ | 1.25 | 2.0 | 2.5 | 2.4 | 3.0 |
| | $L/b = 3.5$ | 1.0 | 1.5 | 2.0 | 1.9 | 2.6 |
| Components/Elements | | | | | | |
| Frame Components Subject to Axial Tension and/or Bending | | 1.0 | 2.5 | 3.0 | 2.5 | 4.0 |
| Frame Components Subject to Axial Compression | | | Force-controlled | | | |
| Wood Piles, Bending and Axial | | 1.2 | 2.5 | 3.0 | — | — |
| Cantilever Pole Structures, Bending and Axial | | 1.2 | 3.0 | 3.5 | — | — |
| Pole Structures With Diagonal Bracing | | 1.0 | 2.5 | 3.0 | — | — |
| Connectors[4] | | | | | | |
| Nails—8d and Larger—Wood to Wood | | 2.0 | 6.0 | 8.0 | 8.0 | 9.0 |
| Nails—8d and Larger—Metal to Wood | | 2.0 | 4.0 | 6.0 | 5.0 | 7.0 |
| Screws—Wood to Wood | | 1.2 | 2.0 | 2.2 | 2.0 | 2.5 |
| Screws—Metal to Wood | | 1.1 | 1.8 | 2.0 | 1.8 | 2.3 |
| Lag Bolts—Wood to Wood | | 1.4 | 2.5 | 3.0 | 2.5 | 3.3 |
| Lag Bolts—Metal to Wood | | 1.3 | 2.3 | 2.5 | 2.4 | 3.0 |
| Machine Bolts—Wood to Wood | | 1.3 | 3.0 | 3.5 | 3.3 | 3.9 |
| Machine Bolts—Metal to Wood | | 1.4 | 2.8 | 3.3 | 3.1 | 3.7 |
| Split Rings and Shear Plates | | 1.3 | 2.2 | 2.5 | 2.3 | 2.7 |

[1]Shear walls shall be permitted to be classified as secondary components or nonstructural components, subject to the limitations of Section 3.2.2.3. Acceptance criteria need not be considered for walls classified as secondary or nonstructural.
[2]Linear interpolation shall be permitted for intermediate values of aspect ratio.
[3]Shear wall components with aspect ratios exceeding maximum listed values shall not be considered effective in resisting lateral loads.
[4]Actions on connectors not listed in this table shall be considered force-controlled.
[5]For diaphragm components with aspect ratios between maximum listed values and 4.0, *m*-factors shall be decreased by linear interpolation between the listed values and 1.0. Diaphragm components with aspect ratios exceeding 4.0 shall not be considered effective in resisting lateral loads.

## 8.4.6 Vertical Wood Siding Shear Walls

### 8.4.6.1 Stiffness

The deflection of vertical wood siding shear walls shall be determined using Eq. 8-1. Properties used to compute shear wall deflection and stiffness shall be based on Section 8.2.2.

### C8.4.6.1 Stiffness

Vertical wood siding has a very low lateral-force-resistance capacity and is very flexible. The strength and stiffness degrade with cyclic loading. These shear walls are suitable only where earthquake shear loads are very low and deflection control is not needed.

### 8.4.6.2 Strength

The expected strength of vertical wood siding shear walls shall be determined in accordance with Section 8.2.2.

### C8.4.6.2 Strength

The strength of vertical wood siding is dependent on the width of the boards, the spacing of the studs, the spacing of blocking, and the size, number, and spacing of the nails. The nail couple method described in the *Western Woods Use Book* (WWPA 1996) can be used to calculate the capacity of vertical wood siding in a manner similar to the method used for horizontal siding.

### 8.4.6.3 Acceptance Criteria

For linear procedures, *m*-factors for use with deformation-controlled actions shall be taken from Table 8-3. For nonlinear procedures, the coordinates of the generalized force–deformation relation, described by Fig. 8-1, and deformation acceptance criteria for primary and secondary components shall be taken from Table 8-4.

**Table 8-4. Modeling Parameters and Numerical Acceptance Criteria for Nonlinear Procedures—Wood Components**

| | | Modeling Parameters | | | Acceptance Criteria[5] | | | | |
| | | | | | IO | Acceptable Deformation Ratio $\Delta/\Delta_y$ | | | |
| | | | | | | Performance Level | | | |
| | | | | | | Component Type | | | |
| | | | | | | Primary | | Secondary | |
| | | $\dfrac{\Delta}{\Delta_y}$ | | Residual Strength Ratio | | | | | |
| | | $d$ | $e$ | $c$ | IO | LS | CP | LS | CP |
| Wood and Light Frame Shear Walls[1] | Height/Width Ratio ($h/b$) | | | | | | | | |
| Horizontal 1-in. × 6-in. Sheathing | $h/b \leq 1.0$ | 5.0 | 6.0 | 0.3 | 2.0 | 4.0 | 5.0 | 5.0 | 6.0 |
| Horizontal 1-in. × 8-in. or 1-in. × 10-in. Sheathing | $h/b \leq 1.0$ | 4.0 | 5.0 | 0.3 | 1.8 | 3.3 | 4.0 | 4.0 | 5.0 |
| Horizontal Wood Siding over Horizontal 1-in. × 6-in. Sheathing | $h/b \leq 1.5$ | 3.0 | 4.0 | 0.2 | 1.5 | 2.5 | 3.0 | 3.0 | 4.0 |
| Horizontal Wood Siding over Horizontal 1-in. × 8-in. or 1-in. × 10-in. Sheathing | $h/b \leq 1.5$ | 2.6 | 3.6 | 0.2 | 1.4 | 2.2 | 2.6 | 2.6 | 3.6 |
| Diagonal 1-in. × 6-in. Sheathing | $h/b \leq 1.5$ | 3.3 | 4.0 | 0.2 | 1.6 | 2.7 | 3.3 | 3.3 | 4.0 |
| Diagonal 1-in. × 8-in. Sheathing | $h/b \leq 1.5$ | 3.1 | 4.0 | 0.2 | 1.5 | 2.6 | 3.1 | 3.1 | 4.0 |
| Horizontal Wood Siding over Diagonal 1-in. × 6-in. Sheathing | $h/b \leq 2.0$ | 2.5 | 3.0 | 0.2 | 1.4 | 2.1 | 2.5 | 2.5 | 3.0 |
| Horizontal Wood Siding over Diagonal 1-in. × 8-in. Sheathing | $h/b \leq 2.0$ | 2.3 | 3.0 | 0.2 | 1.3 | 2.0 | 2.3 | 2.3 | 3.0 |
| Double Diagonal 1-in. × 6-in. Sheathing | $h/b \leq 2.0$ | 2.0 | 2.5 | 0.2 | 1.3 | 1.8 | 2.0 | 2.0 | 2.5 |
| Double Diagonal 1-in. × 8-in. Sheathing | $h/b \leq 2.0$ | 2.0 | 2.5 | 0.2 | 1.3 | 1.8 | 2.0 | 2.0 | 2.5 |
| Vertical 1-in. × 10-in. Sheathing | $h/b \leq 1.0$ | 3.6 | 4.0 | 0.3 | 1.7 | 3.0 | 3.6 | 3.6 | 4.0 |
| Wood Structural Panel Sheathing or Siding[2] | $h/b \leq 2.0$ | 4.5 | 5.5 | 0.3 | 1.9 | 3.6 | 4.5 | 4.5 | 5.5 |
| | $h/b = 3.5$ | 3.0 | 4.0 | 0.2 | 1.5 | 2.5 | 3.0 | 3.0 | 4.0 |
| Stucco on Studs[2] | $h/b \leq 1.0$ | 3.6 | 4.0 | 0.2 | 1.7 | 3.0 | 3.6 | 3.6 | 4.0 |
| | $h/b = 2.0$ | 2.5 | 3.0 | 0.2 | 1.4 | 2.1 | 2.5 | 2.5 | 3.0 |
| Stucco over 1-in. × Horizontal Sheathing | $h/b \leq 2.0$ | 3.5 | 4.0 | 0.2 | 1.6 | 2.9 | 3.5 | 3.5 | 4.0 |
| Gypsum Plaster on Wood Lath | $h/b \leq 2.0$ | 4.6 | 5.0 | 0.2 | 1.9 | 3.7 | 4.6 | 4.6 | 5.0 |
| Gypsum Plaster on Gypsum Lath | $h/b \leq 2.0$ | 5.0 | 6.0 | 0.2 | 2.0 | 4.0 | 5.0 | 5.0 | 6.0 |
| Gypsum Plaster on Metal Lath | $h/b \leq 2.0$ | 4.4 | 5.0 | 0.2 | 1.9 | 3.6 | 4.4 | 4.4 | 5.0 |
| Gypsum Sheathing | $h/b \leq 2.0$ | 5.7 | 6.3 | 0.2 | 2.2 | 4.5 | 5.7 | 5.7 | 6.3 |
| Gypsum Wallboard[2] | $h/b \leq 1.0$ | 5.7 | 6.3 | 0.2 | 2.2 | 4.5 | 5.7 | 5.7 | 6.3 |
| | $h/b = 2.0$ | 4.0 | 5.0 | 0.2 | 1.8 | 3.3 | 4.0 | 4.0 | 5.0 |
| Horizontal 1-in. × 6-in. Sheathing with Cut-In Braces or Diagonal Blocking | $h/b \leq 1.0$ | 4.4 | 5.0 | 0.2 | 1.9 | 3.6 | 4.4 | 4.4 | 5.0 |
| Fiberboard or Particleboard Sheathing | $h/b \leq 1.5$ | 3.8 | 4.0 | 0.2 | 1.7 | 3.1 | 3.8 | 3.8 | 4.0 |
| Diaphragms[3] | Length/Width Ratio ($L/b$) | | | | | | | | |
| Single Straight Sheathing, Chorded | $L/b \leq 2.0$ | 2.5 | 3.5 | 0.2 | 1.4 | 2.1 | 2.5 | 2.5 | 3.5 |
| Single Straight Sheathing, Unchorded | $L/b \leq 2.0$ | 2.0 | 3.0 | 0.3 | 1.3 | 1.8 | 2.0 | 2.0 | 3.0 |
| Double Straight Sheathing, Chorded | $L/b \leq 2.0$ | 2.5 | 3.5 | 0.2 | 1.4 | 2.1 | 2.5 | 2.5 | 3.5 |
| Double Straight Sheathing, Unchorded | $L/b \leq 2.0$ | 2.0 | 3.0 | 0.3 | 1.3 | 1.8 | 2.0 | 2.0 | 3.0 |
| Single Diagonal Sheathing, Chorded | $L/b \leq 2.0$ | 2.5 | 3.5 | 0.2 | 1.4 | 2.1 | 2.5 | 2.5 | 3.5 |
| Single Diagonal Sheathing, Unchorded | $L/b \leq 2.0$ | 2.0 | 3.0 | 0.3 | 1.3 | 1.8 | 2.0 | 2.0 | 3.0 |
| Straight Sheathing over Diagonal Sheathing, Chorded | $L/b \leq 2.0$ | 3.0 | 4.0 | 0.2 | 1.5 | 2.5 | 3.0 | 3.0 | 4.0 |
| Straight Sheathing over Diagonal Sheathing, Unchorded | $L/b \leq 2.0$ | 2.5 | 3.5 | 0.3 | 1.4 | 2.1 | 2.5 | 2.5 | 3.5 |

| | | Modeling Parameters | | | Acceptance Criteria[5] | | | | |
|---|---|---|---|---|---|---|---|---|---|
| | | | | | Acceptable Deformation Ratio $\Delta/\Delta_y$ | | | | |
| | | | | | Performance Level | | | | |
| | | | | | | Component Type | | | |
| | | $\dfrac{\Delta}{\Delta_y}$ | Residual Strength Ratio | | | Primary | | Secondary | |
| | | | | | IO | LS | CP | LS | CP |
| | | $d$ | $e$ | $c$ | IO | LS | CP | LS | CP |
| Double Diagonal Sheathing, Chorded | $L/b \leq 2.0$ | 3.0 | 4.0 | 0.2 | 1.5 | 2.5 | 3.0 | 3.0 | 4.0 |
| Double Diagonal Sheathing, Unchorded | $L/b \leq 2.0$ | 2.5 | 3.5 | 0.2 | 1.4 | 2.1 | 2.5 | 2.5 | 3.5 |
| Wood Structural Panel, Blocked, Chorded[2] | $L/b \leq 3$ | 4.0 | 5.0 | 0.3 | 1.8 | 3.3 | 4.0 | 4.0 | 5.0 |
| | $L/b = 4$ | 3.0 | 4.0 | 0.3 | 1.5 | 2.5 | 3.0 | 3.0 | 4.0 |
| Wood Structural Panel, Unblocked, Chorded[2] | $L/b \leq 3$ | 3.0 | 4.0 | 0.3 | 1.5 | 2.5 | 3.0 | 3.0 | 4.0 |
| | $L/b = 4$ | 2.5 | 3.5 | 0.3 | 1.4 | 2.1 | 2.5 | 2.5 | 3.5 |
| Wood Structural Panel, Blocked, Unchorded[2] | $L/b \leq 2.5$ | 3.0 | 4.0 | 0.3 | 1.5 | 2.5 | 3.0 | 3.0 | 4.0 |
| | $L/b = 3.5$ | 2.5 | 3.5 | 0.3 | 1.4 | 2.1 | 2.5 | 2.5 | 3.5 |
| Wood Structural Panel, Unblocked, Unchorded[2] | $L/b \leq 2.5$ | 2.5 | 3.5 | 0.4 | 1.4 | 2.1 | 2.5 | 2.5 | 3.5 |
| | $L/b = 3.5$ | 2.0 | 3.0 | 0.4 | 1.3 | 1.8 | 2.0 | 2.0 | 3.0 |
| Wood Structural Panel Overlay on Sheathing, Chorded[2] | $L/b \leq 3$ | 3.0 | 4.0 | 0.3 | 1.5 | 2.5 | 3.0 | 3.0 | 4.0 |
| | $L/b = 4$ | 2.5 | 3.5 | 0.3 | 1.4 | 2.1 | 2.5 | 2.5 | 3.5 |
| Wood Structural Panel Overlay on Sheathing, Unchorded[2] | $L/b \leq 2.5$ | 2.5 | 3.5 | 0.4 | 1.4 | 2.1 | 2.5 | 2.5 | 3.5 |
| | $L/b = 3.5$ | 2.0 | 3.0 | 0.4 | 1.3 | 1.8 | 2.0 | 2.0 | 3.0 |
| **Connections[4]** | | | | | | | | | |
| Nails—Wood to Wood | | 7.0 | 8.0 | 0.2 | 2.5 | 5.5 | 7.0 | 7.0 | 8.0 |
| Nails—Metal to Wood | | 5.5 | 7.0 | 0.2 | 2.1 | 4.4 | 5.5 | 5.5 | 7.0 |
| Screws—Wood to Wood | | 2.5 | 3.0 | 0.2 | 1.4 | 2.1 | 2.5 | 2.5 | 3.0 |
| Screws—Wood to Metal | | 2.3 | 2.8 | 0.2 | 1.3 | 2.0 | 2.3 | 2.3 | 2.8 |
| Lag Bolts—Wood to Wood | | 2.8 | 3.2 | 0.2 | 1.5 | 2.4 | 2.8 | 2.8 | 3.2 |
| Lag Bolts—Metal to Wood | | 2.5 | 3.0 | 0.2 | 1.4 | 2.1 | 2.5 | 2.5 | 3.0 |
| Bolts—Wood to Wood | | 3.0 | 3.5 | 0.2 | 1.5 | 2.5 | 3.0 | 3.0 | 3.5 |
| Bolts—Metal to Wood | | 2.8 | 3.3 | 0.2 | 1.5 | 2.4 | 2.8 | 2.8 | 3.3 |

[1]Shear wall components with aspect ratios exceeding maximum listed values shall not be considered effective in resisting lateral loads.

[2]Linear interpolation shall be permitted for intermediate values of aspect ratio.

[3]For diaphragm components with aspect ratios between maximum listed values and 4.0, deformation ratios shall be decreased by linear interpolation between the listed values and 1.0. Diaphragm components with aspect ratios exceeding 4.0 shall not be considered effective in resisting lateral loads.

[4]Actions on connectors not listed in this table shall be considered force-controlled

[5]Primary and secondary component demands shall be within secondary component acceptance criteria where the full backbone curve is explicitly modeled, including strength degradation and residual strength in accordance with Section 3.4.3.2.

### 8.4.6.4 Connections

The presence of connections between parts of the vertical wood siding shear wall assembly and other elements of the lateral-force-resisting system shall be verified. If connections are present, they need not be considered in the analysis conducted in accordance with Chapter 3. In the absence of connections, connections shall be provided in accordance with Section 8.4.1.

### C8.4.6.4 Connections

The load capacity of the vertical siding is low, which makes the capacity of connections between the shear wall and the other elements of less concern (see Section 8.2.2.2.2).

### 8.4.7 Wood Siding over Horizontal Sheathing Shear Walls

#### 8.4.7.1 Stiffness
The deflection of wood siding over horizontal sheathing shear walls shall be determined using Eq. 8-1. Properties used to compute shear wall deflection and stiffness shall be based on Section 8.2.2.

#### C8.4.7.1 Stiffness
Double-layer horizontal sheathed shear walls are stiffer and stronger than single-layer horizontal sheathed shear walls. These shear walls are often suitable for resisting earthquake shear loads that are low to moderate in magnitude. They also provide greater stiffness for deflection control and, thereby, greater damage control.

#### 8.4.7.2 Strength
The expected strength of wood siding over horizontal sheathing shall be determined in accordance with Section 8.2.2.

#### C8.4.7.2 Strength
This capacity is dependent on the width of the boards, the spacing of the studs, the size, number, and spacing of the nails, and the location of joints.

#### 8.4.7.3 Acceptance Criteria
For linear procedures, $m$-factors for use with deformation-controlled actions shall be taken from Table 8-3. For nonlinear procedures, the coordinates of the generalized force–deformation relation, described by Fig. 8-1, and deformation acceptance criteria for primary and secondary components shall be taken from Table 8-4.

#### 8.4.7.4 Connections
The connections between parts of the shear wall assembly and other elements of the lateral-force-resisting system shall be considered in accordance with Section 8.4.1.

### 8.4.8 Wood Siding over Diagonal Sheathing

#### 8.4.8.1 Stiffness
The deflection of these shear walls shall be calculated in accordance with Eq. 8-1. Properties used to compute shear wall deflection and stiffness shall be based on Section 8.2.2.

#### C8.4.8.1 Stiffness
Horizontal wood siding over diagonal sheathing will provide stiff, strong shear walls. These shear walls are often suitable for resisting earthquake shear loads that are moderate in magnitude. They also provide good stiffness for deflection control and damage control.

#### 8.4.8.2 Strength
The expected strength of wood siding over diagonal sheathing shall be determined in accordance with Section 8.2.2.

#### C8.4.8.2 Strength
The capacity of wood siding over diagonal sheathing is dependent on the width of the boards, the spacing of the studs, the size, number, and spacing of the nails, the location of joints, and the boundary conditions.

#### 8.4.8.3 Acceptance Criteria
For linear procedures, $m$-factors for use with deformation-controlled actions shall be taken from Table 8-3. For nonlinear procedures, the coordinates of the generalized force–deformation relation, described by Fig. 8-1, and deformation acceptance criteria for primary and secondary components shall be taken from Table 8-4.

#### 8.4.8.4 Connections
The connections between parts of the shear wall assembly and other elements of the lateral-force-resisting system shall be considered in accordance with Section 8.4.1.

### 8.4.9 Wood Structural Panel Sheathing

#### 8.4.9.1 Stiffness
The deflection of wood structural shear walls at yield shall be determined using Eq. 8-2:

$$\Delta_y = 8v_y h^3/(EAb) + v_y h/(Gt) + 0.75h_{en} + (h/b)d_a \qquad \text{(Eq. 8-2)}$$

where

$v_y$ = shear at yield in the direction under consideration (lb/ft);

$h$ = shear wall height (ft);

$E$ = modulus of elasticity of boundary member (psi);

$A$ = area of boundary member cross section (in.$^2$);

$b$ = shear wall width (ft);

$G$ = modulus of rigidity of wood structural panel (psi);

$t$ = effective thickness of wood structural panel (in.);

$d_a$ = deflection at yield of tie-down anchorage or deflection at load level to anchorage at end of wall, determined by anchorage details and dead load (in.); and

$e_n$ = nail deformation at yield load per nail (in.).
   Values listed are for Structural I panels; multiply
   by 1.2 for all other panel grades;
   = 0.13 for 6d nails at yield;
   = 0.08 for 8d nails at yield;
   = 0.08 for 10d nails at yield.

Properties used to compute shear wall deflection and stiffness shall be based on Section 8.2.2.

### C8.4.9.1 Stiffness

The response of wood structural panel shear walls is dependent on the thickness of the wood structural panels, the height-to-width ($h/b$) ratio, the nailing pattern, and other factors. Values for modulus of rigidity, $G$, and effective thickness, $t$, for various sheathing materials are contained in *Design Capacities of APA Performance Rated Structural-Use Panels* (APA 1995) and *Plywood Design Specification* (APA 1997).

### 8.4.9.2 Strength

The expected strength of wood structural panel shear walls shall be taken as mean maximum strengths obtained experimentally. Expected strengths of wood structural panel shear walls shall be permitted to be based on 1.5 times yield strengths. Yield strengths shall be determined using LRFD procedures contained in AF&PA/ASCE 16 *Standard for Load and Resistance Factor Design (LRFD) for Engineered Wood Construction* (ASCE 1996), except that the resistance factor, $\phi$, shall be taken as unity and expected material properties shall be determined in accordance with Section 8.2.2.

Conversion from tabulated allowable stress values in accordance with Section 8.2.2.5 shall not be permitted for wood structural panel shear walls, but approved allowable stress values for fasteners shall be permitted to be converted in accordance with Section 8.2.2.5 where the strength of a shear wall is computed using principles of mechanics.

### C8.4.9.2 Strength

Shear capacities of wood structural panel shear walls are primarily dependent on the nailing at the plywood panel edges, and the thickness and grade of the plywood.

LRFD-based design values for various configurations are listed in the *LRFD Manual for Engineered Wood Construction* (AF&PA LRFD 1996) and the *NEHRP Recommended Provisions for Seismic Regulations for New Buildings and Other Structures* (BSSC 2000). For tabulated values, some references provide nominal strength and some provide factored strength. It is expected that nominal strength values

(values without a $\phi$ factor) are similar to factored strength value with $\phi = 1.0$.

A method for calculating the capacity of wood structural shear walls based on accepted nail values is provided in Tissell (1993). For this method, use LRFD-based fastener strengths. Due to the differences in load-duration/time-effect factors between the allowable stress and LRFD formats, direct conversion of shear wall tables using the method outlined in Section 8.2.2.5 is not permitted. However, the tabulated LRFD design values, with $\phi = 1$, are intended to be 2.0 times the associated allowable stress design values.

### 8.4.9.3 Acceptance Criteria

For linear procedures, *m*-factors for use with deformation-controlled actions shall be taken from Table 8-3. For nonlinear procedures, the coordinates of the generalized force–deformation relation, described in Eq. 8-1, and deformation acceptance criteria for primary and secondary components shall be taken from Table 8-4.

### 8.4.9.4 Connections

The connections between parts of the shear wall assembly and other elements of the lateral-force-resisting system shall be considered in accordance with Section 8.4.1.

## 8.4.10 Stucco on Studs, Sheathing, or Fiberboard

### 8.4.10.1 Stiffness

The deflection of stucco on studs, sheathing, or fiberboard shear walls shall be determined using Eq. 8-1. Properties used to compute shear wall deflection and stiffness shall be based on Section 8.2.2.

### C8.4.10.1 Stiffness

Stucco is brittle and the lateral-force-resisting capacity of stucco shear walls is low. The walls are stiff until cracking occurs, but the strength and stiffness degrade under cyclic loading. These shear walls are suitable only where earthquake shear loads are low.

### 8.4.10.2 Strength

The expected strength of stucco on studs, sheathing, or fiberboard shall be determined in accordance with Section 8.2.2.

### C8.4.10.2 Strength

This capacity is dependent on the attachment of the stucco netting to the studs and the embedment of the netting in the stucco.

### 8.4.10.3 Acceptance Criteria

For linear procedures, $m$-factors for use with deformation-controlled actions shall be taken from Table 8-3. For nonlinear procedures, the coordinates of the generalized force–deformation relation, described by Fig. 8-1, and deformation acceptance criteria for primary and secondary components shall be taken from Table 8-4.

### 8.4.10.4 Connections

The connection between the stucco netting and the framing shall be investigated. The connections between the shear wall and foundation, and between the shear wall and other elements of the lateral-force-resisting system, shall be considered in accordance with Section 8.4.1.

### C8.4.10.4 Connections

Of less concern is the connection of the stucco to the netting. Unlike plywood, the tensile capacity of the stucco material (Portland cement), rather than the connections, will often govern failure. See Section 8.2.2.2.2.

## 8.4.11 Gypsum Plaster on Wood Lath

### 8.4.11.1 Stiffness

The deflection of gypsum plaster on wood lath shear walls shall be determined using Eq. 8-1. Properties used to compute shear wall deflection and stiffness shall be based on Section 8.2.2.

### C8.4.11.1 Stiffness

Gypsum plaster shear walls are similar to stucco, except their strength is lower. As is the case for stucco, the walls are stiff until failure but the strength and stiffness degrade under cyclic loading. These shear walls are suitable only where earthquake shear loads are very low.

### 8.4.11.2 Strength

The expected strength of gypsum plaster shall be determined in accordance with Section 8.2.2.

### 8.4.11.3 Acceptance Criteria

For linear procedures, $m$-factors for use with deformation-controlled actions shall be taken from Table 8-3. For nonlinear procedures, the coordinates of the generalized force–deformation relation by Fig. 8-1, and deformation acceptance criteria for primary and secondary components shall be taken from Table 8-4.

### 8.4.11.4 Connections

The presence of connections between parts of the shear wall assembly and other elements of the lateral-force-resisting system shall be verified. If connections are present, they need not be considered in the analysis conducted in accordance with Chapter 3. If connections are absent, they shall be provided in accordance with Section 8.4.1.

### C8.4.11.4 Connections

The tensile and bearing capacity of the plaster, rather than the connections, will often govern failure. The relatively low strength of this material makes connections between parts of the shear wall assembly and the other elements of the lateral-force-resisting system of less concern.

## 8.4.12 Gypsum Plaster on Gypsum Lath

### 8.4.12.1 Stiffness

The deflection of gypsum plaster on gypsum lath shear walls shall be determined using Eq. 8-1. Properties used to compute shear wall deflection and stiffness shall be based on Section 8.2.2.

### C8.4.12.1 Stiffness

Gypsum plaster on gypsum lath is similar to gypsum wallboard (see Section 8.4.13).

### 8.4.12.2 Strength

The expected strength of gypsum plaster on gypsum lath shear walls shall be determined in accordance with Section 8.2.2.

### 8.4.12.3 Acceptance Criteria

For linear procedures, $m$-factors for use with deformation-controlled actions shall be taken from Table 8-3. For nonlinear procedures, the coordinates of the generalized force–deformation relation, described by Fig. 8-1, and deformation acceptance criteria for primary and secondary components shall be taken from Table 8-4.

### 8.4.12.4 Connections

The presence of connections between parts of the shear wall assembly and other elements of the lateral-force-resisting system shall be verified. If connections are present, they need not be considered in the analysis conducted in accordance with Chapter 3. If connections are absent, they shall be provided in accordance with Section 8.4.1.

### C8.4.12.4 Connections

The tensile and bearing capacity of the plaster, rather than the connections, will often govern failure. The relatively low strength of this material makes connections between parts of the shear wall assembly and the other elements of the lateral-force-resisting system of less concern.

### 8.4.13 Gypsum Wallboard

*8.4.13.1 Stiffness*

The deflection of gypsum wallboard shear walls shall be determined using Eq. 8-1. Properties used to compute shear wall deflection and stiffness shall be based on Section 8.2.2.

*C8.4.13.1 Stiffness*

Gypsum wallboard has a very low lateral-force-resisting capacity, but is relatively stiff until cracking occurs. The strength and stiffness degrade under cyclic loading. These shear walls are suitable only where earthquake shear loads are very low.

*8.4.13.2 Strength*

The expected strength of gypsum wallboard shear walls shall be determined in accordance with Section 8.2.2.

*C8.4.13.2 Strength*

The default capacity listed in Section Table 8-1 is for typical 7-in. nail spacing of $\frac{1}{2}$-in. or $\frac{5}{8}$-in.-thick panels with 4d or 5d nails. Higher capacities can be used if closer nail spacing, multilayers of gypsum board, and/or the presence of blocking at all panel edges is verified.

*8.4.13.3 Acceptance Criteria*

For linear procedures, *m*-factors for use with deformation-controlled actions shall be taken from Table 8-3. For nonlinear procedures, the coordinates of the generalized force–deformation relation, described by Fig. 8-1, and deformation acceptance criteria for primary and secondary components shall be taken from Table 8-4.

*8.4.13.4 Connections*

The connections between parts of the shear wall assembly and other elements of the lateral-force-resisting system shall be considered in accordance with Section 8.4.1.

### 8.4.14 Gypsum Sheathing

*8.4.14.1 Stiffness*

The deflection of gypsum sheathed shear walls shall be determined using Eq. 8-1. Properties used to compute shear wall deflection and stiffness shall be based on Section 8.2.2.

*C8.4.14.1 Stiffness*

Gypsum sheathing is similar to gypsum wallboard (see Section 8.4.13.1).

*8.4.14.2 Strength*

The expected strength of gypsum wallboard shear walls shall be determined in accordance with Section 8.2.2.

*C8.4.14.2 Strength*

The default capacity listed in Table 8-1 is based on typical 7-in. nail spacing of $\frac{1}{2}$-in. or $\frac{5}{8}$-in.-thick panels with 4d or 5d nails. Higher capacities can be used if closer nail spacing, multilayers of gypsum board, and/or the presence of blocking at all panel edges is verified.

*8.4.14.3 Acceptance Criteria*

For linear procedures, *m*-factors for use with deformation-controlled actions shall be taken from Table 8-3. For nonlinear procedures, the coordinates of the generalized force–deformation relation, described by Fig. 8-1, and deformation acceptance criteria for primary and secondary components shall be taken from Table 8-4.

*8.4.14.4 Connections*

The connections between parts of the shear wall assembly and other elements of the lateral-force-resisting system shall be considered in accordance with Section 8.4.1.

### 8.4.15 Plaster on Metal Lath

*8.4.15.1 Stiffness*

The deflection of plaster on metal lath shear walls shall be determined using Eq. 8-1. Properties used to compute shear wall deflection and stiffness shall be based on Section 8.2.2.

*C8.4.15.1 Stiffness*

Plaster on metal lath is similar to plaster on wood lath, and the lateral-force-resisting capacity of these shear walls is low. The walls are stiff until cracking occurs, but the strength and stiffness degrade under cyclic loading. These shear walls are suitable only where earthquake shear loads are low.

*8.4.15.2 Strength*

The expected strength of plaster on metal lath shear walls shall be determined in accordance with Section 8.2.2.

### 8.4.15.3 Acceptance Criteria

For linear procedures, m-factors for use with deformation-controlled actions shall be taken from Table 8-3. For nonlinear procedures, the coordinates of the generalized force–deformation relation, described by Fig. 8-1, and deformation acceptance criteria for primary and secondary components shall be taken from Table 8-4.

### 8.4.15.4 Connections

The presence of connections between parts of the shear wall assembly and other elements of the lateral-force-resisting system shall be verified. If connections are present, they need not be considered in the analysis conducted in accordance with Chapter 3. If connections are absent, they shall be provided in accordance with Section 8.4.1.

### C8.4.15.4 Connections

The tensile and bearing capacity of the plaster, rather than the connections, will often govern failure. The relatively low strength of this material makes connections between parts of the shear wall assembly and the other elements of the lateral-force-resisting system of less concern.

## 8.4.16 Horizontal Lumber Sheathing with Cut-In Braces or Diagonal Blocking

### 8.4.16.1 Stiffness

The deflection of horizontal lumber sheathing with cut-in braces or diagonal blocking shear walls shall be calculated using Eq. 8-1. Properties used to compute shear wall deflection and stiffness shall be based on Section 8.2.2.

### C8.4.16.1 Stiffness

This assembly is similar to horizontal sheathing without braces, except that the cut-in braces or diagonal blocking provide higher stiffness at initial loads. After the braces or blocking fail (at low loads), the behavior of the wall is the same as with horizontal sheathing without braces. The strength and stiffness degrade under cyclic loading.

### 8.4.16.2 Strength

The expected strength of horizontal sheathing or siding shall be determined in accordance with Section 8.2.2.

### 8.4.16.3 Acceptance Criteria

For linear procedures, m-factors for use with deformation-controlled actions shall be taken from

Table 8-3. For nonlinear procedures, the coordinates of the generalized force–deformation relation, described by Fig. 8-1, and deformation acceptance criteria for primary and secondary components shall be taken from Table 8-4.

### 8.4.16.4 Connections

The connections between the parts of the shear wall assembly and other elements of the lateral-force-resisting system shall be considered in accordance with Section 8.4.1.

### C8.4.16.4 Connections

The capacity and ductility of these connections will often determine the failure mode as well as the capacity of the assembly. Ductile connections with sufficient capacity will give acceptable performance (see Section 8.2.2.2.2).

## 8.4.17 Fiberboard or Particleboard Sheathing

### 8.4.17.1 Stiffness

For structural particleboard sheathing, see Section 8.4.9. The deflection of shear walls sheathed in nonstructural particleboard shall be determined using Eq. 8-1. Properties used to compute shear wall deflection and stiffness shall be based on Section 8.2.2. Fiberboard sheathing shall not be considered a structural element for resisting seismic loads.

### C8.4.17.1 Stiffness

Fiberboard sheathing is very weak, lacks stiffness, and is unable to resist lateral loads. Particleboard comes in two varieties: one is similar to structural panels, the other (nonstructural) is slightly stronger than gypsum board but more brittle. Nonstructural particleboard should only be used where earthquake loads are very low.

### 8.4.17.2 Strength

The expected strength of structural particleboard shall be based on Section 8.4.9. The strength of nonstructural fiberboard or particleboard sheathed walls shall be determined in accordance with Section 8.2.2.

### C8.4.17.2 Strength

Fiberboard has very low strength and is therefore not considered a structural element for resisting seismic loads.

### 8.4.17.3 Acceptance Criteria

For linear procedures, m-factors for use with deformation-controlled actions shall be taken from

Table 8-3. For nonlinear procedures, the coordinates of the generalized force–deformation relation, described by Fig. 8-1, and deformation acceptance criteria for primary and secondary components shall be taken from Table 8-4.

### 8.4.17.4 Connections

The connections between parts of structural particleboard shear wall assemblies and other elements of the lateral-force-resisting system shall be considered in accordance with Section 8.4.1.

The presence of connections between parts of nonstructural particleboard shear wall assemblies and other elements of the lateral-force-resisting system shall be verified. If connections are present, they need not be considered in the analysis conducted in accordance with Chapter 3. If connections are absent, they shall be provided in accordance with Section 8.4.1.

### C8.4.17.4 Connections

The capacity and ductility of the connections in structural particleboard shear walls will often determine the failure mode as well as the capacity of the assembly. Ductile connections with sufficient capacity will give acceptable performance. The tensile and bearing capacity of the nonstructural particleboard, rather than the connections, will often govern failure. The relatively low strength of this material makes connections between parts of the shear wall assembly and the other elements of the lateral-force-resisting system of less concern.

### 8.4.18 Light Gage Metal Frame Shear Walls

#### 8.4.18.1 Plaster on Metal Lath

The criteria for plaster on metal lath shall be based on Section 8.4.15.

#### 8.4.18.2 Gypsum Wallboard

The criteria for gypsum wallboard shall be based on Section 8.4.13.

#### 8.4.18.3 Wood Structural Panels

The criteria for wood structural panels shall be based on Section 8.4.9. The expected strength values of fasteners shall be calculated in accordance with Section 8.2.2.5, based on approved data. The expected strength of the wood structural panels shall be adjusted to account for differences in strength values of fasteners into light gage metal studs rather than wood studs.

## 8.5 WOOD DIAPHRAGMS

### 8.5.1 General

The expected strength of wood diaphragm assemblies, $Q_{CE}$, shall be determined in accordance with Sections 8.5.3 through 8.5.10. The expected strength, $Q_{CE}$, of braced horizontal diaphragm systems shall be determined in accordance with Section 8.5.11.

The effects of openings in wood diaphragms shall be considered. Chords and collectors shall be added to provide sufficient load capacity around openings to meet the strength requirements for the diaphragm.

Connections between diaphragms and other components including shear walls, diaphragm ties, collectors, cross ties, and out-of-plane anchors shall be considered in accordance with Section 8.3.3, and designed for forces calculated in accordance with Chapter 3.

### C8.5.1 General

The behavior of horizontal wood diaphragms is influenced by the type of sheathing, size and amount of fasteners, existence of perimeter chord or flange members, and the ratio of span length to width of the diaphragm.

The presence of any but small openings in wood diaphragms will cause a reduction in the stiffness and strength of the diaphragm due to a reduced length of diaphragm available to resist lateral forces. Special analysis techniques and detailing are required at the openings. The presence or addition of chord members around the openings will reduce the loss in stiffness of the diaphragm and limit damage in the area of the openings. See *Guidelines for the Design of Horizontal Wood Diaphragms*, ATC-7 (ATC 1981) and Tissell and Elliott (1997) for a discussion of the effects of openings in wood diaphragms.

The presence of chords at the perimeter of a diaphragm will significantly reduce the diaphragm deflection due to bending, and increase the stiffness of the diaphragm over that of an unchorded diaphragm. However, the increase in stiffness due to chords in a single straight-sheathed diaphragm is minimal due to the flexible nature of these diaphragms.

### 8.5.2 Types of Wood Diaphragms

#### 8.5.2.1 Existing Wood Diaphragms

8.5.2.1.1 Single Straight Sheathing Single straight-sheathed diaphragms shall include diaphragms with sheathing laid perpendicular to the framing members.

*C8.5.2.1.1 Single Straight Sheathing* Typically, single straight-sheathed diaphragms consist of 1-in. × sheathing laid perpendicular to the framing members; 2-in. × or 3-in. × sheathing may also be present. The sheathing serves the dual purpose of supporting gravity loads and resisting shear forces in the diaphragm. Most often, 1-in. × sheathing is nailed with 8d or 10d nails, with two or more nails per sheathing board at each support. Shear forces perpendicular to the direction of the sheathing are resisted by the nail couple. Shear forces parallel to the direction of the sheathing are transferred through the nails in the supporting joists or framing members below the sheathing joints.

*8.5.2.1.2 Double Straight Sheathing* Double straight-sheathed diaphragms shall include diaphragms with one layer of sheathing laid perpendicular to the framing members and a second layer of sheathing laid either perpendicular or parallel to the first layer, where both layers of sheathing are fastened to the framing members.

*C8.5.2.1.2 Double Straight Sheathing* Construction of double straight-sheathed diaphragms is the same as that for single straight-sheathed diaphragms, except that an upper layer of straight sheathing is laid over the lower layer of sheathing. The upper sheathing can be placed either perpendicular or parallel to the lower layer of sheathing. If the upper layer of sheathing is parallel to the lower layer, the board joints are usually offset sufficiently that nails at joints in the upper layer of sheathing are driven into a common sheathing board below, with sufficient edge distance. The upper layer of sheathing is nailed to the framing members through the lower layer of sheathing.

*8.5.2.1.3 Single Diagonal Sheathing* Single diagonally sheathed diaphragms shall include diaphragms with sheathing laid at approximately a 45-degree angle and connected to the framing members.

*C8.5.2.1.3 Single Diagonal Sheathing* Typically, 1-in. × sheathing is laid at an approximate 45-degree angle to the framing members. In some cases 2-in. × sheathing may also be used. The sheathing supports gravity loads and resists shear forces in the diaphragm. Commonly, 1-in. × sheathing is nailed with 8d nails, with two or more nails per board at each support. The recommended nailing for diagonally sheathed diaphragms is published in the *Western Woods Use Book* (WWPA 1996) and the *NEHRP Recommended Provisions for Seismic Regulations for New Buildings and Other Structures* (BSSC 2000). The shear capacity

of the diaphragm is dependent on the size and quantity of the nails at each sheathing board.

*8.5.2.1.4 Diagonal Sheathing with Straight Sheathing or Flooring Above* Diagonal sheathing with straight sheathing or flooring above shall include diaphragms with sheathing laid at a 45-degree angle to the framing members, with a second layer of straight sheathing or wood flooring laid on top of the diagonal sheathing at a 90-degree angle to the framing members.

*C8.5.2.1.4 Diagonal Sheathing with Straight Sheathing or Flooring Above* Typically, these consist of a lower layer of 1-in. × diagonal sheathing laid at a 45-degree angle to the framing members, with a second layer of straight sheathing or wood flooring laid on top of the diagonal sheathing at a 90-degree angle to the framing members. Both layers of sheathing support gravity loads and resist shear forces in the diaphragm. Sheathing boards are commonly connected with two or more 8d nails per board at each support.

*8.5.2.1.5 Double Diagonal Sheathing* Double diagonally sheathed diaphragms shall include diaphragms with one layer of sheathing laid at a 45-degree angle to the framing members and a second layer of sheathing laid at a 90-degree angle to the first layer.

*C8.5.2.1.5 Double Diagonal Sheathing* Typically, double diagonally sheathed diaphragms consist of a lower layer of 1-in. × diagonal sheathing with a second layer of 1-in. × diagonal sheathing laid at a 90-degree angle to the lower layer. The sheathing supports gravity loads and resists shear forces in the diaphragm. The sheathing is commonly nailed with 8d nails, with two or more nails per board at each support. The recommended nailing for double diagonally sheathed diaphragms is published in the *Western Woods Use Book* (WWPA 1996).

*8.5.2.1.6 Wood Structural Panel Sheathing* Wood structural panel-sheathed diaphragms shall include diaphragms with wood structural panel, or other wood structural panels as defined in this standard, fastened to the framing members.

*C8.5.2.1.6 Wood Structural Panel Sheathing* Typically, these consist of wood structural panels, such as wood structural panel or oriented strand board, placed on framing members and nailed in place. Different grades and thicknesses of wood structural panels are commonly used, depending on requirements for gravity load support and shear capacity. Edges at the ends of the wood structural panels are usually supported by the framing members. Edges at the sides of the panels can be blocked or unblocked. In some cases, tongue and groove wood structural panels are used. Nailing patterns and nail size can vary greatly. Nail spacing is commonly in the range of 3 to 6 in. on center at the supported and blocked edges of the panels, and 10 to 12 in. on center at the panel infield. Staples are sometimes used to attach the wood structural panels.

*8.5.2.1.7 Braced Horizontal Diaphragms* Braced horizontal diaphragms shall include diaphragms with a horizontal truss system at the floor or roof level of the building.

*C8.5.2.1.7 Braced Horizontal Diaphragms* Typically, these consist of "X" rod bracing and wood struts forming a horizontal truss system at the floor or roof levels of the building. The "X" bracing usually consists of steel rods drawn taut by turnbuckles or nuts. The struts usually consist of wood members, which may or may not be part of the gravity-load-bearing system of the floor or roof. The steel rods function as tension members in the horizontal truss, while the struts function as compression members. Truss chords (similar to diaphragm chords) are needed to resist bending in the horizontal truss system.

### 8.5.2.2 Enhanced Wood Diaphragms
Enhanced wood diaphragms shall include existing diaphragms rehabilitated by an approved method.

### C8.5.2.2 Enhanced Wood Diaphragms
Possible rehabilitation methods for wood diaphragms are described in Sections 8.5.2.2.1 through 8.5.2.2.3.

*C8.5.2.2.1 Wood Structural Panel Overlays on Straight or Diagonal Sheathing* Existing sheathed diaphragms may be overlaid with new wood structural panels. Nails or staples may be used to connect the new structural panels to the existing diaphragms. Nails should be of sufficient length to provide the required embedment into framing members below the sheathing.

These diaphragms typically consist of new wood structural panels placed over existing straight or diagonal sheathing and nailed or stapled to the existing framing members through the existing sheathing. If the new overlay is nailed to the existing framing members only—without nailing at the panel edges perpendicular to the framing—the response of the new overlay will be similar to that of an unblocked wood structural panel diaphragm.

If a stronger and stiffer diaphragm is desired, the joints of the new wood structural panel overlay should be placed parallel to the joints of the existing sheathing, with the overlay nailed or stapled to the existing sheathing. The edges of the new wood structural panels should be offset from the joints in the existing sheathing below by a sufficient distance that the new nails may be driven into the existing sheathing without splitting the sheathing. If the new panels are nailed at all edges as described above, the response of the new overlay will be similar to that of a blocked wood structural panel diaphragm. As an alternative, new blocking may be installed below all panel joints perpendicular to the existing framing members.

Because the joints of the overlay and the joints of the existing sheathing may not be offset consistently without cutting the panels, it may be advantageous to place the wood structural panel overlay at a 45-degree angle to the existing sheathing. If the existing diaphragm is straight-sheathed, the new overlay should be placed at a 45-degree angle to the existing sheathing and joists. If the existing diaphragm is diagonally sheathed, the new wood structural panel overlay should be placed perpendicular to the existing joists at a 45-degree angle to the diagonal sheathing. Nails should be driven into the existing sheathing with sufficient edge distance to prevent splitting of the existing sheathing. At boundaries, nails should be of sufficient length to penetrate the sheathing into the framing below. New structural panel overlays shall be connected to shear walls or vertical bracing elements to ensure the effectiveness of the added panel.

Care should be exercised where placing new wood structural panel overlays on existing diaphragms. The changes in stiffness and dynamic characteristics of the diaphragm may have negative effects by causing increased forces in other components or elements. The increased stiffness and the associated increase in dynamic forces may not be desirable in some diaphragms for certain performance levels.

*C8.5.2.2.2 Wood Structural Panel Overlays on Existing Wood Structural Panels* Existing wood structural panel diaphragms may be overlaid with new wood structural panels. Panel joints should be offset, or the overlay should be placed at a 45-degree angle to the existing wood structural panels.

The placement of a new overlay over an existing diaphragm should follow the same construction methods and procedures as those used for straight-sheathed and diagonally sheathed diaphragms (see Section C8.5.2.2.1).

*C8.5.2.2.3 Increased Attachment* The nailing or attachment of the existing sheathing to the supporting framing may be increased. Nailing or attachment to the supporting framing should be increased and blocking for the diaphragm at the wood structural panel joints should be added.

For straight-sheathed diaphragms, the increase in shear capacity will be minimal. Double straight-sheathed diaphragms with minimal nailing in the upper or both layers of sheathing may be enhanced significantly by adding new nails or staples to the existing diaphragm. The same is true for diaphragms that are single diagonally sheathed, double diagonally sheathed, or single diagonally sheathed with straight sheathing or flooring.

In some cases, increased nailing at the wood structural panel infield may also be required. If the required shear capacity or stiffness is greater than that which can be provided by increased attachment, a new overlay on the existing diaphragm may be required to provide the desired enhancement.

### 8.5.2.3 New Wood Diaphragms

*8.5.2.3.1 New Wood Structural Panel Sheathing* New wood structural panel sheathed diaphragms shall include new wood structural panels connected to new framing members, or connected to existing framing members after existing sheathing has been removed.

*C8.5.2.3.1 New Wood Structural Panel Sheathing* Typically, these consist of wood structural panels—such as wood structural panel or oriented strand board—nailed or stapled to existing framing members after existing sheathing has been removed. Different grades and thicknesses of wood structural panels can be used, depending on the requirements for gravity load support and diaphragm shear capacity. In most cases, the panels are placed with the long dimension perpendicular to the framing members, and panel edges at the ends of the panels are supported by, and nailed to, the framing members. Edges at the sides of the panels can be blocked or unblocked, depending on the shear capacity and stiffness required in the new diaphragm. Wood structural panels can be placed in various patterns as shown in the *LRFD Manual for Engineered Wood Construction* (AF&PA LRFD 1996) and the *NEHRP Recommended Provisions for Seismic Regulations for New Buildings and Other Structures* (BSSC 2000).

*8.5.2.3.2 New Single-Diagonal Sheathing* New single-diagonally sheathed wood diaphragms shall include

new sheathing laid at approximately a 45-degree angle and connected to the existing framing members.

*8.5.2.3.3 New Double-Diagonal Sheathing* New double-diagonally sheathed wood diaphragms shall include diaphragms with new sheathing laid at approximately a 45-degree angle to the existing framing members with a second layer of sheathing laid at approximately a 90-degree angle to the first layer, where both layers shall be connected to the framing members.

*8.5.2.3.4 New Braced Horizontal Diaphragms* New braced horizontal diaphragms shall include a new horizontal truss system attached to the existing framing at the floor or roof level of the building.

*C8.5.2.3.4 New Braced Horizontal Diaphragms* Because new horizontal truss systems will induce new forces on existing framing members, it may be more economical to design floor or roof sheathing as a diaphragm. This eliminates the potential need to strengthen wood members at the compression struts. Braced horizontal diaphragms are more feasible where sheathing cannot provide sufficient shear capacity, or where diaphragm openings reduce the shear capacity of the diaphragm and additional shear capacity is needed.

### 8.5.3 Single Straight Sheathing

#### 8.5.3.1 Stiffness

The deflection of straight-sheathed diaphragms shall be calculated using Eq. 8-3:

$$\Delta_y = v_y L/(2G_d) \qquad \text{(Eq. 8-3)}$$

where

$G_d$ = diaphragm shear stiffness from Table 8-2 (lb/in.);

$L$ = diaphragm span, distance between shear walls or collectors (in.);

$v_y$ = shear per unit length at yield in the direction under consideration (lb/in.); and

$\Delta_y$ = calculated diaphragm deflection at yield (in.).

Properties used to compute diaphragm deflection and stiffness shall be based on Section 8.2.2.

#### C8.5.3.1 Stiffness

Straight-sheathed diaphragms are characterized by high flexibility with a long period of vibration. These diaphragms are suitable for low shear conditions where control of diaphragm deflections is not needed to attain the desired performance levels.

### 8.5.3.2 Strength

The expected strength of straight-sheathed diaphragms shall be determined in accordance with Section 8.2.2.

### C8.5.3.2 Strength

The expected capacity of straight-sheathed diaphragms is dependent on the size, number, and spacing between the nails at each sheathing board, and the spacing of the supporting framing members. The shear capacity of straight-sheathed diaphragms can be calculated using the nail-couple method. See *Guidelines for the Design of Horizontal Wood Diaphragms*, ATC-7 (ATC 1981) for a discussion of calculating the shear capacity of straight-sheathed diaphragms.

### 8.5.3.3 Acceptance Criteria

For linear procedures, *m*-factors for use with deformation-controlled actions shall be taken from Table 8-3. For nonlinear procedures, the coordinates of the generalized force–deformation relation, described by Fig. 8-1, and deformation acceptance criteria shall be taken from Table 8-4.

### C8.5.3.3 Acceptance Criteria

Deformation acceptance criteria will largely depend on the allowable deformations for other structural and nonstructural components and elements that are laterally supported by the diaphragm. Allowable deformations must also be consistent with the permissible damage state of the diaphragm.

### 8.5.3.4 Connections

Connections between diaphragms and shear walls and other vertical elements shall be considered in accordance with Section 8.5.1.

### C8.5.3.4 Connections

The load capacity of connections between diaphragms and shear walls or other vertical elements, as well as diaphragm chords and shear collectors, is critical.

### 8.5.4 Double Straight Sheathing

### 8.5.4.1 Stiffness

The deflection of double straight-sheathed diaphragms shall be calculated using Eq. 8-3. Properties used to compute diaphragm deflection and stiffness shall be based on Section 8.2.2.

### C8.5.4.1 Stiffness

The double-sheathed system will provide a significant increase in stiffness over a single straight-sheathed diaphragm, but very little test data are available on the stiffness and strength of these diaphragms. Both layers of straight sheathing must have sufficient nailing, and the joints of the top layer must be either offset or perpendicular to the bottom layer.

### 8.5.4.2 Strength

The expected strength of double straight-sheathed diaphragms shall be determined in accordance with Section 8.2.2.

### C8.5.4.2 Strength

The strength and stiffness of double straight-sheathed diaphragms is highly dependent on the nailing of the upper layer of sheathing. If the upper layer has minimal nailing, the increase in strength and stiffness over a single straight-sheathed diaphragm may be slight. If the upper layer of sheathing has nailing similar to that of the lower layer of sheathing, the increase in strength and stiffness will be significant.

### 8.5.4.3 Acceptance Criteria

For linear procedures, *m*-factors for use with deformation-controlled actions shall be taken from Table 8-3. For nonlinear procedures, the coordinates of the generalized force–deformation relation, described by Fig. 8-1, and deformation acceptance criteria shall be taken from Table 8-4.

### 8.5.4.4 Connections

Connections between diaphragms and shear walls and other vertical elements shall be considered in accordance with Section 8.5.1.

### 8.5.5 Single Diagonal Sheathing

### 8.5.5.1 Stiffness

The deflection of single diagonally sheathed diaphragms shall be calculated using Eq. 8-3. Properties used to compute diaphragm deflection and stiffness shall be based on Section 8.2.2.

### C8.5.5.1 Stiffness

Single diagonally sheathed diaphragms are significantly stiffer than straight-sheathed diaphragms, but are still quite flexible.

### 8.5.5.2 Strength

The expected strength for diagonally sheathed wood diaphragms with chords shall be determined in accordance with Section 8.2.2.

### C8.5.5.2 Strength

Diagonally sheathed diaphragms are usually capable of resisting moderate shear loads.

Because the diagonal sheathing boards function in tension and compression to resist shear forces in the diaphragm, and the boards are placed at a 45-degree angle to the chords at the ends of the diaphragm, the component of the force in the sheathing boards that is perpendicular to the axis of the end chords will create a bending force in the end chords. If the shear in diagonally sheathed diaphragms is limited to approximately 300 lb/ft or less, bending forces in the end chords are usually neglected. If shear forces exceed 300 lb/ft, the end chords should be designed or reinforced to resist bending forces from the sheathing. See *Guidelines for the Design of Horizontal Wood Diaphragms,* ATC-7 (ATC 1981) for methods of calculating the shear capacity of diagonally sheathed diaphragms.

### 8.5.5.3 Acceptance Criteria

For linear procedures, *m*-factors for use with deformation-controlled actions shall be taken from Table 8-3. For nonlinear procedures, the coordinates of the generalized force–deformation relation, described by Fig. 8-1, and deformation acceptance criteria shall be taken from Table 8-4.

### 8.5.5.4 Connections

Connections between diaphragms and shear walls and other vertical elements shall be considered in accordance with Section 8.5.1.

## 8.5.6 Diagonal Sheathing with Straight Sheathing or Flooring Above

### 8.5.6.1 Stiffness

The deflection of diagonally sheathed diaphragms with straight sheathing or flooring above shall be calculated using Eq. 8-3. Properties used to compute diaphragm deflection and stiffness shall be based on Section 8.2.2.

### C8.5.6.1 Stiffness

Straight sheathing or flooring over diagonal sheathing provides a significant increase in stiffness over single-sheathed diaphragms. The increased stiffness of these diaphragms may make them suitable where Life Safety or Immediate Occupancy Structural Performance Levels are desired.

### 8.5.6.2 Strength

The expected strength of diagonally sheathed diaphragms with straight sheathing or flooring above shall be determined in accordance with Section 8.2.2.

### C8.5.6.2 Strength

Shear capacity is dependent on the nailing of the diaphragm. The strength and stiffness of diagonally sheathed diaphragms with straight sheathing above is highly dependent on the nailing of both layers of sheathing. Both layers of sheathing should have at least two 8d common nails per board at each support.

### 8.5.6.3 Acceptance Criteria

For linear procedures, *m*-factors for use with deformation-controlled actions shall be taken from Table 8-3. For nonlinear procedures, the coordinates of the generalized force–deformation relation, described by Fig. 8-1, and deformation acceptance criteria shall be taken from Table 8-4.

### 8.5.6.4 Connections

Connections between diaphragms and shear walls and other vertical elements shall be considered in accordance with Section 8.5.1.

## 8.5.7 Double Diagonal Sheathing

### 8.5.7.1 Stiffness

The deflection of double diagonally sheathed diaphragms shall be calculated using Eq. 8-3. Properties used to compute diaphragm deflection and stiffness shall be based on Section 8.3.2.

### C8.5.7.1 Stiffness

Double diagonally sheathed diaphragms have greater stiffness than diaphragms with single diagonal sheathing. The response of these diaphragms is similar to the response of diagonally sheathed diaphragms with straight sheathing overlays.

The increased stiffness of these diaphragms may make them suitable where Life Safety or Immediate Occupancy Structural Performance Levels are desired.

### 8.5.7.2 Strength

The expected strength of double diagonally sheathed wood diaphragms shall be determined in accordance with Section 8.2.2.

### C8.5.7.2 Strength

Shear capacity is dependent on the nailing of the diaphragm, but these diaphragms are usually suitable for moderate to high shear loads.

Shear capacities are similar to those of diagonally sheathed diaphragms with straight-sheathing overlays. The sheathing boards in both layers of sheathing should be nailed with at least two 8d common nails at each support. The presence of a double layer of diagonal sheathing will eliminate the bending forces that single diagonally sheathed diaphragms impose on the chords at the ends of the diaphragm. As a result, the bending capacity of the end chords does not have an effect on the shear capacity and stiffness of the diaphragm.

### 8.5.7.3 Acceptance Criteria

For linear procedures, $m$-factors for use with deformation-controlled actions shall be taken from Table 8-3. For nonlinear procedures, the coordinates of the generalized force–deformation, described by Fig. 8-1, and deformation acceptance criteria shall be taken from Table 8-4.

### 8.5.7.4 Connections

Connections between diaphragms and shear walls and other vertical elements shall be considered in accordance with Section 8.5.1.

## 8.5.8 Wood Structural Panel Sheathing

### 8.5.8.1 Stiffness

The deflection of blocked and chorded wood structural panel diaphragms with constant nailing across the diaphragm length shall be determined using Eq. 8-4:

$$\Delta_y = 5v_yL^3/(8EAb) + v_yL/(4Gt)$$
$$+ 0.188Le_n + \Sigma(\Delta_cX)/2b \quad \text{(Eq. 8-4)}$$

where

$A$ = area of diaphragm chords cross section (in.²);
$b$ = diaphragm width (ft);
$E$ = modulus of elasticity of diaphragm chords (psi);
$e_n$ = nail deformation at yield load per nail (in.). Values listed are for Structural I panels; multiply by 1.2 for all other panel grades;
  = 0.13 for 6d nails at yield;
  = 0.08 for 8d nails at yield;
  = 0.08 for 10d nails at yield;
$G$ = modulus of rigidity of wood structural panels (psi);
$L$ = diaphragm span, distance between shear walls or collectors (ft);
$t$ = effective thickness of wood structural panel for shear (in.);
$v_y$ = shear at yield in the direction under consideration (lb/ft);
$\Delta_y$ = calculated deflection of diaphragm at yield (in.); and
$\Sigma(\Delta_cX)$ = sum of individual chord-splice slip values on both sides of the diaphragm, each multiplied by its distance to the nearest support.

Alternatively, a more rigorous calculation of diaphragm deflection based on rational engineering principles shall be permitted.

The deflection of blocked and chorded wood structural panel diaphragms with variable nailing across the diaphragm length shall be determined using Eq. 8-5:

$$\Delta_y = 5v_yL^3/(8EAb) + v_yL/(4Gt)$$
$$+ 0.376Le_n + \Sigma(\Delta_cX)/(2b) \quad \text{(Eq. 8-5)}$$

Alternatively, a more rigorous calculation of diaphragm deflection based on rational engineering principles shall be permitted.

The deflection of unblocked diaphragms shall be calculated using Eq. 8-3. Properties used to compute diaphragm deflection and stiffness shall be based on Section 8.2.2.

### C8.5.8.1 Stiffness

The response of wood structural panel diaphragms is dependent on the thickness of the wood structural panels, the length-to-width ($L/b$) ratio, nailing pattern, and presence of chords in the diaphragm, as well as other factors. Values for modulus rigidity, $G$, and effective thickness, $t$, for various sheathing materials are contained in *Design Capacities of APA Performance Rated Structural-Use Panels* (APA 1995) and *Plywood Design Specification* (APA 1997).

In most cases the area of the diaphragm chord equals the area of the continuous wood (or steel) member to which the sheathing is attached. For buildings with wood diaphragms and concrete or masonry walls, however, the area of the diaphragm chord is more difficult to identify and engineering judgment is required. The tension area of the diaphragm chord on both edges of the diaphragm should be used for deflection calculations. Generally, this is conservative as it results in a larger calculated deflection. Use of the tension area of the diaphragm chord may not yield conservative results, however, where calculating the period of the building using Eq. 3-8.

The term $\Delta_cX$ is determined by multiplying the assumed diaphragm chord slip at a single chord splice, $\Delta_c$, by the distance, $X$, from diaphragm chord splice to the nearest support (shear wall).

An alternate constant that can be used in the nail slip contribution term where panel nailing is not uniform is provided in Appendix C of the *Diaphragms and Shear Walls Design/Construction Guide* (APA 2001).

Example calculations of diaphragm deflection are provided in Design Example A.1 of the *Structural Use Panel Shear Wall and Diaphragm Supplement*

(AF&PA ASD 2001) and *Design of Wood Structures* (Breyer et al. 1999).

### 8.5.8.2 Strength

The expected strength of wood structural panel diaphragms shall be taken as mean maximum strengths obtained experimentally. Expected strengths shall be permitted to be based on 1.5 times yield strengths of wood structural panel diaphragms. Yield strengths shall be determined using LRFD procedures contained in AF&PA/ASCE 16 *Standard for Load and Resistance Factor Design (LRFD) for Engineered Wood Construction* (ASCE 1996), except that the resistance factor, $\phi$, shall be taken as unity and expected material properties shall be determined in accordance with Section 8.2.2.

Conversion for tabulated allowable stress values in accordance with Section 8.2.2.5 shall not be permitted for wood structural panel diaphragms, but approved allowable stress values for fasteners shall be permitted to be converted in accordance with Section 8.2.2.5 where the strength of a shear wall is computed using principles of mechanics.

The expected shear capacity of unchorded diaphragms shall be calculated by multiplying the values given for chorded diaphragms by 0.60.

### C8.5.8.2 Strength

Shear capacities of wood structural panel diaphragms are primarily dependent on the nailing at the wood structural panel edges, and the thickness and grade of the wood structural panel in the diaphragm.

LRFD-based design values for various configurations are listed in the *LRFD Manual for Engineered Wood Construction* (AF&PA LRFD 1996) and the *NEHRP Recommended Provisions for Seismic Regulations for New Buildings and Other Structures* (BSSC 2000). A method for calculating the capacity of wood structural panel diaphragms based on accepted nail values and panel shear strength is provided in Tissell and Elliott (1997). For this method, use LRFD-based fastener strengths. Due to the differences in load duration/time effect factors between the allowable stress and LRFD formats, direct conversion of diaphragm tables using the method outlined in Section 8.2.2.5 is not permitted. However, the tabulated LRFD design values, with $\phi = 1$, are intended to be 2.0 times the associated allowable stress design values.

### 8.5.8.3 Acceptance Criteria

For linear procedures, *m*-factors for use with deformation-controlled actions shall be taken from Table 8-3. For nonlinear procedures, the coordinates of the generalized force–deformation relation, described by Fig. 8-1, and deformation acceptance criteria shall be taken from Table 8-4.

### 8.5.8.4 Connections

Connections between diaphragms and shear walls and other vertical elements shall be considered in accordance with Section 8.5.1.

## 8.5.9 Wood Structural Panel Overlays on Straight or Diagonal Sheathing

### 8.5.9.1 Stiffness

Placement of the new wood structural panel overlay shall be consistent with Section 8.5.2.2.

The deflection of wood structural panel overlays on straight or diagonally sheathed diaphragms shall be calculated using Eq. 8-3.

### C8.5.9.1 Stiffness

The stiffness of existing straight-sheathed diaphragms can be increased significantly by placing a new wood structural panel overlay over the existing diaphragm. The stiffness of existing diagonally sheathed diaphragms and wood structural panel diaphragms will be increased, but not in proportion to the stiffness increase for straight-sheathed diaphragms.

Depending on the nailing of the new overlay, the response of the diaphragm may be similar to that of a blocked or an unblocked diaphragm.

The increased stiffness of these diaphragms may make them suitable where Life Safety or Immediate Occupancy Structural Performance Levels are desired.

### 8.5.9.2 Strength

Strength of wood structural panel overlays shall be determined in accordance with Section 8.3.2.2. It shall be permitted to take the expected strength of wood structural panel overlays as the value for the corresponding wood structural panel diaphragm without the existing sheathing below, computed in accordance with Section 8.5.8.2.

### 8.5.9.3 Acceptance Criteria

For linear procedures, *m*-factors for use with deformation-controlled actions shall be taken from Table 8-3. For nonlinear procedures, the coordinates of the generalized force–deformation relation, described by Fig. 8-1, and deformation acceptance criteria shall be taken from Table 8-4.

### 8.5.9.4 Connections

Connections between diaphragms and shear walls and other vertical elements shall be considered in accordance with Section 8.5.1.

## 8.5.10 Wood Structural Panel Overlays on Existing Wood Structural Panel Sheathing

### *8.5.10.1 Stiffness*

Diaphragm deflection shall be calculated in accordance with Eq. 8-3 or using accepted principles of mechanics. Nails in the upper layer of wood structural panel shall have sufficient embedment in the framing to meet the requirements of AF&PA/ASCE 16 *Standard for Load and Resistance Factor Design (LRFD) for Engineered Wood Construction* (ASCE 1996).

### *C8.5.10.1 Stiffness*

According to Tissell and Elliott (1997), Eq. 8-4 is not applicable to two-layer diaphragms, presumably due to the difficulty in estimating the combined nail slip. Diaphragm deflection may be estimated using principles of mechanics that include consideration of nail slip, blocking, and the embedment of nails into the framing.

### *8.5.10.2 Strength*

Expected strength shall be calculated based on the combined two layers of wood structural panel sheathing, with the strength of the overlay limited to 75% of the values calculated in accordance with Section 8.5.8.2.

### *8.5.10.3 Acceptance Criteria*

For linear procedures, *m*-factors for use with deformation-controlled actions shall be taken from Table 8-3. For nonlinear procedures, the coordinates of the generalized force–deformation relation, described by Fig. 8-1, and deformation acceptance criteria shall be taken from Table 8-4.

### *8.5.10.4 Connections*

Connections between diaphragms and shear walls and other vertical elements shall be considered in accordance with Section 8.5.1.

## 8.5.11 Braced Horizontal Diaphragms

Braced horizontal diaphragms shall be considered in accordance with Section 8.7.1.

Connections between members of the horizontal bracing system and shear walls or other vertical elements shall be considered in accordance with Section 8.5.1.

## 8.6 WOOD FOUNDATIONS

### 8.6.1 Types of Wood Foundations

Types of wood foundations include wood piling, wood footings, and pole structures. Wood piling shall include friction or end-bearing piles that resist only vertical loads.

### C8.6.1 Types of Wood Foundations

1. **Wood Piling**. Wood piles are generally used with a concrete pile cap and are usually keyed into the base of the concrete cap. The piles are usually treated with preservatives. Piles are classified as either friction- or end-bearing piles. Piles are generally not able to resist uplift loads because of the manner in which they are attached to the pile cap. The piles may be subjected to lateral loads from seismic loading, which are resisted by bending of the piles. The analysis of pile bending is generally based on a pinned connection at the top of the pile, and fixity of the pile at some depth established by the geotechnical engineer. However, it should be evaluated with consideration for the approximate nature of the original assumption of the depth to point of fixity. Where battered piles are present, the lateral loads can be resisted by the horizontal component of the axial load.

2. **Wood Footings**. Wood grillage footings, sleepers, skids, and pressure-treated all-wood foundations can be encountered in existing structures. These foundations are highly susceptible to deterioration. The seismic resistance of wood footings is generally very low; they are essentially dependent on friction between the wood and soil for their performance.

3. **Pole Structures**. Pole structures resist lateral loads by acting as cantilevers fixed in the ground, with the lateral load considered to be applied perpendicular to the pole axis. It is possible to design pole structures to have moment-resisting capacity at floor and roof levels by the use of knee braces or trusses. Pole structures are frequently found on sloping sites. The varying unbraced lengths of the poles generally affect the stiffness and performance of the structure, and can result in unbalanced loads to the various poles along with significant torsional distortion, which must be investigated and evaluated. Additional horizontal and diagonal braces can be used to reduce the flexibility of tall poles or reduce the torsional eccentricity of the structure.

### 8.6.2 Analysis, Strength, and Acceptance Criteria of Wood Foundations

The expected strength of wood piles shall be computed in accordance with Section 8.3.2.2. Lateral deflection of piles under seismic loads shall be calculated based on an assumed point of fixity. Unless rigidly connected to the pile cap, wood piles shall be taken as pinned at the top.

Flexure and axial loads in wood piles shall be considered deformation-controlled. The *m*-factors shall be taken from Table 8-3.

Wood footings shall be thoroughly investigated for the presence of deterioration. Acceptability of soils below wood footings shall be determined in accordance with Chapter 4.

Component and connection strength of pole structures shall be based on Section 8.2. Pole structures shall be modeled as cantilever elements and analyzed in accordance with Chapter 3.

Flexure and axial loads in pole structures shall be considered deformation-controlled. The *m*-factors shall be taken from Table 8-3. Where concentrically braced diagonals are added to enhance the capacity of the pole structure, reduced *m*-factors taken from Table 8-3 shall be used.

### C8.6.2 Analysis, Strength, and Acceptance Criteria of Wood Foundations

The strength of the components, elements, and connections of a pole structure are the same as for a conventional structure.

### 8.6.3 Rehabilitation Measures

Wood foundations not meeting the acceptance criteria for the selected Rehabilitation Objective shall be rehabilitated in accordance with Section 8.3.4. Wood foundations exhibiting signs of deterioration shall be rehabilitated or replaced.

### C8.6.3 Rehabilitation Measures

Wood footings showing signs of deterioration may be replaced with reinforced concrete footings. Wood pole structures can be rehabilitated with the installation of diagonal braces or other supplemental lateral-force-resisting elements. Structures supported on wood piles may be rehabilitated by the installation of additional piles.

## 8.7 OTHER WOOD ELEMENTS AND COMPONENTS

### 8.7.1 General

Wood elements and components, other than shear walls, diaphragms, and foundations, shall be considered in accordance with this section. Where an assembly includes wood components and steel rods, the rods shall be considered in accordance with applicable provisions of Chapter 5.

### C8.7.1 General

Other wood elements include knee-braced frames, rod-braced frames, and braced horizontal diaphragms, among other systems.

Knee-braced frames produce moment-resisting joints by the addition of diagonal members between columns and beams. The resulting "semi-rigid" frame resists lateral loads. The moment-resisting capacity of knee-braced frames varies widely. The controlling part of the assembly is usually the connection; however, bending of members can be the controlling feature of some frames. Once the capacity of the connection is determined, members can be checked and the capacity of the frame can be determined by statics. Particular attention should be given to the beam-column connection. Additional tensile forces may be developed in this connection due to knee-brace action under vertical loads.

Similar to knee-braced frames, the connections of rods to timber framing will usually govern the capacity of the rod-braced frame. Typically, the rods act only in tension. Once the capacity of the connection is determined, the capacity of the frame can be determined by statics.

Braced horizontal diaphragms are described in Section 8.5.2.1.7.

### 8.7.1.1 Stiffness

The stiffness and deflection of wood elements other than shear walls, diaphragms, and foundations shall be determined based on a mathematical model or by a test program for the assembly considering the configuration, stiffness, and interconnection of the individual components approved by the authority having jurisdiction.

### 8.7.1.2 Strength

The capacities of individual components, including connections, shall be determined in accordance with Section 8.3.2.

### C8.7.1.2 Strength

The strength of wood elements is dependent on the strength of the individual components that comprise the assembly. In many cases the capacity of the connections between components will be the limiting factor in the strength of the assembly.

### 8.7.1.3 Acceptance Criteria

Design actions shall be compared with design capacities in accordance with Section 3.4.2.2. Connections shall be considered in accordance with Section 8.3.3. Axial tension and axial tension with bending shall be considered deformation-controlled. Axial compression and connections between steel rods and wood components shall be considered force-controlled. The *m*-factors for deformation-controlled actions shall be taken from Table 8-3 for component actions listed. The *m*-factors for deformation-controlled component actions not included in Table 8-3 shall be established in accordance with Section 2.8. For nonlinear procedures, coordinates of the generalized force–deformation relation described by Fig. 8-1, and deformation acceptance criteria shall be taken from Table 8-4.

### C8.7.1.3 Acceptance Criteria

Deformation acceptance criteria will largely depend on the allowable deformations for other structural and nonstructural components that are supported by the element. Allowable deformations must also be consistent with the desired performance level. Actions on connection types that do not appear in Table 8-3 (e.g., truss plates) are force-controlled.

## 9.0 SEISMIC ISOLATION AND ENERGY DISSIPATION

## 9.1 SCOPE

This chapter sets forth requirements for the Systematic Rehabilitation of buildings using seismic isolation and energy dissipation systems. Section 9.2 provides analysis and design criteria for seismic isolation systems. Section 9.3 provides analysis and design criteria for passive energy dissipation systems.

Components and elements in buildings with seismic isolation and energy dissipation systems shall also comply with the requirements of Chapters 1 through 8 and 11 of this standard, unless modified by the requirements of this chapter.

## C9.1 SCOPE

The basic form and formulation of requirements for seismic isolation and energy dissipation systems have been established and coordinated with the Rehabilitation Objectives, target Building Performance Levels, and seismic ground shaking hazard criteria of Chapter 1 and the linear and nonlinear procedures of Chapter 3.

Criteria for modeling the stiffness, strength, and deformation capacities of conventional structural components of buildings with seismic isolation or energy dissipation systems are given in Chapters 5 through 8 and Chapter 10.

Limited guidance for other special seismic systems, including active control systems, hybrid active

and passive systems, and tuned mass and liquid dampers, is provided in this chapter.

Special seismic protective systems should be evaluated as possible rehabilitation strategies based on the Rehabilitation Objectives established for the building.

Seismic isolation and energy dissipation systems are viable design strategies that have already been used for seismic rehabilitation of a number of buildings. Other special seismic protective systems—including active control, hybrid combinations of active and passive energy devices, and tuned mass and liquid dampers—may also provide practical solutions in the near future. These systems are similar in that they enhance performance during an earthquake by modifying the building's response characteristics.

Seismic isolation and energy dissipation systems will not be appropriate design strategies for most buildings, particularly buildings that have only Limited Rehabilitation Objectives. In general, these systems will be most applicable to the rehabilitation of buildings whose owners desire superior earthquake performance and can afford the special costs associated with the design, fabrication, and installation of seismic isolators and/or energy dissipation devices. These costs are typically offset by the reduced need for stiffening and strengthening measures that would otherwise be required to meet Rehabilitation Objectives.

Seismic isolation and energy dissipation systems are relatively new and sophisticated concepts that require more extensive design and detailed analysis than do most conventional rehabilitation schemes. Similarly, design (peer) review is required for all rehabilitation schemes that use either seismic isolation or energy dissipation systems.

Seismic isolation and energy dissipation systems include a wide variety of concepts and devices. In most cases, these systems and devices will be implemented with some additional conventional strengthening of the structure; in all cases they will require evaluation of existing building components. As such, this chapter supplements the requirements of other chapters of this document with additional criteria and methods of analysis that are appropriate for buildings rehabilitated with seismic isolators and/or energy dissipation devices.

Seismic isolation is increasingly being considered for historic buildings that are free-standing and have a basement or bottom space of no particular historic significance. In selecting such a solution, special consideration should be given to the possibility that historic or archaeological resources may be present at the site. If historic or archaeological resources are present at the site, the guidance of the State Historic Preservation Officer should be obtained in a timely manner. Isolation is also often considered for essential facilities, to protect valuable contents, and on buildings with a complete but insufficiently strong lateral-force–resisting system.

Conceptually, isolation reduces response of the superstructure by "decoupling" the building from the ground. Typical isolation systems reduce forces transmitted to the superstructure by lengthening the period of the building and adding some amount of damping. Added damping is an inherent property of most isolators, but may also be provided by supplemental energy dissipation devices installed across the isolation interface. Under favorable conditions, the isolation system reduces drift in the superstructure by a factor of at least two—and sometimes by as much as factor of five—from that which would occur if the building were not isolated. Accelerations are also reduced in the structure, although the amount of reduction depends on the force-deflection characteristics of the isolators and may not be as significant as the reduction of drift. Reduction of drift in the superstructure protects structural components and elements, as well as nonstructural components sensitive to drift-induced damage. Reduction of acceleration protects nonstructural components that are sensitive to acceleration-induced damage.

Passive energy dissipation devices add damping (and sometimes stiffness) to the building. A wide variety of passive energy dissipation devices are available, including fluid viscous dampers, viscoelastic materials, and hysteretic devices. Ideally, energy dissipation devices dampen earthquake excitation of the structure that would otherwise cause higher levels of response and damage to components of the building. Under favorable conditions, energy dissipation devices reduce drift of the structure by a factor of about two to three (if no stiffness is added) and by larger factors if the devices also add stiffness to the structure. Energy dissipation devices will also reduce force in the structure—provided the structure is responding elastically—but would not be expected to reduce force in structures that are responding beyond yield.

Active control systems sense and resist building motion, either by applying external force or by modifying structural properties of active components (e.g., so-called smart braces). Tuned mass or liquid dampers modify properties and add damping to key building modes of vibration.

Special seismic systems, such as isolation or energy dissipation systems, should be considered early in the design process and be based on the

Rehabilitation Objectives established for the building (Chapter 2). Whether a special seismic system is found to be the "correct" design strategy for building rehabilitation will depend primarily on the performance required at the specified level of earthquake demand. In general, special seismic systems will be found to be more attractive as a rehabilitation strategy for buildings that have more stringent Rehabilitation Objectives (i.e., higher levels of performance and more severe levels of earthquake demand). Table C9-1 provides some simple guidance on the performance levels for which isolation and energy dissipation systems should be considered as possible design strategies for building rehabilitation.

Table C9-1 suggests that isolation systems should be considered for achieving the Immediate Occupancy Structural Performance Level and the Operational Nonstructural Performance Level. Conversely, isolation will likely not be an appropriate design strategy for achieving the Collapse Prevention Structural Performance Level. In general, isolation systems provide significant protection to the building structure, nonstructural components, and contents, but at a cost that precludes practical application where the budget and Rehabilitation Objectives are modest.

Energy dissipation systems should be considered in a somewhat broader context than isolation systems. For the taller buildings (where isolation systems may not be feasible), energy dissipation systems should be considered as a design strategy where performance goals include the Damage Control Performance Range. Conversely, certain energy dissipation devices are quite economical and might be practical for performance goals that address only Limited Safety. In general, however, energy dissipation systems are more likely to be an appropriate design strategy where the desired Structural Performance Level is Life Safety, or

**Table C9-1. Applicability of Isolation and Energy Dissipation Systems**

| Performance Level | Performance Range | Isolation | Energy Dissipation |
|---|---|---|---|
| Operational | Damage Control | Very Likely | Limited |
| Immediate Occupancy | | Likely | Likely |
| Life Safety | Limited Safety | Limited | Likely |
| Collapse Prevention | | Not Practical | Limited |

perhaps Immediate Occupancy. Other objectives may also influence the decision to use energy dissipation devices, since these devices can also be useful for control of building response due to small earthquakes, wind, or mechanical loads.

## 9.2 SEISMIC ISOLATION SYSTEMS

### 9.2.1 General Requirements

Seismic isolation systems using seismic isolators, classified as either elastomeric or sliding, as defined in Section 9.2.2, shall comply with the requirements of Section 9.2. Properties of seismic isolation systems shall be based on Section 9.2.2. Seismic isolation systems shall be designed and analyzed in accordance with Section 9.2.3. Linear and nonlinear analyses shall be performed, as required by Section 9.2.3, in accordance with Sections 9.2.4 and 9.2.5, respectively. Nonstructural components shall be rehabilitated in accordance with Section 9.2.6. Additional requirements for seismic isolation systems as defined in Section 9.2.7 shall be met. Seismic isolation systems shall be reviewed and tested in accordance with Sections 9.2.8 and 9.2.9, respectively.

The seismic isolation system shall include wind-restraint and tie-down systems, if such systems are required by this standard. The isolation system also shall include supplemental energy dissipation devices, if such devices are used to transmit force between the structure above the isolation system and the structure below the isolation system.

For seismically isolated structures, the coefficients $C_0$, $C_1$, $C_2$, and $J$ defined in Chapter 3, shall be taken as 1.0

### C9.2.1 General Requirements

Analysis methods and design criteria for seismic isolation systems are based on criteria for the Rehabilitation Objectives of Chapter 1.

The methods described in this section augment the analysis requirements of Chapter 3. The analysis methods and other criteria of this section are based largely on FEMA 302, *NEHRP Recommended Provisions for Seismic Regulations for New Buildings and Other Structures* (FEMA 1997).

Seismic isolation has typically been used as a rehabilitation strategy that enhances the performance of the building above that afforded by conventional stiffening and strengthening schemes. Seismic isolation rehabilitation projects have targeted performance at least equal to, and commonly exceeding, the Basic Safety Objective of this standard, effectively achieving

a target Building Performance Level of Immediate Occupancy or better.

A number of buildings rehabilitated with seismic isolators have been historic. For these projects, seismic isolation reduced the extent and intrusion of seismic modifications on the historical fabric of the building that would otherwise be required to meet desired performance levels.

### 9.2.2 Mechanical Properties and Modeling of Seismic Isolation Systems

#### 9.2.2.1 General

Seismic isolators shall be classified as either elastomeric or sliding. Elastomeric isolators shall include any one of the following: high-damping rubber bearings (HDR), low-damping rubber bearings (RB), or low-damping rubber bearings with a lead core (LRB). Sliding isolators shall include flat assemblies or have a curved surface, such as the friction-pendulum system (FPS). Rolling systems shall be characterized as a subset of sliding systems. Rolling isolators shall be flat assemblies or have a curved or conical surface, such as the ball and cone system (BNC). Isolators that cannot be classified as either elastomeric or sliding are not addressed in this standard.

#### C9.2.2.1 General

A seismic isolation system is the collection of all individual seismic isolators (and separate wind restraint and tie-down devices, if such devices are used to meet the requirements of this standard). Seismic isolation systems may be composed entirely of one type of seismic isolator, a combination of different types of seismic isolators, or a combination of seismic isolators acting in parallel with energy dissipation devices (i.e., a hybrid system).

Elastomeric isolators are typically made of layers of rubber separated by steel shims.

#### 9.2.2.2 Mechanical Properties of Seismic Isolators

*9.2.2.2.1 Elastomeric Isolators* Force–deformation response properties shall be established for elastomeric isolators taking into consideration axial–shear interaction, bilateral deformation, load history including the effects of scragging of virgin elastomeric isolators, temperature, and other environmental loads and aging effects over the design life of the isolator.

For mathematical modeling of isolators, mechanical characteristics based on analysis or available material test properties shall be permitted. For design, mechanical characteristics shall be based on tests of isolator prototypes in accordance with Section 9.2.9.

*C9.2.2.2.1 Elastomeric Isolators* Elastomeric bearings represent a common means for introducing flexibility into an isolated structure. They consist of thin layers of natural rubber that are vulcanized and bonded to steel plates. Natural rubber exhibits a complex mechanical behavior, which can be described simply as a combination of viscoelastic and hysteretic behavior. Low-damping natural rubber bearings exhibit essentially linearly elastic and linearly viscous behavior at large shear strains. The effective damping is typically less than or equal to 0.07 for shear strains in the range of 0 to 2.0.

Lead-rubber bearings are generally constructed of low-damping natural rubber with a preformed central hole into which a lead core is press-fitted. Under lateral deformation, the lead core deforms in almost pure shear, yields at low levels of stress (approximately 8 to 10 MPa in shear at normal temperature), and produces hysteretic behavior that is stable over many cycles. Unlike mild steel, lead recrystallizes at normal temperature (about 20 °C), so that repeated yielding does not cause fatigue failure. Lead-rubber bearings generally exhibit characteristic strength that ensures rigidity under service loads. Figure C9-1 shows an idealized force–displacement relation of a lead-rubber bearing. The characteristic strength, $Q$, is related to the lead plug area, $A_p$, and the shear yield stress of lead, $\sigma_{YL}$:

$$Q = A_p \sigma_{YL} \qquad \text{(Eq. C9-1)}$$

The post-yield stiffness, $k_p$, is typically higher than the shear stiffness of the bearing without the lead core:

$$k_p = \frac{A_r G f_L}{\Sigma t} \qquad \text{(Eq. C9-2)}$$

where

$A_r$ = bonded rubber area;
$\Sigma$ = total rubber thickness;

$G$ = shear modulus of rubber (typically computed at shear strain of 0.5); and
$f_L$ = a factor larger than unity.

Typically, $f_L$ is 1.15, and the elastic stiffness ranges between 6.5 to 10 times the post-yield stiffness.

The behavior of lead-rubber bearings may be represented by a bilinear hysteretic model. Computer programs *3D-BASIS* (Nagarajaiah et al. 1991; Reinhorn et al. 1994; Tsopelas et al. 1994b), and *ETABS*, Version 8 (CSI, 2003) have the capability of modeling hysteretic behavior for isolators. These models typically require definition of three parameters, namely, the post-yield stiffness $k_p$, the yield force $F_y$, and the yield displacement $D_y$. For lead-rubber bearings in which the elastic stiffness is approximately equal to 6.5 $k_p$, the yield displacement can be estimated as:

$$D_y = \frac{Q}{5.5 k_p} \qquad \text{(Eq. C9-3)}$$

The yield force is then given by:

$$F_y = Q + k_p D_y \qquad \text{(Eq. C9-4)}$$

High-damping rubber bearings are made of specially compounded rubber that exhibits effective damping between 0.10 and 0.20 of critical. The increase in effective damping of high-damping rubber is achieved by the addition of chemical compounds that may also affect other mechanical properties of rubber. Figure C9-2 shows representative force–displacement loops of a high-damping rubber bearing under scragged conditions.

Scragging is the process of subjecting an elastomeric bearing to one or more cycles of large-amplitude displacement. The scragging process modi-

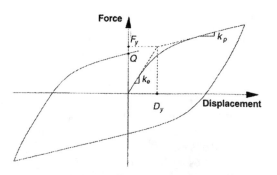

**FIGURE C9-1. Idealized Hysteretic Force–Displacement Relation of a Lead-Rubber Bearing.**

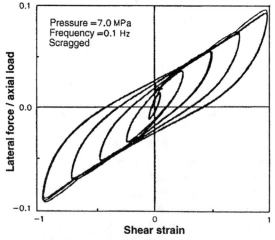

**FIGURE C9-2. Force–Displacement Loops of a High-Damping Rubber Bearing.**

fies the molecular structure of the elastomer and results in more stable hysteresis at strain levels lower than that to which the elastomer was scragged. Although it is usually assumed that the scragged properties of an elastomer remain unchanged with time, recent studies by Cho and Retamal (1993) and Murota et al. (1994) suggest that partial recovery of unscragged properties is likely. The extent of this recovery is dependent on the elastomer compound.

Mathematical models capable of describing the transition between virgin and scragged properties of high-damping rubber bearings are not yet available. It is appropriate in this case to perform multiple analyses with stable hysteretic models and obtain bounds on the dynamic response. A smooth, bilinear hysteretic model that is capable of modeling the behavior depicted in Fig. C9-1 is appropriate for such analyses, as long as the peak shear strain is below the stiffening limit of approximately 1.5 to 2.0, depending on the rubber compound. Beyond this strain limit many elastomers exhibit stiffening behavior, with tangent stiffness approximately equal to twice the tangent stiffness prior to initiation of stiffening. For additional information, refer to Tsopelas and Constantinou (1994a).

To illustrate the calculations of parameters from prototype bearings test data, Fig. C9-3 shows experimentally determined properties of the high-damping rubber bearings, for which loops are shown in Fig. C9-2. The properties identified are the tangent shear modulus, $G$, and the effective damping ratio, $\beta_{eff}$ (described by Eq. C9-18), which is now defined for a single bearing rather than the entire isolation system), under scragged conditions. With reference to Fig. C9-1, $G$ is related to the post-yielding stiffness $k_p$.

$$k_p = \frac{GA}{\Sigma t}$$  (Eq. C9-5)

where $A$ is the bonded rubber area. The results of Fig. C9-3 demonstrate that the tangent shear modulus and equivalent damping ratio are only marginally affected by the frequency of loading and the bearing pressure, within the indicated range for the tested elastomer. Different conclusions may be drawn from the testing of other high-damping rubber compounds.

The parameters of the bilinear hysteretic model may be determined by use of the mechanical properties $G$ and $\beta_{eff}$ at a specific shear strain, such as the strain corresponding to the design displacement $D$. The post-yield stiffness $k_p$ is determined from Eq. C9-5, whereas the characteristic strength, $Q$, can be determined as:

$$Q = \frac{\pi \beta_{eff} k_p D^2}{(2 - \pi \beta_{eff})D - 2D_y}$$  (Eq. C9-6)

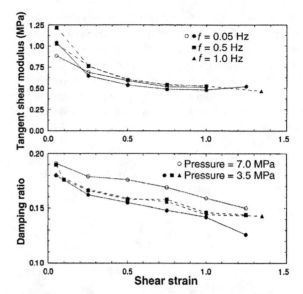

FIGURE C9-3. Tangent Shear Modulus and Effective Damping Ratio of High-Damping Rubber Bearing.

where $D_y$ is the yield displacement. The yield displacement is generally not known a priori. However, experimental data suggest that $D_y$ is approximately equal to 0.05 to 0.1 times the total rubber thickness, $\Sigma t$. With the yield displacement approximately determined, the model can be completely defined by determining the yield force (Eq. C9-4). It should be noted that the characteristic strength may be alternatively determined from the effective stiffness, $k_{eff}$ (Eq. C9-17), of the bearing, as follows:

$$Q = \frac{\pi \beta_{eff} k_{eff} D^2}{2(D - D_y)}$$  (Eq. C9-7)

The effective stiffness is a more readily determined property than the post-yielding stiffness. The effective stiffness is commonly used to obtain the effective shear modulus, $G_{eff}$, defined as:

$$G_{eff} = \frac{k_{eff} \Sigma t}{A}$$  (Eq. C9-8)

The behavior of the bearing for which the force–displacement loops are shown in Fig. C9-2 is now analytically constructed using the mechanical properties at a shear strain of 1.0 and a bearing pressure of 7.0 MPa. These properties are $G_{eff} = 0.50$ MPa and $\beta_{eff} = 0.16$. With the bonded area and total thickness of rubber known, and assuming $D_y = 0.1\Sigma t$, a bilinear hysteretic model was defined and implemented in the program 3D-BASIS. The simulated loops are shown in Fig. C9-4, where it may be observed that the calculated hysteresis loop at shear strain of 1.0 agrees well with the corresponding experimental hysteresis loop. However, at lower peak shear strain the analytical loops have a constant characteristic strength, whereas the experimental loops have a characteristic strength dependent on the shear strain amplitude. Nevertheless, the analytical model will likely produce acceptable results where the design parameters are based on the mechanical properties at a strain corresponding to the design displacement.

Elastomeric bearings have finite vertical stiffness that affects the vertical response of the isolated structure. The vertical stiffness of an elastomeric bearing may be obtained from

$$k_v = \frac{E_c A}{\Sigma t} \qquad \text{(Eq. C9-9)}$$

where $E_c$ is the compression modulus. Although a number of approximate empirical relations have been proposed for the calculation of the compression modulus, the correct expression for circular bearings is:

$$E_c = \left( \frac{1}{6G_{eff}S^2} + \frac{4}{3K} \right)^{-1} \qquad \text{(Eq. C9-10)}$$

(Kelly 1993) where $K$ is the bulk modulus (typically assumed to have a value of 2,000 MPa) and $S$ is the shape factor, which is defined as the ratio of the loaded area to the bonded perimeter of a single rubber layer. For a circular bearing of bonded diameter $\phi$ and rubber layer thickness $t$, the shape factor is given by:

$$S = \frac{\phi}{4t} \qquad \text{(Eq. C9-11)}$$

Seismic elastomeric bearings are generally designed with a large shape factor, typically 12 to 20.

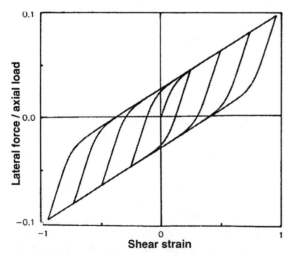

**FIGURE C9-4. Analytical Force–Displacement Loops of High-Damping Rubber Bearing.**

Considering an elastomeric bearing design with $S = 15$, $G_{eff} = 1$ MPa, and $K = 2,000$ MPa, the ratio of vertical stiffness (Eq. C9-9) to effective horizontal stiffness (Eq. C9-8) is approximately equal to 700. Thus, the vertical period of vibration of a structure on elastomeric isolation bearings will be about 26 times (i.e., $\sqrt{700}$) less than the horizontal period, on the order of 0.1 sec. This value of vertical period provides potential for amplification of the vertical ground acceleration by the isolation system. The primary effect of this amplification is to change the vertical load on the bearings, which may need to be considered for certain design applications.

Another consideration in the design of seismically isolated structures with elastomeric bearings is reduction in height of a bearing with increasing lateral deformation (Kelly 1993). While this reduction of height is typically small, it may be important where elastomeric bearings are combined with other isolation components that are vertically rigid (such as sliding bearings). In addition, incompatibilities in vertical displacements may lead to a redistribution of loads.

*9.2.2.2.2 Sliding Isolators* Force–deformation response properties shall be established for sliding isolators, taking into consideration contact pressure, rate of loading or velocity, bilateral deformation, temperature, contamination, and other environmental loads and aging effects over the design life of the isolator.

Mechanical characteristics for use in mathematical models shall be based on analysis and available material test properties. Verification of isolator properties used for design shall be based on tests of isolator prototypes in accordance with Section 9.2.9.

*C9.2.2.2.2 Sliding Isolators* Sliding bearings will tend to limit the transmission of force to an isolated structure to a predetermined level. While this is desirable, the lack of significant restoring force can result in significant variations in the peak displacement response, and can result in permanent offset displacements. To avoid these undesirable features, sliding bearings are typically used in combination with a restoring force mechanism.

The lateral force developed in a sliding bearing can be defined as:

$$F = \frac{N}{R} U + \mu_s N \operatorname{sgn}(\dot{U}) \qquad \text{(Eq. C9-12)}$$

where

$U =$ displacement;
$\dot{U} =$ sliding velocity;
$R =$ radius of curvature of sliding surface;

$\mu_s$ = coefficient of sliding friction;

$N$ = normal load on bearing; and

$\text{sgn}(\dot{U})$ = sign of sliding velocity vector; $+1$ or $-1$.

The normal load consists of the gravity load, $W$, the effect of vertical ground acceleration, $\ddot{U}_v$, and the additional seismic load due to overturning moment, $P_s$:

$$N = W\left(1 + \frac{\ddot{U}_v}{g} + \frac{P_s}{W}\right) \qquad \text{(Eq. C9-13)}$$

The first term in Eq. C9-13 denotes the restoring force component, and the second term describes the friction force. For flat sliding bearings, the radius of curvature is infinite, so the restoring force term in Eq. C9-13 vanishes. For a spherical sliding surface (Zayas et al. 1987), the radius of curvature is constant, so the bearing exhibits a linear restoring force; that is, under constant gravity load the stiffness is equal to $W/R_o$, where $R_o$ is the radius of the spherical sliding surface. Where the sliding surface takes a conical shape, the restoring force is constant. Figure C9-5 shows idealized force–displacement loops of sliding bearings with flat, spherical, and conical surfaces

Sliding bearings with either a flat or single curvature spherical sliding surface are typically made of polytetrafluoroethylene (PTFE) or PTFE-based composites in contact with polished stainless steel. The shape of the sliding surface allows large contact areas that, depending on the materials used, are loaded to average bearing pressures in the range of 7 to 70 MPa. For interfaces with shapes other than flat or spherical, the load needs to be transferred through a bearing as illustrated in Fig. C9-5 for the conical sliding surface.

Such an arrangement typically results in a very low coefficient of friction.

For bearings with large contact area, and in the absence of liquid lubricants, the coefficient of friction depends on a number of parameters, of which the three most important are the composition of the sliding interface, bearing pressure, and velocity of sliding. For interfaces composed of polished stainless steel in contact with PTFE or PTFE-based composites, the coefficient of sliding friction may be described by:

$$\mu_s = f_{max} - (f_{max} - f_{min})\exp(-a|\dot{U}|) \qquad \text{(Eq. C9-14)}$$

where parameters $f_{min}$ and $f_{max}$ describe the coefficient of friction at small and large velocities of sliding and under constant pressure, respectively, all as depicted in Fig. C9-6. Parameters $f_{max}$, $f_{min}$, and $a$ depend on the bearing pressure, although only the dependency of $f_{max}$ on pressure is of practical significance.

A good approximation to the experimental data (Constantinou et al. 1993) is

$$f_{max} = f_{max\,o} - (f_{max\,o} - f_{max\,p})\tanh \varepsilon p \qquad \text{(Eq. C9-15)}$$

where the physical significance of parameters $f_{maxo}$ and $f_{maxp}$ is as illustrated in Fig. C9-6. The term $p$ is the instantaneous bearing pressure, which is equal to the normal load $N$ computed by Eq. C9-13, divided by the contact area; and $\varepsilon$ is a parameter that controls the variation of $f_{max}$ with pressure.

Figure C9-6 illustrates another feature of sliding bearings. On initiation of motion, the coefficient of friction exhibits a static or breakaway value, $\mu_B$, which is typically higher than the minimum value $f_{min}$. To demonstrate frictional properties, Fig. C9-6 shows the relation between bearing pressure and the friction

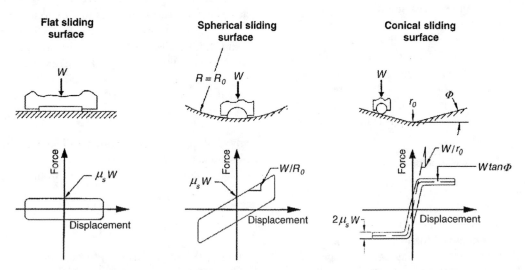

**FIGURE C9-5. Idealized Force–Displacement Loops of Sliding Bearings.**

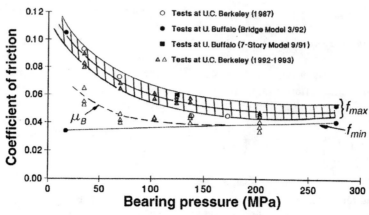

**FIGURE C9-6. Coefficient of Friction of PTFE-Based Composite in Contact with Polished Stainless Steel at Normal Temperature.**

coefficients $f_{max}$, $\mu_B$, and $f_{min}$ of a PTFE-based composite material in contact with polished stainless steel at normal temperature. These data were compiled from testing of bearings in four different testing programs (Soong and Constantinou 1994).

Combined elastomeric-sliding isolation systems have been used in buildings in the United States. Japanese engineers have also used elastomeric bearings in combination with mild steel components that are designed to yield in strong earthquakes and enhance the energy dissipation capability of the isolation system (Kelly 1988). These mild steel components exhibit either elasto-plastic behavior or bilinear hysteretic behavior with low post-yielding stiffness. Moreover, fluid viscous energy dissipation devices have been used in combination with elastomeric bearings. The behavior of fluid viscous devices is described in Section 9.3.3.2.3.

Hybrid seismic isolation systems composed of elastomeric and sliding bearings should be modeled taking into account the likely significant differences in the relationships between vertical displacement as a function of horizontal displacement. The use of elastomeric and sliding isolators in close proximity to one another under vertically stiff structural framing elements (e.g., reinforced concrete shear walls) may be problematic and could result in significant redistributions of gravity loads.

### 9.2.2.3 Modeling of Isolators

*9.2.2.3.1 General* If the mechanical characteristics of a seismic isolator are dependent on axial load (due to gravity, earthquake overturning effects, and vertical earthquake shaking), rate of loading (velocity), bilateral deformation, temperature, or aging, then upper- and lower-bound values of stiffness and damping shall be

used in multiple analyses of the model to determine the range and sensitivity of response to design parameters.

*9.2.2.3.2 Linear Models* The restoring force, $F$, of an isolator shall be calculated as the product of effective stiffness, $k_{eff}$, and response displacement, $D$:

$$F = k_{eff}D \qquad \text{(Eq. 9-1)}$$

The effective stiffness, $k_{eff}$, of an isolator shall be calculated from test data using Eq. 9-12. The area enclosed by the force–displacement hysteresis loop shall be used to calculate the effective damping, $\beta_{eff}$, of an isolator using Eq. 9-13. Effective stiffness and effective damping shall be evaluated at all response displacements of design interest.

*C9.2.2.3.2 Linear Models* Linear procedures use effective stiffness, $k_{eff}$, and effective damping, $\beta_{eff}$, to characterize nonlinear properties of isolators.

For linear procedures [see FEMA 274, Section C9.2.3 (FEMA 1997)], the seismic isolation system can be represented by an equivalent linearly elastic model. The force in a seismic isolation device is calculated as:

$$F = k_{eff}D \qquad \text{(Eq. C9-16)}$$

where all terms are as defined in Section 9.2.2.3.2 of this standard. The effective stiffness of the seismic isolation device may be calculated from test data as follows:

$$k_{eff} = \frac{|F^+| + |F^-|}{|\Delta^+| + |\Delta^-|} \qquad \text{(Eq. C9-17)}$$

Figure C9-7 illustrates the physical significance of the effective stiffness.

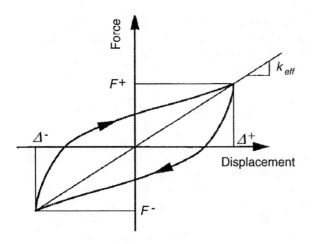

**Hysteretic behavior**

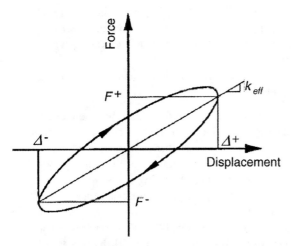

**Viscoelastic behavior**

**FIGURE C9-7. Definition of Effective Stiffness of Seismic Isolation Devices.**

Analysis by a linear method requires that either each seismic isolator or groups of seismic isolators be represented by linear springs of either stiffness, $k_{eff}$, or the combined effective stiffness of each group. The energy dissipation capability of an isolation system is generally represented by effective damping. Effective damping is amplitude-dependent and calculated at design displacement, $D$, as follows:

$$\beta_{eff} = \frac{1}{2\pi}\left[\frac{\Sigma E_D}{K_{eff}D^2}\right] \qquad \text{(Eq. C9-18)}$$

where $\Sigma E_D$ is the sum of the areas of the hysteresis loops of all isolators, and $K_{eff}$ is the sum of the effective stiffnesses of all seismic isolation devices. Both the area of the hysteresis loops and the effective stiffness are determined at the design displacement, $D$.

The application of Eqs. C9-16 through Eq. C9-18 to the design of isolation systems is complicated if the effective stiffness and loop area depend on axial load. Multiple analyses are then required to establish bounds on the properties and response of the isolators. For example, sliding isolation systems exhibit such dependencies as described in Section C9.2.2.2.2. To account for these effects, the following procedure is proposed.

1. In sliding isolation systems, the relation between horizontal force and vertical load is substantially linear (see Eq. C9-16). Accordingly, the net effect of overturning moment on the mechanical behavior of a group of bearings is small and can be neglected.

   Al-Hussaini et al. (1994) provided experimental results that demonstrate this behavior up to the point of imminent bearing uplift. Similar results are likely for elastomeric bearings.

2. The effect of vertical ground acceleration is to modify the load on the isolators. If it is assumed that the building is rigid in the vertical direction, and axial forces due to overturning moments are absent, the axial loads can vary between $W(1 - \dot{U}/g)$ and $W(1 + \dot{U}/g)$, where $\dot{U}$ is the peak vertical ground acceleration. However, recognizing that horizontal and vertical ground motion components are likely not correlated unless in the near field, it is appropriate to use a combination rule that uses only a fraction of the peak vertical ground acceleration. Based on the use of 50% of the peak vertical ground acceleration, maximum and minimum axial loads on a given isolator may be defined as:

$$N_C = W(1 \pm 0.20S_{DS}) \qquad \text{(Eq. C9-19)}$$

where the plus sign gives the maximum value and the minus sign gives the minimum value. Equation C9-19 is based on the assumption that the short-period spectral response parameter, $S_{DS}$, is 2.5 times the peak value of the vertical ground acceleration. For analysis for the Maximum Considered Earthquake, the axial load should be determined from:

$$N_C = W(1 \pm 0.20S_{MS}) \qquad \text{(Eq. C9-20)}$$

Equations C9-19 and C9-20 should be used with caution if the building is located in the near field of a major active fault. In this instance, expert advice should be sought regarding correlation of horizontal and vertical ground motion components.

Load $N_C$ represents a constant load on isolators, which can be used for determining the effective stiffness and area of the hysteresis loop. To obtain these properties, the characteristic strength $Q$ (see Fig. C9-7) is needed. For sliding isolators, $Q$ can be taken as equal to $f_{max}N_C$, where $f_{max}$ is determined at the bearing pressure corresponding to load $N_C$. For example, for a sliding bearing with spherical sliding surface of radius $R_O$ (see Fig. C9-5), the effective stiffness and area of the loop at the design displacement $D$ are:

$$k_{eff} = \left( \frac{1}{R_O} + \frac{f_{max}}{D} \right) N_C \qquad \text{(Eq. C9-21)}$$

$$\text{Loop Area} = 4f_{max}N_C D \qquad \text{(Eq. C9-22)}$$

*9.2.2.3.3 Nonlinear Models* The nonlinear force-deflection properties of isolators shall be explicitly modeled if nonlinear procedures are used.

The inelastic (hysteretic) response of the isolators shall represent damping. Additional viscous damping shall not be included in the model unless supported by rate-dependent tests of isolators.

*C9.2.2.3.3 Nonlinear Models* For dynamic nonlinear time-history analysis, the seismic isolation components should be explicitly modeled. FEMA 274 Sections C9.2.2.2 through C9.2.2.4 (FEMA 1997) present relevant information. Where uncertainties exist, and where aspects of behavior cannot be modeled, multiple analyses should be performed in order to establish bounds on the dynamic response.

For simplified nonlinear analysis, each seismic isolation component can be modeled by an appropriate ra e-independent hysteretic model. Elastomeric bearings may be modeled as bilinear hysteretic components as described in FEMA 274, Section C9.2.2.2. Sliding bearings may also be modeled as bilinear hysteretic components with characteristic strength (see Fig. C9-5) given by

$$Q = f_{max}N_C \qquad \text{(Eq. C9-23)}$$

where $N_C$ is determined by either Eq. C9-19 or C9-20, and $f_{max}$ is the coefficient of sliding friction at the appropriate sliding velocity. The post-yield stiffness can then be determined as:

$$k_p = \frac{N_C}{R} \qquad \text{(Eq. C9-24)}$$

where $R$ is as defined in FEMA 274 Section C9.2.2.2.B. The yield displacement $D_y$ in a bilinear hysteretic model of a sliding bearing should be very small, perhaps on the order of 2 mm. Alternatively, a bilinear hysteretic model for sliding bearings may be defined to have an elastic stiffness that is at least 100 times larger than the post-yield stiffness, $k_p$.

Isolation devices that exhibit viscoelastic behavior as shown in Fig. C9-7 should be modeled as linearly elastic components with effective stiffness $k_{eff}$ as determined by Eq. C9-21.

### 9.2.2.4 Isolation System and Superstructure Modeling

*9.2.2.4.1 General* Mathematical models of the isolated building, including the isolation system, the lateral-force-resisting system of the superstructure, other structural components and elements, and connections between the isolation system and the structure, shall meet the requirements of Chapters 2 and 3 and Sections 9.2.2.4.2 and 9.2.2.4.3.

*9.2.2.4.2 Isolation System Model* The isolation system shall be modeled using deformation characteristics developed and verified by test in accordance with the requirements of Section 9.2.9.

The isolation system shall be modeled with sufficient detail to:

1. Account for the spatial distribution of isolator units;
2. Calculate translation, in both horizontal directions, and torsion of the structure above the isolation interface, considering the most disadvantageous location of mass eccentricity;
3. Assess overturning/uplift forces on individual isolators;
4. Account for the effects of vertical load, bilateral load, and/or the rate of loading, if the force deflection properties of the isolation system are dependent on one or more of these factors.
5. Assess forces due to P-D moments; and
6. Account for nonlinear components. Isolation systems with nonlinear components include systems that do not meet the criteria of Section 9.2.3.3.1, Item 2.

*9.2.2.4.3 Superstructure Model* The maximum displacement of each floor, the total design displacement, and the total maximum displacement across the isolation system shall be calculated using a model of the isolated building that incorporates the force-deformation characteristics of nonlinear components.

Calculation of design forces and displacements in primary components of the lateral-force-resisting system using linearly elastic models of the isolated structure shall be permitted if both of the following criteria are met:

1. Pseudo-elastic properties assumed for nonlinear isolation system components are based on the

maximum effective stiffness of the isolation system; and

2. The lateral-force-resisting system remains linearly elastic for the earthquake demand level of interest.

A lateral-force-resisting system that meets both of the following criteria may be classified as linearly elastic:

1. For all deformation-controlled actions, Eq. 3-20 is satisfied using an *m*-factor equal to 1.0; and
2. For all force-controlled actions, Eq. 3-21 is satisfied.

### 9.2.3 General Criteria for Seismic Isolation Design

#### 9.2.3.1 General
The design, analysis, and testing of the isolation system shall be based on the requirements of this section.

#### C9.2.3.1 General
Criteria for the seismic isolation of buildings are divided into two sections:
Rehabilitation of the building; and
Design, analysis, and testing of the isolation system.

*9.2.3.1.1 Stability of the Isolation System* The stability of the vertical load-carrying components of the isolation system shall be verified by analysis and test, as required by Section 9.2.9, for a lateral displacement equal to the total maximum displacement computed in accordance with Section 9.2.4.3.5 or Section 9.2.5.1.2, or for the maximum displacement allowed by displacement-restraint devices, if such devices are part of the isolation system.

*9.2.3.1.2 Configuration Requirements* The isolated building shall be classified as regular or irregular, as defined in Section 2.4.1.1, based on the structural configuration of the structure above the isolation system.

#### 9.2.3.2 Ground Shaking Criteria
Ground shaking criteria for the Design Earthquake and the Maximum Considered Earthquake shall be established in accordance with Section 1.6 as modified by this section. The design Earthquake Hazard Level shall be user-specified and shall be permitted to be chosen equal to the BSE-1 Earthquake Hazard Level. The Maximum Considered Earthquake shall be taken equal to the BSE-2 Earthquake Hazard Level.

*9.2.3.2.1 User-Specified Design Earthquake* For the Design Earthquake, the following ground shaking criteria shall be established:

1. Short-period spectral response acceleration parameter, $S_{XS}$ and spectral response acceleration parameter at 1.0 second, $S_{X1}$, in accordance with Section 1.6.1.4;
2. Five-percent-damped response spectrum of the design earthquake (where a response spectrum is required for linear procedures by Section 9.2.3.3.2, or to define acceleration time histories); and
3. At least three acceleration time histories compatible with the design earthquake spectrum (where acceleration time histories are required for nonlinear procedures by Section 9.2.3.3.3).

*9.2.3.2.2 Maximum Considered Earthquake* For the BSE-2, the following ground shaking criteria shall be established:

1. Short period spectral response acceleration parameter, $S_{XS}$, and spectral response acceleration parameter at 1.0 sec, $S_{X1}$, in accordance with Section 1.6.1.4.
2. Five-percent-damped site-specific response spectrum of the BSE-2 (where a response spectrum is required for linear procedures by Section 9.2.3.3.2, or to define acceleration time histories); and
3. At least three acceleration time histories compatible with the BSE-2 spectrum (where acceleration time histories are required for nonlinear procedures by Section 9.2.3.3.3).

#### 9.2.3.3 Selection of Analysis Procedure

*9.2.3.3.1 Linear Procedures* Linear procedures shall be permitted for design of seismically isolated buildings, provided the following criteria are met:

1. The building is located on Soil Profile Type A, B, C, or D; or E if $S_1 \geq 0.6$ for BSE-2;
2. The isolation system meets all of the following criteria:
   2.1. The effective stiffness of the isolation system at the design displacement is greater than one-third of the effective stiffness at 20% of the design displacement;
   2.2. The isolation system is capable of producing a restoring force as specified in Section 9.2.7.2.4;
   2.3. The isolation system has force-deflection properties that are independent of the rate of loading;

2.4. The isolation system has force-deflection properties that are independent of vertical load and bilateral load;

2.5. Where considering analysis procedures, for the BSE-2, the isolation system does not limit BSE-2 displacement to less than the ratio of the design spectral response acceleration at one second ($S_{X1}$) for the BSE-2 to that for the Design Earthquake times the total design displacement; and

3. The structure above the isolation system exhibits global elastic behavior for the earthquake motions under consideration.

*9.2.3.3.2 Response Spectrum Analysis* Response spectrum analysis shall be used for design of seismically isolated buildings that meet any of the following criteria:

1. The building is over 65 ft (19.8 m) in height;
2. The effective period of the structure, $T_M$, is greater than three seconds;
3. The effective period of the isolated structure, $T_D$, is less than or equal to three times the elastic, fixed-base period of the structure above the isolation system; or
4. The structure above the isolation system is irregular in configuration.

*9.2.3.3.3 Nonlinear Procedures* Nonlinear procedures shall be used for design of seismic-isolated buildings for which any of the following conditions apply:

1. The structure above the isolation system is nonlinear for the earthquake motions under consideration; and
2. The isolation system does not meet all of the criteria of Section 9.2.3.3.1.

Nonlinear acceleration time-history analysis shall be performed for the design of seismically isolated buildings for which conditions (1) and (2) apply.

### C9.2.3.3 Selection of Analysis Procedure

Linear procedures include prescriptive formulas and Response Spectrum Analysis. Linear procedures based on formulas (similar to the seismic-coefficient equation required for design of fixed-base buildings) prescribe peak lateral displacement of the isolation system, and define "minimum" design criteria that may be used for design of a very limited class of isolated structures (without confirmatory dynamic analyses). These simple formulas are useful for preliminary design and provide a means of expeditious review of more complex calculations.

Response Spectrum Analysis is recommended for design of isolated structures that have either (1) a tall or otherwise flexible superstructure, or (2) an irregular superstructure. For most buildings, Response Spectrum Analysis will not predict significantly different displacements of the isolation system than those calculated by prescriptive formulas, provided both calculations are based on the same effective stiffness and damping properties of the isolation system. The real benefit of Response Spectrum Analysis is not in the prediction of isolation system response but, rather, in the calculation and distribution of forces in the superstructure. Response Spectrum Analysis permits the use of more detailed models of the superstructure that better estimate forces and deformations of components and elements considering flexibility and irregularity of the structural system.

Nonlinear procedures include the Nonlinear Static Procedure (NSP) and the Nonlinear Dynamic Procedure (NDP). The NSP is a static pushover procedure and the NDP is based on nonlinear time-history analysis. The NSP or the NDP is required for isolated structures that do not have essentially linearly elastic superstructures (during BSE-2 demand). In this case, the superstructure would be modeled with nonlinear components.

Time-history analysis is required for isolated structures on very soft soil (i.e., Soil Profile Type E where shaking is strong, or Soil Profile Type F) that could shake the building with a large number of cycles of long-period motion, and for buildings with isolation systems that are best characterized by nonlinear models. Such isolation systems include:

1. Systems with more than about 30% effective damping (because high levels of damping can significantly affect higher-mode response of the superstructure);
2. Systems that lack significant restoring force (because these systems may not stay centered during earthquake shaking);
3. Systems that are expected to exceed the sway-space clearance with adjacent structures (because impact with adjacent structures could impose large demands on the superstructure); and
4. Systems that are rate- or load-dependent (because their properties will vary during earthquake shaking).

For the types of isolation systems described above, appropriate nonlinear properties must be used to model isolators. Linear properties could be used to model the superstructure, provided the superstructure's response is essentially linearly elastic for BSE-2 demand.

The restrictions placed on the use of linear procedures effectively suggest that nonlinear procedures be used for virtually all isolated buildings. However, lower-bound limits on isolation system design displacement and force are specified by this standard as a percentage of the demand prescribed by the linear formulas, even where dynamic analysis is used as the basis for design. These lower-bound limits on key design attributes ensure consistency in the design of isolated structures and serve as a "safety net" against gross under design.

### 9.2.4 Linear Procedures

#### 9.2.4.1 General

Seismically isolated buildings for which linear analysis procedures are selected based on the criteria of Section 9.2.3.3 shall be designed and constructed to resist the earthquake displacements and forces specified in this section, at a minimum.

#### 9.2.4.2 Deformation Characteristics of the Isolation System

The deformation characteristics of the isolation system shall be based on tests performed in accordance with Section 9.2.9.

The deformation characteristics of the isolation system shall explicitly include the effects of the wind-restraint and tie-down systems, and supplemental energy dissipation devices, if such systems and devices are used to meet the design requirements of this standard.

#### 9.2.4.3 Minimum Lateral Displacements

*9.2.4.3.1 Design Displacement* The isolation system shall be designed and constructed to withstand, as a minimum, lateral earthquake displacements that act in the direction of each of the main horizontal axes of the structure in accordance with Eq. 9-2:

$$D_D = \left[\frac{g}{4\pi^2}\right]\frac{S_{X1}T_D}{B_{D1}} \qquad \text{(Eq. 9-2)}$$

where $S_{X1}$ is evaluated for the Design Earthquake.

*9.2.4.3.2 Effective Period at the Design Displacement* The effective period, $T_D$, of the isolated building at the design displacement shall be determined using the deformation characteristics of the isolation system in accordance with Eq. 9-3:

$$T_D = 2\pi\sqrt{\frac{W}{K_{D\min}g}} \qquad \text{(Eq. 9-3)}$$

*9.2.4.3.3 Maximum Displacement* The maximum displacement of the isolation system, $D_M$, in the most critical direction of horizontal response shall be calculated in accordance with Eq. 9-4:

$$D_M = \left[\frac{g}{4\pi^2}\right]\frac{S_{X1}T_M}{B_{M1}} \qquad \text{(Eq. 9-4)}$$

where $S_{X1}$ is evaluated for the BSE-2.

*9.2.4.3.4 Effective Period at the Maximum Displacement* The effective period, $T_M$, of the isolated building at the maximum displacement shall be determined using the deformation characteristics of the isolation system in accordance with Eq. 9-5:

$$T_M = 2\pi\sqrt{\frac{W}{K_{M\min}g}} \qquad \text{(Eq. 9-5)}$$

*9.2.4.3.5 Total Displacement* The total design displacement, $D_{TD}$, and the total maximum displacement, $D_{TM}$, of components of the isolation system shall include additional displacement due to actual and accidental torsion calculated considering the spatial distribution of the effective stiffness of the isolation system at the design displacement and the most disadvantageous location of mass eccentricity.

The total design displacement, $D_{TD}$, and the total maximum displacement, $D_{TM}$, of components of an isolation system with a uniform spatial distribution of effective stiffness at the design displacement shall be taken as not less than that prescribed by Eqs. 9-6 and 9-7:

$$D_{TD} = D_D\left[1 + y\frac{12e}{b^2 + d^2}\right] \qquad \text{(Eq. 9-6)}$$

$$D_{TM} = D_M\left[1 + y\frac{12e}{b^2 + d^2}\right] \qquad \text{(Eq. 9-7)}$$

A value for the total maximum displacement, $D_{TM}$, less than the value prescribed by Eq. 9-7, but not less than 1.1 times $D_M$, shall be permitted, provided the isolation system is shown by calculation to be configured to resist torsion.

#### 9.2.4.4 Minimum Lateral Forces

*9.2.4.4.1 Isolation System and Structural Components and Elements at or below the Isolation System* The isolation system, the foundation, and all other structural components and elements below the isolation system shall be designed and constructed to withstand a minimum lateral seismic force, $V_b$, prescribed by Eq. 9-8:

$$V_b = K_{D\,max}D_D \qquad \text{(Eq. 9-8)}$$

*9.2.4.4.2 Structural Components and Elements above the Isolation System* The components and elements above the isolation system shall be designed and constructed to resist a minimum lateral seismic force, $V_s$, equal to the value of $V_b$, prescribed by Eq. 9-8.

*9.2.4.4.3 Limits on $V_s$* The value of $V_s$ shall be taken as not less than the following:

1. The base shear corresponding to the design wind load; and
2. The lateral seismic force required to fully activate the isolation system factored by 1.5.

*C9.2.4.4.3 Limits on $V_s$* Examples of lateral seismic forces required to fully activate the isolation system include the yield level of a softening system, the ultimate capacity of a sacrificial wind-restraint system, or the break-away friction level of a sliding system.

*9.2.4.4.4 Vertical Distribution of Force* The total force, $V_s$, shall be distributed over the height of the structure above the isolation interface in accordance with Eq. 9-9:

$$F_x = \frac{V_s w_x h_x}{\sum_{i}^{n} w_i h_i} \qquad \text{(Eq. 9-9)}$$

At each level designated as $x$, the force $F_x$ shall be applied over the area of the building in accordance with the weight, $w_x$, distribution at that level, $h_x$. Response of structural components and elements shall be calculated as the effect of the force $F_x$ applied at the appropriate levels above the base.

### 9.2.4.5 Response Spectrum Analysis

*9.2.4.5.1 Earthquake Input* The Design Earthquake spectrum shall be used to calculate the total design displacement of the isolation system and the lateral forces and displacements of the isolated building. The BSE-2 spectrum shall be used to calculate the total maximum displacement of the isolation system.

*9.2.4.5.2 Modal Damping* Response spectrum analysis shall be performed, using a damping value for isolated modes equal to the effective damping of the isolation system, or 30% of critical, whichever is less. The damping value assigned to higher modes of response shall be consistent with the material type and stress level of the superstructure.

*9.2.4.5.3 Combination of Earthquake Directions* Response spectrum analysis used to determine the total design displacement and total maximum displacement shall include simultaneous excitation of the model by 100% of the most critical direction of ground motion, and not less than 30% of the ground motion in the orthogonal axis. The maximum displacement of the isolation system shall be calculated as the vector sum of the two orthogonal displacements.

*9.2.4.5.4 Scaling of Results* If the total design displacement determined by response spectrum analysis is found to be less than the value of $D_{TD}$ prescribed by Eq. 9-6, or if the total maximum displacement determined by response spectrum analysis is found to be less than the value of $D_{TM}$ prescribed by Eq. 9-7, then all response parameters, including component actions and deformations, shall be adjusted by the greater of the following:

1. $D_{TD}$/Design displacement determined by response spectrum analysis, or
2. $D_{TM}$/Maximum displacement determined by response spectrum analysis.

### 9.2.4.6 Design Forces and Deformations

Components and elements of the building shall be designed for forces and displacements estimated by linear procedures using the acceptance criteria of Section 3.4.2.2, except that deformation-controlled components and elements shall be designed using component $m$-factors equal to or less than 1.5.

### 9.2.5 Nonlinear Procedures

Seismically isolated buildings evaluated using nonlinear procedures shall be represented by three-dimensional models that incorporate the nonlinear characteristics of both the isolation system and the structure above the isolation system.

### 9.2.5.1 Nonlinear Static Procedure

*9.2.5.1.1 General* The Nonlinear Static Procedure (NSP) for seismically isolated buildings shall be based on the criteria of Section 3.3.3, except that the target displacement and pattern of applied lateral load shall be based on the criteria given in Sections 9.2.5.1.2 and 9.2.5.1.3, respectively.

*9.2.5.1.2 Target Displacement* In each principal direction, the building model shall be pushed to the Design Earthquake target displacement, $D'_D$, and to the BSE-2 target displacement, $D'_M$, as defined by Eqs. 9-10 and 9-11:

$$D'_D = \frac{D_D}{\sqrt{1 + \left(\dfrac{T_e}{T_D}\right)^2}} \qquad \text{(Eq. 9-10)}$$

$$D'_M = \frac{D_M}{\sqrt{1 + \left(\dfrac{T_e}{T_M}\right)^2}} \qquad \text{(Eq. 9-11)}$$

where $T_e$ is the effective period of the structure above the isolation interface on a fixed base as prescribed by Eq. 3-14. The target displacements, $D'_D$ and $D'_M$, shall be evaluated at a control node that is located at the center of mass of the first floor above the isolation interface.

*9.2.5.1.3 Lateral Load Pattern* The pattern of applied lateral load shall be proportional to the distribution of the product of building mass and the deflected shape of the isolated mode of response at the target displacement.

### 9.2.5.2 Nonlinear Dynamic Procedure

*9.2.5.2.1 General* The NDP for seismically isolated buildings shall be based on the nonlinear procedure requirements of Section 3.3.4, except that results shall be scaled for design based on the criteria given in the following section.

*9.2.5.2.2 Scaling of Results* If the design displacement determined by time-history analysis is less than the $D'_D$ value of prescribed by Eq. 9-10, or if the maximum displacement determined by response spectrum analysis is found to be less than the value of $D'_M$ prescribed by Eq. 9-11, then all response parameters, including component actions and deformations, shall be adjusted by the greater of the following:

1. $D'_D$/Design displacement determined by time history analysis, or
2. $D'_M$/Maximum displacement determined by time history analysis.

### 9.2.5.3 Design Forces and Deformations
Components and elements of the building shall be designed for the forces and deformations estimated by nonlinear procedures using the acceptance criteria of Section 3.4.3.2.

### 9.2.6 Nonstructural Components

### 9.2.6.1 General
Permanent nonstructural components and the attachments to them shall be designed to resist seismic forces and displacements as given in this section and the applicable requirements of Chapter 11.

### 9.2.6.2 Forces and Displacements

*9.2.6.2.1 Components and Elements at or above the Isolation Interface* Nonstructural components, or portions thereof, that are at or above the isolation interface shall be designed to resist a total lateral seismic force equal to the maximum dynamic response of the element or component under consideration.

**EXCEPTION:** Design of elements of seismically isolated structures and nonstructural components, or portions thereof, to resist the total lateral seismic force as required for conventional fixed-base buildings by Chapter 11, shall be permitted.

*9.2.6.2.2 Components and Elements that Cross the Isolation Interface* Nonstructural components, or portions thereof, that cross the isolation interface shall be designed to withstand the total maximum (horizontal) displacement and maximum vertical displacement of the isolation system at the total maximum (horizontal) displacement. Components and elements that cross the isolation interface shall not restrict displacement of the isolated building or otherwise compromise the Rehabilitation Objectives of the building.

*9.2.6.2.3 Components and Elements below the Isolation Interface* Nonstructural components, or portions thereof, that are below the isolation interface shall be designed and constructed in accordance with the requirements of Chapter 11.

### 9.2.7 Detailed System Requirements

### 9.2.7.1 General
The isolation system and the structural system shall comply with the detailed system requirements specified in Section 9.2.7.2 and 9.2.7.3, respectively.

### 9.2.7.2 Isolation System

*9.2.7.2.1 Environmental Conditions* In addition to the requirements for vertical and lateral loads induced by wind and earthquake, the isolation system shall be designed with consideration given to other environmental conditions, including aging effects, creep, fatigue, operating temperature, and exposure to moisture or damaging substances.

*9.2.7.2.2 Wind Forces* Isolated buildings shall resist design wind loads at all levels above the isolation interface in accordance with the applicable wind design provisions. At the isolation interface, a wind-restraint system shall be provided to limit lateral displacement in the isolation system to a value equal to that required between floors of the structure above the isolation interface.

*9.2.7.2.3 Fire Resistance* Fire resistance rating for the isolation system shall be consistent with the requirements of columns, walls, or other such components of the building.

*9.2.7.2.4 Lateral Restoring Force* The isolation system shall be configured to produce either a restoring force such that the lateral force at the total design displacement is at least $0.025W$ greater than the lateral force at 50% of the total design displacement, or a restoring force of not less than $0.05W$ at all displacements greater than 50% of the total design displacement.

**EXCEPTION:** The isolation system need not be configured to produce a restoring force, as required above, provided the isolation system is capable of remaining stable under full vertical load and accommodating a total maximum displacement equal to the greater of either 3.0 times the total design displacement or $36S_{XI}$ in., where $S_{XI}$ is calculated for the BSE-2.

*9.2.7.2.5 Displacement Restraint* Configuration of the isolation system to include a displacement restraint that limits lateral displacement due to the BSE-2 to less than the ratio of the design spectral response acceleration parameter at 1 sec ($S_{XI}$) for the BSE-2 to that for the Design Earthquake times the total design displacement shall be permitted, provided that the seismically isolated building is designed in accordance with the following criteria where more stringent than the requirements of Section 9.2.3:

1. BSE-2 response is calculated in accordance with the dynamic analysis requirements of Section 9.2.5, explicitly considering the nonlinear characteristics of the isolation system and the structure above the isolation system;
2. The ultimate capacity of the isolation system, and structural components and elements below the isolation system, shall exceed the force and displacement demands of the BSE-2;

3. The structure above the isolation system is checked for stability and ductility demand of the BSE-2; and
4. The displacement restraint does not become effective at a displacement less than 0.75 times the total design displacement, unless it is demonstrated by analysis that earlier engagement does not result in unsatisfactory performance.

*9.2.7.2.6 Vertical Load Stability* Each component of the isolation system shall be designed to be stable under the full maximum vertical load, $1.2Q_D + Q_L + |Q_E|$, and the minimum vertical load, $0.8Q_D - |Q_E|$, at a horizontal displacement equal to the total maximum displacement. The earthquake vertical load on an individual isolator unit, $Q_E$, shall be based on peak building response due to the BSE-2.

*9.2.7.2.7 Overturning* The factor of safety against global structural overturning at the isolation interface shall be not less than 1.0 for required load combinations. All gravity and seismic loading conditions shall be investigated. Seismic forces for overturning calculations shall be based on the BSE-2, and the vertical restoring force shall be based on the building's weight, $W$, above the isolation interface.

Local uplift of individual components and elements shall be permitted, provided the resulting deflections do not cause overstress or instability of the isolator units or other building components and elements. A tie-down system to limit local uplift of individual components and elements shall be permitted, provided that the seismically isolated building is designed in accordance with the following criteria where more stringent than the requirements of Section 9.2.3:

1. BSE-2 response is calculated in accordance with the dynamic analysis requirements of Section 9.2.5, explicitly considering the nonlinear characteristics of the isolation system and the structure above the isolation system;
2. The ultimate capacity of the tie-down system exceeds the force and displacement demands of the BSE-2; and
3. The isolation system is designed and shown by test to be stable (Section 9.2.9.2.4) for BSE-2 loads that include additional vertical load due to the tie-down system.

*9.2.7.2.8 Inspection and Replacement* Access for inspection and replacement of all components and elements of the isolation system shall be provided.

*9.2.7.2.9 Manufacturing Quality Control* A manufacturing quality control testing program for isolator units shall be established by the design professional.

### 9.2.7.3 Structural System

*9.2.7.3.1 Horizontal Distribution of Force* A horizontal diaphragm or other structural components and elements shall provide continuity above the isolation interface. The diaphragm or other structural components and elements shall have adequate strength and ductility to transmit forces (due to nonuniform ground motion) calculated in accordance with this section from one part of the building to another, and have sufficient stiffness to effect rigid diaphragm response above the isolation interface.

*9.2.7.3.2 Building Separations* Separations between the isolated building and surrounding retaining walls or other fixed obstructions shall be not less than the total maximum displacement.

### 9.2.8 Design Review

#### 9.2.8.1 General

A review of the design of the isolation system and related test programs shall be performed by an independent engineering team, including persons experienced in seismic analysis methods and the theory and application of seismic isolation.

#### 9.2.8.2 Isolation System

Isolation system design review shall include the following:

1. Site-specific seismic criteria, including site-specific spectra and ground motion time history, and all other design criteria developed specifically for the project;
2. Preliminary design, including the determination of the total design and total maximum displacement of the isolation system, and the lateral force design level;
3. Isolation system prototype testing in accordance with Section 9.2.9;
4. Final design of the isolated building and supporting analyses; and
5. Isolation system quality control testing in accordance with Section 9.2.7.2.9.

### 9.2.9 Isolation System Testing and Design Properties

#### 9.2.9.1 General

The deformation characteristics and damping values of the isolation system used in the design and analysis of seismically isolated structures shall be based on the following tests of a selected sample of the components prior to construction.

The isolation system components to be tested shall include isolators and components of the wind-restraint system and supplemental energy dissipation devices if such components and devices are used in the design.

The tests specified in this section establish design properties of the isolation system, and shall not be considered as satisfying the manufacturing quality control testing requirements of Section 9.2.7.2.9.

#### 9.2.9.2 Prototype Tests

*9.2.9.2.1 General* Prototype tests shall be performed separately on two full-sized specimens of each type and size of isolator of the isolation system. The test specimens shall include components of the wind-restraint system, as well as individual isolators, if such components are used in the design. Supplementary energy dissipation devices shall be tested in accordance with Section 9.3.8. Specimens tested shall not be used for construction unless approved by the engineer responsible for the structural design.

*9.2.9.2.2 Record* For each cycle of tests, the force–deflection and hysteretic behavior of the test specimen shall be recorded.

*9.2.9.2.3 Sequence and Cycles* The following sequence of tests shall be performed for the prescribed number of cycles at a vertical load equal to the average $Q_D + 0.5Q_L$ on all isolators of a common type and size:

Twenty fully reversed cycles of loading at a lateral force corresponding to the wind design force;

Three fully reversed cycles of loading at each of the following displacements: $0.25D_D$, $0.50D_D$, $1.0D_D$, and $1.0D_M$;

Three fully reversed cycles at the total maximum displacement, $1.0D_{TM}$; and

$30S_{X1}/S_{XS}B_{D1}$, but not less than 10, fully reversed cycles of loading at the design displacement, $1.0D_D$. $S_{X1}$ and $S_{XS}$ shall be evaluated for the Design Earthquake.

*9.2.9.2.4 Vertical Load-Carrying Isolators* If an isolator is also a vertical-load-carrying component, then Item 2 of the sequence of cyclic tests specified in Section 9.2.9.2.3 shall be performed for two additional vertical load cases:

1. $1.2Q_D + 0.5Q_L + |Q_E|$; and
2. $0.8Q_D - |Q_E|$

where $D$, $L$, and $E$ refer to dead, live, and earthquake loads, respectively. $Q_D$ and $Q_L$ are as defined in Section 3.2.8. The vertical test load on an individual isolator unit shall include the load increment $Q_E$ due to earthquake overturning, and shall be equal to or greater than the peak earthquake vertical force response corresponding to the test displacement being evaluated. In these tests, the combined vertical load shall be taken as the typical or average downward force on all isolators of a common type and size.

*9.2.9.2.5 Isolators Dependent on Loading Rates* If the force-deflection properties of the isolators are dependent on the rate of loading, then each set of tests specified in Sections 9.2.9.2.3 and 9.2.9.2.4 shall be performed dynamically at a frequency equal to the inverse of the effective period, $T_D$, of the isolated structure.

**EXCEPTION**: If reduced-scale prototype specimens are used to quantify rate-dependent properties of isolators, the reduced-scale prototype specimens shall be of the same type and material and be manufactured with the same processes and quality as full-scale prototypes, and shall be tested at a frequency that represents full-scale prototype loading rates.

The force-deflection properties of an isolator shall be considered to be dependent on the rate of loading if there is greater than a plus or minus 10% difference in the effective stiffness at the design displacement (1) where tested at a frequency equal to the inverse of the effective period of the isolated structure, and (2) where tested at any frequency in the range of 0.1 to 2.0 times the inverse of the effective period of the isolated structure.

*9.2.9.2.6 Isolators Dependent on Bilateral Load* If the force-deflection properties of the isolators are dependent on bilateral load, then the tests specified in Sections 9.2.9.2.3 and 9.2.9.2.5 shall be augmented to include bilateral load at the following increments of the total design displacement: 0.25 and 1.0; 0.50 and 1.0; 0.75 and 1.0; and 1.0 and 1.0.

**EXCEPTION**: If reduced-scale prototype specimens are used to quantify bilateral-load-dependent properties, then such scaled specimens shall be of the same type and material, and manufactured with the same processes and quality as full-scale prototypes.

The force-deflection properties of an isolator shall be considered to be dependent on bilateral load, if the bilateral and unilateral force-deflection properties have greater than a plus or minus 15% difference in effective stiffness at the design displacement.

*9.2.9.2.7 Maximum and Minimum Vertical Load* Isolators that carry vertical load shall be statically tested for the maximum and minimum vertical load, at the total maximum displacement. In these tests, the combined vertical loads of $1.2Q_D + 1.0Q_L + |Q_E|$ shall be taken as the maximum vertical force, and the combined vertical load of $0.8Q_{\bar{D}}\,|Q_E|$ shall be taken as the minimum vertical force, on any one isolator of a common type and size. The earthquake vertical load on an individual isolator, $Q_E$, shall be based on peak building response due to the BSE-2.

*9.2.9.2.8 Sacrificial Wind-Restraint Systems* If a sacrificial wind-restraint system is part of the isolation system, then the ultimate capacity shall be established by testing in accordance with this section.

*9.2.9.2.9 Testing Similar Units* Prototype tests need not be performed if an isolator unit, where compared to another tested unit, complies with the following criteria:

1. Is of similar dimensional characteristics;
2. Is of the same type and materials; and
3. Is fabricated using identical manufacturing and quality control procedures

The testing exemption shall be approved by the review team specified in Section 9.2.8.

### 9.2.9.3 Determination of Force-Deflection Characteristics

The force-deflection characteristics of the isolation system shall be based on the cyclic load testing of isolator prototypes specified in Section 9.2.9.2.3.

As required, the effective stiffness of an isolator unit, $k_{eff}$, shall be calculated for each cycle of deformation by Eq. 9-12:

$$k_{eff} = \frac{|F^+| + |F^-|}{|\Delta^+| + |\Delta^-|} \qquad \text{(Eq. 9-12)}$$

where $F^+$ and $F^-$ are the positive and negative forces at positive and negative test displacements, $\Delta^+$ and $\Delta^-$, respectively.

As required, the effective damping of an isolator unit, $\beta_{eff}$, shall be calculated for each cycle of deformation by Eq. 9-13:

$$\beta_{eff} = \frac{2}{\pi}\left[ \frac{E_{Loop}}{k_{eff}(|\Delta^+| + |\Delta^-|)^2} \right] \qquad \text{(Eq. 9-13)}$$

where the energy dissipated per cycle of loading, $E_{Loop}$, and the effective stiffness, $k_{eff}$, are based on test displacements, $D^+$ and $D^-$.

#### 9.2.9.4 System Adequacy

The performance of the test specimens shall be assessed as adequate if the following conditions are satisfied:

1. The force–deflection plots of all tests specified in Section 9.2.9.2 have a non-negative incremental force-carrying capacity;
2. For each increment of test displacement specified in Section 9.2.9.2.3, Item 2, and for each vertical load case specified in Section 9.2.9.2.3, the following criteria are met:
   2.1. There is no greater than a plus or minus 15% difference between the effective stiffness at each of the three cycles of test and the average value of effective stiffness for each test specimen;
   2.2. There is no greater than a 15% difference in the average value of effective stiffness of the two test specimens of a common type and size of the isolator unit over the required three cycles of test;
3. For each specimen there is no greater than a plus or minus 20% change in the initial effective stiffness of each test specimen over the $30S_{XI}/S_{XS}B_{DI}$, but not less than 10, cycles of the test specified in Section 9.2.9.2.3, Item 3. $S_{XI}$ and $S_{XS}$ shall be evaluated for the Design Earthquake;
4. For each specimen there is no greater than a 20% decrease in the initial effective damping over the $30S_{XI}/S_{XS}B_{DI}$, but not less than 10, cycles of the test specified in Section 9.2.9.2.3, Item 4. $S_{XI}$ and $S_{XS}$ shall be evaluated for the Design Earthquake;
5. All specimens of vertical-load-carrying components of the isolation system remain stable at the total maximum displacement for static load as prescribed in Section 9.2.9.2.6; and
6. The effective stiffness and effective damping of test specimens fall within the limits specified by the engineer responsible for structural design.

#### 9.2.9.5 Design Properties of the Isolation System

*9.2.9.5.1 Maximum and Minimum Effective Stiffness* At the design displacement, the maximum and minimum effective stiffness of the isolation system, $K_{Dmax}$ and $K_{Dmin}$, shall be based on the cyclic tests of Section 9.2.9.2 and calculated by Eqs. 9-14 and 9-15:

$$K_{Dmax} = \frac{\Sigma|F_D^+|_{max} + \Sigma|F_D^-|_{max}}{2D_D} \quad \text{(Eq. 9-14)}$$

$$K_{Dmin} = \frac{\Sigma|F_D^+|_{min} + \Sigma|F_D^-|_{min}}{2D_D} \quad \text{(Eq. 9-15)}$$

At the maximum displacement, the maximum and minimum effective stiffness of the isolation system shall be based on cyclic tests of Section 9.2.9.2 and calculated by Eqs. 9-16 and 9-17:

$$K_{Mmax} = \frac{\Sigma|F_M^+|_{max} + \Sigma|F_M^-|_{max}}{2D_M} \quad \text{(Eq. 9-16)}$$

$$K_{Mmin} = \frac{\Sigma|F_M^+|_{max} + \Sigma|F_M^-|_{max}}{2D_M} \quad \text{(Eq. 9-17)}$$

*9.2.9.5.2 Effective Damping* At the design displacement, the effective damping of the isolation system, $\beta_D$, shall be based on the cyclic tests of Section 9.2.9.2 and calculated by Eq. 9-18:

$$\beta_D = \frac{1}{2\pi}\left\lfloor \frac{\Sigma E_D}{K_{Dmax}D_D^2} \right\rfloor \quad \text{(Eq. 9-18)}$$

In Eq. 9-18, the total energy dissipated in the isolation system per displacement cycle, $\Sigma E_D$, shall be taken as the sum of the energy dissipated per cycle in all isolators measured at test displacements, $\Delta^+$ and $\tilde{\Delta}$, that are equal in magnitude to the design displacement, $D_D$.

At the maximum displacement, the effective damping of the isolation system, $\beta_M$, shall be based on the cyclic tests of Section 9.2.9.2 and calculated by Eq. 9-19:

$$\beta_M = \frac{1}{2\pi}\left\lfloor \frac{\Sigma E_M}{K_{Mmax}D_M^2} \right\rfloor \quad \text{(Eq. 9-19)}$$

In Eq. 9-19, the total energy dissipated in the isolation system per displacement cycle, $\Sigma E_D$, shall be taken as the sum of the energy dissipated per cycle in all isolators measured at test displacements, $\Delta^+$ and $\Delta^-$, that are equal in magnitude to the maximum displacement, $D_M$.

### 9.3 PASSIVE ENERGY DISSIPATION SYSTEMS

#### 9.3.1 General Requirements

Passive energy dissipation systems classified as either displacement-dependent, velocity-dependent, or other, as defined in Section 9.3.3, shall comply with the requirements of Section 9.3. Linear and nonlinear analyses shall be performed, as required, in accordance with Section 9.3.4 and 9.3.5, respectively. Additional requirements for passive energy dissipation systems, as defined in Section 9.3.6, shall be met. Passive energy dissipation systems shall be reviewed and tested in accordance with Sections 9.3.7 and 9.3.8, respectively.

The energy dissipation devices shall be designed with consideration given to environmental conditions including wind, aging effects, creep, fatigue, ambient temperature, operating temperature, and exposure to moisture or damaging substances.

The mathematical model of the rehabilitated building shall include the plan and vertical distribution of the energy dissipation devices. Analyses shall account for the dependence of the devices on excitation frequency, ambient and operating temperature, velocity, sustained loads, and bilateral loads. Multiple analyses of the building shall be conducted to bound the effects of each varying mechanical characteristic of the devices.

Energy dissipation devices shall be capable of sustaining larger displacements for displacement-dependent devices and larger velocities for velocity-dependent devices than the maximum calculated for the BSE-2 in accordance with the following criteria:

1. If four or more energy dissipation devices are provided in a given story of a building in one principal direction of the building, with a minimum of two devices located on each side of the center of stiffness of the story in the direction under consideration, all energy dissipation devices shall be capable of sustaining displacements equal to 130% of the maximum calculated displacement in the device in the BSE-2. A velocity-dependent device as described in Section 9.3.3 shall be capable of sustaining the force associated with a velocity equal to 130% of the maximum calculated velocity for that device in the BSE-2; and

2. If fewer than four energy dissipation devices are provided in a given story of a building in one principal direction of the building, or fewer than two devices are located on each side of the center of stiffness of the story in the direction under consideration, all energy dissipation devices shall be capable of sustaining displacements equal to 200% of the maximum calculated displacement in the device in the BSE-2. A velocity-dependent device shall be capable of sustaining the force associated with a velocity equal to 200% of the maximum calculated velocity for that device in the BSE-2.

The components and connections transferring forces between the energy dissipation devices shall be designed to remain linearly elastic for the forces described in Items 1 or 2 above.

### C9.3.1 General Requirements

The increase in displacement (and velocity) capacity is dependent on the level of redundancy in the supplemental damping system.

Passive energy dissipation is an emerging technology that enhances the performance of the building by adding damping (and in some cases, stiffness) to the building. The primary use of energy dissipation devices is to reduce earthquake displacement of the structure. Energy dissipation devices will also reduce force in the structure—provided the structure is responding elastically—but would not be expected to reduce force in structures that are responding beyond yield.

For most applications, energy dissipation provides an alternative approach to conventional stiffening and strengthening schemes, and would be expected to achieve comparable performance levels. In general, these devices would be expected to be good candidates for projects that have a target Building Performance Level of Life Safety or perhaps Immediate Occupancy, but would be expected to have only limited applicability to projects with a target Building Performance Level of Collapse Prevention.

Other objectives may also influence the decision to use energy dissipation devices since these devices can also be useful for control of building response due to small earthquakes, wind, or mechanical loads. The analysis procedures set forth in this standard are approximate. Roof displacements calculated using the linear and nonlinear procedures are likely to be more accurate than the corresponding estimates of story drift and relative velocity between adjacent stories.

Accordingly, this standard requires that energy dissipation devices be capable of sustaining larger displacements (and velocities for velocity-dependent devices) than the maxima calculated by analysis in the BSE-2. Recognizing that the response of a building frame incorporating four or more devices in each principal direction in each story will be more reliable than a frame with fewer devices in each principal direction, the increase in displacement (and velocity) capacity is dependent on the level of redundancy in the supplemental damping system. The increased force shall be used to design the framing that supports the energy dissipation devices-reflecting the objective of keeping the device support framing elastic in the BSE-2. The increases in force and displacement capacity listed in this standard are based on the judgment of the authors.

### 9.3.2 Implementation of Energy Dissipation Devices

Energy dissipation devices shall be implemented in accordance with requirements specified in Chapters 1 through 3 but as modified in the subsequent sections of this chapter.

## 9.3.3 Modeling of Energy Dissipation Devices

Displacement-dependent devices shall include devices that exhibit either rigid-plastic (friction devices), bilinear (metallic yielding devices), or trilinear hysteresis. The response of displacement-dependent devices shall be independent of velocity and frequency of excitation. Velocity-dependent devices shall include solid and fluid viscoelastic devices, and fluid viscous devices. Devices not classified as displacement- or velocity-dependent shall be classified as "other."

Models of the energy dissipation system shall include the stiffness of structural components that are part of the load path between energy dissipation devices and the ground and whose flexibility affects the performance of the energy dissipation system, including components of the foundation, braces that work in series with the energy dissipation devices, and connections between braces and the energy dissipation devices.

Energy dissipation devices shall be modeled as described in the following subsections, unless approved methods are used.

## C9.3.3 Modeling of Energy Dissipation Devices

Examples of "other" devices include shape-memory alloys (superelastic effect), friction-spring assemblies with recentering capability, and fluid-restoring, force-damping devices.

### 9.3.3.1 Displacement-Dependent Devices

A displacement-dependent device shall have a force–displacement relationship that is a function of the relative displacement between each end of the device. The response of a displacement-dependent device shall be independent of the relative velocity between each end of the device and frequency of excitation.

Displacement-dependent devices shall be modeled in sufficient detail to capture their force–displacement response, and their dependence, if any, on axial–shear–flexure interaction, or bilateral deformation response.

For evaluating the response of a displacement-dependent device from testing data, the force in a displacement-dependent device shall be calculated in accordance with Eq. 9-20:

$$F = k_{eff}D \qquad \text{(Eq. 9-20)}$$

where the effective stiffness, $k_{eff}$, of the device is calculated in accordance with Eq. 9-21:

$$k_{eff} = \frac{|F^+| + |F^-|}{|D^+| + |D^-|} \qquad \text{(Eq. 9-21)}$$

The forces in the device, $F^+$ and $F^-$, shall be evaluated at displacements $D^+$ and $D^-$, respectively.

### 9.3.3.2 Velocity-Dependent Devices

#### 9.3.3.2.1 Solid Viscoelastic Devices
Solid viscoelastic devices shall be modeled using a spring and dashpot in parallel (Kelvin model). The spring and dashpot constants selected shall capture the frequency and temperature dependence of the device consistent with fundamental frequency of the rehabilitated building ($f_1$), and the operating temperature range. If the cyclic response of a viscoelastic solid device cannot be captured by single estimates of the spring and dashpot constants, the response of the rehabilitated building shall be estimated by multiple analyses of the building frame, using limiting upper- and lower-bound values for the spring and dashpot constants.

The force in a viscoelastic device shall be determined in accordance with Eq. 9-22:

$$F = k_{eff}D + C\dot{D} \qquad \text{(Eq. 9-22)}$$

where $C$ is the damping coefficient for the viscoelastic device, $D$ is the relative displacement between each end of the device, $\dot{D}$ is the relative velocity between each end of the device, and $k_{eff}$ is the effective stiffness of the device calculated in accordance with Eq. 9-23:

$$k_{eff} = \frac{|F^+| + |F^-|}{|D^+| + |D^-|} = K' \qquad \text{(Eq. 9-23)}$$

where $K'$ is the storage stiffness.

The damping coefficient for the device shall be calculated in accordance with Eq. 9-24:

$$C = \frac{W_D}{\pi \omega_1 D_{ave}^2} = \frac{K''}{\omega_1} \qquad \text{(Eq. 9-24)}$$

where $K''$ is the loss stiffness, the angular frequency $\omega_1$ is equal to $2\pi f_1$, $D_{ave}$ is the average of the absolute values of displacements $D^+$ and $D^-$, and $W_D$ is the area enclosed by one complete cycle of the force–displacement response of the device.

#### C9.3.3.2.1 Solid Viscoelastic Devices
The cyclic response of viscoelastic solids is generally dependent on the frequency and amplitude of the motion and the operating temperature (including temperature rise due to excitation).

#### 9.3.3.2.2 Fluid Viscoelastic Devices
Fluid viscoelastic devices shall be modeled using a spring and dashpot in series (Maxwell model). The spring and dashpot constants selected shall capture the frequency and temperature dependence of the device consistent with fundamental frequency of the rehabilitated building

($f_1$), and the operating temperature range. If the cyclic response of a viscoelastic fluid device cannot be captured by single estimates of the spring and dashpot constants, the response of the rehabilitated building shall be estimated by multiple analyses of the building frame, using limiting upper- and lower-bound values for the spring and dashpot constants.

*C9.3.3.2.2 Fluid Viscoelastic Devices* The cyclic response of viscoelastic fluid devices is generally dependent on the frequency and amplitude of the motion and the operating temperature (including temperature rise due to excitation).

*9.3.3.2.3 Fluid Viscous Devices* Linear fluid viscous dampers exhibiting stiffness in the frequency range $0.5 f_1$ to $2.0 f_1$ shall be modeled as a fluid viscoelastic device.

In the absence of stiffness in the frequency range $0.5 f_1$ to $2.0 f_1$, the force in the fluid viscous device shall be computed in accordance with Eq. 9-25:

$$F = C_0 |\dot{D}|^\alpha \, \text{sgn}(\dot{D}) \qquad \text{(Eq. 9-25)}$$

where $C_0$ is the damping coefficient for the device, $\alpha$ is the velocity exponent for the device, $\dot{D}$ is the relative velocity between each end of the device, and sgn is the signum function that, in this case, defines the sign of the relative velocity term.

### 9.3.3.3 Other Types of Devices

Energy dissipation devices not classified as either displacement-dependent or velocity-dependent shall be modeled using approved methods. Such models shall accurately describe the force–velocity–displacement response of the device under all sources of loading including gravity, seismic, and thermal.

### C9.3.3.3 Other Types of Devices

Other energy dissipating devices, such as those having hysteresis of the type shown in Fig. C9-8, require modeling techniques different from those described above. Tsopelas and Constantinou (1994a), Nims et al. (1993), and Pekcan et al. (1995) describe analytical models for some of these devices.

### 9.3.4 Linear Procedures

Linear procedures shall be permitted only if the following criteria are met:

1. The framing system exclusive of the energy dissipation devices remains linearly elastic for the selected Earthquake Hazard Level after the effects of added damping are considered;
2. The effective damping afforded by the energy dissipation does not exceed 30% of critical in the fundamental mode;
3. The secant stiffness of each energy dissipation device, calculated at the maximum displacement in the device, is included in the mathematical model of the rehabilitated building;
4. Where evaluating the regularity of a building, the energy dissipation devices are included in the mathematical model; and
5. Higher mode effects are not significant as defined in Section 2.4.2.1.

### 9.3.4.1 Linear Static Procedure

*9.3.4.1.1 Displacement-Dependent Devices* Use of the Linear Static Procedure (LSP) shall be permitted to analyze displacement-dependent energy dissipation devices, provided that, in addition to the requirements

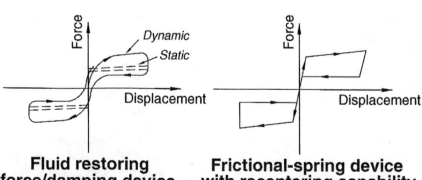

**Fluid restoring force/damping device**

**Frictional-spring device with recentering capability**

**FIGURE C9-8. Idealized Force–Displacement Loops of Energy Dissipation Devices with Recentering Capability.**

of Section 9.3.4, the following requirements are satisfied:

1. The ratio of the maximum resistance in each story, in the direction under consideration, to the story shear demand calculated using Eqs. 3-10 and 3-11, shall range between 80% and 120% of the average value of the ratio for all stories. The maximum story resistance shall include the contributions from all components, elements, and energy dissipation devices; and
2. The maximum resistance of all energy dissipation devices in a story, in the direction under consideration, shall not exceed 50% of the resistance of the remainder of the framing where said resistance is calculated at the displacements anticipated in the BSE-2. Aging and environmental effects shall be considered in calculating the maximum resistance of the energy dissipation devices

The pseudo-lateral load of Eq. 3-10 shall be reduced by the damping modification factors of Section 1.6.1.5.1 to account for the energy dissipation (damping) afforded by the energy dissipation devices. The damping effect shall be calculated in accordance with Eq. 9-26:

$$\beta_{eff} = \beta + \frac{\sum_j W_j}{4\pi W_k} \qquad \text{(Eq. 9-26)}$$

where $\beta$ is the damping in the framing system and shall be set equal to 0.05 unless modified in Section 1.6.1.4, $W_j$ shall be taken as the work done by device $j$ in one complete cycle corresponding to floor displacements $\delta_i$, the summation extends over all devices $j$, and $W_k$ is the maximum strain energy in the frame, determined using Eq. 9-27:

$$W_k = \frac{1}{2} \sum_i F_i \delta_i \qquad \text{(Eq. 9-27)}$$

where $F_i$ shall be taken as the inertia force at floor level $i$ and the summation extends over all floor levels.

### 9.3.4.1.2 Velocity-Dependent Devices

Use of the LSP shall be permitted to analyze velocity-dependent energy dissipation devices, provided that in addition to the requirements of Section 9.3.4, the following requirements are satisfied:

1. The maximum resistance of all energy dissipation devices in a story in the direction under consideration shall not exceed 50% of the resistance of the remainder of the framing where said resistance is calculated at the displacements anticipated in the BSE-2. Aging and environmental effects shall be considered in calculating the maximum resistance of the energy dissipation devices; and

2. The pseudo-lateral load of Eq. 3-10 shall be reduced by the damping modification factors of Section 1.6.1.5.1 to account for the energy dissipation (damping) afforded by the energy dissipation devices. The damping effect shall be calculated in accordance with Eq. 9-28:

$$\beta_{eff} = \beta + \frac{\sum_j W_j}{4\pi W_k} \qquad \text{(Eq. 9-28)}$$

where $\beta$ is the damping in the structural frame and shall be set equal to 0.05 unless modified in Section 1.6.1.5.3, $W_j$ shall be taken as the work done by device $j$ in one complete cycle corresponding to floor displacements $\delta_i$, the summation extends over all devices $j$, and $W_k$ is the maximum strain energy in the frame, determined using Eq. 9-27.

The work done by linear viscous device $j$ in one complete cycle of loading shall be calculated in accordance with Eq. 9-29:

$$W_j = \frac{2\pi^2}{T} C_j \delta_{rj}^2 \qquad \text{(Eq. 9-29)}$$

where $T$ is the fundamental period of the rehabilitated building including the stiffness of the velocity-dependent devices, $C_j$ is the damping constant for device $j$, and $\delta_{rj}$ is the relative displacement between the ends of device $j$ along the axis of device $j$.

Calculation of effective damping in accordance with Eq. 9-30 rather than Eq. 9-28 shall be permitted for linear viscous devices:

$$\beta_{eff} = \beta + \frac{T \sum_j C_j \cos^2 \theta_j \phi_{rj}^2}{4\pi \sum_i \left(\frac{w_i}{g}\right) \phi_i^2} \qquad \text{(Eq. 9-30)}$$

where $\theta_j$ is the angle of inclination of device $j$ to the horizontal, $\phi_{rj}$ is the first mode relative displacement between the ends of device $j$ in the horizontal direction, $w_i$ is the reactive weight of floor level $i$, $f_i$ is the first mode displacement at floor level $i$, and other terms are as defined above.

### 9.3.4.1.3 Design Actions

The design actions for components of the rehabilitated building shall be calculated in three distinct stages of deformation as follows. The maximum action shall be used for design.

1. **At the stage of maximum drift.** The lateral forces at each level of the building shall be calculated using Eq. 3-11, where $V^*$ is the modified equivalent base shear.
2. **At the stage of maximum velocity and zero drift.** The viscous component of force in each energy

dissipation device shall be calculated by Eq. 9-22 or 9-25, where the relative velocity $D$ is given by $2\pi f_1 D$, where $D$ is the relative displacement between the ends of the device calculated at the stage of maximum drift. The calculated viscous forces shall be applied to the mathematical model of the building at the points of attachment of the devices and in directions consistent with the deformed shape of the building at maximum drift. The horizontal inertia forces at each floor level of the building shall be applied concurrently with the viscous forces so that the horizontal displacement of each floor level is zero.

3. **At the stage of maximum floor acceleration.** Design actions in components of the rehabilitated building shall be determined as the sum of actions determined at the stage of maximum drift times $CF_1$, and actions determined at the stage of maximum velocity times $CF_2$, where

$$CF_1 = \cos[\tan^{-1}(2\beta_{eff})] \qquad \text{(Eq. 9-31)}$$

$$CF_2 = \sin[\tan^{-1}(2\beta_{eff})] \qquad \text{(Eq. 9-32)}$$

in which $\beta_{eff}$ is defined by either Eq. 9-28 or Eq. 9-30.

### 9.3.4.2 Linear Dynamic Procedure

If the Linear Dynamic Procedure (LDP) is selected based on the requirements of Section 9.2.3.3 and Section 2.4, the LDP of Section 3.3.2.2 shall be followed unless explicitly modified by this section.

Use of the response spectrum method of the LDP shall be permitted where the effective damping in the fundamental mode of the rehabilitated building, in each principal direction, does not exceed 30% of critical.

#### 9.3.4.2.1 Displacement-Dependent Devices

Application of the LDP for the analysis of rehabilitated buildings incorporating displacement-dependent devices shall comply with the restrictions set forth in Section 9.3.4.1.1.

For analysis by the Response Spectrum Method, modification of the 5%-damped response spectrum shall be permitted to account for the damping afforded by the displacement-dependent energy dissipation devices. The 5%-damped acceleration spectrum shall be reduced by the modal-dependent damping modification factor, $B$, either $B_s$ or $B_1$, for periods in the vicinity of the mode under consideration; the value of $B$ will be different for each mode of vibration. The damping modification factor in each significant mode shall be determined in accordance with Section 1.6.1.5.1 and the calculated effective damping in that mode. The effective damping shall be determined using a procedure similar to that described in Section 9.3.4.1.1.

If the maximum base shear force calculated by dynamic analysis is less than 80% of the modified equivalent base shear of Section 9.3.4.1, component and element actions and deformations shall be proportionally increased to correspond to 80% of the modified equivalent base shear.

#### 9.3.4.2.2 Velocity-Dependent Devices

For analysis by the Response Spectrum Method, modification of the 5%-damped response spectrum shall be permitted to account for the damping afforded by the velocity-dependent energy dissipation devices. The 5%-damped acceleration spectrum shall be reduced by the modal-dependent damping modification factor, $B$, either $B_s$ or $B_1$, for periods in the vicinity of the mode under consideration; note that the value of $B$ will be different for each mode of vibration. The damping modification factor in each significant mode shall be determined in accordance with Section 1.6.1.5.1 and the calculated effective damping in that mode.

The effective damping in the $m$-th mode of vibration ($\beta_{eff-m}$) shall be calculated in accordance with Eq. 9-33:

$$\beta_{eff-m} = \beta_m + \frac{\sum_j W_{mj}}{4\pi W_{mk}} \qquad \text{(Eq. 9-33)}$$

where $\beta_m$ is the $m$-th mode damping in the building frame, $W_{mj}$ is work done by device $j$ in one complete cycle corresponding to modal floor displacements $\delta_{mi}$, and $W_{mk}$ is the maximum strain energy in the frame in the $m$-th mode, determined using Eq. 9-34:

$$W_{mk} = \frac{1}{2} \sum_i F_{mi} \delta_{mi} \qquad \text{(Eq. 9-34)}$$

where $F_{mi}$ is the $m$-th mode horizontal inertia force at floor level $i$ and $\delta_{mi}$ is the $m$-th mode horizontal displacement at floor level $i$. The work done by linear viscous device $j$ in one complete cycle of loading in the $m$-th mode may be calculated in accordance with Eq. 9-35:

$$W_{mj} = \frac{2\pi^2}{T_m} C_j \delta_{mrj}^2 \qquad \text{(Eq. 9-35)}$$

where $T_m$ is the $m$-th mode period of the rehabilitated building including the stiffness of the velocity-dependent devices, $C_j$ is the damping constant for device $j$, and $\delta_{mrj}$ is the $m$-th mode relative displacement between the ends of device $j$ along the axis of device $j$.

In addition to direct application of the Response Spectrum Method in accordance with this section to obtain member actions at maximum drift, member actions at maximum velocity and maximum acceleration in each significant mode shall be determined using the procedure described in Sections 9.3.4.1.2. The combination factors $CF_1$ and $CF_2$ shall be determined based on Eqs. 9-31 and 9-32 using $\beta_{eff\,m}$ for the $m$-th mode.

If the maximum base shear force calculated by dynamic analysis is less than 80% of the modified equivalent base shear of Section 9.3.4.1, component and element actions and deformations shall be proportionally increased to correspond to 80% of the modified equivalent base shear.

## 9.3.5 Nonlinear Procedures

### 9.3.5.1 Nonlinear Static Procedure

If the Nonlinear Static Procedure (NSP) is selected based on the requirements of Section 9.2.3.3 and Section 2.4, the NSP of Section 3.3.3 shall be followed unless explicitly modified by this section.

The nonlinear mathematical model of the rehabilitated building shall include the nonlinear force-velocity-displacement characteristics of the energy dissipation devices explicitly, and the mechanical characteristics of the components supporting the devices. Stiffness characteristics shall be consistent with the deformations corresponding to the target displacement and a frequency equal to the inverse of period $T_e$, as defined in Section 3.3.3.2.

The nonlinear mathematical model of the rehabilitated building shall include the nonlinear force-velocity-displacement characteristics of the energy dissipation devices, and the mechanical characteristics of the components supporting the devices. Energy dissipation devices with stiffness and damping characteristics that are dependent on excitation frequency and/or temperature shall be modeled with characteristics consistent with (1) the deformations expected at the target displacement, and (2) a frequency equal to the inverse of the effective period.

Equation 3-15 shall be used to calculate the target displacement.

### C9.3.5.1 Nonlinear Static Procedure

**Benefits of Adding Energy Dissipation Devices.** The benefit of adding displacement-dependent energy dissipation devices is recognized in this standard by the increase in building stiffness afforded by such devices, and the reduction in target displacement associated with the reduction in $T_e$. The alternative NSP uses a different strategy to calculate the target displacement and explicitly recognizes the added damping provided by the energy dissipation devices.

The benefits of adding velocity-dependent energy dissipation devices are recognized by the increase in stiffness and equivalent viscous damping in the building frame. For most velocity-dependent devices, the primary benefit will result from the added viscous damping. Higher-mode damping forces in the energy dissipation devices must be evaluated regardless of the NSP used.

*9.3.5.1.1 Displacement-Dependent Devices* The stiffness characteristics of the energy dissipation devices shall be included in the mathematical model.

*9.3.5.1.2 Velocity-Dependent Devices* The target displacement and the spectral acceleration in Eq. 3-15 shall be reduced to account for the damping added by the velocity-dependent energy dissipation devices. The calculation of the damping effect shall be calculated in accordance with Eq. 9-36:

$$\beta_{eff} = \beta + \frac{\sum_j W_j}{4\pi W_k} \qquad \text{(Eq. 9-36)}$$

where $\beta$ is the damping in the structural frame and shall be set equal to 0.05 unless modified in Section 1.6.1.5, $W_j$ shall be taken as the work done by device $j$ in one complete cycle corresponding to floor displacements $\delta_j$, the summation extends over all devices $j$, and $W_k$ is the maximum strain energy in the frame, determined using Eq. 9-27.

The work done by device $j$ in one complete cycle of loading shall be calculated based on Eq. 9-37:

$$W_j = \frac{2\pi^2}{T_{ss}} C_j \delta_{rj}^2 \qquad \text{(Eq. 9-37)}$$

where $T_{ss}$ is the secant fundamental period of the rehabilitated building including the stiffness of the velocity-dependent devices (if any), calculated using Eq. 3-14 but replacing the effective stiffness ($K_e$) with the secant stiffness ($K_s$) at the target displacement as shown in Fig. 9-1; $C_j$ is the damping constant for device $j$; and $\delta_{rj}$ is the relative displacement between the ends of device $j$ along the axis of device $j$ at a roof displacement corresponding to the target displacement.

The acceptance criteria of Section 3.4.3 shall apply to buildings incorporating energy dissipation devices. Checking for displacement-controlled actions shall use deformations corresponding to the target displacement. Checking for force-controlled actions shall use component actions calculated for three limit states: maximum drift, maximum velocity, and maximum acceleration. Maximum actions shall be used for design. Higher-mode effects shall be explicitly evaluated.

*C9.3.5.1.2 Velocity-Dependent Devices* The use of Eq. 9-36 will generally capture the maximum displacement of the building.

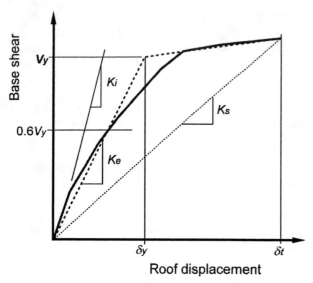

**FIGURE 9-1. Calculation of Secant Stiffness, $K_s$.**

### 9.3.5.2 Nonlinear Dynamic Procedure

If the NDP is selected based on the requirements of Section 9.2.3.3 and Section 2.4, a nonlinear time-history analysis shall be performed as required by Section 3.3.4.2, except as modified by this section. The mathematical model shall account for both the plan and vertical spatial distribution of the energy dissipation devices in the rehabilitated building. If the energy dissipation devices are dependent on excitation frequency, operating temperature (including temperature rise due to excitation), deformation (or strain), velocity, sustained loads, and bilateral loads, such dependence shall be accounted for in the analysis by assuming upper- and lower-bound properties to bound the solution.

The viscous forces in velocity-dependent energy dissipation devices shall be included in the calculation of design actions and deformations. Substitution of viscous effects in energy dissipation devices by global structural damping for nonlinear time-history analysis shall not be permitted.

### C9.3.5.2 Nonlinear Dynamic Procedure

If energy dissipation devices are dependent on loading frequency, operating temperature (including temperature rise due to excitation), deformation (or strain), velocity, sustained loads, and bilateral loads, such dependence should be accounted for in the non-linear time-history analysis. One way to account for variations in the force–deformation response of energy dissipation devices is to perform multiple analyses of the rehabilitated building using the likely bounding response characteristics of the energy dissipation

devices. The design of the rehabilitated building, including the energy dissipation devices, should be based on the maximum responses computed from the multiple analyses.

The viscous forces (if any) developed in the seismic framing system should be accounted for in the analysis and design of the seismic framing system. Evaluation of member action histories should be based on nodal displacements (operating on member stiffness matrices) and nodal velocities (operating on member damping matrices).

Key to the acceptable response of a rehabilitated building incorporating energy dissipation devices is the stable response of the energy dissipation devices. The forces and deformations in the energy dissipation devices that develop during the Design Earthquake should be demonstrated to be adequate by prototype testing in accordance with Section 9.3.8.

### 9.3.6 Detailed Systems Requirements

#### 9.3.6.1 General

The energy dissipation system and the remainder of the lateral-force-resisting system shall comply with the detailed systems requirements specified in this section.

#### 9.3.6.2 Operating Temperature

The analysis of a rehabilitated building shall account for variations in the force–displacement response of the energy dissipation devices due to variation in the ambient temperature and temperature rise due to earthquake cyclic excitation. Multiple analysis shall be performed to bound the seismic response of the building during the Design Earthquake, and develop limits for defining the acceptable response of the prototype devices and production devices.

#### C9.3.6.2 Operating Temperature

The force–displacement response of an energy dissipation device will generally be dependent on ambient temperature and temperature rise due to cyclic or earthquake excitation.

#### 9.3.6.3 Environmental Conditions

In addition to the requirements for vertical and lateral loads induced by wind and earthquake actions, the energy dissipation devices shall be designed with consideration given to other environmental conditions, including aging effects, creep, fatigue, ambient temperature, and exposure to moisture and damaging substances.

#### 9.3.6.4 Wind Forces

The fatigue life of energy dissipation devices, or components thereof, including seals in a fluid viscous device, shall be investigated and shown to be adequate for the design life of the devices. Devices subject to failure by low-cycle fatigue shall resist wind forces in the linearly elastic range.

#### 9.3.6.5 Inspection and Replacement

Access for inspection and replacement of the energy dissipation devices shall be provided.

#### 9.3.6.6 Manufacturing Quality Control

A manufacturing quality control plan for production of energy dissipation devices shall be established by the design professional. This plan shall include descriptions of the manufacturing processes, inspection procedures, and testing necessary to ensure quality control of production devices.

#### 9.3.6.7 Maintenance

The design professional shall establish a maintenance and testing schedule for energy dissipation devices to ensure reliable response of the devices over the design life of the devices. The degree of maintenance and testing shall reflect the established in-service history of the devices.

### 9.3.7 Design Review

#### 9.3.7.1 General

Design review of all rehabilitated buildings incorporating energy dissipation devices shall be performed in accordance with the requirements of Section 2.7, unless modified by the requirements of this section. Design review of the energy dissipation system and related test programs shall be performed by an independent engineering review panel that includes persons experienced in seismic analysis and the theory and application of energy dissipation methods.

The following items shall be included in the design review:

1. Preliminary design including sizing of the devices;
2. Prototype testing conducted in accordance with Section 9.3.8.2;
3. Final design of the rehabilitated building and supporting analyses; and
4. Manufacturing quality control program for the energy dissipation devices.

### 9.3.8 Required Tests of Energy Dissipation Devices

#### 9.3.8.1 General

The force–displacement relations and damping values assumed in the design of the energy dissipation system shall be confirmed by the tests conducted in accordance with this section prior to production of devices for construction. If tests conducted in accordance with this section precede the design phase of a project, the results of the testing program shall be used for the design.

The tests specified in this section shall be conducted to: (1) confirm the force–displacement properties of the energy dissipation devices assumed for design, and (2) demonstrate the robustness of individual devices to extreme seismic excitation. These tests shall not be considered as satisfying the manufacturing quality control (production) plan of Section 9.3.6.6.

The design professional shall provide explicit acceptance criteria for the effective stiffness and damping values established by the prototype tests. These criteria shall reflect the values assumed in design, account for likely variations in material properties, and provide limiting response values outside of which devices will be rejected.

The design professional shall provide explicit acceptance criteria for the effective stiffness and damping values established by the production tests of Section 9.3.6.6. The results of the prototype tests shall form the basis of the acceptance criteria for the production tests unless an alternate basis is established by the design professional in the specification. Such acceptance criteria shall recognize the influence of loading history on the response of individual devices by requiring production testing of devices prior to prototype testing.

The fabrication and quality control procedures used for all prototype and production devices shall be identical. These procedures shall be approved by the design professional prior to the fabrication of prototype devices.

#### 9.3.8.2 Prototype Tests

9.3.8.2.1 General The following prototype tests shall be performed separately on two full-sized devices of each type and size used in the design. If approved by the design professional, selection of representative sizes of each type of device shall be permitted for prototype testing, rather than each type and size, provided that the fabrication and quality control procedures are identical for each type and size of devices used in the rehabilitated building.

Test specimens shall not be used for construction unless approved in writing by the design professional.

*9.3.8.2.2 Data Recording* The force–deflection relationship for each cycle of each test shall be recorded electronically.

*9.3.8.2.3 Sequence and Cycles of Testing* For the following minimum test sequence, each energy dissipation device shall be loaded to simulate the gravity loads on the device as installed in the building and the extreme ambient temperatures anticipated:

1. Each device shall be loaded with the number of cycles expected in the design wind storm, but not less than 2,000 fully-reversed cycles of load (displacement-dependent and viscoelastic devices) or displacement (viscous devices) at amplitudes expected in the design wind storm, at a frequency equal to the inverse of the fundamental period of the rehabilitated building.
   **EXCEPTION:** Devices not subject to wind-induced forces or displacements need not be subjected to these tests.
2. Each device shall be loaded with 20 fully reversed cycles at the displacement in the energy dissipation device corresponding to the BSE-2, at a frequency equal to the inverse of the fundamental period of the rehabilitated building.
   **EXCEPTION:** Testing methods for energy dissipation devices other than those noted above shall be permitted, provided that: (1) equivalency between the proposed method and cyclic testing can be demonstrated; (2) the proposed method captures the dependence of the energy dissipation device response to ambient temperature, frequency of loading, and temperature rise during testing; and (3) the proposed method is approved by the design professional.

*C9.3.8.2.3 Sequence and Cycles of Testing* Energy dissipation devices should not form part of the gravity-load-resisting system, but may be required to support some gravity load.

*9.3.8.2.4 Devices Dependent on Velocity and/or Frequency of Excitation* If the force-deformation properties of the energy dissipation devices at any displacement less than or equal to the total design displacement change by more than 15% for changes in testing frequency from $0.5 f_1$ to $2.0 f_1$, the preceding tests shall be performed at frequencies equal to $0.5 f_1$, $f_1$, and $2.0 f_1$.

**EXCEPTION:** If reduced-scale prototypes are used to quantify the rate-dependent properties of energy dissipation devices, the reduced-scale prototypes shall be of the same type and materials—and manufactured with the same processes and quality control procedures—as full-scale prototypes, and tested at a similitude-scaled frequency that represents the full-scale loading rates.

*9.3.8.2.5 Devices Dependent on Bilateral Displacement* If the energy dissipation devices are subjected to bilateral deformation, the preceding tests shall be made at both zero bilateral displacement, and peak lateral displacement in the BSE-2.

**EXCEPTION:** If reduced-scale prototypes are used to quantify the bilateral displacement properties of the energy dissipation devices, the reduced-scale prototypes shall be of the same type and materials, and manufactured with the same processes and quality control procedures, as full-scale prototypes, and tested at similitude-scaled displacements that represent the full-scale displacements.

*9.3.8.2.6 Testing Similar Devices* Energy dissipation devices that are (1) of similar size, identical materials, internal construction, and static and dynamic internal pressures (if any), and (2) fabricated with identical internal processes and manufacturing quality control procedures, and that have been previously tested by an independent laboratory in the manner described above need not be tested, provided that:

1. All pertinent testing data are made available to, and are approved by, the design professional;
2. The manufacturer can substantiate the similarity of the previously tested devices to the satisfaction of the design professional; and
3. The submission of data from a previous testing program is approved in writing by the design professional.

### 9.3.8.3 Determination of Force-Displacement Characteristics

The force-displacement characteristics of an energy dissipation device shall be based on the cyclic load and displacement tests of prototype devices specified in Section 9.3.8.2.

As required, the effective stiffness ($k_{eff}$) of an energy dissipation device with stiffness shall be calculated for each cycle of deformation in accordance with Eq. 9-38:

$$k_{eff} = \frac{|F^-| + |F^+|}{|\Delta^-| + |\Delta^+|} \qquad \text{(Eq. 9-38)}$$

where forces $F+$ and $\tilde{F}$ shall be calculated at displacements $\Delta+$ and $\tilde{\Delta}$, respectively. The effective stiffness of an energy dissipation device shall be established at the test displacements given in Section 9.3.8.2.3.

The equivalent viscous damping of an energy dissipation device ($\beta_{eff}$) exhibiting stiffness shall be calculated for each cycle of deformation based on Eq. 9-39:

$$\beta_{eff} = \frac{1}{2\pi} \frac{W_D}{k_{eff}\Delta_{ave}^2} \qquad \text{(Eq. 9-39)}$$

where $k_{eff}$ shall be calculated in accordance with Eq. 9-38, and $W_D$ shall be taken as the area enclosed by one complete cycle of the force–displacement response for a single energy dissipation device at a prototype test displacement ($\Delta_{ave}$) equal to the average of the absolute values of displacements $\Delta^+$ and $\tilde{\Delta}$.

### 9.3.8.4 System Adequacy

The performance of a prototype device shall be considered adequate if all of the following conditions are satisfied:

1. The force–displacement curves for the tests in Sction 9.3.8.2.3 have nonnegative incremental force-carrying capacities.
   **EXCEPTION:** Energy dissipation devices that exhibit velocity-dependent behavior need not comply with this requirement.
2. Within each test of Section 9.3.8.2.3, the effective stiffness ($k_{eff}$) of a prototype energy dissipation device for any one cycle does not differ by more than plus or minus 15% from the average effective stiffness as calculated from all cycles in that test.
   **EXCEPTIONS:** (1) The 15% limit may be increased by the design professional in the specification, provided that the increased limit has been demonstrated by analysis to not have a deleterious effect on the response of the rehabilitated building; and (2) fluid viscous energy dissipation devices, and other devices that do not have effective stiffness, need not comply with this requirement.
3. Within each test of Section 9.3.8.2.3, the maximum force and minimum force at zero displacement for a prototype device for any one cycle does not differ by more than plus or minus 15% from the average maximum and minimum forces as calculated from all cycles in that test.
   **EXCEPTION:** The 15% limit may be increased by the design professional in the specification, provided that the increased limit has been demonstrated by analysis to not have a deleterious effect on the response of the rehabilitated building.

4. Within each test of Section 9.3.8.2.3, the area of the hysteresis loop ($W_D$) of a prototype energy dissipation device for any one cycle does not differ by more than plus or minus 15% from the average area of the hysteresis curve as calculated from all cycles in that test.
   **EXCEPTION:** The 15% limit may be increased by the design professional in the specification, provided that the increased limit has been demonstrated by analysis to not have a deleterious effect on the response of the rehabilitated building.
5. For displacement-dependent devices, the average effective stiffness, average maximum and minimum force at zero displacement, and average area of the hysteresis loop ($W_D$), calculated for each test in the sequence described in Section 9.3.8.2.3, shall fall within the limits set by the design professional in the specification. The area of the hysteresis loop at the end of cyclic testing shall not differ by more than plus or minus 15% from the average area of the 20 test cycles.
6. For velocity-dependent devices, the average maximum and minimum force at zero displacement, effective stiffness (for viscoelastic devices only), and average area of the hysteresis loop ($W_D$), calculated for each test in the sequence described in Section 9.3.8.2.3, shall fall within the limits set by the design professional in the specification.

## 9.4 OTHER RESPONSE CONTROL SYSTEMS

The analysis and design of other response control systems shall be reviewed by an independent engineering review panel in accordance with the requirements of Section 9.3.7. This revicw panel shall be selected by the owner prior to the development of the preliminary design.

## C9.4 OTHER RESPONSE CONTROL SYSTEMS

Response control strategies other than base isolation (Section 9.2) and passive energy dissipation (Section 9.3) systems have been proposed. Dynamic vibration absorption and active control systems are two such response control strategies. Although both dynamic vibration absorption and active control systems have been implemented to control the wind-induced vibration of buildings, the technology is not sufficiently mature and the necessary hardware is not sufficiently robust to warrant the preparation of general guidelines for the implementation of other response control systems.

## 10.0 SIMPLIFIED REHABILITATION

### 10.1 SCOPE

This chapter sets forth requirements for the rehabilitation of buildings using the Simplified Rehabilitation Method. Section 10.2 outlines the procedure of the Simplified Rehabilitation Method. Section 10.3 specifies actions for correction of deficiencies using the Simplified Rehabilitation Method.

### C10.1 SCOPE

The Simplified Rehabilitation Method is intended primarily for use on a select group of simple buildings.

The Simplified Rehabilitation Method only applies to buildings that fit into one of the Model Building Types and conform to the limitations of Table 10-1, which sets the standard for simple, regularly configured buildings defined in Table 10-2. Building regularity is an important consideration in the application of the method. Regularity is determined by checklist statements addressing building configuration issues. The Simplified Rehabilitation Method may be used if an evaluation shows no deficiencies with regard to regularity. Buildings that have configuration irregularities (as determined by an ASCE 31 Tier 1 or Tier 2 Evaluation) (ASCE 2002) may use this Simplified Rehabilitation Method to achieve the Life Safety Building Performance Level only if the resulting rehabilitation work eliminates all significant vertical and horizontal irregularities and results in a building with a complete seismic lateral-force-resisting load path.

The technique described in this chapter is one of the two rehabilitation methods defined in Section 2.3. It is to be used only by a design professional and only in a manner consistent with this standard. Consideration must be given to all aspects of the rehabilitation process, including the development of appropriate as-built information, proper design of rehabilitation techniques, and specification of appropriate levels of quality assurance.

"Simplified Rehabilitation" reflects a level of analysis and design that (1) is appropriate for small, regular buildings and buildings that do not require advanced analytical procedures; and (2) achieves the Life Safety Performance Level for the BSE-1 Earthquake Hazard Level as defined in Chapter 1, but does not necessarily achieve the Basic Safety Objective (BSO).

FEMA 178, the *NEHRP Handbook for the Seismic Evaluation of Existing Buildings*, a nationally

applicable method for seismic evaluation of buildings (FEMA 1992), was the basis for the Simplified Rehabilitation Method in FEMA 273 (FEMA 1997) and in this standard. FEMA 178 is based on the historic behavior of buildings in past earthquakes and the success of current code provisions in achieving the Life Safety Building Performance Level. It is organized around a set of common construction styles called model buildings.

Since the preliminary version of FEMA 178 was completed in the late 1980s, new information has become available and has been incorporated into ASCE 31, which is an updated version of FEMA 178. This information includes additional Model Building Types, eight new evaluation statements for potential deficiencies, a reorganization of the procedure to clearly state the intended three-tier approach, and new analysis techniques that parallel those of FEMA 273. ASCE 31 is the basis of the Simplified Rehabilitation Method in this standard.

The Simplified Rehabilitation Method may yield a more conservative result than the Systematic Method because of a variety of simplifying assumptions.

### 10.2 PROCEDURE

Simplified Rehabilitation Objectives, subject to the limitations of Section 2.3.1, shall be permitted to satisfy Limited Rehabilitation Objectives, as described in Sections 1.4.3. Reduced Rehabilitation shall be performed in accordance with Section 10.2.1. Partial Rehabilitation shall be performed in accordance with Section 10.2.2.

#### 10.2.1 Procedure for Reduced Rehabilitation

Where the Simplified Rehabilitation Method is used to achieve Reduced Rehabilitation, each of the following steps shall be completed:

1. The building shall be classified as one of the Model Building Types listed in Table 10-1 and defined in Table 10-2;
2. A Tier 1 and a Tier 2 Seismic Evaluation of the building in its existing state shall be performed for the Life Safety Building Performance Level in accordance with ASCE 31 (ASCE 2002), except that the spectral response acceleration parameters shall be defined in accordance with Section 1.6.1.2 of this standard. For any other differences between this standard and the ASCE 31 procedures, the ASCE 31 procedures shall govern;
3. The deficiencies identified by the ASCE 31 Evaluation conducted in Step 2 shall be ranked

from highest to lowest priority based on the extent of nonconformance and the significance of the deficiency;

4. Rehabilitation measures shall be developed in accordance with Section 10.3. The proposed rehabilitation scheme shall be designed such that all deficiencies identified by the ASCE 31 Evaluation of Step 2 are eliminated;

5. A complete Tier 1 and Tier 2 Evaluation of the building in its proposed rehabilitated state shall be performed in accordance with ASCE 31, except that the spectral response acceleration parameters shall be defined in accordance with Section 1.6.1.2 of this standard. For any other differences between this standard and the ASCE 31 procedures, the ASCE 31 procedures shall govern;

### Table 10-1. Limitations on Use of the Simplified Rehabilitation Method

| Model Building Type[2] | Maximum Building Height in Stories by Seismic Level1 for Use of the Simplified Rehabilitation Method | | |
|---|---|---|---|
| | Low | Moderate | High |
| **Wood Frame** | | | |
| Light (W1) | 3 | 3 | 2 |
| Multistory Multi-Unit Residential (W1A) | 3 | 3 | 2 |
| Commercial and Industrial (W2) | 3 | 3 | 2 |
| **Steel Moment Frame** | | | |
| Stiff Diaphragm (S1) | 6 | 4 | 3 |
| Flexible Diaphragm (S1A) | 4 | 4 | 3 |
| **Steel Braced Frame** | | | |
| Stiff Diaphragm (S2) | 6 | 4 | 3 |
| Flexible Diaphragm (S2A) | 3 | 3 | 3 |
| **Steel Light Frame (S3)** | 2 | 2 | 2 |
| **Steel Frame with Concrete Shear Walls (S4)** | 6 | 4 | 3 |
| **Steel Frame with Infill Masonry Shear Walls** | | | |
| Stiff Diaphragm (S5) | 3 | 3 | n.p. |
| Flexible Diaphragm (S5A) | 3 | 3 | n.p. |
| **Concrete Moment Frame (C1)** | 3 | n.p. | n.p. |
| **Concrete Shear Walls** | | | |
| Stiff Diaphragm (C2) | 6 | 4 | 3 |
| Flexible Diaphragm (C2A) | 3 | 3 | 3 |
| **Concrete Frame with Infill Masonry Shear Walls** | | | |
| Stiff Diaphragm (C3) | 3 | n.p. | n.p. |
| Flexible Diaphragm (C3A) | 3 | n.p. | n.p. |
| **Precast/Tilt-Up Concrete Shear Walls** | | | |
| Flexible Diaphragm (PC1) | 3 | 2 | 2 |
| Stiff Diaphragm (PC1A) | 3 | 2 | 2 |
| **Precast Concrete Frame** | | | |
| With Shear Walls (PC2) | 3 | 2 | n.p. |
| Without Shear Walls (PC2A) | n.p. | n.p. | n.p. |
| **Reinforced Masonry Bearing Walls** | | | |
| Flexible Diaphragm (RM1) | 3 | 3 | 3 |
| Stiff Diaphragm (RM2) | 6 | 4 | 3 |
| **Unreinforced Masonry Bearing Walls** | | | |
| Flexible Diaphragm (URM) | 3 | 3 | 2 |
| Stiff Diaphragm (URMA) | 3 | 3 | 2 |

n.p., Use of Simplified Rehabilitation Method shall not be permitted.

[1]Seismic levels shall be as defined in Section 1.6.3.

[2]Buildings with different types of flexible diaphragms shall be permitted to be considered as having flexible diaphragms. Multistory buildings having stiff diaphragms at all levels except the roof shall be permitted to be considered as having stiff diaphragms. Buildings having both flexible and stiff diaphragms, or having diaphragm systems that are neither flexible nor stiff, in accordance with this chapter, shall be rehabilitated using the Systematic Method.

## Table 10-2. Description of Model Building Types

### Building Type 1—Wood Light Frame

W1:     These buildings are single or multiple family dwellings of one or more stories in height. Building loads are light and the framing spans are short. Floor and roof framing consists of wood joists or rafters on wood studs spaced no more than 24 in. apart. The first-floor framing is supported directly on the foundation, or is raised up on cripple studs and post-and-beam supports. The foundation consists of spread footings constructed on concrete, concrete masonry block, or brick masonry in older construction. Chimneys, where present, consist of solid brick masonry, masonry veneer, or wood frame with internal metal flues. Lateral forces are resisted by wood frame diaphragms and shear walls. Floor and roof diaphragms consist of straight or diagonal lumber sheathing, tongue-and-groove planks, oriented strand board, or wood structural panel. Shear walls consist of straight or lumber sheathing, plank siding, oriented strand board, wood structural panel, stucco, gypsum board, particle board, or fiber board. Interior partitions are sheathed with plaster or gypsum board.

W1A:    These buildings are multistory, similar in construction to W1 buildings, but have plan areas on each floor of greater than 3,000 sf. Older construction often has open front garages at the lowest story.

### Building Type 2—Wood Frames, Commercial and Industrial

W2:     These buildings are commercial or industrial buildings with a floor area of 5,000 sf or more. There are few, if any, interior walls. The floor and roof framing consists of wood or steel trusses, glulam or steel beams, and wood posts or steel columns. Lateral forces are resisted by wood diaphragms and exterior stud walls sheathed with wood structural panel, oriented strand board, stucco, plaster, straight or diagonal wood sheathing, or braced with rod bracing. Wall openings for storefronts and garages, where present, are framed by post-and-beam framing.

### Building Type 3—Steel Moment Frames

S1:     These buildings consist of a frame assembly of steel beams and steel columns. Floor and roof framing consists of cast-in-place concrete slabs or metal deck with concrete fill supported on steel beams, open web joists, or steel trusses. Lateral forces are resisted by steel moment frames that develop their stiffness through rigid or semi-rigid beam–column connections. Where all connections are moment-resisting connections, the entire frame participates in lateral-force resistance. Where only selected connections are moment-resisting connections, resistance is provided along discrete frame lines. Columns are oriented so that each principal direction of the building has columns resisting forces in strong axis bending. Diaphragms consist of concrete or metal deck with concrete fill and are stiff relative to the frames. Where the exterior of the structure is concealed, walls consist of metal panel curtain walls, glazing, brick masonry, or precast concrete panels. Where the interior of the structure is finished, frames are concealed by ceilings, partition walls, and architectural column furring. Foundations consist of concrete-spread footings or deep pile foundations.

S1A:    These buildings are similar to S1 buildings except that diaphragms consist of wood framing, untopped metal deck, or metal deck with lightweight insulating concrete, poured gypsum, or similar nonstructural topping and are flexible relative to the frames.

### Building Type 4—Steel Braced Frames

S2:     These buildings have a frame of steel columns, beams, and braces. Braced frames develop resistance to lateral forces by the bracing action of the diagonal members. The braces induce forces in the associated beams and columns such that all components work together in a manner similar to a truss, with all component stresses being primarily axial. Where the braces do not completely triangulate the panel, some of the members are subjected to shear and flexural stresses; eccentrically braced frames are one such case. Diaphragms transfer lateral loads to braced frames. The diaphragms consist of concrete or metal deck with concrete fill and are stiff relative to the frames.

S2A:    These buildings are similar to S2 buildings except that diaphragms consist of wood framing, untopped metal deck, or metal deck with lightweight insulating concrete, poured gypsum, or similar nonstructural topping and are flexible relative to the frames.

### Building Type 5—Steel Light Frames

S3:     These buildings are pre-engineered and prefabricated with transverse rigid steel frames. They are one story in height. The roof and walls consist of lightweight metal, fiberglass or cementitious panels. The frames are designed for maximum efficiency and the beams and columns consist of tapered, built-up sections with thin plates. The frames are built in segments and assembled in the field with bolted or welded joints. Lateral forces in the transverse direction are resisted by the rigid frames. Lateral forces in the longitudinal direction are resisted by wall panel shear components or rod bracing. Diaphragm forces are resisted by untopped metal deck, roof panel shear components, or a system of tension-only rod bracing.

**Building Type 6—Steel Frames with Concrete Shear Walls**

S4: These buildings consist of a frame assembly of steel beams and steel columns. The floors and roof consist of cast-in-place concrete slabs or metal deck with or without concrete fill. Framing consists of steel beams, open web joists, or steel trusses. Lateral forces are resisted by cast-in-place concrete shear walls. These walls are bearing walls where the steel frame does not provide a complete vertical support system. In older construction, the steel frame is designed for vertical loads only. In modern dual systems, the steel moment frames are designed to work together with the concrete shear walls in proportion to their relative rigidity. In the case of a dual system, the walls shall be evaluated under this building type and the frames shall be evaluated under S1 or S1A, Steel Moment Frames. The steel frame may provide a secondary lateral-force-resisting system depending on the stiffness of the frame and the moment capacity of the beam–column connections.

**Building Type 7—Steel Frame with Infill Masonry Shear Walls**

S5: This is an older type of building construction that consists of a frame assembly of steel beams and steel columns. The floors and roof consist of cast-in-place concrete slabs or metal deck with concrete fill. Framing consists of steel beams, open web joists, or steel trusses. Walls consist of infill panels constructed of solid clay brick, concrete block, or hollow clay tile masonry. Infill walls may completely encase the frame members, and present a smooth masonry exterior with no indication of the frame. The seismic performance of this type of construction depends on the interaction between the frame and infill panels. The combined behavior is more like a shear wall structure than a frame structure. Solidly infilled masonry panels form diagonal compression struts between the intersections of the frame members. If the walls are offset from the frame and do not fully engage the frame members, the diagonal compression struts will not develop. The strength of the infill panel is limited by the shear capacity of the masonry bed joint or the compression capacity of the strut. The post-cracking strength is determined by an analysis of a moment frame that is partially restrained by the cracked infill.

S5A: These buildings are similar to S5 buildings except that diaphragms consist of wood sheathing or untopped metal deck, or have large aspect ratios and are flexible relative to the walls.

**Building Type 8—Concrete Moment Frames**

C1: These buildings consist of a frame assembly of cast-in-place concrete beams and columns. Floor and roof framing consists of cast-in-place concrete slabs, concrete beams, one-way joists, two-way waffle joists, or flat slabs. Lateral forces are resisted by concrete moment frames that develop their stiffness through monolithic beam–column connections. In older construction or in levels of low seismicity, the moment frames may consist of the column strips of two-way flat slab systems. Modern frames in levels of high seismicity have joint reinforcing, closely spaced ties, and special detailing to provide ductile performance. This detailing is not present in older construction. Foundations consist of concrete-spread footings, mat foundations, or deep pile foundations.

**Building Type 9—Concrete Shear Wall Buildings**

C2: These buildings have floor and roof framing that consists of cast-in-place concrete slabs, concrete beams, one-way joists, two-way waffle joists, or flat slabs. Floors are supported on concrete columns or bearing walls. Lateral forces are resisted by cast-in-place concrete shear walls. In older construction, shear walls are lightly reinforced, but often extend throughout the building. In more recent construction, shear walls occur in isolated locations and are more heavily reinforced with concrete slabs and are stiff relative to the walls. Foundations consist of concrete-spread footings, mat foundations, or deep pile foundations.

C2A: These buildings are similar to C2 buildings except that diaphragms consist of wood sheathing, or have large aspect ratios and are flexible relative to the walls.

**Building Type 10—Concrete Frame with Infill Masonry Shear Walls**

C3: This is an older type of building construction that consists of a frame assembly of cast-in-place concrete beams and columns. The floors and roof consist of cast-in-place concrete slabs and are stiff relative to the walls. Walls consist of infill panels constructed of solid clay brick, concrete block, or hollow clay tile masonry. The seismic performance of this type of construction depends on the interaction between the frame and the infill panels. The combined behavior is more like a shear wall structure than a frame structure. Solidly infilled masonry panels form diagonal compression struts between the intersections of the frame members. If the walls are offset from the frame and do not fully engage the frame members, the diagonal compression struts will not develop. The strength of the infill panel is limited by the shear capacity of the masonry bed joint or the compression capacity of the strut. The post-cracking strength is determined by an analysis of a moment frame that is partially restrained by the cracked infill. The shear strength of the concrete columns, after racking of the infill, may limit the semi-ductile behavior of the system. .

C3A: These buildings are similar to C3 buildings except that diaphragms consists of wood sheathing or untopped metal deck, or have large aspect ratios and are flexible relative to the walls.

*continued*

**TABLE 10-2. (Continued)**

**Building Type 11—Precast/Tilt-Up Concrete Shear Wall Buildings**

PC1:     These buildings have precast concrete perimeter wall panels that are cast on site and tilted into place. Floor and roof framing consists of wood joists, glulam beams, steel beams, or open web joists. Framing is supported on interior steel columns and perimeter concrete bearing walls. The floors and roof consist of wood sheathing or untopped metal deck. Lateral forces are resisted by the precast concrete perimeter wall panels. Wall panels may be solid, or have large window and door openings which cause the panels to behave more as frames than as shear walls. In older construction, wood framing is attached to the walls with wood ledgers. Foundations consist of concrete-spread footings or deep pile foundations.

PC1A:    These buildings are similar to PC1 buildings except that diaphragms consist of precast components, cast-in-place concrete, or metal deck with concrete fill, and are stiff relative to the walls.

**Building Type 12—Precast Concrete Frames**

PC2:     These buildings consist of a frame assembly of precast concrete girders and columns with the presence of shear walls. Floor and roof framing consists of precast concrete planks, tees, or double-tees supported on precast concrete girders and columns. Lateral forces are resisted by precast or cast-in-place concrete shear walls. Diaphragms consist of precast components interconnected with welded inserts, cast-in-place closure strips, or reinforced concrete topping slabs.

PC2A:    These buildings are similar to PC2 buildings except that concrete shear walls are not present. Lateral forces are resisted by precast concrete moment frames that develop their stiffness through beam–column joints rigidly connected by welded inserts or cast-in-place concrete closures. Diaphragms consist of precast components interconnected with welded inserts, cast-in-place closure strips, or reinforced concrete topping slabs.

**Building Type 13—Reinforced Masonry Bearing Wall Buildings with Flexible Diaphragms**

RM1:     These buildings have bearing walls that consist of reinforced brick or concrete block masonry. The floor and roof framing consists of steel or wood beams and girders or open web joists, and are supported by steel, wood, or masonry columns. Lateral forces are resisted by the reinforced brick or concrete block masonry shear walls. Diaphragms consist of straight or diagonal wood sheathing, wood structural panel, or untopped metal deck, and are flexible relative to the walls. Foundations consist of brick or concrete-spread footings or deep foundations.

**Building Type 14—Reinforced Masonry Bearing Wall Buildings with Stiff Diaphragms**

RM2:     These building are similar to RM1 buildings except that the diaphragms consist of metal deck with concrete fill, precast concrete planks, tees, or double-tees, with or without a cast-in-place concrete topping slab, and are stiff relative to the walls. The floor and roof framing is supported on interior steel or concrete frames or interior reinforced masonry walls.

**Building Type 15—Unreinforced Masonry Bearing Wall Buildings**

URM:     These buildings have perimeter bearing walls that consist of unreinforced clay brick, stone, or concrete masonry. Interior bearing walls, where present, also consist of unreinforced clay brick, stone, or concrete masonry. In older construction, floor and roof framing consists of straight or diagonal lumber sheathing supported by wood joists, which, in turn, are supported on posts and timbers. In more recent construction, floors consist of structural panel or wood structural panel sheathing rather than lumber sheathing. The diaphragms are flexible relative to the walls. Where they exist, ties between the walls and diaphragms consist of anchors or bent steel plates embedded in the mortar joints and attached to framing. Foundations consist of brick or concrete-spread footings, or deep foundations.

URMA:    These buildings are similar to URM buildings except that the diaphragms are stiff relative to the unreinforced masonry walls and interior framing. In older construction or large, multistory buildings, diaphragms consist of cast-in-place concrete. In levels of low seismicity, more recent construction consists of metal deck and concrete fill supported on steel framing.

---

6. Rehabilitation measures for architectural, mechanical, and electrical components shall be developed in accordance with Chapter 11 for the Life Safety Nonstructural Performance Level at the BSE-1 Earthquake Hazard Level; and

7. Construction documents, including drawings, specifications, and a quality assurance plan, shall be developed as defined in Chapter 2.

**10.2.2 Procedure for Partial Rehabilitation**

Where the Simplified Rehabilitation Method is used to achieve Partial Rehabilitation, Steps 1, 2, 3, 5, and 7 of Section 10.2.1 shall be completed. Steps 4 and 6 of Section 10.2.1 shall be completed only as they apply to the deficiencies being addressed as part of the Partial Rehabilitation.

**C10.2 PROCEDURE**

The basis of the Simplified Rehabilitation Method is the ASCE 31 (ASCE 2002) procedure. There are intentional differences between the provisions of this standard and ASCE 31 with regard to site class amplification factors, seismicity, and design earthquake, among other issues. The Earthquake Hazard Level defined in ASCE 31 is taken as two-thirds of the Maximum Considered Earthquake (MCE) for simplicity and conservatism in the evaluation procedures. However, rehabilitation to the BSE-1 Earthquake Hazard Level in accordance with this standard is the traditional level of safety taken as the lesser of the 10%/50-year or two-thirds of the MCE.

For simple buildings with specific deficiencies, it is possible and advisable to prioritize the rehabilitation measures. This is often done where the construction has limited funding or must take place while the building is occupied. In both cases, it is preferable to correct the worst deficiency first.

Potential deficiencies are ranked in Tables C10-1 through C10-19; items in these tables are ordered roughly from highest priority at the top to lowest at the bottom, although this can vary widely in individual cases. Tables C10-1 through C10-19 are presented at the end of this Commentary section.

ASCE 31 lists specific deficiencies both by Model Building Type and by association with each building system. Tables C10-1 through C10-19 of this standard further group deficiencies by general characteristics. For example, the deficiency listing "Diaphragm Stiffness/Strength," includes deficiencies related to the type of sheathing used, diaphragm span, and lack of

### Table C10-1. W1: Wood Light Frame

Typical Deficiencies

Load Path
    Redundancy
    Vertical Irregularities
Shear Walls in Wood Frame Buildings
    Shear Stress
    Openings
    Wall Detailing
    Cripple Walls
    Narrow Wood Shear Walls
    Stucco Shear Walls
    Gypsum Wallboard or Plaster Shear Walls
Diaphragm Openings
    Diaphragm Stiffness/Strength
    Spans
    Diaphragm Continuity
Anchorage to Foundations
    Condition of Foundations
    Geologic Site Hazards
Condition of Wood

### Table C10-2. W1A: Multi-Story, Multi-Unit, Wood Frame Construction

Typical Deficiencies

Load Path
    Redundancy
    Vertical Irregularities
Shear Walls in Wood Frame Buildings
    Shear Stress
    Openings
    Wall Detailing
    Cripple Walls
    Narrow Wood Shear Walls
    Stucco Shear Walls
    Gypsum Wallboard or Plaster Shear Walls
Diaphragm Openings
    Diaphragm Stiffness/Strength
    Spans
    Diaphragm Continuity
Anchorage to Foundations
    Condition of Foundations
    Geologic Site Hazards
Condition of Wood

blocking. Table C10-20 provides a complete cross-reference for sections in this standard and in ASCE 31 and is presented at the end of this Commentary section.

Within the table for each Model Building Type, typical deficiencies are ranked from most critical at the top of each deficiency group to least critical at the bottom. For example, in Table C10-12, in a precast/tilt-up concrete shear wall with flexible diaphragm (PC1) building, the lack of positive gravity frame connections (e.g., of girders to posts by sheet metal hardware or bolts) has a greater potential to lower the building's performance (a partial collapse of the roof structure supported by the beam) than a deficiency in lateral forces on foundations (e.g., poor reinforcing in the footings).

The ranking was based on the following characteristics of each deficiency group:

1. Most critical
    1.1. Building systems: those with a discontinuous load path and little redundancy.
    1.2. Building components: those with low strength and low ductility.
2. Intermediate
    2.1. Building systems: those with a discontinuous load path but substantial redundancy.
    2.2. Building components: those with substantial strength but low ductility.

3. Least critical
  3.1. Building systems: those with a substantial load path but little redundancy.
  3.2. Building components: those with low strength but substantial ductility.

The intent of Tables C10-1 through C10-19 is to guide the design professional in accomplishing a Partial Rehabilitation Objective. For example, if the foundation is strengthened in a PC1 building but a poor girder/wall connection is left alone, relatively little has been done to improve the expected performance of the building. Considerable professional judgment must be used where evaluating a structure's unique behavior and determining which deficiencies should be strengthened and in what order.

As a rule, the resulting rehabilitated building must be one of the Model Building Types. For example, adding concrete shear walls to concrete shear wall buildings or adding a complete system of concrete shear walls to a concrete frame building meets this requirement. Steel bracing may be used to strengthen wood or unreinforced masonry (URM) construction. For large buildings, it is advisable to explore several rehabilitation strategies and compare alternative ways

### Table C10-3. W2: Wood—Commercial, and Industrial

Typical Deficiencies

Load Path
  Redundancy
  Vertical Irregularities
Shear Walls in Wood Frame Buildings
  Shear Stress
  Openings
  Wall Detailing
  Cripple Walls
  Narrow Wood Shear Walls
  Stucco Shear Walls
  Gypsum Wallboard or Plaster Shear Walls
Diaphragm Openings
  Diaphragm Stiffness/Strength
  Sheathing
  Unblocked Diaphragms
  Spans
  Span-to-Depth Ratio
  Diaphragm Continuity
  Chord Continuity
Anchorage to Foundations
  Condition of Foundations
  Geologic Site Hazards
Condition of Wood

### Table C10-4. S1 and S1A: Steel Moment Frames with Stiff or Flexible Diaphragms

Typical Deficiencies

Load Path
  Redundancy
  Vertical Irregularities
  Plan Irregularities
  Adjacent Buildings
  Uplift at Pile Caps
Steel Moment Frames
  Drift Check
  Frame Concerns
  Strong Column–Weak Beam
  Connections
Re-entrant Corners
  Diaphragm Openings
  Diaphragm Stiffness/Strength
Diaphragm/Frame Shear Transfer
Anchorage to Foundations
  Condition of Foundations
  Overturning
  Lateral Loads
  Geologic Site Hazards
Condition of Steel

of eliminating deficiencies. A Tier 1 and Tier 2 Evaluation of the proposed rehabilitated state is performed to verify the proposed rehabilitation design.

For a Limited Rehabilitation Objective, the deficiencies identified by the ASCE 31 Evaluation of Step 2 should be mitigated in order of priority based on the ranking performed in Step 3.

A complete evaluation of the building should confirm that the strengthening of any one component or system has not merely shifted the deficiency to another.

Specific application of the Systematic Rehabilitation Method is needed to achieve the BSO. The total strength of the building should be sufficient, and the ability of the building to experience the predicted maximum displacement without partial or complete collapse must be established.

If only a Partial Rehabilitation or Limited Rehabilitation Objective is intended, deficiencies should be corrected in priority order and in a way that will facilitate fulfillment of the requirements of a higher objective at a later date. Care must be taken to ensure that a Partial Rehabilitation effort does not make the building's overall performance worse by unintentionally channeling failure to a more critical component.

**Table C10-5. S2 and S2A: Steel Braced Frames with Stiff or Flexible Diaphragms**

Typical Deficiencies

Load Path
    Redundancy
    Vertical Irregularities
    Plan Irregularities
    Uplift at Pile Caps
Stress Level
    Stiffness of Diagonals
    Chevron or K-Bracing
    Braced Frame Connections
Re-entrant Corners
    Diaphragm Openings
    Diaphragm Stiffness/Strength
Diaphragm/Frame Shear Transfer
Anchorage to Foundations
    Condition of Foundations
    Overturning
    Lateral Loads
    Geologic Site Hazards
Condition of Steel

**Table C10-6. S3: Steel Light Frames**

Typical Deficiencies

Load Path
    Redundancy
    Vertical Irregularities
    Plan Irregularities
Steel Moment Frames
    Frame Concerns
Masonry Shear Walls
    Infill Walls
Steel Braced Frames
    Stress Level
    Braced Frame Connections
Re-entrant Corners
    Diaphragm Openings
Diaphragm/Frame Shear Transfer
    Wall Panels and Cladding
    Light Gage Metal, Plastic, or Cementitious Roof Panels
Anchorage to Foundations
    Condition of Foundations
    Geologic Site Hazards
Condition of Steel

## 10.3 CORRECTION OF DEFICIENCIES

For Simplified Rehabilitation, deficiencies identified by an ASCE 31 Evaluation shall be mitigated by implementing rehabilitation measures in accordance with this standard. The resulting building, including strengthening measures, shall comply with the requirements of ASCE 31, except that the spectral response acceleration parameters shall be defined in accordance with Section 1.6.1.2 of this standard. The rehabilitated building shall conform to one of the Model Building Types contained in Table 10-1, except that steel bracing in wood or unreinforced masonry buildings shall be permitted.

The Simplified Rehabilitation Method shall only be used to achieve Limited Rehabilitation Objectives. To achieve the Life Safety Building Performance Level (3-C) at the BSE-1 Earthquake Hazard Level, all deficiencies identified by an ASCE 31 Evaluation shall be corrected to meet the ASCE 31 criteria except that the spectral response acceleration parameters shall be defined in accordance with Section 1.6.1.2 of this standard. To achieve a Partial Rehabilitation Objective, only selected deficiencies need to be corrected.

To achieve the Basic Safety Objective, the Simplified Rehabilitation Method is not permitted, and deficiencies shall be corrected in accordance with the Systematic Rehabilitation Method of Section 2.3.

## C10.3 CORRECTION OF DEFICIENCIES

Implementing a rehabilitation scheme that mitigates all of a building's ASCE 31 (ASCE 2002) deficiencies using the Simplified Rehabilitation Method does not, in and of itself, achieve the Basic Safety Objective or any Enhanced Rehabilitation Objective as defined in Chapter 2 since the rehabilitated building may not meet the Collapse Prevention Structural Performance Level for the BSE-2 Earthquake Hazard Level. If the goal is to attain the Basic Safety Objective as described in Chapter 2 or other Enhanced Rehabilitation Objectives, this can be accomplished using the Systematic Rehabilitation Method defined in Chapter 2.

Suggested rehabilitation measures are listed by deficiency in the following sections.

### C10.3.1 Building Systems

#### C10.3.1.1 Load Path
Load path discontinuities can be mitigated by adding components to complete the load path. This may require adding new, well-founded shear walls or frames to fill gaps in existing shear walls or frames that are not carried continuously to the foundation. Alternatively, it may require the addition of components throughout the building to pick up loads from

**Table C10-7. S4: Steel Frames with Concrete Shear Walls**

Typical Deficiencies

Load Path
  Redundancy
  Vertical Irregularities
  Plan Irregularities
  Uplift at Pile Caps
Cast-in-Place Concrete Shear Walls
  Shear Stress
  Overturning
  Coupling Beams
  Boundary Component Detailing
  Wall Reinforcement
Re-entrant Corners
  Diaphragm Openings
  Diaphragm Stiffness/Strength
Diaphragm/Wall Shear Transfer
Anchorage to Foundations
  Condition of Foundations
  Overturning
  Lateral Loads
  Geologic Site Hazards
Condition of Steel
  Condition of Concrete

**Table C10-8. S5, S5A: Steel Frames with Infill Masonry Shear Walls and Stiff or Flexible Diaphragms**

Typical Deficiencies

Load Path
  Redundancy
  Vertical Irregularities
  Plan Irregularities
  Uplift at Pile Caps
Frames Not Part of the Lateral-Force-Resisting System
  Complete Frames
Masonry Shear Walls
  Reinforcing in Masonry Walls
  Shear Stress
  Reinforcing at Openings
  Unreinforced Masonry Shear Walls
  Proportions, Solid Walls
  Infill Walls
Re-entrant Corners
  Diaphragm Openings
  Diaphragm Stiffness/Strength
  Span-to-Depth Ratio
Diaphragm/Wall Shear Transfer
  Anchorage for Normal Forces
Anchorage to Foundations
  Condition of Foundations
  Overturning
  Lateral Loads
  Geologic Site Hazards
Condition of Steel
  Quality of Masonry

diaphragms that have no path into existing vertical elements [ASCE 31, Section 4.3.1 (ASCE 2002)].

### C10.3.1.2 Redundancy

The most prudent rehabilitation strategy for a building without redundancy is to add new lateral-force-resisting elements in locations where the failure of a few components will cause an instability in the building. The added lateral-force-resisting elements should be of the same stiffness as the elements they are supplementing. It is not generally satisfactory just to strengthen a nonredundant element (such as by adding cover plates to a slender brace), because its failure would still result in an instability [ASCE 31, Sections 4.4.1.1.1, 4.4.2.1.1, 4.4.3.1.1, and 4.4.4.1.1 (ASCE 2002)].

### C10.3.1.3 Vertical Irregularities

New vertical lateral-force-resisting elements can be provided to eliminate the vertical irregularity. For weak stories, soft stories, and vertical discontinuities, new elements of the same type can be added as needed. Mass and geometric discontinuities must be evaluated and strengthened based on the Systematic Rehabilitation Method, if required by Chapter 2 [ASCE 31, Sections 4.3.2.4–4.3.2.5 (ASCE 2002)].

### C10.3.1.4 Plan Irregularities

The effects of plan irregularities that create torsion can be eliminated with the addition of lateral-force-resisting bracing elements that will support all major diaphragm segments in a balanced manner. Although it is possible in some cases to allow the irregularity to remain and instead strengthen those structural components that are overstressed by its existence, this does not directly address the problem and will require the use of the Systematic Rehabilitation Method [ASCE 31, Section 4.3.2.6 (ASCE 2002)].

### C10.3.1.5 Adjacent Buildings

Stiffening elements (typically braced frames or shear walls) can be added to one or both buildings to reduce the expected drifts to acceptable levels. With separate structures in a single building complex, it may be possible to tie them together structurally to force them to respond as a single structure. The relative stiffnesses of each and the resulting force interactions must be determined to ensure that additional deficiencies are not created. Pounding can also be eliminated by demolishing a portion of one building to increase the separation [ASCE 31, Section 4.3.1.2 (ASCE 2002)].

### C10.3.1.6 Uplift at Pile Caps

Typically, deficiencies in the load path at the pile caps are not a life safety concern. However, if the design professional has determined that there is a strong possibility of a life safety hazard due to this deficiency, piles and pile caps may be modified, supplemented, repaired, or in the most severe condition, replaced in their entirety. Alternatively, the building system may be rehabilitated such that the pile caps are protected [ASCE 31, Section 4.6.3.10 (ASCE 2002)].

### C10.3.1.7 Deflection Compatibility

Vertical lateral-force-resisting elements can be added to decrease the drift demands on the columns, or the ductility of the columns can be increased. Jacketing the columns with steel or concrete is one approach to increase their ductility [ASCE 31, Section 4.4.1.6.2 (ASCE 2002)].

### Table C10-9. C1: Concrete Moment Frames

Typical Deficiencies

Load Path
  Redundancy
  Vertical Irregularities
  Plan Irregularities
  Adjacent Buildings
  Uplift at Pile Caps
  Deflection Compatibility
Concrete Moment Frames
  Quick Checks, Frame and Nonductile Detail Concerns
  Precast Moment Frame Concerns
  Frames Not Part of the Lateral-Force-Resisting System
  Captive Columns
Re-entrant Corners
  Diaphragm Openings
  Diaphragm Stiffness/Strength
Diaphragm/Frame Shear Transfer
  Precast Connections
Anchorage to Foundations
  Condition of Foundations
  Overturning
  Lateral Loads
  Geologic Site Hazards
Condition of Concrete

### Table C10-10. C2, C2A: Concrete Shear Walls with Stiff or Flexible Diaphragms

Typical Deficiencies

Load Path
  Redundancy
  Vertical Irregularities
  Plan Irregularities
  Uplift at Pile Caps
  Deflection Compatibility
Frames Not Part of the Lateral-Force-Resisting System
  Captive Columns
Cast-in-Place Concrete Shear Walls
  Shear Stress
  Overturning
  Coupling Beams
  Boundary Component Detailing
  Wall Reinforcement
Re-entrant Corners
  Diaphragm Openings
  Diaphragm Stiffness/Strength
  Sheathing
Diaphragm/Wall Shear Transfer
Anchorage to Foundations
  Condition of Foundations
  Overturning
  Lateral Loads
  Geologic Site Hazards
Condition of Concrete

## C10.3.2 Moment Frames

### C10.3.2.1 Steel Moment Frames

*C10.3.2.1.1 Drift* The most direct mitigation approach is to add properly placed and distributed stiffening elements—new moment frames, braced frames, or shear walls—that can reduce the story drifts to acceptable levels. Alternatively, the addition of energy dissipation devices to the system may reduce the drift, though these are outside the scope of the Simplified Rehabilitation Method [ASCE 31, Section 4.4.1.3.1 (ASCE 2002)].

*C10.3.2.1.2 Frames* Noncompact members can be eliminated by adding appropriate steel plates. Eliminating or properly reinforcing large member penetrations will develop the demanded strength and deformations. Lateral bracing in the form of new steel components can be added to reduce member unbraced lengths to within the limits prescribed. Stiffening elements (e.g., braced frames, shear walls, or additional moment frames) can be added throughout the building to reduce the expected frame demands [ASCE 31,

### Table C10-11. C3, C3A: Concrete Frames with Infill Masonry Shear Walls and Stiff or Flexible Diaphragms

Typical Deficiencies

Load Path
    Redundancy
    Vertical Irregularities
    Plan Irregularities
    Uplift at Pile Caps
    Deflection Compatibility
Frames Not Part of the Lateral-Force-Resisting System
    Complete Frames
Masonry Shear Walls
    Reinforcing in Masonry Walls
    Shear Stress
    Reinforcing at Openings
    Unreinforced Masonry Shear Walls
    Proportions, Solid Walls
    Infill Walls
Re-entrant Corners
    Diaphragm Openings
    Diaphragm Stiffness/Strength
    Span-to-Depth Ratio
Diaphragm/Wall Shear Transfer
    Anchorage for Normal Forces
Anchorage to Foundations
    Condition of Foundations
    Overturning
    Lateral Loads
    Geologic Site Hazards
Condition of Concrete
    Quality of Masonry

### Table C10-12. PC1: Precast/Tilt-Up Concrete Shear Walls with Flexible Diaphragms

Typical Deficiencies

Load Path
    Redundancy
    Vertical Irregularities
    Plan Irregularities
    Deflection Compatibility
Diaphragm/Wall Shear Transfer
    Anchorage for Normal Forces
    Girder/Wall Connections
    Stiffness of Wall Anchors
Precast Concrete Shear Walls
    Panel-to-Panel Connections
Wall Openings
    Collectors
Re-entrant Corners
    Cross Ties
    Diaphragm Openings
    Diaphragm Stiffness/Strength
    Sheathing
    Unblocked Diaphragms
    Span-to-Depth Ratio
    Chord Continuity
Anchorage to Foundations
    Condition of Foundation
    Overturning
    Lateral Loads
    Geologic Site Hazards
Condition of Concrete

Sections 4.4.1.3.7, 4.4.1.3.8, and 4.4.1.3.10 (ASCE 2002)].

*C10.3.2.1.3 Strong Column-Weak Beam* Steel plates can be added to increase the strength of the steel columns to beyond that of the beams to eliminate this issue. Stiffening elements (e.g., braced frames, shear walls, or additional moment frames) can be added throughout the building to reduce the expected frame demands [ASCE 31, Section 4.4.1.3.6 (ASCE 2002)].

*C10.3.2.1.4 Connections* Adding a stiffer lateral-force-resisting system (e.g., braced frames or shear walls) can reduce the expected rotation demands. Connections can be modified by adding flange cover plates, vertical ribs, haunches, or brackets, or removing beam flange material to initiate yielding away from the connection location (e.g., via a pattern of drilled holes or the cutting out of flange material). Partial penetration splices, which may become more

vulnerable for conditions where the beam-column connections are modified to be more ductile, can be modified by adding plates and/or welds. Adding continuity plates alone is not likely to enhance the connection performance significantly [ASCE 31, Sections 4.4.1.3.3–4.4.1.3.5, and 4.4.1.3.9 (ASCE 2002)].

Moment-resisting connection capacity can be increased by adding cover plates or haunches, or using other techniques as stipulated in FEMA 351 (FEMA 2000).

### *C10.3.2.2 Concrete Moment Frames*

*C10.3.2.2.1 Frame and Nonductile Detail Concerns* Adding properly placed and distributed stiffening elements such as shear walls or braced frames will fully supplement the moment frame system with a new lateral-force-resisting system. For eccentric joints, columns and/or beams may be jacketed to reduce the effective eccentricity. Jackets may also be provided for shear-critical columns.

## Table C10-13. PC1A: Precast/Tilt-Up Concrete Shear Walls with Stiff Diaphragms

Typical Deficiencies

Load Path
  Redundancy
  Vertical Irregularities
  Plan Irregularities
Precast Concrete Shear Walls
  Panel-to-Panel Connections
  Wall Openings
  Collectors
Re-entrant Corners
  Diaphragm Openings
  Diaphragm Stiffness/Strength
Diaphragm/Wall Shear Transfer
  Anchorage for Normal Forces
  Girder/Wall Connections
Anchorage to Foundations
  Condition of Foundations
  Overturning
  Lateral Loads
  Geologic Site Hazards
Condition of Concrete

It must be verified that this new system sufficiently reduces the frame shears and story drifts to acceptable levels [ASCE 31, Section 4.4.1.4 (ASCE 2002)].

*C10.3.2.2.2 Precast Moment Frames* Precast concrete frames without shear walls may not be addressed under the Simplified Rehabilitation Method (see Table 10-1). Where shear walls are present, the precast connections must be strengthened sufficiently to meet the ASCE 31 (ASCE 2002) requirements.

The development of a competent load path is extremely critical in these buildings. If the connections have sufficient strength so that yielding will first occur in the members rather than in the connections, the building should be evaluated as a shear wall system Type C2 (ASCE 31, Section 4.4.1.5).

### C10.3.2.3 Frames Not Part of the Lateral-Force-Resisting System

*C10.3.2.3.1 Complete Frames* Complete frames of steel or concrete form a complete vertical load-carrying system.

Incomplete frames are essentially bearing wall systems. The wall must be strengthened to resist the combined gravity/seismic loads or new columns added to complete the gravity load path [ASCE 31, Section 4.4.1.6.1 (ASCE 2002)].

## Table C10-14. PC2: Precast Concrete Frames with Shear Walls

Typical Deficiencies

Load Path
  Redundancy
  Vertical Irregularities
  Plan Irregularities
  Uplift at Pile Caps
  Deflection Compatibility
Concrete Moment Frames
  Precast Moment Frame Concerns
Cast-in-Place Concrete Shear Walls
  Shear Stress
  Overturning
  Coupling Beams
  Boundary Component Detailing
  Wall Reinforcement
Re-entrant Corners
  Cross Ties
  Diaphragm Openings
  Diaphragm Stiffness/Strength
Diaphragm/Wall Shear Transfer
  Anchorage for Normal Forces
  Girder/Wall Connections
  Precast Connections
Anchorage to Foundations
  Condition of Foundations
  Overturning
  Lateral Loads
  Geologic Site Hazards
Condition of Concrete

*C10.3.2.3.2 Captive Columns* Columns may be jacketed with steel or concrete such that they can resist the expected forces and drifts. Alternatively, the expected story drifts can be reduced throughout the building by infilling openings or adding shear walls [ASCE 31, Section 4.4.1.4.5 (ASCE 2002)].

### C10.3.3 Shear Walls

#### C10.3.3.1 Cast-in-Place Concrete Shear Walls

*C10.3.3.1.1 Shearing Stress* New shear walls can be provided and/or the existing walls can be strengthened to satisfy seismic demand criteria. New and strengthened walls must form a complete, balanced, and properly detailed lateral-force-resisting system for the building. Special care is needed to ensure that the connection of the new walls to the existing diaphragm is appropriate and of sufficient strength such that yielding will first occur in the wall. All shear walls must have sufficient shear and overturning resistance to

**Table C10-15. PC2A: Precast Concrete Frames without Shear Walls**

Typical Deficiencies

Load Path
  Redundancy
  Vertical Irregularities
  Plan Irregularities
  Adjacent Buildings
  Uplift at Pile Caps
  Deflection Compatibility
Concrete Moment Frames
  Precast Moment Frame Concerns
  Frames Not Part of the Lateral-Force-Resisting System
  Short Captive Columns
Re-entrant Corners
  Diaphragm Openings
  Diaphragm Stiffness/Strength
Diaphragm/Frame Shear Transfer
  Precast Connections
Anchorage to Foundations
  Condition of Foundations
  Overturning
  Lateral Loads
  Geologic Site Hazards
Condition of Concrete

**Table C10-16. RM1: Reinforced Masonry Bearing Wall Buildings with Flexible Diaphragms**

Typical Deficiencies

Load Path
  Redundancy
  Vertical Irregularities
  Plan Irregularities
Diaphragm/Wall Shear Transfer
  Anchorage for Normal Forces
  Stiffness of Wall Anchors
Masonry Shear Walls
  Reinforcing in Masonry Walls
  Shear Stress
  Reinforcing at Openings
Re-entrant Corners
  Cross Ties
  Diaphragm Openings
  Diaphragm Stiffness/Strength
  Sheathing
  Unblocked Diaphragms
  Span-to-Depth Ratio
Anchorage to Foundations
  Condition of Foundations
  Geologic Site Hazards
Quality of Masonry

meet the ASCE 31 load criteria [ASCE 31, Section 4.4.2.2.1 (ASCE 2002)].

*C10.3.3.1.2 Overturning* Lengthening or adding shear walls can reduce overturning demands; increasing the length of footings will capture additional building dead load [ASCE 31, Section 4.4.2.2.4 (ASCE 2002)].

*C10.3.3.1.3 Coupling Beams* To eliminate the need to rely on the coupling beam, the walls may be strengthened as required. The beam should be jacketed only as a means of controlling debris. If possible, the opening that defines the coupling beam should be infilled [ASCE 31, Section 4.4.2.2.3 (ASCE 2002)].

*C10.3.3.1.4 Boundary Component Detailing* Splices may be improved by welding bars together after exposing them. The shear transfer mechanism can be improved by adding steel studs and jacketing the boundary components. [ASCE 31, Sections 4.4.2.2.5, 4.4.2.2.8, and 4.4.2.2.9 (ASCE 2002)].

*C10.3.3.1.5 Wall Reinforcement* Shear walls can be strengthened by infilling openings, or by thickening the walls [FEMA 172, Section 3.2.1.2 (FEMA 1992)] and ASCE 31, Sections 4.4.2.2.2 and 4.4.2.2.6 (ASCE 2002)].

### C10.3.3.2 Precast Concrete Shear Walls

*C10.3.3.2.1 Panel-to-Panel Connections* Appropriate Simplified Rehabilitation solutions are outlined in FEMA 172, Section 3.2.2.3 (FEMA 1992) and ASCE 31, Section 4.4.2.3.5 (ASCE 2002).

    Interpanel connections with inadequate capacity can be strengthened by adding steel plates across the joint, or by providing a continuous wall by exposing the reinforcing steel in the adjacent units and providing ties between the panels and patching with concrete. Providing steel plates across the joint is typically the most cost-effective approach, although care must be taken to ensure adequate anchor bolt capacity by providing adequate edge distances [FEMA 172, Section 3.2.2 (FEMA 1992)].

*C10.3.3.2.2 Wall Openings* Infilling openings or adding shear walls in the plane of the open bays can reduce demand on the connections and eliminate frame action [ASCE 31, Section 4.4.2.3.3 (ASCE 2002)].

*C10.3.3.2.3 Collectors* Upgrading the concrete section and/or the connections (e.g., exposing the existing connection, adding confinement ties, increasing embedment) can increase strength and/or ductility.

## Table C10-17. RM2: Reinforced Masonry Bearing Wall Buildings with Stiff Diaphragms

Typical Deficiencies

Load Path
  Redundancy
  Vertical Irregularities
  Plan Irregularities
Masonry Shear Walls
  Reinforcing in Masonry Walls
  Shear Stress
  Reinforcing at Openings
Re-entrant Corners
  Diaphragm Openings
  Diaphragm Stiffness/Strength
Diaphragm/Wall Shear Transfer
  Anchorage for Normal Forces
Anchorage to Foundations
  Condition of Foundations
  Geologic Site Hazards
Quality of Masonry

Alternative load paths for lateral forces can be provided, and shear walls can be added to reduce demand on the existing collectors [ASCE 31, Section 4.4.2.3.4 (ASCE 2002)].

### C10.3.3.3 Masonry Shear Walls

*C10.3.3.3.1 Reinforcing in Masonry Walls*
Nondestructive methods should be used to locate reinforcement, and selective demolition used if necessary to determine the size and spacing of the reinforcing. If it cannot be verified that the wall is reinforced in accordance with the minimum requirements, then the wall should be assumed to be unreinforced and therefore must be supplemented with new walls, or the procedures for URM should be followed [ASCE 31, Section 4.4.2.4.2 (ASCE 2002)].

*C10.3.3.3.2 Shearing Stress* To meet the lateral force requirements of ASCE 31 (ASCE 2002), new walls can be provided or the existing walls can be strengthened as needed. New and strengthened walls must form a complete, balanced, and properly detailed lateral-force-resisting system for the building. Special care is needed to ensure that the connection of the new walls to the existing diaphragm is appropriate and of sufficient strength to deliver the actual lateral loads or force yielding in the wall. All shear walls must have sufficient shear and overturning resistance [ASCE 31, Section 4.4.2.4.1 (ASCE 2002)].

## Table C10-18. URM: Unreinforced Masonry Bearing Wall Buildings with Flexible Diaphragms

Typical Deficiencies

Load Path
  Redundancy
  Vertical Irregularities
  Plan Irregularities
  Adjacent Buildings
Diaphragm/Wall Shear Transfer
  Anchorage for Normal Forces
  Stiffness of Wall Anchors
Masonry Shear Walls
  Unreinforced Masonry Shear Walls
  Properties, Solid Walls
Re-entrant Corners
  Cross Ties
  Diaphragm Openings
  Diaphragm Stiffness/Strength
  Sheathing
  Unblocked Diaphragms
  Span-to-Depth Ratio
Anchorage to Foundations
  Condition of Foundations
  Geologic Site Hazards
Quality of Masonry

*C10.3.3.3.3 Reinforcing at Openings* The presence and location of reinforcing steel at openings may be established using nondestructive or destructive methods at selected locations to verify the size and location of the reinforcing, or using both methods. Reinforcing must be provided at all openings as required to meet the ASCE 31 criteria. Steel plates may be bolted to the surface of the section as long as the bolts are sufficient to yield the steel plate [ASCE 31, Section 4.4.2.4.3 (ASCE 2002)].

*C10.3.3.3.4 Unreinforced Masonry Shear Walls*
Openings in the lateral-force-resisting walls should be infilled as needed to meet the ASCE 31 stress check. If supplemental strengthening is required, it should be designed using the Systematic Rehabilitation Method as defined in Chapter 2. Walls that do not meet the masonry lay-up requirements should not be considered as lateral-force-resisting elements and shall be specially supported for out-of-plane loads [ASCE 31, Sections 4.4.2.5.1 and 4.4.2.5.3 (ASCE 2002)].

*C10.3.3.3.5 Proportions of Solid Walls* Walls with insufficient thickness should be strengthened either by increasing the thickness of the wall or by adding a well-detailed strong-back system. The thickened wall must be detailed in a manner that fully interconnects

**Table C10-19. URMA: Unreinforced Masonry Bearing Walls Buildings with Stiff Diaphragms**

Typical Deficiencies

Load Path
  Redundancy
  Vertical Irregularities
  Plan Irregularities
  Adjacent Buildings
Masonry Shear Walls
  Unreinforced Masonry Shear Walls
  Properties, Solid Walls
Re-entrant Corners
  Diaphragm Openings
  Diaphragm Stiffness/Strength
Diaphragm/Wall Shear Transfer
  Anchorage for Normal Forces
Anchorage to Foundations
  Condition of Foundations
  Geologic Site Hazards
Quality of Masonry

the wall over its full height. The strong-back system must be designed for strength, connected to the structure in a manner that it: (1) develops the full yield strength of the strong-back, and (2) connects to the diaphragm in a manner that distributes the load into the diaphragm and has sufficient stiffness to ensure that the components will perform in a compatible and acceptable manner. The stiffness of the bracing should limit the out-of-plane deflections to acceptable levels such as L/600 to L/900 [ASCE 31, Sections 4.4.2.4.4 and 4.4.2.5.2 (ASCE 2002)].

*C10.3.3.3.6 Infill Walls* The partial infill wall should be isolated from the boundary columns to avoid a "short column" effect, except where it can be shown that the column is adequate. In sizing the gap between the wall and the columns, the anticipated story drift must be considered. The wall must be positively restrained against out-of-plane failure by either bracing the top of the wall or installing vertical girts. These bracing components must not violate the isolation of the frame from the infill [ASCE 31, Section 4.4.2.6 (ASCE 2002)].

### C10.3.3.4 Shear Walls in Wood Frame Buildings

*C10.3.3.4.1 Shear Stress* Walls may be added or existing openings filled. Alternatively, the existing walls and connections can be strengthened. The walls should be distributed across the building in a balanced man-

ner to reduce the shear stress for each wall. Replacing heavy materials such as tile roofing with lighter materials will also reduce shear stress [ASCE 31, Section 4.4.2.7.1 (ASCE 2002)].

*C10.3.3.4.2 Openings* Local shear transfer stresses can be reduced by distributing the forces from the diaphragm. Chords and/or collector members can be provided to collect and distribute shear from the diaphragm to the shear wall or bracing [FEMA 172, Figure 3.7.1.3 (FEMA 1992)]. Alternatively, the opening can be closed off by adding a new wall with wood structural panel sheathing [ASCE 31, Section 4.4.2.7.8 (ASCE 2002)].

*C10.3.3.4.3 Wall Detailing* If the walls are not bolted to the foundation or if the bolting is inadequate, bolts can be installed through the sill plates at regular intervals [FEMA 172, Figure 3.8.1.2a (FEMA 1992)]. If the crawl space is not deep enough for vertical holes to be drilled through the sill plate, the installation of connection plates or angles may be a practical alternative (FEMA 172, Figure 3.8.1.2b). Sheathing and additional nailing can be added where walls lack proper nailing or connections. Where the existing connections are inadequate, adding clips or straps will deliver lateral loads to the walls and to the foundation sill plate [ASCE 31, Section 4.4.2.7.9 (ASCE 2002)].

*C10.3.3.4.4 Cripple Walls* Where bracing is inadequate, new wood structural panel sheathing can be added to the cripple wall studs. The top edge of the wood structural panel is nailed to the floor framing and the bottom edge is nailed into the sill plate [FEMA 172, Figure 3.8.1.3 (FEMA 1992)]. Verify that the cripple wall does not change height along its length (stepped top of foundation). If it does, the shorter portion of the cripple wall will carry the majority of the shear and significant torsion will occur in the foundation. Added wood structural panel sheathing must have adequate strength and stiffness to reduce torsion to an acceptable level. Also, it should be verified that the sill plate is properly anchored to the foundation. If anchor bolts are lacking or insufficient, additional anchor bolts should be installed. Blocking and/or framing clips may be needed to connect the cripple wall bracing to the floor diaphragm or the sill plate [ASCE 31, Section 4.4.2.7.7 (ASCE 2002)].

*C10.3.3.4.5 Narrow Wood Shear Walls* Where narrow shear walls lack capacity, they should be replaced with shear walls with a height-to-width aspect ratio of 2:1

**Table C10-20. Cross-Reference Between This Standard and ASCE 31[1] Deficiency Reference Numbers**

| ASCE 31 | | ASCE 41 | |
|---------|--------------|---------|--------------|
| **Section** | **Section Heading** | **Section** | **Section Heading** |
| 4.3.1.1 | **Load Path** | C10.3.1.1 | Load Path |
| 4.3.1.2 | **Adjacent Buildings** | C10.3.1.5 | Adjacent Buildings |
| 4.3.1.3 | Mezzanines | C10.3.6.8 | Mezzanine Connections |
| 4.3.2 | **Configuration** | C10.3.1 | Building Systems |
| 4.3.2.1 | Weak Story | C10.3.1.3 | Vertical Irregularities |
| 4.3.2.2 | Soft Story | C10.3.1.3 | Vertical Irregularities |
| 4.3.2.3 | Geometry | C10.3.1.3 | Vertical Irregularities |
| 4.3.2.4 | Vertical Discontinuities | C10.3.1.3 | Vertical Irregularities |
| 4.3.2.5 | Mass | C10.3.1.3 | Vertical Irregularities |
| 4.3.2.6 | Torsion | C10.3.1.4 | Plan Irregularities |
| 4.3.3 | **Condition of Materials** | C10.3.8 | Evaluation of Materials and Conditions |
| 4.3.3.1 | Deterioration of Wood | C10.3.8.2 | Condition of Wood |
| 4.3.3.2 | Wood Structural Panel Shear Wall Fasteners | C10.3.8.3 | Wood Structural Panel Shear Wall Fasteners |
| 4.3.3.3 | Deterioration of Steel | C10.3.8.4 | Condition of Steel |
| 4.3.3.4 | Deterioration of Concrete | C10.3.8.5 | Condition of Concrete |
| 4.3.3.5 | Post-Tensioning Anchors | C10.3.8.6 | Post-Tensioning Anchors |
| 4.3.3.6 | Precast Concrete Walls | C10.3.8.5 | Condition of Concrete |
| 4.3.3.7 | Masonry Units | C10.3.8.7 | Quality of Masonry |
| 4.3.3.8 | Masonry Joints | C10.3.8.7 | Quality of Masonry |
| 4.3.3.9 | Concrete Wall Cracks | C10.3.8.5 | Condition of Concrete |
| 4.3.3.10 | Reinforced Masonry Wall Cracks | C10.3.8.7 | Quality of Masonry |
| 4.3.3.11 | Unreinforced Masonry Wall Cracks | C10.3.8.7 | Quality of Masonry |
| 4.3.3.12 | Cracks in Infill Walls | C10.3.8.7 | Quality of Masonry |
| 4.3.3.13 | Cracks in Boundary Columns | C10.3.8.5 | Condition of Concrete |
| 4.4.1.1.1 | Redundancy | C10.3.1.2 | Redundancy |
| 4.4.1.2 | **Moment Frames with Infill Walls** | C10.3.3 | Shear Walls |
| 4.4.1.2.1 | Interfering Walls | C10.3.3.3.6 | Infill Walls |
| 4.4.1.3 | **Steel Moment Frames** | C10.3.2.1 | Steel Moment Frames |
| 4.4.1.3.1 | Drift Check | C10.3.2.1.1 | Drift |
| 4.4.1.3.2 | Axial Stress Check | C10.3.2.1.2 | Frames |
| 4.4.1.3.3 | Moment-Resisting Connections | C10.3.2.1.4 | Connections |
| 4.4.1.3.4 | Panel Zones | C10.3.2.1.4 | Connections |
| 4.4.1.3.5 | Column Splices | C10.3.2.1.2 | Frames |
| 4.4.1.3.6 | Strong Column–Weak Beam | C10.3.2.1.3 | Strong Column–Weak Beam |
| 4.4.1.3.7 | Compact Members | C10.3.2.1.2 | Frames |
| 4.4.1.3.8 | Beam Penetration | C10.3.2.1.2 | Frames |
| 4.4.1.3.9 | Girder Flange Continuity | C10.3.2.1.4 | Connections |
| 4.4.1.3.10 | Out-of-Plane Bracing | C10.3.2.1.2 | Frames |
| 4.4.1.3.11 | Bottom Flange Bracing | C10.3.2.1.2 | Frames |
| 4.4.1.4 | **Concrete Moment Frames** | C10.3.2.2 | Concrete Moment Frames |
| 4.4.1.4.1 | Shear Stress Check | C10.3.2.2.1 | Frame and Nonductile Detail Concerns |
| 4.4.1.4.2 | Axial Stress Check | C10.3.2.2.1 | Frame and Nonductile Detail Concerns |
| 4.4.1.4.3 | Flat Slab Frames | C10.3.2.2.1 | Frame and Nonductile Detail Concerns |
| 4.4.1.4.4 | Prestressed Frame Elements | C10.3.2.2.1 | Frame and Nonductile Detail Concerns |
| 4.4.1.4.5 | Captive Columns | C10.3.2.3.2 | Captive Columns |
| 4.4.1.4.6 | No Shear Failures | C10.3.2.2.1 | Frame and Nonductile Detail Concerns |
| 4.4.1.4.7 | Strong Column–Weak Beam | C10.3.2.2.1 | Frame and Nonductile Detail Concerns |
| 4.4.1.4.8 | Beam Bars | C10.3.2.2.1 | Frame and Nonductile Detail Concerns |
| 4.4.1.4.9 | Column–Bar Splices | C10.3.2.2.1 | Frame and Nonductile Detail Concerns |
| 4.4.1.4.10 | Beam–Bar Splices | C10.3.2.2.1 | Frame and Nonductile Detail Concerns |
| 4.4.1.4.11 | Column–Tie Spacing | C10.3.2.2.1 | Frame and Nonductile Detail Concerns |
| 4.4.1.4.12 | Stirrup Spacing | C10.3.2.2.1 | Frame and Nonductile Detail Concerns |

*continued*

**TABLE 10-20. (Continued)**

| ASCE 31 | | ASCE 41 | |
|---|---|---|---|
| **Section** | **Section Heading** | **Section** | **Section Heading** |
| 4.4.1.4.13 | Joint Reinforcing | C10.3.2.2.1 | Frame and Nonductile Detail Concerns |
| 4.4.1.4.14 | Joint Eccentricity | C10.3.2.2.1 | Frame and Nonductile Detail Concerns |
| 4.4.1.4.15 | Stirrup and Tie Hooks | C10.3.2.2.1 | Frame and Nonductile Detail Concerns |
| 4.4.1.5 | **Precast Moment Frames** | C10.3.2.2.2 | Precast Moment Frames |
| 4.4.1.5.1 | Precast Connection Check | C10.3.2.2.2 | Precast Moment Frames |
| 4.4.1.5.2 | Precast Frames | C10.3.2.2.2 | Precast Moment Frames |
| 4.4.1.5.3 | Precast Connections | C10.3.6.5 | Precast Connections |
| 4.4.1.6 | **Frames Not Part of the Lateral-Force-Resisting System** | C10.3.2.3 | Frames Not Part of the Lateral-Force-Resisting System |
| 4.4.1.6.1 | Complete Frames | C10.3.2.3.1 | Complete Frames |
| 4.4.1.6.2 | Deflection Compatibility | C10.3.1.7 | Deflection Compatibility |
| 4.4.1.6.3 | Flat Slabs | C10.3.2.2.1 | Frame and Nonductile Detail Concerns |
| 4.4.2.1.1 | Redundancy | C10.3.1.2 | Redundancy |
| 4.4.2.2 | **Concrete Shear Walls** | C10.3.3 | Shear Walls |
| 4.4.2.2.1 | Shear Stress Check | C10.3.3.1.1 | Shearing Stress |
| 4.4.2.2.2 | Reinforcing Steel | C10.3.3.1.5 | Wall Reinforcement |
| 4.4.2.2.3 | Coupling Beams | C10.3.3.1.3 | Coupling Beams |
| 4.4.2.2.4 | Overturning | C10.3.3.1.2 | Overturning |
| 4.4.2.2.5 | Confinement Reinforcing | C10.3.3.1.4 | Boundary Component Detailing |
| 4.4.2.2.6 | Reinforcing at Openings | C10.3.3.1.5 | Wall Reinforcement |
| 4.4.2.2.7 | Wall Thickness | C10.3.3.1.1 | Shearing Stress |
| 4.4.2.2.8 | Wall Connections | C10.3.3.1.4 | Boundary Component Detailing |
| 4.4.2.2.9 | Column Splices | C10.3.3.1.4 | Boundary Component Detailing |
| 4.4.2.3 | **Precast Concrete Shear Walls** | C10.3.3.2 | Precast Concrete Shear Walls |
| 4.4.2.3.1 | Shear Stress Check | C10.3.3.1.1 | Shearing Stress |
| 4.4.2.3.2 | Reinforcing Steel | C10.3.3.1.5 | Wall Reinforcement |
| 4.4.2.3.3 | Wall Openings | C10.3.3.2.2 | Wall Openings |
| 4.4.2.3.4 | Corner Openings | C10.3.3.2.3 | Collectors |
| 4.4.2.3.5 | Panel-to-Panel Connections | C10.3.3.2.1 | Panel-to-Panel Connections |
| 4.4.2.3.6 | Wall Thickness | C10.3.3.1.1 | Shearing Stress |
| 4.4.2.4 | **Reinforced Masonry Shear Walls** | C10.3.3.3 | Masonry Shear Walls |
| 4.4.2.4.1 | Shear Stress Check | C10.3.3.3.2 | Shearing Stress |
| 4.4.2.4.2 | Reinforcing Steel | C10.3.3.3.1 | Reinforcing in Masonry Walls |
| 4.4.2.4.3 | Reinforcing at Openings | C10.3.3.3.3 | Reinforcing at Openings |
| 4.4.2.4.4 | Proportions | C10.3.3.3.5 | Proportions of Solid Walls |
| 4.4.2.5 | **Unreinforced Masonry Shear Walls** | C10.3.3.3.4 | Unreinforced Masonry Shear Walls |
| 4.4.2.5.1 | Shear Stress Check | C10.3.3.3.4 | Unreinforced Masonry Shear Walls |
| 4.4.2.5.2 | Proportions | C10.3.3.3.5 | Proportions of Solid Walls |
| 4.4.2.5.3 | Masonry Lay-Up | C10.3.3.3.4 | Unreinforced Masonry Shear Walls |
| 4.4.2.6 | **Infill Walls in Frames** | C10.3.3.3.6 | Infill Walls |
| 4.4.2.6.1 | Wall Connections | C10.3.3.3.6 | Infill Walls |
| 4.4.2.6.2 | Proportions | C10.3.3.3.5 | Proportions of Solid Walls |
| 4.4.2.6.3 | Solid Walls | C10.3.3.3.5 | Proportions of Solid Walls |
| 4.4.2.6.4 | Infill Walls | C10.3.3.3.6 | Infill Walls |
| 4.4.2.7 | **Walls in Wood Frame Buildings** | C10.3.3.4 | Shear Walls in Wood Frame Buildings |
| 4.4.2.7.1 | Shear Stress Check | C10.3.3.4.1 | Shear Stress |
| 4.4.2.7.2 | Stucco (Exterior Plaster) Shear Walls | C10.3.3.4.6 | Stucco Shear Walls |
| 4.4.2.7.3 | Gypsum Wallboard or Plaster Shear Walls | C10.3.3.4.7 | Gypsum Wallboard or Plaster Shear Walls |
| 4.4.2.7.4 | Narrow Wood Shear Walls | C10.3.3.4.5 | Narrow Wood Shear Walls |
| 4.4.2.7.5 | Walls Connected Through Floors | C10.3.3.4.3 | Wall Detailing |
| 4.4.2.7.6 | Hillside Site | C10.3.3.4.4 | Cripple Walls |
| 4.4.2.7.7 | Cripple Walls | C10.3.3.4.4 | Cripple Walls |
| 4.4.2.7.8 | Openings | C10.3.3.4.2 | Openings |
| 4.4.2.7.9 | Hold-Down Anchors | C10.3.3.4.3 | Wall Detailing |

| ASCE 31 | | ASCE 41 | |
|---|---|---|---|
| **Section** | **Section Heading** | **Section** | **Section Heading** |
| 4.4.3.1 | **Braced Frames** | C10.3.4 | Steel Braced Frames |
| 4.4.3.1.1 | Redundancy | C10.3.1.2 | Redundancy |
| 4.4.3.1.2 | Axial Stress Check | C10.3.4.1 | System Concerns |
| 4.4.3.1.3 | Column Splices | C10.3.4.4 | Braced Frame Connections |
| 4.4.3.1.4 | Slenderness of Diagonals | C10.3.4.2 | Stiffness of Diagonals |
| 4.4.3.1.5 | Connection Strength | C10.3.4.4 | Braced Frame Connections |
| 4.4.3.1.6 | Out-of-Plane Bracing | C10.3.4.1 | System Concerns |
| 4.4.3.2 | **Concentrically Braced Frames** | C10.3.4 | Steel Braced Frames |
| 4.4.3.2.1 | K-Bracing | C10.3.4.3 | Chevron or K-Bracing |
| 4.4.3.2.2 | Tension-Only Braces | C10.3.4.2 | Stiffness of Diagonals |
| 4.4.3.2.3 | Chevron Bracing | C10.3.4.3 | Chevron or K-Bracing |
| 4.4.3.2.4 | Concentrically Braced Frame Joints | C10.3.4.4 | Braced Frame Connections |
| 4.4.3.3 | **Eccentrically Braced Frames** | C10.3.4.1 | System Concerns |
| 4.5 | **Diaphragms** | C10.3.5 | Diaphragms |
| 4.5.1.1 | Diaphragm Continuity | C10.3.5.4.5 | Diaphragm Continuity |
| 4.5.1.2 | Cross Ties | C10.3.5.2 | Cross Ties |
| 4.5.1.3 | Roof Chord Continuity | C10.3.5.4.6 | Chord Continuity |
| 4.5.1.4 | Openings at Shear Walls | C10.3.5.3 | Diaphragm Openings |
| 4.5.1.5 | Openings at Braced Frames | C10.3.5.3 | Diaphragm Openings |
| 4.5.1.6 | Openings at Exterior Masonry Shear Walls | C10.3.5.3 | Diaphragm Openings |
| 4.5.1.7 | Plan Irregularities | C10.3.5.1 | Re-entrant Corners |
| 4.5.1.8 | Diaphragm Reinforcing at Openings | C10.3.5.3 | Diaphragm Openings |
| 4.5.2 | **Wood Diaphragms** | C10.3.5 | Diaphragms |
| 4.5.2.1 | Straight Sheathing | C10.3.5.4.1 | Board Sheathing |
| 4.5.2.2 | Spans | C10.3.5.4.3 | Spans |
| 4.5.2.3 | Unblocked Diaphragms | C10.3.5.4.2 | Unblocked Diaphragms |
| | | C10.3.5.4.4 | Span-to-Depth Ratio |
| 4.5.3 | Metal Deck Diaphragms | C10.3.5 | Diaphragms |
| 4.5.3.1 | Non-Concrete Filled Diaphragms | C10.3.5 | Diaphragms |
| 4.5.4 | **Concrete Diaphragms** | C10.3.5 | Diaphragms |
| 4.5.5 | **Precast Concrete Diaphragms** | C10.3.5 | Diaphragms |
| 4.5.5.1 | Topping Slab | C10.3.5 | Diaphragms |
| 4.5.6 | **Horizontal Bracing** | C10.3.5 | Diaphragms |
| 4.5.7.1 | **Other Diaphragms** | C10.3.5 | Diaphragms |
| 4.6.1 | **Anchorage for Normal Forces** | C10.3.6.3 | Anchorage for Normal Forces |
| 4.6.1.1 | Wall Anchorage | C10.3.6.3 | Anchorage for Normal Forces |
| 4.6.1.2 | Wood Ledgers | C10.3.6.3 | Anchorage for Normal Forces |
| 4.6.1.3 | Precast Panel Connections | C10.3.6.3 | Anchorage for Normal Forces |
| 4.6.1.4 | Stiffness of Wall Anchors | C10.3.6.3 | Anchorage for Normal Forces |
| 4.6.2 | **Shear Transfer** | C10.3.6 | Connections |
| 4.6.2.1 | Transfer to Shear Walls | C10.3.6.1 | Diaphragm/Wall Shear Transfer |
| 4.6.2.2 | Transfer to Steel Frames | C10.3.6.2 | Diaphragm/Frame Shear Transfer |
| 4.6.2.3 | Topping Slab to Walls or Frames | C10.3.6.1 | Diaphragm/Wall Shear Transfer |
| 4.6.3 | **Vertical Components** | C10.3.7.1 | Anchorage to Foundations |
| 4.6.3.1 | Steel Columns | C10.3.7.1 | Anchorage to Foundations |
| 4.6.3.2 | Concrete Columns | C10.3.7.1 | Anchorage to Foundations |
| 4.6.3.3 | Wood Posts | C10.3.7.1 | Anchorage to Foundations |
| 4.6.3.4 | Wood Sills | C10.3.7.1 | Anchorage to Foundations |
| 4.6.3.5 | Foundation Dowels | C10.3.7.1 | Anchorage to Foundations |
| 4.6.3.6 | Shear-Wall-Boundary Columns | C10.3.7.1 | Anchorage to Foundations |
| 4.6.3.7 | Precast Wall Panels | C10.3.7.1 | Anchorage to Foundations |
| 4.6.3.8 | Wall Panels | C10.3.7.1 | Anchorage to Foundations |
| 4.6.3.9 | Wood Sill Bolts | C10.3.7.1 | Anchorage to Foundations |
| 4.6.3.10 | Uplift at Pile Caps | C10.3.7.4 | Lateral Loads |

*continued*

**TABLE 10-20. (Continued)**

| ASCE 31 | | ASCE 41 | |
|---|---|---|---|
| **Section** | **Section Heading** | **Section** | **Section Heading** |
| 4.6.4 | **Interconnection of Elements** | C10.3.6 | Connections |
| 4.6.4.1 | Girder/Column Connection | C10.3.6.4 | Girder-Wall Connections |
| 4.6.4.2 | Girders | C10.3.6.4 | Girder-Wall Connections |
| 4.6.4.3 | Corbel Bearing | C10.3.6.4 | Girder-Wall Connections |
| 4.6.4.4 | Corbel Connections | C10.3.6.4 | Girder-Wall Connections |
| 4.6.5 | **Panel Connections** | C10.3.6 | Connections |
| 4.6.5.1 | Roof Panels | C10.3.6.7 | Light Gage Metal, Plastic, or Cementitious Roof Panels |
| 4.6.5.2 | Wall Panels | C10.3.6.6 | Wall Panels and Cladding |
| 4.6.5.3 | Roof Panel Connections | C10.3.6.7 | Light Gage Metal, Plastic, or Cementitious Roof Panels |
| 4.7.1 | **Geologic Site Hazards** | C10.3.7 | Foundations and Geologic Hazards |
| 4.7.1.1 | Liquefaction | C10.3.7.5 | Geologic Site Hazards |
| 4.7.1.2 | Slope Failure | C10.3.7.5 | Geologic Site Hazards |
| 4.7.1.3 | Surface Fault Rupture | C10.3.7.5 | Geologic Site Hazards |
| 4.7.2 | **Condition of Foundations** | C10.3.7.2 | Condition of Foundations |
| 4.7.2.1 | Foundation Performance | C10.3.7.2 | Condition of Foundations |
| 4.7.2.2 | Deterioration | C10.3.7.2 | Condition of Foundations |
| 4.7.3 | **Capacity of Foundations** | C10.3.7 | Foundations and Geologic Hazards |
| 4.7.3.1 | Pole Foundations | C10.3.7.4 | Lateral Loads |
| 4.7.3.2 | Overturning | C10.3.7.3 | Overturning |
| 4.7.3.3 | Ties between Foundation Elements | C10.3.7.4 | Lateral Loads |
| 4.7.3.4 | Deep Foundations | C10.3.7.4 | Lateral Loads |
| 4.7.3.5 | Sloping Sites | C10.3.7.4 | Lateral Loads |

[1]ASCE 31 (2002). *Seismic Evaluation of Existing Buildings,* American Society of Civil Engineers, Reston, Virginia.

or less. These replacement walls must have sufficient strength, including being adequately connected to the diaphragm and sufficiently anchored to the foundation for shear and overturning forces [ASCE 31, Section 4.4.2.7.4 (ASCE 2002)].

*C10.3.3.4.6 Stucco Shear Walls* For strengthening or repair, the stucco should be removed, a wood structural panel shear wall added, and new stucco applied. The wood structural panel should be the manufacturer's recommended thickness for the installation of stucco. The new stucco should be installed in accordance with building code requirements for waterproofing. Walls should be sufficiently anchored to the diaphragm and foundation [ASCE 31, Section 4.4.2.7.2 (ASCE 2002)].

*C10.3.3.4.7 Gypsum Wallboard or Plaster Shear Walls* Plaster and gypsum wallboard can be removed and replaced with structural panel shear wall as required, and the new shear walls covered with gypsum wallboard [ASCE 31, Section 4.4.2.7.3 (ASCE 2002)].

### C10.3.4 Steel Braced Frames

#### C10.3.4.1 System Concerns

If the strength of the braced frames is inadequate, more braced bays or shear wall panels can be added. The resulting lateral-force-resisting system must form a well-balanced system of braced frames that do not fail at their joints, are properly connected to the floor diaphragms, and whose failure mode is yielding of braces rather than overturning [ASCE 31, Sections 4.4.3.1.1 and 4.4.3.1.2 (ASCE 2002)].

#### C10.3.4.2 Stiffness of Diagonals

Diagonals with inadequate stiffness should be strengthened using supplemental steel plates, or replaced with a larger and/or different type of section. Global stiffness can be increased by the addition of braced bays or shear wall panels [(ASCE 31, Sections 4.4.3.1.3 and 4.4.3.2.2 (ASCE 2002)].

#### C10.3.4.3 Chevron or K-Bracing

Columns or horizontal girts can be added as needed to support the tension brace when the compression brace buckles, or the bracing can be revised to another system throughout the building. The beam components can be strengthened with cover plates to provide them with the capacity to fully develop the unbalanced forces created by tension brace yielding [ASCE 31, Sections 4.4.3.2.1 and 4.4.3.2.3 (ASCE 2002)].

### C10.3.4.4 Braced Frame Connections

Column splices or other braced frame connections can be strengthened by adding plates and welds to ensure that they are strong enough to develop the connected components. Connection eccentricities that reduce component capacities can be eliminated, or the components can be strengthened to the required level by the addition of properly placed plates. Demands on the existing elements can be reduced by adding braced bays or shear wall panels [ASCE 31, Sections 4.4.3.1.4 and 4.4.3.1.5 (ASCE 2002)].

## C10.3.5 Diaphragms

### C10.3.5.1 Re-Entrant Corners

New chords with sufficient strength to resist the required force can be added at the re-entrant corner. If a vertical lateral-force-resisting element exists at the re-entrant corner, a new collector component should be installed in the diaphragm to reduce tensile and compressive forces at the re-entrant corner. The same basic materials used in the diaphragm should be used for the chord [ASCE 31, Section 4.5.1.7 (ASCE 2002)].

### C10.3.5.2 Cross Ties

New cross ties and wall connections can be added to resist the required out-of-plane wall forces and distribute these forces through the diaphragm. New strap plates and/or rod connections can be used to connect existing framing members together so they function as a cross tie in the diaphragm [ASCE 31, Section 4.5.1.2 (ASCE 2002)].

### C10.3.5.3 Diaphragm Openings

New diaphragm ties or chords can be added around the perimeter of existing openings to distribute tension and compression forces along the diaphragm. The existing sheathing should be nailed to the new diaphragm ties or chords. In some cases it may also be necessary to: (1) increase the shear capacity of the diaphragm adjacent to the opening by overlaying the existing diaphragm with a wood structural panel, or (2) decrease the demand on the diaphragm by adding new vertical elements near the opening [ASCE 31,

Sections 4.5.1.4 through 4.5.1.6 and 4.5.1.8 (ASCE 2002)].

### C10.3.5.4 Diaphragm Stiffness/Strength

*C10.3.5.4.1 Board Sheathing* Where the diaphragm does not have at least two nails through each board into each of the supporting members, and the lateral drift and/or shear demands on the diaphragm are not excessive, the shear capacity and stiffness of the diaphragm can be increased by adding nails at the sheathing boards. This method of upgrade is most often suitable in areas of low seismicity. In other cases, a new wood structural panel should be placed over the existing straight sheathing, and the joints of the wood structural panels placed so they are near the center of the sheathing boards or at a 45-degree angle to the joints between sheathing boards [FEMA 172, Section 3.5.1.2 (FEMA 1992); ATC-7 (ATC 1981), and ASCE 31, Section 4.5.2.1 (ASCE 2002)].

*C10.3.5.4.2 Unblocked Diaphragm* The shear capacity of unblocked diaphragms can be improved by adding new wood blocking and nailing at the unsupported panel edges. Placing a new wood structural panel over the existing diaphragm will increase the shear capacity. Both of these methods will require the partial or total removal of existing flooring or roofing to place and nail the new overlay or nail the existing panels to the new blocking. Strengthening of the diaphragm is usually not necessary at the central area of the diaphragm where shear is low. In certain cases where the design loads are low, it may be possible to increase the shear capacity of unblocked diaphragms with sheet metal plates stapled on the underside of the existing wood panels. These plates and staples must be designed for all related shear and torsion caused by the details related to their installation [ASCE 31, Section 4.5.2.3 (ASCE 2002)].

*C10.3.5.4.3 Spans* New vertical elements can be added to reduce the diaphragm span. The reduction of the diaphragm span will also reduce the lateral deflection and shear demand in the diaphragm. However, adding new vertical elements will result in a different distribution of shear demands. Additional blocking, nailing, or other rehabilitation measures may need to be provided at these areas [FEMA 172, Section 3.4 (FEMA 1992) and ASCE 31, Section 4.5.2.2 (ASCE 2002)].

*C10.3.5.4.4 Span-to-Depth Ratio* New vertical elements can be added to reduce the diaphragm span-to-depth ratio. The reduction of the diaphragm span-to-depth ratio will also reduce the lateral deflection and shear demand in the diaphragm. Typical construction details and methods are discussed in FEMA 172, Section 3.4 (FEMA 1992).

*C10.3.5.4.5 Diaphragm Continuity* The diaphragm discontinuity should in all cases be eliminated by adding new vertical elements at the diaphragm offset or the expansion joint [FEMA 172, Section 3.4 (FEMA 1992)]. In some cases, special details may be used to transfer shear across an expansion joint—while still allowing the expansion joint to function—thus eliminating a diaphragm discontinuity [ASCE 31, Section 4.5.1.1 (ASCE 2002)].

*C10.3.5.4.6 Chord Continuity* If members such as edge joists, blocking, or wall top plates have the capacity to function as chords but lack connection, adding nailed or bolted continuity splices will provide a continuous diaphragm chord. New continuous steel or wood chord members can be added to the existing diaphragm where existing members lack sufficient capacity or no chord exists. New chord members can be placed at either the underside or topside of the diaphragm. In some cases, new vertical elements can be added to reduce the diaphragm span and stresses on any existing chord members [FEMA 172, Section 3.5.1.3 (FEMA 1992) and ATC-7 (ATC 1981)]. New chord connections should not be detailed such that they are the weakest component in the chord [ASCE 31, Section 4.5.1.3 (ASCE 2002)].

## C10.3.6 Connections

### C10.3.6.1 Diaphragm/Wall Shear Transfer

Collector members, splice plates, and shear transfer devices can be added as required to deliver collector forces to the shear wall. Adding shear connectors from the diaphragm to the wall and/or to the collectors will transfer shear. See FEMA 172, Section 3.7 for Wood Diaphragms, 3.7.2 for concrete diaphragms, 3.7.3 for poured gypsum, and 3.7.4 for metal deck diaphragms (FEMA 1992) and ASCE 31, Sections 4.6.2.1 and 4.6.2.3 (ASCE 2002).

### C10.3.6.2 Diaphragm/Frame Shear Transfer

Adding collectors and connecting the framing will transfer loads to the collectors. Connections can be provided along the collector length and at the collector-to-frame connection to withstand the calculated forces. See FEMA 172, Sections 3.7.5 and 3.7.6 (FEMA 1992) and ASCE 31, Sections 4.6.2.2 and 4.6.2.3 (ASCE 2002).

### C10.3.6.3 Anchorage for Normal Forces

To account for inadequacies identified by ASCE 31, wall anchors can be added. Complications that may result from inadequate anchorage include cross-grain tension in wood ledgers or failure of the diaphragm-to-wall connection due to: (1) insufficient strength, number, or stability of anchors; (2) inadequate embedment of anchors; (3) inadequate development of anchors and straps into the diaphragm; and (4) deformation of anchors and their fasteners that permit diaphragm boundary connection pullout, or cross-grain tension in wood ledgers.

Existing anchors should be tested to determine load capacity and deformation potential, including fastener slip, according to the requirements in ASCE 31. Special attention should be given to the testing procedure to maintain a high level of quality control. Additional anchors should be provided as needed to supplement those that fail the test, as well as those needed to meet the ASCE 31 criteria. The quality of the rehabilitation depends greatly on the quality of the performed tests [ASCE 31, Sections 4.6.1.1 through 4.6.1.5 (ASCE 2002)].

### C10.3.6.4 Girder–Wall Connections

The existing reinforcing must be exposed, and the connection modified as necessary. For out-of-plane loads, the number of column ties can be increased by jacketing the pilaster or, alternatively, by developing a second load path for the out-of-plane forces. Bearing length conditions can be addressed by adding bearing extensions. Frame action in welded connections can be mitigated by adding shear walls [ASCE 31, Section 4.6.4.1 (ASCE 2002)].

### C10.3.6.5 Precast Connections

The connections of chords, ties, and collectors can be upgraded to increase strength and/or ductility, providing alternative load paths for lateral forces. Upgrading can be achieved by such methods as adding confinement ties or increasing embedment. Shear walls can be added to reduce the demand on connections [ASCE 31, Section 4.4.1.5.3 (ASCE 2002)].

### C10.3.6.6 Wall Panels and Cladding

It may be possible to improve the connection between the panels and the framing. If architectural or occupancy conditions warrant, the cladding can be replaced with a new system. The building can be stiffened with the addition of shear walls or braced frames to reduce the drifts in the cladding components [ASCE 31, Section 4.8.4.6 (ASCE 2002)].

### C10.3.6.7 Light Gage Metal, Plastic, or Cementitious Roof Panels

It may be possible to improve the connection between the roof and the framing. If architectural or occupancy conditions warrant, the roof diaphragm can be replaced with a new one. Alternatively, a new diaphragm may be added using rod braces or wood structural panel above or below the existing roof, which remains in place [ASCE 31, Section 4.6.5.1 (ASCE 2002)].

### C10.3.6.8 Mezzanine Connections

Diagonal braces, moment frames, or shear walls can be added at or near the perimeter of the mezzanine where bracing elements are missing, so that a complete and balanced lateral-force-resisting system is provided that meets the requirements of ASCE 31 [ASCE 31, Section 4.3.1.3 (ASCE 2002)].

## C10.3.7 Foundations and Geologic Hazards

### C10.3.7.1 Anchorage to Foundations

For wood walls, expansion anchors or epoxy anchors can be installed by drilling through the wood sill to the concrete foundation. Similarly, steel columns and wood posts can be anchored to concrete slabs or footings, using expansion anchors and clip angles. If the concrete or masonry walls and columns lack dowels, a concrete curb can be installed adjacent to the wall or column by drilling dowels and installing anchors into the wall that lap with dowels installed in the slab or footing. However, this curb can cause significant architectural problems. Alternatively, steel angles may be used with drilled anchors. The anchorage of shear wall boundary components can be challenging due to very high concentrated forces [ASCE 31, Sections 4.6.3.2 through 4.6.3.9 (ASCE 2002)].

### C10.3.7.2 Condition of Foundations

All deteriorated and otherwise damaged foundations should be strengthened and repaired using the same materials and style of construction. Some conditions of material deterioration can be mitigated in the field, including patching of spalled concrete. Pest infestation or dry rot of wood piles can be very difficult to correct and often require full replacement. The deterioration of these components may have implications that extend beyond seismic safety and must be considered in the rehabilitation [ASCE 31, Sections 4.7.2.1 and 4.7.2.2 (ASCE 2002)].

### C10.3.7.3 Overturning

Existing foundations can be strengthened as needed to resist overturning forces. Spread footings may be enlarged or additional piles, rock anchors, or piers may be added to deep foundations. It may also be possible to use grade beams or new wall elements to spread out overturning loads over a greater distance. Adding new lateral-load-resisting elements will reduce overturning effects of existing elements [ASCE 31, Section 4.7.3.2 (ASCE 2002)].

### C10.3.7.4 Lateral Loads

As with overturning effects, the correction of lateral load deficiencies in the foundations of existing buildings is expensive and may not be justified by more realistic analysis procedures. For this reason, the Systematic Rehabilitation Method is recommended for these cases [ASCE 31, Sections 4.7.3.1, 4.7.3.3 through 4.7.3.5 (ASCE 2002)].

### C10.3.7.5 Geologic Site Hazards

Site hazards other than ground shaking should be considered. Rehabilitation of structures subject to life safety hazards from ground failures is impractical unless site hazards can be mitigated to the point where acceptable performance can be achieved. Not all ground failures need necessarily be considered as life safety hazards. For example, in many cases liquefaction beneath a building does not pose a life safety hazard; however, related lateral spreading can result in collapse of buildings with inadequate foundation strength. For this reason, the liquefaction potential and the related consequences should be thoroughly investigated for sites that do not satisfy the ASCE 31 requirements. Further information on the evaluation of site hazards is provided in Chapter 4 of this standard [ASCE 31, Sections 4.7.1.1 through 4.7.1.3 (ASCE 2002)].

## C10.3.8 Evaluation of Materials and Conditions

### C10.3.8.1 General

Proper evaluation of the existing conditions and configuration of the existing building structure is an important aspect of the Simplified Rehabilitation Method. As Simplified Rehabilitation is often concerned with specific deficiencies in a particular structural system, the evaluation may either be focused on affected structural elements and components, or be comprehensive and include the complete structure. If the degree of existing damage or deficiencies in a structure has not been established, the evaluation shall consist of an inspection of gravity- and lateral-force-resisting systems in accordance with ASCE 31 (ASCE 2002). This inspection should include the following:

1. Verify existing data (e.g., accuracy of drawings);
2. Develop other needed data (e.g., measure and sketch building if necessary);
3. Verify the vertical and lateral systems;
4. Check the condition of the building;
5. Look for special conditions and anomalies;
6. Address the evaluation statements and goals during the inspection; and
7. Perform material tests that are justified through a weighing of the cost of destructive testing and the cost of corrective work.

The materials testing and evaluation methods of this standard should not be used for Simplified Rehabilitation except those required for all new work specified in the Construction Quality Assurance Plan.

### C10.3.8.2 Condition of Wood

An inspection should be conducted to grade the existing wood and verify physical condition, using techniques from Section C10.3.8.1. Any damage or deterioration and its source must be identified. Wood that is significantly damaged due to splitting, decay, aging, or other phenomena must be removed and replaced. Localized problems can be eliminated by adding new appropriately sized reinforcing components extending beyond the damaged area and connecting to undamaged portions. Additional connectors between components should be provided to correct any discontinuous load paths. It is necessary to verify that any new reinforcing components or connectors will not be exposed to similar deterioration or damage [ASCE 31, Section 4.3.3.1 (ASCE 2002)].

### C10.3.8.3 Wood Structural Panel Shear Wall Fasteners

Where visual inspection determines that extensive overdriving of fasteners exists in greater than 20% of the installed connectors, the fasteners and shear panels can generally be repaired through addition of a new same-sized fastener for every two overdriven fasteners. To avoid splitting because of closely spaced nails, it may be necessary to predrill to 90% of the nail shank diameter for installation of new nails. For other conditions, such as cases where the addition of new connectors is not possible or where component damage is suspected, further investigation shall be conducted using the guidance of Section C10.3.8.1 [ASCE 31, Section 4.3.3.2 (ASCE 2002)].

### C10.3.8.4 Condition of Steel

Should visual inspection or testing conducted in accordance with Section C10.3.8.1 reveal the presence of steel component or connection deterioration, further evaluation is needed. The source of the damage shall be identified and mitigated to preserve the remaining structure. In areas of significant deterioration, restoration of the material cross section can be performed by the addition of plates or other reinforcing techniques. When sizing reinforcements, the design professional shall consider the effects of existing stresses in the original structure, load transfer, and strain compatibility. The demands on the deteriorated steel elements and components may also be reduced through careful addition of bracing or shear wall panels [ASCE 31, Section 4.3.3.3 (ASCE 2002)].

### C10.3.8.5 Condition of Concrete

Should visual inspections or testing conducted in accordance with Section C10.3.8.1 reveal the presence of concrete component or reinforcing steel deterioration, further evaluation is needed. The source of the damage shall be identified and mitigated to preserve the remaining structure. Existing deteriorated material, including reinforcing steel, shall be removed to the limits defined by testing; reinforcing steel in good condition shall be cleaned and left in place for splicing purposes as appropriate. Cracks in otherwise sound material shall be evaluated to determine cause, and repaired as necessary using techniques appropriate to the source and activity level [ASCE 31, Section 4.3.3.4 (ASCE 2002)]. FEMA 306 (FEMA 1998), FEMA 307 (FEMA 1998), and FEMA 308 (FEMA 1998) can be used as a source of further information on evaluation and repair of concrete wall buildings.

### C10.3.8.6 Post-Tensioning Anchors

Prestressed concrete systems may be adversely affected by cyclic deformations produced by earthquake motion. One rehabilitation process that may be considered is to add stiffness to the system. Another concern for these systems is the adverse effects of tendon corrosion. A thorough visual inspection of prestressed systems shall be performed to verify absence of concrete cracking or spalling, staining from embedded tendon corrosion, or other signs of damage along the tendon spans and at anchorage zones. If degradation is observed or suspected, more detailed evaluations will be required as indicated in Chapter 6. Rehabilitation of these systems, except for local anchorage repair, should be in accordance with the Systematic Rehabilitation provisions in this standard. Professionals with special prestressed concrete construction expertise should also be consulted for further interpretation of damage [ASCE 31, Section 4.3.3.5 (ASCE 2002)].

### C10.3.8.7 Quality of Masonry

Should visual inspections or testing conducted in accordance with Section C10.3.8.1 reveal the presence of masonry components or construction deterioration, further evaluation is needed. Certain damage such as degraded mortar joints or simple cracking may be rehabilitated through repointing or rebuilding. If the wall is repointed, care should be taken to ensure that the new mortar is compatible with the existing masonry units and mortar, and that suitable wetting is performed. The strength of the new mortar is critical to load-carrying capacity and seismic performance. Significant degradation should be treated as specified in Chapter 7 of this standard [ASCE 31, Sections 4.3.3.7, 4.3.3.8, 4.3.3.10, 4.3.3.11, and 4.3.3.12)]. FEMA 306 (FEMA 1998), FEMA 307 (FEMA 1998), and FEMA 308 (FEMA 1998) can be used as a source of further information on evaluation and repair of masonry wall buildings.

## 11.0 ARCHITECTURAL, MECHANICAL, AND ELECTRICAL COMPONENTS

### 11.1 SCOPE

This chapter sets forth requirements for the seismic rehabilitation of existing architectural, mechanical, and electrical components and systems that are permanently installed in, or are an integral part of, a building system. Procedures of this chapter are applicable to both the Simplified and Systematic Rehabilitation Methods. Requirements are provided for nonstructural components that are rehabilitated to the Immediate Occupancy, Life Safety, and Hazards Reduced Nonstructural Performance Levels. The requirements for Operational Performance shall be as approved by the authority having jurisdiction.

Sections 11.2, 11.3, 11.4, and 11.5 provide requirements for condition assessment, component evaluation, Rehabilitation Objectives, and structural–nonstructural interaction. Section 11.6 defines acceleration- and deformation-sensitive components. Section 11.7 specifies procedures for determining design forces and deformations on nonstructural components. Section 11.8 identifies rehabilitation methods. Sections 11.9, 11.10, and 11.11 specify evaluation and acceptance criteria for architectural components; mechanical, electrical, and plumbing (MEP) systems; and other equipment.

New nonstructural components installed in existing buildings shall conform to the requirements of this standard. New nonstructural components designed to Life Safety or Hazards Reduced Performance Levels may be designed using the requirements of similar components for new buildings.

### C11.1 SCOPE

The assessment process necessary to make a final determination of which nonstructural components are to be rehabilitated is not part of this standard, but the subject is discussed briefly in Section 11.3.

The core of this chapter is contained in Table 11-1, which provides:

A list of nonstructural components subject to the Hazards Reduced, Life Safety and Immediate Occupancy requirements of this standard;

Rehabilitation requirements related to the level of seismicity and Hazards Reduced, Life Safety, and Immediate Occupancy Performance Levels. Requirements for Operational Performance are not included in this standard. References that may be used to seismically qualify equipment and systems to achieve Operational Performance for some nonstructural components are provided in C1.5.2.1. Requirements for Hazards Reduced Performance will generally be based on the requirements for Life Safety Performance, so separate evaluation procedures and acceptance criteria have not been provided; and

Identification of the required evaluation procedure (Analytical or Prescriptive).

Section 11.4 provides general requirements and discussion of Rehabilitation Objectives, Performance Levels, and Performance Ranges as they pertain to nonstructural components. Criteria for means of egress are not specifically included in this standard.

Section 11.5 briefly discusses structural–nonstructural interaction, and Section 11.6 provides general requirements for acceptance criteria for acceleration-sensitive and deformation-sensitive components, and those sensitive to both kinds of response.

Section 11.7 provides sets of equations for a simple "default" force analysis, as well as an extended analysis method that considers a number of additional factors. This section defines the Analytical Procedure for determining drift ratios and relative displacement, and outlines general requirements for the Prescriptive Procedure.

Section 11.8 notes the general ways in which nonstructural rehabilitation is carried out.

Sections 11.9, 11.10, and 11.11 provide the rehabilitation criteria for each component category identified in Table 11-1. For each component, the following information is given:

1. Definition and scope;
2. Component behavior and rehabilitation concepts;

3. Acceptance criteria; and
4. Evaluation requirements.

## 11.2 PROCEDURE

Nonstructural components shall be rehabilitated by completing the following steps:

1. The Rehabilitation Objectives shall be established in accordance with Section 11.4, which includes selection of a Nonstructural Performance Level and Earthquake Hazard level. The level of seismicity shall be determined in accordance with Section 1.6.3. A target Building Performance Level that includes Nonstructural Performance Not Considered need not comply with the provisions of this chapter;
2. A walk-through and condition assessment shall be performed in accordance with Sections 11.2.1 and 11.2.2;
3. Analysis and rehabilitation requirements for the selected Nonstructural Performance Level and appropriate level of seismicity shall be determined for nonstructural components using Table 11-1. "Yes" indicates that rehabilitation shall be required if the component does not meet applicable acceptance criteria specified in Section 11.3.2;
4. Interaction between structural and nonstructural components shall be considered in accordance with Section 11.5;
5. The classification of each type of nonstructural component shall be determined in accordance with Section 11.6;
6. Evaluation shall be conducted in accordance with Section 11.7 using the procedure specified in Table 11-1. The acceptability of bracing components and connections between nonstructural components and the structure shall be determined in accordance with Section 11.3.2; and
7. Nonstructural components not meeting the requirements of the selected Nonstructural Performance Level shall be rehabilitated in accordance with Section 11.8.

## C11.2 PROCEDURE

Where Hazards Reduced Performance is used, the engineer should consider the location of nonstructural components relative to areas of public occupancy. The authority having jurisdiction should be consulted to establish the areas of the building for which nonstruc-

tural hazards will be considered. Other nonstructural components, such as those designated by the owner also should be included in those that are evaluated.

### 11.2.1 Condition Assessment

A condition assessment of nonstructural components shall be performed as part of the nonstructural rehabilitation process. As a minimum, this assessment shall determine the following:

1. The presence and configuration of each type of nonstructural component and its attachment to the structure;
2. The physical condition of each type of nonstructural component and whether or not degradation is present;
3. The presence of nonstructural components that potentially influence overall building performance; and
4. The presence of other nonstructural components whose failure could affect the performance of the nonstructural component being considered.

### C11.2.1 Condition Assessment

For the purpose of visual observation, nonstructural component types should be based on the general types listed in Table 11-1. Further distinction can be made where difference in structural configuration of the component or its bracing exists.

Seismic interactions between nonstructural components and systems can have a profound influence on the performance of these systems. Where appropriate, the condition assessment should include an interaction review. A seismic interaction involves two components—a source and a target. An interaction source is the component or structure that could fail or displace and interact with another component. An interaction target is component that is being impacted, sprayed, or spuriously activated. For an interaction to affect a component, it must be credible and significant. A credible interaction is one that can take place. For example, the fall of a ceiling panel located overhead from a motor control center is a credible interaction because the falling panel can reach and impact the motor control center (MCC). The target (the MCC) is said to be within the zone of influence of the source (the ceiling panel). A significant interaction is one that can result in damage to the target. For example, the fall of a light fixture on a 20-in. steel pipe may be credible (the light fixture being above the pipe) but may not be significant (the light fixture will not damage the steel pipe). An important aspect of the interaction review is engineering judgment, because only credible and significant sources of interaction should be considered in the condition assessment.

### 11.2.2 Sample Size

Direct visual inspection shall be performed on each type of nonstructural component in the building as follows:

1. If detailed drawings are available, at least one sample of each type of nonstructural component shall be observed. If no deviations from the drawings exist, the sample shall be considered representative of installed conditions. If deviations are observed, then at least 10% of all occurrences of the component shall be observed; and

2. If detailed drawings are not available, at least three samples of each type of nonstructural component shall be observed. If no deviations among the three components are observed, the sample shall be considered representative of installed conditions. If deviations are observed, at least 20% of all occurrences of the component shall be observed.

## 11.3 HISTORICAL AND COMPONENT EVALUATION CONSIDERATIONS

### 11.3.1 Historical Information

Available construction documents, equipment specification and data, and as-built information shall be obtained as specified in Section 2.2. Data on nonstructural components and equipment shall be collected to estimate the year of manufacture or installation of nonstructural components to justify selection of rehabilitation approaches and techniques based on available historical information, prevailing codes, and assessment of existing conditions.

### C11.3.1 Historical Information

The architectural, mechanical, and electrical components and systems of a historic building may be historically significant, especially if they are original to the building, very old, or innovative. Historic buildings may also contain hazardous materials, such as lead pipes and asbestos, that may or may not pose a hazard depending on their location, condition, use or abandonment, containment, and/or disturbance during the rehabilitation.

### C11.3.1.1 Background

Prior to the 1961 *Uniform Building Code* and the 1964 Alaska earthquake, architectural components and mechanical and electrical systems for buildings had typically been designed with little, if any, regard to stability when subjected to seismic forces. By the time of the 1971 San Fernando earthquake, it became clear that damage to nonstructural components could result in serious casualties, severe building functional impairment, and major economic losses, even where structural damage was not significant (Lagorio 1990). This historical perspective presents the background for the development of building code provisions, together with a historical review of professional and construction practices related to the seismic design and construction of nonstructural components. Since the 1964 Alaska earthquake, the poor performance of nonstructural components has been identified in earthquake reconnaissance reports. Subsequent editions of the *Uniform Building Code* (ICBO 1994), as well as California and federal codes and laws have increased both the scope and strictness of nonstructural seismic provisions in an attempt to achieve better performance. Table C11-1 and Table C11-2 provide a comprehensive list of nonstructural hazards that have been observed in these earthquakes.

The following quote, taken from statements made after the Alaska earthquake, characterizes the hazard nonstructural components pose to building occupants:

> "If, during an earthquake, [building occupants] must exit through a shower of falling light fixtures and ceilings, maneuver through shifting and toppling furniture and equipment, stumble down dark corridors and debris-laden stairs, and then be met at the street by falling glass, veneers, or facade components, then the building cannot be described as a safe structure." (Ayres and Sun, 1973a)

In reviewing the design and construction of architectural nonstructural components in this century, four general phases can be distinguished.

**A. Phase 1: 1900 to 1920s**

Buildings featured monumental classical architecture, generally with a steel frame structure using stone facing with a backing of unreinforced masonry and concrete. Interior partitions were of unreinforced hollow clay tile or brick unit masonry, or wood partitions with wood lath and plaster. These buildings had natural ventilation systems with hot water radiators (later, forced-air), and surface- or pendant-mounted incandescent light fixtures.

**B. Phase 2: 1930s to 1950s**

Buildings were characterized by poured-in-place reinforced concrete or steel frame structures, employing columns and (in California) limited exterior and interior shear walls. Windows were large and horizontal. Interior partitions of unreinforced hollow clay tile or concrete block unit masonry, or light wood frame

**Table 11-1. Nonstructural Components: Applicability of Hazards Reduced, Life Safety and Immediate Occupancy Requirements and Methods of Analysis**

| Component Type | Performance Level | | | | | |
| | | Seismicity | | | | |
| | | High and Moderate Seismicity | | Low Seismicity | | |
| | IO | LS | HR | LS | HR | Evaluation Procedure |
|---|---|---|---|---|---|---|
| **Architectural (Section 11.9)** | | | | | | |
| 1. **Exterior Wall Components** | | | | | | |
| Adhered Veneer | Yes | Yes | Yes[15] | No | No | F/D |
| Anchored Veneer | Yes | Yes | Yes[15] | No | No | F/D |
| Glass Blocks | Yes | Yes | Yes[15] | No | No | F/D |
| Prefabricated Panels | Yes | Yes | Yes[15] | Yes | Yes[15] | F/D |
| Glazed Exterior Wall Systems | Yes | Yes | Yes[15] | Yes | Yes[15] | F/D/PR |
| 2. **Partitions** | | | | | | |
| Heavy | Yes | Yes | Yes[15] | No | No | F/D |
| Light | Yes | No | No | No | No | F/D |
| Glazed | Yes | Yes | Yes[15] | Yes | Yes[15] | F/D/PR |
| 3. **Interior Veneers** | | | | | | |
| Stone, Including Marble | Yes | Yes[18] | Yes[15] | No | No | F/D |
| 4. **Ceilings** | | | | | | |
| Directly Applied to Structure | Yes | No[13] | No[15] | No | No | F |
| Dropped Furred Gypsum Board | Yes | No | No | No | No | F |
| Suspended Lath and Plaster | Yes | Yes | Yes[15] | No | No | F |
| Suspended Integrated Ceiling | Yes | No[11] | No | No[11] | No | PR |
| 5. **Parapets and Appendages** | Yes | Yes | Yes[15] | Yes | Yes | F[1] |
| 6. **Canopies and Marquees** | Yes | Yes | Yes[15] | Yes | Yes | F |
| 7. **Chimneys and Stacks** | Yes | Yes | Yes[15] | No | No | F[2] |
| 8. **Stairs** | Yes | Yes | No | Yes | No | * |
| **Mechanical Equipment (Section 11.10)** | | | | | | |
| 1. **Mechanical Equipment** | | | | | | |
| Boilers, Furnaces, Pumps, and Chillers | Yes | Yes | No | Yes | No | F |
| General Mfg. and Process Machinery | Yes | No[3] | No | No | No | F |
| HVAC Equipment, Vibration-Isolated | Yes | No[3] | No | No | No | F |
| HVAC Equipment, Non-Vibration-Isolated | Yes | No[3] | No | No | No | F |
| HVAC Equipment, Mounted In-Line with Ductwork | Yes | No[3] | No | No | No | PR |
| 2. **Storage Vessels and Water Heaters** | | | | | | |
| Structurally Supported Vessels (Category 1) | Yes | No[3] | No | No | No | Note[4] |
| Flat-Bottom Vessels (Category 2) | Yes | No[3] | No | No | No | Note[5] |
| 3. **Pressure Piping** | Yes | Yes | No | No | No | Note[5] |
| 4. **Fire Suppression Piping** | Yes | Yes | No | No | No | PR |
| 5. **Fluid Piping, not Fire Suppression** | | | | | | |
| Hazardous Materials | Yes | Yes | Yes[12] | Yes | Yes[12] | PR/F/D |
| Nonhazardous Materials | Yes[14] | No | No | No | No | PR/F/D |
| 6. **Ductwork** | Yes | No[6] | No | No | No | PR |
| **Electrical And Communications (Section 11.10)** | | | | | | |
| 1. **Electrical and Communications Equipment** | Yes | No[7] | No | No | No | F |
| 2. **Electrical and Communications Distribution Equipment** | Yes | No[8] | No | No | No | PR |
| 3. **Light Fixtures** | | | | | | |
| Recessed | No | No | No | No | No | PR[17] |
| Surface-Mounted | No | No | No | No | No | PR[17] |
| Integrated Ceiling | Yes | Yes | Yes[15] | No | No | PR |
| Pendant | Yes | No[9] | No | No | No | F/PR |

| Component Type | Performance Level | | | | | Evaluation Procedure |
|---|---|---|---|---|---|---|
| | | Seismicity | | | | |
| | | High and Moderate Seismicity | | Low Seismicity | | |
| | IO | LS | HR | LS | HR | |
| **Furnishings and Interior Equipment (Section 11.11)** | | | | | | |
| 1. **Storage Racks** | Yes | Yes[10] | Yes[16] | No | No | F |
| 2. **Bookcases** | Yes | Yes | No | No | No | F |
| 3. **Computer Access Floors** | Yes | No | No | No | No | PR/FD |
| 4. **Hazardous Materials Storage** | Yes | Yes | No[12] | No[12] | No[12] | PR |
| 5. **Computer and Communication Racks** | Yes | No | No | No | No | PR/F/D |
| 6. **Elevators** | Yes | Yes | No | No | No | F/D/PR |
| 7. **Conveyors** | Yes | No | No | No | No | F/D/PR |

[1]Rehabilitation of unreinforced masonry parapets not over 4 ft in height by the Prescriptive Design Concept shall be permitted.

[2]Rehabilitation of residential masonry chimneys by the Prescriptive Design Concept shall be permitted.

[3]Equipment type 1 or 2 that is 6 ft or more in height, equipment type 3, equipment forming part of an emergency power system, and gas-fired equipment in occupied or unoccupied space shall be rehabilitated to the Life Safety Nonstructural Performance Level in areas of High Seismicity. In areas of Moderate Seismicity, this equipment need not be considered. Refer to Section 11.10.1.1 for equipment type designations.

[4]Rehabilitation of residential water heaters with capacity less than 100 gal by the Prescriptive Procedure shall be permitted. Other vessels shall meet the force provisions of Sections 11.7.3 or 11.7.4.

[5]Rehabilitation of vessels or piping systems according to Prescriptive Standards shall be permitted. Storage vessels shall meet the force provisions of Sections 11.7.3 or 11.7.4. Piping shall meet drift provisions of Section 11.7.5 and the force provisions of Sections 11.7.3 or 11.7.4.

[6]Ductwork that conveys hazardous materials, exceeds 6 sf in cross-sectional area, or is suspended more than 12 in. from top of duct to supporting structure at any support point shall meet the requirements of the selected Rehabilitation Objective.

[7]Equipment that is 6 ft or more in height, weighs over 20 lbs., or forms part of an emergency power and/or communication system shall meet the Life Safety Nonstructural Performance Level.

[8]Equipment that forms part of an emergency lighting, power, and/or communication system shall meet the Life Safety Nonstructural Performance Level.

[9]Fixtures that exceed 20 lbs. per support shall meet the Life Safety Nonstructural Performance Level.

[10]Rehabilitation shall not be required for storage racks in unoccupied spaces.

[11]Panels that exceed 2 lbs/sf, or for which Enhanced Rehabilitation Objectives have been selected, shall meet the Life Safety Nonstructural Performance Level.

[12]Where material is in close proximity to occupancy such that leakage could cause an immediate life safety threat, the requirements of the selected Rehabilitation Objective shall be met.

[13]Plaster ceilings on metal or wood lath over 10 sf in area shall meet the Life Safety Nonstructural Performance Level.

[14]Unbraced pressure pipes with a 2-in. or larger diameter and suspended more than 12 in. from the top of the pipe to the supporting structure at any support point shall meet the requirements of the selected Rehabilitation Objective.

[15]Where heavy nonstructural components are located in areas of public occupancy or egress, the components shall meet the Life Safety Nonstructural Performance Level.

[16]Storage racks in areas of public assembly shall meet the requirements of the selected Rehabilitation Objective.

[17]Evaluation for the presence of an adequate attachment shall be checked as described in Section 11.10.9.3.

[18]In areas of Moderate Seismicity, interior veneers of ceramic tile need not be considered.

Key:

HR, Hazards Reduced Nonstructural Performance Level; LS, Life Safety Nonstructural Performance Level; IO, Immediate Occupancy Nonstructural Performance Level; PR, Use of the Prescriptive Procedure of Section 11.7.2 shall be permitted; F, the Analytical Procedure of Section 11.7.1 shall be implemented and a force analysis shall be performed in accordance with Sections 11.7.3 or 11.7.4; F/D, the Analytical Procedure of Section 11.7.1 shall be implemented and a force and deformation analysis shall be performed in accordance with Sections 11.7.4 and 11.7.5, respectively.

*Individual components shall be rehabilitated as required.

**Table C11-1. Nonstructural Architectural Component Seismic Hazards**

| Component | Principal Concerns |
|---|---|
| Suspended ceilings | Dropped acoustical tiles, perimeter damage, separation of runners and cross-runners |
| Plaster ceilings | Collapse, local spalling |
| Cladding | Falling from building, damaged panels and connections, broken glass |
| Ornamentation | Damage leading to a falling hazard |
| Plaster and gypsum board walls | Cracking |
| Demountable partitions | Collapse |
| Raised access floors | Collapse, separation between modules |
| Recessed light fixtures and HVAC diffusers | Dropping out of suspended ceilings |
| Unreinforced masonry walls and partitions | Parapet and wall collapse and spalling, partitions debris and falling hazard |

**Table C11-2. Mechanical and Electrical Equipment Seismic Hazards**

| Equipment/Component | Principal Concerns |
|---|---|
| Boilers | Sliding, broken gas/fuel and exhaust lines, broken/bent steam and relief lines |
| Chillers | Sliding, overturning, loss of function, leaking refrigerant |
| Emergency generators | Failed vibration isolation mounts; broken fuel, signal, and power lines, loss of function, broken exhaust lines |
| Fire pumps | Anchorage failure, misalignment between pump and motor, broken piping |
| On-site water storage | Tank or vessel rupture, pipe break |
| Communications equipment | Sliding, overturning, or toppling leading to loss of function |
| Main transformers | Sliding, oil leakage, bushing failure, loss of function |
| Main electrical panels | Sliding or overturning, broken or damaged conduit or electrical bus |
| Elevators (traction) | Counterweights out of guide rails, cables out of sheaves, dislodged equipment |
| Other fixed equipment | Sliding or overturning, loss of function or damage to adjacent equipment |
| Ducts | Collapse, separation, leaking, fumes |
| Piping | Breaks, leaks |

partitions with plaster, were gradually replaced by gypsum. Suspended ceilings and fluorescent lights arrived, generally surface- or pendant-mounted. Air conditioning (cooling) was introduced and HVAC systems became more complex, with increased demands for duct space.

**C. Phase 3: 1950s to 1960s**

This phase saw the advent of simple rectangular metal or reinforced concrete frame structures ("International Style"), and metal and glass curtain walls with a variety of opaque claddings (porcelain enamel, ceramic tile, concrete, cement plaster). Interior partitions became primarily metal studs and gypsum board. Proprietary suspended ceilings were developed using wire-hung metal grids with infill of acoustic panels, lighting fixtures, and air diffusion units. HVAC systems increased in size, requiring large mechanical rooms and increased above-ceiling space for ducts. Sprinklers and more advanced electrical control systems were introduced, and more HVAC equipment was spring-mounted to prevent transmission of motor vibration.

**D. Phase 4: 1960s to Present**

This period saw the advent of exterior precast concrete and, in the 1980s, glass-fiber-reinforced concrete (GFRC) cladding. Interior partition systems of metal studs and gypsum board, demountable partitions, and suspended ceiling systems become catalog proprietary items. The evolution of the late 1970s architectural style ("Post-Modern") resulted in less-regular forms and much more interior and exterior decoration, much of it accomplished by nonstructural components: assemblies of glass, metal panel, GFRC, and natural stone cladding for the exteriors, and use of gypsum board for exaggerated structural concealment and form-making in interiors. Suspended ceilings and HVAC systems changed little, but the advent of office landscaping often reduced floor-to-ceiling partitions to almost nothing in general office space. Starting in the 1980s, the advent of the "smart" office greatly increased electrical and communications needs and the use of raised floors, and increased the need for the mechanical and electrical systems to remain functional after earthquakes.

### C11.3.1.2 Background to Mechanical and Electrical Considerations

Prior to the 1964 Alaska earthquake, mechanical and electrical systems for buildings had been designed with little, if any, regard to stability when subjected to seismic forces. The change in design from the heavily structured and densely partitioned structures of the pre-war era, with their simple mechanical, electrical and lighting systems, to the light frame and curtain wall, gypsum board and integrated ceiling buildings of the 1950s and onward, had been little reflected in the seismic building codes. The critical yet fragile nature of the new nonstructural systems was not fully realized, except for nuclear power plant design and other special-purpose, high-risk structures. Equipment supports were generally designed for gravity loads only, and attachments to the structure itself were often deliberately designed to be flexible to allow for vibration isolation or thermal expansion.

Few building codes, even in regions with a history of seismic activity, have contained provisions governing the behavior of mechanical and electrical systems until relatively recently. One of the earliest references to seismic bracing can be found in NFPA 13, *Standard for the Installation of Sprinkler Systems* (NFPA 2002). This pamphlet has been updated periodically since 1896, and seismic bracing requirements have been included since 1947. Piping systems for building sprinklers are static and do not require vibration isolation. They do, however, require flexibility where the service piping enters the building. The issue of protecting flexibly mounted piping was not studied until after the 1964 Alaska earthquake.

The designers of building mechanical systems must also address the seismic restraints required for emergency generators, fire protection pumps, and plumbing systems that are vital parts of an effective fire suppression system.

Studies published following the 1971 San Fernando earthquake all indicated that buildings that sustained only minor structural damage became uninhabitable and hazardous to life due to failures of mechanical and electrical systems.

### C11.3.1.3 HVAC Systems

A study by Ayres and Sun (1973b) clearly identified the need to anchor tanks and equipment that did not require vibration isolation, and to provide lateral restraints on equipment vibration isolation devices. Some of these suggested corrective measures are now incorporated into manufactured products. The HVAC system designers had to become aware of the earthquake-induced forces on the system's components and the need for seismic restraints to limit damage; they also had to understand the requirements for the suspension and bracing of ceilings and light fixtures because of their adjacency to and interaction with the HVAC system components.

To provide technical guidance to HVAC system designers and installers, the Sheet Metal Industry Fund of Los Angeles published its first manual, *Guidelines for Seismic Restraints of Mechanical Systems* (Sheet Metal Industry Fund, 1976). This manual was updated in 1982 with assistance from the Plumbing and Piping Industry Council (PPIC). The most recent manual, *Seismic Restraint Guidelines for Mechanical Equipment* (SMACNA 1991), is designed for use in California as well as other locations with lower seismic hazard levels.

Secondary effects of earthquakes (fires, explosions, and hazardous materials releases resulting from damaged mechanical and electrical equipment) have only recently been considered. In addition, the potential danger of secondary damage from falling architectural and structural components, which could inflict major damage to adjacent equipment and render it unusable, needs to be carefully assessed.

These secondary effects can represent a considerable hazard to the building, its occupants, and its contents. Steam and hot water boilers and other pressure vessels can release fluids at hazardous temperatures. Mechanical systems often include piping systems filled with flammable, toxic, or noxious substances, such as ammonia or other refrigerants. Some of the nontoxic halogen refrigerants used in air-conditioning apparatus can be converted to a poisonous gas (phosgene) upon contact with open flame. Hot parts of disintegrating boilers, such as portions of the burner and firebrick, are at high enough temperatures to ignite combustible materials with which they might come in contact.

### 11.3.2 Component Evaluation

Nonstructural components shall be evaluated to achieve the Rehabilitation Objective selected in accordance with Section 1.4. Analysis and rehabilitation requirements for the Hazards Reduced, Life Safety, and Immediate Occupancy Nonstructural Performance Levels for the appropriate level of seismicity shall be as specified in Table 11-1. Design forces shall be calculated in accordance with Section 11.7.3 or 11.7.4, and design deformations shall be calculated in accordance with Section 11.7.5. Analysis and rehabilitation requirements for the Hazards Reduced Nonstructural Performance Level shall follow the requirements for the Life Safety Nonstructural Performance Level. Analysis and rehabilitation requirements for the Operational Nonstructural Performance Level shall be based on approved codes.

Acceptance criteria for nonstructural components being evaluated to the Life Safety and Immediate Occupancy Nonstructural Performance Levels shall be based on criteria listed in Sections 11.9 through 11.11. Forces on bracing and connections for nonstructural components calculated in accordance with Section 11.7 shall be compared to capacities using strength design procedures. Acceptance criteria for the Life Safety Nonstructural Performance Level shall be used for nonstructural components being evaluated to the Hazards Reduced Nonstructural Performance Level. For nonstructural components being evaluated to the Operational Nonstructural Performance Level, approved acceptance criteria shall be used.

### C11.3.2 Component Evaluation

The Hazards Reduced Nonstructural Performance Level applies only to high-hazard components as specified in Section 1.5.2.4 and Table 11-1. Life Safety Nonstructural Performance Level criteria—or other approved criteria—should be used for the Hazards Reduced Nonstructural Performance Level. Criteria for the Operational Nonstructural Performance Level has not been developed to date. Evaluation, rehabilitation, and acceptance criteria for the Immediate Occupancy Nonstructural Performance Level may be used for the Operational Nonstructural Performance Level if more appropriate data are not available.

Forces on nonstructural components calculated in accordance with Section 11.7 are at a strength design level. Where allowable stress values are available for proprietary products used as bracing for nonstructural

components, these values shall be factored up to strength design levels. In the absence of manufacturer's data on strength values, allowable stress values can be increased by a factor of 1.4 to obtain strength design values.

Where nonstructural components are evaluated using Hazards Reduced Nonstructural Performance Level, the force level associated with Life Safety Nonstructural Performance in Section 11.7 should be used. In many instances, if bracing of the nonstructural component exists, or if it is rehabilitated, there would not be a substantial justification for evaluating or rehabilitating the component using a force level or acceptance criteria less stringent than Life Safety. However, in cases where it is not considered critical or feasible, the engineer may, with appropriate approval, evaluate or rehabilitate the nonstructural component using a criterion that is less stringent than Life Safety.

In cases where the Basic Safety Objective is not required—such as where the Limited Safety Performance Range applies—there may be more latitude in the selection of components or criteria for nonstructural rehabilitation.

A suggested general procedure for developing a mitigation plan for the rehabilitation of nonstructural components is as follows:

1. It is assumed that the building has been evaluated in a feasibility phase, using a procedure such as that described in ASCE 31 (ASCE 2002). For nonstructural components, use of this procedure will have provided a broad list of deficiencies that are generally, but not specifically, related to a Rehabilitation Objective. Issues related to other objectives and possible nonstructural components not discussed in ASCE 31, as well as issues raised by nonstructural rehabilitation unaccompanied by structural rehabilitation (e.g., planning, cost-benefit) are outlined in this commentary, and references are provided for more detailed investigation;

2. The decision is made to rehabilitate the building, either structurally, nonstructurally, or both;

3. From Chapter 1 of this standard, the designer reviews Rehabilitation Objectives and, in concert with the authority having jurisdiction, determines the objective. Alternatively, the objective may have been already defined in an ordinance or other policy;

4. Following a decision on the Rehabilitation Objective, which includes the Nonstructural Performance Level or Range as well as ground motion criteria, the designer consults Chapter 11 of this standard;

5. Using Chapter 11, the designer prepares a definitive list of nonstructural components that are within the scope of the rehabilitation, based on the selected Nonstructural Performance Level and an assessment of component condition. For the Life Safety Nonstructural Performance Level and, to some extent, the Immediate Occupancy Nonstructural Performance Level, Chapters 2 and 11 of this standard specify requirements. However, for other levels and ranges, there is a need to evaluate and prioritize. From the list of nonstructural components within the project scope, a design assessment is made to determine if the component requires rehabilitation and, from Table 11-1, the rehabilitation Analysis Method (Analytical or Prescriptive) for each component or component group is determined;

6. For those components that do not meet the criteria, an appropriate analysis and design procedure is undertaken, with the aim of bringing the component into compliance with the criteria appropriate to the Nonstructural Performance Level or Range and the ground motion criteria; and

7. Nonstructural rehabilitation design documents are prepared.

## 11.4 REHABILITATION OBJECTIVES AND PERFORMANCE LEVELS

Rehabilitation Objectives that include performance levels for nonstructural components shall be established in accordance with Section 1.4. The level of seismicity shall be determined in accordance with Section 1.6.3.

## C11.4 REHABILITATION OBJECTIVES AND PERFORMANCE LEVELS

The nonstructural Rehabilitation Objective may be the same as the Structural Rehabilitation Objective, or it may differ. For the Basic Safety Objective (BSO), structural and nonstructural requirements specified in this standard must be met.

This standard is also intended to be applicable to the situation where nonstructural—but not structural—components are to be rehabilitated. Rehabilitation that is restricted to the nonstructural components will typically fall within the Limited Safety Nonstructural Performance Range unless the structure is already determined to meet a specified Rehabilitation Objective. To qualify for any Rehabilitation Objective

higher than Limited Safety, consideration of structural behavior is necessary to properly take into account loads on nonstructural components generated by inertial forces or deformations imposed by the structure.

### C11.4.1 Regional Seismicity and Nonstructural Components

Requirements for the rehabilitation of nonstructural components relating to the three Seismic Levels—High, Moderate, and Low—are shown in Table 11-1 and noted in each section, where applicable. In general, for levels of low seismicity, certain nonstructural components have no rehabilitation requirements with respect to the Life Safety Nonstructural Performance Level. Rehabilitation of these components, particularly where rehabilitation is simple, may nevertheless be desirable for damage control and property loss reduction.

### C11.4.2 Means of Egress: Escape and Rescue

Preservation of egress is accomplished primarily by ensuring that the most hazardous nonstructural components are replaced or rehabilitated. The items listed in Table 11-1 for achieving the Life Safety Nonstructural Performance Level show that typical requirements for maintaining egress will, in effect, be accomplished if the egress-related components are addressed. These would include the following items listed in ASCE 31 (ASCE 2002).

1. Walls around stairs, elevator enclosures, and corridors are not hollow clay tile or unreinforced masonry;
2. Stair enclosures do not contain any piping or equipment except as required for life safety;
3. Veneers, cornices, and other ornamentation above building exits are well-anchored to the structural system; and
4. Parapets and canopies are anchored and braced to prevent collapse and blockage of building exits.

Beyond this, the following list describes some conditions that might be commonly recognized as representing major obstruction; the building should be inspected to see whether these or any similar hazardous conditions exist. If so, their replacement or rehabilitation should be included in the rehabilitation plan:

1. Partitions taller than 6 ft and weighing more than 5 lbs/sf, if collapse of the entire partition—rather than cracking—is the expected mode of failure and if egress would be impeded;

2. Ceilings, soffits, or any ceiling or decorative ceiling component weighing more than 2 lbs/sf, if it is expected that large areas (pieces measuring 10 sf or larger) would fall;

3. Potential for falling ceiling-located light fixtures or piping; diffusers and ductwork, speakers and alarms, and other objects located higher than 42 in. off the floor;

4. Potential for falling debris weighing more than 100 lbs that, if it fell in an earthquake, would obstruct a required exit door or other component, such as a rescue window or fire escape; and

5. Potential for jammed doors or windows required as part of an exit path—including doors to individual offices, rest rooms, and other occupied spaces.

Of these, the first four are also taken care of in the Life Safety Nonstructural Performance Level requirement. The last condition is very difficult to eliminate with any assurance, except for low levels of shaking in which structural drift and deformation will be minimal, and the need for escape and rescue correspondingly slight.

## 11.5 STRUCTURAL–NONSTRUCTURAL INTERACTION

### 11.5.1 Response Modification

Nonstructural components shall be included in the mathematical model of the building in accordance with the requirements of Section 3.2.2.3. Nonstructural components included in the mathematical model of the building shall be evaluated for forces and deformations imposed by the structure, computed in accordance with Chapter 3.

### 11.5.2 Base Isolation

In a base-isolated structure, nonstructural components located at or above the isolation interface shall comply with the requirements in Section 9.2.6.2.1. Nonstructural components that cross the isolation interface shall comply with the requirements of Section 9.2.6.2.2. Nonstructural components located below the isolation interface shall comply with the requirements of this chapter.

## 11.6 CLASSIFICATION OF ACCELERATION-SENSITIVE AND DEFORMATION-SENSITIVE COMPONENTS

Nonstructural components shall be classified based on their response sensitivity as follows:

1. Nonstructural components that are sensitive to and subject to damage from inertial loading shall be classified as *acceleration-sensitive* components;

2. Nonstructural components that are sensitive and subject to damage imposed by drift or deformation of the structure shall be classified as *deformation-sensitive*; and

3. Nonstructural components that are sensitive to both inertial loading and drift and deformation of the structure shall be classified as *deformation-sensitive*.

## C11.6 CLASSIFICATION OF ACCELERATION-SENSITIVE AND DEFORMATION-SENSITIVE COMPONENTS

Classification of acceleration-sensitive or deformation-sensitive components are discussed, where necessary, in each component section (Sections 11.9, 11.10, and 11.11). Table C11-3 summarizes the sensitivity of non-structural components listed in Table 11-1, and identifies which are of primary or secondary concern. The guiding principle for deciding whether a component requires a force analysis, as defined in Section 11.7, is that analysis of inertial loads generated within the component is necessary to properly consider the component's seismic behavior. The guiding principle for deciding whether a component requires a drift analysis, as defined in Section 11.7, is that analysis of drift is necessary to properly consider the component's seismic behavior.

Glazing or other components that can hazardously fail at a drift ratio less than 0.01 (depending on installation details) or components that can undergo greater distortion without hazardous failure resulting—for example, typical gypsum board partitions—should be considered.

**Use of Drift Ratio Values as Acceptance Criteria.** The data on drift ratio values related to damage states are limited, and the use of single median drift ratio values as acceptance criteria must cover a broad range of actual conditions. It is therefore suggested that the limiting drift values shown in this chapter be used as a guide for evaluating the probability of a given damage state for a subject building, but not be used as absolute acceptance criteria. At higher Nonstructural Performance Levels, it is likely that the criteria for nonstructural deformation-sensitive components may control the structural rehabilitation design. These criteria should be regarded as a flag for the careful evaluation of structural–nonstructural interaction and consequent damage states, rather than the required imposition of absolute acceptance criteria

### Table C11-3. Nonstructural Components: Response Sensitivity

| Component | Sensitivity | |
|---|---|---|
| | Acceleration | Deformation |
| **Architectural (Section 11.9)** | | |
| 1. **Exterior Skin** | S[2] | P[1] |
|    Adhered Veneer | S | P |
|    Anchored Veneer | S | P |
|    Glass Blocks | S | P |
|    Prefabricated Panels | S | P |
|    Glazing Systems | S | P |
| 2. **Partitions** | | |
|    Heavy | S | P |
|    Light | S | P |
| 3. **Interior Veneers** | S | P |
|    Stone, Including Marble | S | P |
|    Ceramic Tile | S | P |
| 4. **Ceilings** | | |
|    Directly Applied to Structure | P | |
|    Dropped Furred Gypsum Board | P | |
|    Suspended Lath and Plaster | S | P |
|    Suspended Integrated Ceiling | S | P |
| 5. **Parapets and Appendages** | P | |
| 6. **Canopies and Marquees** | P | |
| 7. **Chimneys and Stacks** | P | |
| 8. **Stairs** | P | S |
| **Mechanical Equipment (Section 11.10)** | | |
| 1. **Mechanical Equipment** | P | |
|    Boilers and Furnaces | P | |
|    General Mfg. and Process Machinery | P | |
|    HVAC Equipment, Vibration-Isolated | P | |
|    HVAC Equipment, Non-Vibration-Isolated | P | |
|    HVAC Equipment, Mounted In-Line with Ductwork | P | |
| 2. **Storage Vessels and Water Heaters** | | |
|    Structurally Supported Vessels (Category 1) | P | |
|    FlatBottom Vessels (Category 2) | P | |
| 3. **Pressure Piping** | P | S |
| 4. **Fire Suppression Piping** | P | S |
| 5. **Fluid Piping, not Fire Suppression** | | |
|    Hazardous Materials | P | S |
|    Nonhazardous Materials | P | S |
| 6. **Ductwork** | P | S |

[1]P, Primary response
[2]S, Secondary response

that might require costly redesign of the structural rehabilitation.

## 11.7 EVALUATION PROCEDURES

One of the following evaluation procedures for nonstructural components shall be selected based on the requirements of Table 11-1:

1. Analytical Procedure; or
2. Prescriptive Procedure.

### 11.7.1 Analytical Procedure

Where the Prescriptive Procedure is not permitted based on Table 11-1, forces and deformations on nonstructural components shall be calculated as follows:

1. If a force analysis only is permitted by Table 11-1 and either the Hazards Reduced or Life Safety Nonstructural Performance Level is selected, then use of the default equations given in Section 11.7.3 shall be permitted to calculate seismic design forces on nonstructural components;

2. If a force analysis only is permitted by Table 11-1 and a Nonstructural Performance Level higher than Life Safety is selected, then the default equations of Section 11.7.3 do not apply, and seismic design forces shall be calculated in accordance with Section 11.7.4; and

3. If both force and deformation analysis are required by Table 11-1, then seismic design forces shall be calculated in accordance with Section 11.7.4 and drift ratios or relative displacements shall be calculated in accordance with Section 11.7.5. The deformation and associated drift ratio of the structural component(s) to which the deformation-sensitive nonstructural component is attached shall be determined in accordance with Chapter 3; or

4. Alternatively, the calculation of seismic design forces and deformations in accordance with Section 11.7.6 shall be permitted.

### C11.7.1 Analytical Procedure

For nonstructural components, the Analytical Procedure, which consists of the default equation and general equation approaches, is applicable to any case. The Prescriptive Procedure is limited by Table 11-1 to specified combinations of seismicity and component type for compliance with the Life Safety Nonstructural Performance Level.

### 11.7.2 Prescriptive Procedure

Where the Prescriptive Procedure is permitted in Table 11-1, the characteristics of the nonstructural component shall be compared with characteristics as specified in approved codes.

### C11.7.2 Prescriptive Procedure

A Prescriptive Procedure consists of published standards and references that describe the design concepts and construction features that must be present for a given nonstructural component to be seismically protected. No engineering calculations are required in a Prescriptive Procedure, although in some cases an engineering review of the design and installation is required.

Suggested references for prescriptive requirements are listed in the commentary of the "Component Behavior and Rehabilitation Concepts" subsection of Sections 11.9 through 11.11 for each component type.

### 11.7.3 Force Analysis: Default Equations

Calculation of seismic design forces on nonstructural components using the following default Eqs. 11-1 and 11-2 shall be permitted in accordance with Section 11.7.1. Horizontal seismic design forces shall be

computed using Eq. 11-1. Where specifically required in Sections 11.9, 11.10, and 11.11, vertical seismic forces for horizontal cantilever components shall be determined using Eq. 11-2. Vertical seismic forces for all other components shall be determined using Eq. 11-3.

$$F_p = 1.6 S_{XS} W_P \qquad \text{(Eq. 11-1)}$$

$$F_{pv} = \frac{2}{3} F_p \qquad \text{(Eq. 11-2)}$$

$$F_{pv} \text{ (minimum)} = \pm 0.2 S_{XS} W_p \qquad \text{(Eq. 11-3)}$$

where

$F_p$ = component seismic design force applied horizontally at the center of gravity of the component or distributed according to the mass distribution of the component;

$F_{pv}$ = component seismic design force applied vertically at the center of gravity of the component or distributed according to the mass distribution of the component;

$S_{XS}$ = spectral response acceleration parameter at short periods for any Earthquake Hazard Level and any damping determined in accordance with Section 1.6.1.4 or 1.6.2.1; and

$W_p$ = component operating weight.

### 11.7.4 Force Analysis: General Equations

#### 11.7.4.1 Horizontal Seismic Forces

*11.7.4.1.1 Life Safety and Hazards Reduced Nonstructural Performance Levels* Where default equations of Section 11.7.3 do not apply, horizontal seismic design forces on nonstructural components shall be determined in accordance with Eq. 11-4.

$$F_p = \frac{0.4 a_p S_{XS} I_p W_p \left(1 + \frac{2x}{h}\right)}{R_p} \qquad \text{(Eq. 11-4)}$$

$F_p$ calculated in accordance with Eq. 11-4 shall be based on the stiffness of the component and ductility of its bracing and anchorage, but it need not exceed the default value of $F_p$ calculated in accordance with Eq. 11-1 and shall not be less than $F_p$ computed in accordance with Eq. 11-5.

$$F_p \text{ (minimum)} = 0.3 S_{XS} I_p W_p \qquad \text{(Eq. 11-5)}$$

where

$a_p$ = component amplification factor from Table 11-2;

$F_p$ = component seismic design force applied horizontally at the center of gravity of the component and distributed according to the mass distribution of the component;

$S_{XS}$ = spectral response acceleration parameter at short periods for any Earthquake Hazard Level and any damping determined in accordance with Section 1.6.1.4 or 1.6.2.1;

$h$ = average roof elevation of structure, relative to grade elevation;

$I_p$ = component performance factor (1.0 shall be used for the Life Safety and Hazards Reduced Nonstructural Performance Levels);

$R_p$ = component response modification factor from Table 11-2; and

$x$ = elevation in structure of the center of gravity of the component relative to grade elevation.

### 11.7.4.1.2 Immediate Occupancy Nonstructural Performance Level

Seismic design forces for nonstructural components being evaluated to the Immediate Occupancy Nonstructural Performance Level shall be evaluated considering the dynamic characteristics of the building and the nonstructural component. The fundamental period of vibration of the nonstructural component ($T_p$) in each direction shall be estimated using Eq. 11-5a.

$$T_p = 2\pi \sqrt{\frac{W_p}{K_p g}} \qquad \text{(Eq. 11-5a)}$$

where

$T_p$ = component fundamental period;

$W_p$ = component operating weight;

$g$ = gravitational acceleration; and

$K_p$ = approximate stiffness of the support system of the component, its bracing, and its attachment, determined in terms of load per unit deflection at the center of gravity of the component.

Nonstructural seismic design forces shall be calculated based on Eq. 11-5b.

$$F_p = \frac{I_p a_p A_x W_p}{R_p} \qquad \text{(Eq. 11-5b)}$$

where

$I_p$ = component performance factor = 1.5;

$a_p$ = component amplification factor determined based on the dynamic interaction between the nonstructural component and the building vibrational characteristics; in lieu of a rigorous analysis, the value of $a_p$ may be obtained from Table 11-2;

$R_p$ = component response modification factor from Table 11-2; and

$A_x$ = story acceleration at level x calculated based on a linear dynamic analysis of the building in accordance with Section 3.3.2; in lieu of a rigorous analysis, the value of $A_x$ may be obtained using Eq. 11-5c.

$$A_x = 0.4 S_{XS} \left( 1 + \frac{2x}{h} \right) \qquad \text{(Eq. 11-5c)}$$

where

$S_{XS}$ = the 5% damped spectral response acceleration parameter at short periods for a given Earthquake Hazard Level determined in accordance with Section 1.6.1.4 or 1.6.2.1;

$h$ = average roof elevation of structure, relative to grade elevation; and

$x$ = elevation in structure of the center of gravity of the component relative to grade elevation.

### C11.7.4.1 Horizontal Seismic Forces

Seismic forces for nonstructural components are generated based on three effects: the ground acceleration at the base of the building, the ratio of the floor acceleration at the location of the nonstructural component to the ground acceleration, and the dynamic amplification due to resonance between the nonstructural component and the building response. Equation 11-4 provides an estimate of the horizontal acceleration of a nonstructural component. The peak ground acceleration is calculated as 0.4 times the short period response acceleration ($S_{XS}$).

The ratio of the floor acceleration at the location of the nonstructural component is based on a linearly increasing variation of acceleration over the height of the building. The term $(1 + 2x/h)$ is used to calculate this variation based on a linear variation of floor accelerations over the height of the building and is based on an assumed first-mode response of a building with uniform stiffness and mass. For buildings that have significant higher-mode response, this linearly increasing assumption may overestimate the acceleration at floors below the roof. A linear dynamic analysis using a response spectrum can be used as an alternate method of estimating the variation of floor accelerations.

The $a_p$ factor provides an estimate of the dynamic amplification due to the resonance of response of the nonstructural component with one of the modes of vibration of the building. Table 11-2 provides an estimate of this amplification for most nonstructural components. In Table 11-2, components assumed to be rigid are assigned an $a_p$ value of 1 and components assumed to be flexible are assigned an $a_p$ value of 2.5.

### Table 11-2. Nonstructural Component Amplification and Response Modification Factors

| Architectural Component or Component (Section 11.9) | $a_p{}^1$ | $R_p{}^4$ |
|---|---|---|
| Interior nonstructural walls and partitions[2] | | |
|     Plain masonry walls | 1.0 | 1.5 |
|     All other walls and partitions | 1.0 | 2.5 |
| Cantilever components, unbraced or braced (to structural frame) below their centers of mass | | |
|     Parapets and cantilevered interior nonstructural walls | 2.5 | 2.5 |
|     Chimneys and stacks where laterally supported by structures | 2.5 | 2.5 |
| Cantilever components, braced (to structural frame) above their centers of mass | | |
|     Parapets | 1.0 | 2.5 |
|     Chimneys and stacks | 1.0 | 2.5 |
|     Exterior nonstructural walls[2] | 1.0 | 2.5 |
| Exterior nonstructural wall components and connections[2] | | |
|     Wall component | 1.0 | 2.5 |
|     Body of wall–panel connections | 1.0 | 2.5 |
|     Fasteners of the connecting system | 1.25 | 1.0 |
| Veneer | | |
|     High-deformability components and attachments | 1.0 | 2.5 |
|     Low-deformability components and attachments | 1.0 | 1.5 |
| Penthouses (except where framed by an extension of the building frame) | 2.5 | 3.5 |
| Ceilings | | |
|     All | 1.0 | 2.5 |
| Cabinets | | |
|     Storage cabinets and laboratory equipment | 1.0 | 2.5 |
| Storage Racks[5] | 2.5 | 4.0 |
| Access floors | | |
|     Special access floors | 1.0 | 2.5 |
|     All other | 1.0 | 1.5 |
| Appendages and ornamentation | 2.5 | 2.5 |
| Signs and billboards | 2.5 | 2.5 |
| Other rigid components | | |
|     High-deformability components and attachments | 1.0 | 3.5 |
|     Limited-deformability components and attachments | 1.0 | 2.5 |
|     Low-deformability components and attachments | 1.0 | 1.5 |
| Other flexible components | | |
|     High-deformability components and attachments | 2.5 | 3.5 |
|     Limited-deformability components and attachments | 2.5 | 2.5 |
|     Low-deformability components and attachments | 2.5 | 1.5 |
| **Mechanical and Electrical Components (Section 11.10)** | | |
| Air-side HVAC, fans, air handlers, air conditioning units, cabinet heaters, air distribution boxes, and other mechanical components constructed of sheet metal framing. | 2.5 | 3.0 |
| Wet-side HVAC, boilers, furnaces, atmospheric tanks and bins, chillers, water heaters, heat exchangers, evaporators, air separators, manufacturing or process equipment, and other mechanical components constructed of high-deformability materials. | 1.0 | 2.5 |
| Engines, turbines, pumps, compressors, and pressure vessels not supported on skirts and not within the scope of Section 9.14. | 1.0 | 2.5 |
| Skirt-supported pressure vessels not within the scope of Section 9.14. | 2.5 | 2.5 |
| Elevator and escalator components. | 1.0 | 2.5 |

| Architectural Component or Component (Section 11.9) | $a_p{}^1$ | $R_p{}^4$ |
|---|---|---|
| Generators, batteries, inverters, motors, transformers, and other electrical components constructed of high-deformability materials. | 1.0 | 2.5 |
| Motor control centers, panel boards, switch gear, instrumentation cabinets, and other components constructed of sheet metal framing. | 2.5 | 3.0 |
| Communication equipment, computers, instrumentation, and controls. | 1.0 | 2.5 |
| Roof-mounted chimneys, stacks, cooling and electrical towers laterally braced below their center of mass. | 2.5 | 3.0 |
| Roof-mounted chimneys, stacks, cooling and electrical towers laterally braced above their center of mass. | 1.0 | 2.5 |
| Lighting fixtures. | 1.0 | 1.5 |
| Other mechanical or electrical components. | 1.0 | 1.5 |
| **Vibration-Isolated Components and Systems**[3] | | |
| Components and systems isolated using neoprene components and neoprene isolated floors with built-in or separate elastomeric snubbing devices or resilient perimeter stops. | 2.5 | 2.5 |
| Spring-isolated components and systems and vibration isolated floors closely restrained using built-in or separate elastomeric snubbing devices or resilient perimeter stops. | 2.5 | 2.0 |
| Internally isolated components and systems. | 2.5 | 2.0 |
| Suspended vibration isolated equipment including in-line duct devices and suspended internally isolated components. | 2.5 | 2.5 |
| **Distribution Systems** | | |
| Piping in accordance with ASME B31[6], including in-line components with joints made by welding or brazing | 2.5 | 12.0 |
| Piping in accordance with ASME B31, including in-line components, constructed of high- or limited-deformability materials, with joints made by threading, bonding, compression couplings, or grooved couplings. | 2.5 | 6.0 |
| Piping and tubing not in accordance with ASME B31, including in-line components, constructed of high-deformability materials, with joints made by welding or brazing. | 2.5 | 9.0 |
| Piping and tubing not in accordance with ASME B31, including in-line components, constructed of high- or limited-deformability materials, with joints made by threading, bonding, compression couplings, or grooved couplings | 2.5 | 4.5 |
| Piping and tubing constructed of low-deformability materials, such as cast iron, glass, and nonductile plastics. | 2.5 | 3.0 |
| Ductwork, including in-line components, constructed of high-deformability materials, with joints made by welding or brazing. | 2.5 | 9.0 |
| Ductwork, including in-line components, constructed of high- or limited-deformability materials with joints made by means other than welding or brazing. | 2.5 | 6.0 |
| Ductwork, including in-line components, constructed of low-deformability materials, such as cast iron, glass, and nonductile plastics. | 2.5 | 3.0 |
| Electrical conduit, bus ducts, rigidly mounted cable trays, and plumbing. | 1.0 | 2.5 |
| Manufacturing or process conveyors (nonpersonnel). | 2.5 | 3.0 |
| Suspended cable trays. | 2.5 | 6.0 |
| **Furnishings and Interior Equipment (Section 11.11)** | | |
| Storage racks[5] | 2.5 | 4 |
| Bookcases | 1 | 3 |
| Computer access floors | 1 | 3 |
| Hazardous materials storage | 2.5 | 1 |
| Computer and communications racks | 2.5 | 6 |

*Continued*

**Table 11-2. (Continued)**

| Architectural Component or Component (Section 11.9) | $a_p{}^1$ | $R_p{}^4$ |
|---|---|---|
| Elevators | 1 | 3 |
| Conveyors | 2.5 | 3 |

[1]A lower value for ap is permitted where justified by detailed dynamic analyses. The value for $a_p$ shall be not be less than 1.0. The value of $a_p$ equal to 1.0 is for rigid components and rigidly attached components. The value of $a_p$ equal to 2.5 is for flexible components and flexibly attached components.

[2]Where flexible diaphragms provide lateral support for concrete or masonry walls or partitions, the design forces for anchorage to the diaphragm shall be as specified in Sec. 2.6.7.1.

[3]Components mounted on vibration isolators shall have a bumper restraint or snubber in each horizontal direction. The design force shall be taken as $2F_p$ if the nominal clearance (air gap) between the equipment support frame and restraint is greater than $\frac{1}{4}$ in. If the nominal clearance specified on the construction documents is not greater than $\frac{1}{4}$ in., the design force may be taken as $F_p$.

[4]The value of $R_p$ used to determine the forces in the connected part shall not exceed 1.5 unless the component anchorage is governed by the strength of a ductile steel component.

[5]Storage racks over 6 ft in height shall be designed in accordance with the provisions of Section 11.11.1.

[6]American Society of Mechanical Engineers (ASME B31). (2000). *Code for Pressure Piping*, New York.

A period of vibration of 0.06 sec is used to distinguish between rigid and flexible components. The engineer should verify that the $a_p$ value used is appropriate for the actual component and its support system.

For many buildings, the primary mode of vibration in each direction will have the most influence on the dynamic amplification of nonstructural components. For buildings with primary mode periods greater than 1 sec, the second or third mode of vibration may also cause some dynamic amplification.

Equation 11-5c provides a slightly revised form of Equation 11-5b for use where checking nonstructural components for Immediate Occupancy Performance Level. In Equation 11-5b, the factor $(a_p)$ is defined as the dynamic amplification factor considering resonance of the nonstructural component with one of the modes of the building. The intent is to consider this dynamic amplification effect for nonstructural components for Immediate Occupancy Performance Level. Guidelines for considering this effect are provided in the *Tri-Services Seismic Design for Buildings*, TM5-809-10 and *Seismic Design Guidelines for Essential Buildings*, TM5-809-10-1 (Dept. of the Army, Navy, and Air Force 1986). Other approved procedures could also be used. It is permissible to use the $a_p$ factors from Table 11-2.

Equation 11-5c also provides a factor $A_x$, which represents the floor accelerations. The intent is that a linear dynamic analysis of the building be performed to determine the actual story accelerations based on the ground motion considered for a sufficient number of modes of vibration for the range of periods of vibration of the nonstructural components to be designed. The modal story accelerations can be combined using standard modal combination procedures. Linear dynamic analysis procedures are considered sufficiently accurate for estimating the story accelerations since buildings checked for Immediate Occupancy Performance Level are expected to behave nearly elastically for the design earthquake.

### Eq. 11-11.7.4.2 *Vertical Seismic Forces*

Where the default equations of Section 11.7.3 do not apply, and where specifically required by Sections 11.9, 11.10, and 11.11, vertical seismic design forces on nonstructural components shall be determined in accordance with Eq. 11-6.

$$F_{pv} = \frac{0.27 a_p S_{XS} I_p W_p}{R_p} \qquad \text{(Eq. 11-6)}$$

$F_p$ calculated in accordance with Eq. 11-6 need not exceed $F_p$ calculated in accordance with Eq. 11-2 and shall not be less than $F_{pv}$ (minimum) computed in accordance with Eq. 11-7.

$$F_{pv}(\text{minimum}) = 0.2\, S_{XS} I_p W_p \qquad \text{(Eq. 11-7)}$$

where

$F_{pv}$ = component seismic design force applied vertically at the center of gravity of the component or distributed according to the mass distribution of the component.

All other terms in Eqs. 11-6 and 11-7 shall be as defined in Section 11.7.4.1.

### 11.7.5 Deformation Analysis

Where nonstructural components are anchored by connection points at different levels $x$ and $y$ on the same building or structural system, drift ratios $(D_r)$ shall be calculated in accordance with Eq. 11-8.

$$D_r = (\delta_{xA} - \delta_{yA}) / (X - Y) \qquad \text{(Eq. 11-8)}$$

where

$D_r$ = drift ratio;
$X$ = height of upper support attachment at level $x$ as measured from grade; and
$Y$ = height of lower support attachment at level $y$ as measured from grade.

Where nonstructural components are anchored by connection points on separate buildings or structural systems at the same level $x$, relative displacements ($D_p$) shall be calculated in accordance with Eq. 11-9.

$$D_p = |\delta_{xA}| + |\delta_{xB}| \qquad \text{(Eq. 11-9)}$$

where

$D_p$ = relative seismic displacement;
$\delta_{xA}$ = deflection at building level $x$ of Building A, determined by analysis as defined in Chapter 3;
$\delta_{yA}$ = deflection at building level $y$ of Building A, determined by analysis as defined in Chapter 3; and
$\delta{xB}$ = deflection at building level $x$ of Building B, determined by analysis as defined in Chapter 3 or equal to 0.03 times the height $X$ of level $x$ above grade or as determined using other approved approximate procedures.

The effects of seismic displacements shall be considered in combination with displacements caused by other loads that are present.

### 11.7.6 Other Procedures

Other approved procedures shall be permitted to determine the maximum acceleration of the building at each component support and the maximum drift ratios or relative displacements between two supports of an individual component.

### C11.7.6 Other Procedures

Linear and nonlinear procedures may be used to calculate the maximum acceleration of each component support and the story drifts of the building, taking into account the location of the component in the building. Consideration of the flexibility of the component, and the possible amplification of the building roof and floor accelerations and displacements in the component, would require the development of roof and floor response spectra or acceleration time histories at the nonstructural support locations, derived from the dynamic response of the structure. If the resulting floor spectra are less than demands calculated in accordance with Sections 11.7.3 and 11.7.4, it may be advantageous to use this procedure.

Relative displacements between component supports are difficult to calculate, even with the use of acceleration time histories, because the maximum displacement of each component support at different levels in the building might not occur at the same time during the building response.

Guidelines for these dynamic analyses for nonstructural components are given in Chapter 6 of *Seismic Design Guidelines for Essential Buildings*, a supplement to TM5-809-10.1 (Dept. of the Army, Navy, and Air Force 1986).

These other analytical procedures are considered too complex for the rehabilitation of nonessential building nonstructural components for Immediate Occupancy and Life Safety Nonstructural Performance Levels.

Recent research (Drake and Bachman 1995) has shown that the analytical procedures in Sections 11.7.3 and 11.7.4, which are based on FEMA 302 (FEMA 1997) analytical procedures, provide an upper bound for the seismic forces on nonstructural components.

## 11.8 REHABILITATION APPROACHES

Nonstructural rehabilitation shall be accomplished by approved methods based on the classification of the nonstructural component and the performance level desired for the nonstructural component.

1. For the rehabilitation of nonstructural components that are acceleration-sensitive for Hazards Reduced or Life Safety Performance Levels, the rehabilitation approach shall provide for position retention. Position retention shall be defined as providing bracing, anchorage, attachment, or other approved methods to prevent the nonstructural component from becoming dislodged during earthquake shaking.
2. The rehabilitation of nonstructural components for Immediate Occupancy Performance Level shall provide for position retention. In addition, the rehabilitation of mechanical and electrical components shall prevent damage to the components that will affect the occupancy of the building.
3. For the rehabilitation of nonstructural components that are deformation sensitive, the rehabilitation approach shall provide for sufficient deformation capability for the nonstructural components to allow the nonstructural component to undergo the calculated deformation while maintaining position retention.

## C11.8 REHABILITATION APPROACHES

A general set of alternate methods is available for the rehabilitation of nonstructural components. These are briefly outlined in this section with examples to clarify the intent. However, the choice of rehabilitation technique and its design is the responsibility of the design professional, and use of alternative approaches to those noted below or otherwise customarily in use is acceptable, provided that it can be shown to the satisfaction of the building official that the acceptance criteria are met.

For Hazards Reduced and Life Safety Performance Levels, most nonstructural components that are acceleration-sensitive should be rehabilitated considering position retention. Nonstructural components that are drift-sensitive should be rehabilitated to allow for imposed deformation. Nonstructural components that are drift-sensitive need not be designed to prevent damage to the nonstructural component or its attachments provided that stability of the component is maintained. Components that are both acceleration-sensitive in one direction and drift-sensitive in the other direction should be rehabilitated considering both effects.

### C11.8.1 Replacement

Replacement involves the complete removal of the component and its connections, and its replacement by new components (e.g., the removal of exterior cladding panels, the installation of new connections, and installation of new panels). As with structural components, the installation of new nonstructural components as part of a seismic rehabilitation project should be the same as for new construction.

### C11.8.2 Strengthening

Strengthening involves additions to the component to improve its strength to meet the required force levels (e.g., additional members might be welded to a support to prevent buckling).

### C11.8.3 Repair

Repair involves the repair of any damaged parts or members of the component to enable the component to meet its acceptance criteria (e.g., some corroded attachments for a precast concrete cladding system might be repaired and replaced without removing or replacing the entire panel system).

### C11.8.4 Bracing

Bracing involves the addition of members and attachments that brace the component internally or to the building structure. A suspended ceiling system might be rehabilitated by the addition of diagonal wire bracing and vertical compression struts.

### C11.8.5 Attachment

Attachment refers to methods that are primarily mechanical, such as bolting, by which nonstructural components are attached to the structure or other supporting components. Typical attachments are the bolting of items of mechanical equipment to a reinforced concrete floor or base. Supports and attachments for mechanical and electrical equipment should be designed according to accepted engineering principles. The following guidelines are recommended:

1. Attachments and supports transferring seismic loads should be constructed of materials suitable for the application, and designed and constructed in accordance with a nationally recognized standard;
2. Attachments embedded in concrete should be suitable for cyclic loads;
3. Rod hangers may be considered seismic supports if the length of the hanger from the supporting structure is 12 in. or less. Rod hangers should not be constructed in a manner that would subject the rod to bending moments;
4. Seismic supports should be constructed so that support engagement is maintained;
5. Friction clips should not be used for anchorage attachment;
6. Expansion anchors should not be used for mechanical equipment rated over 10 hp, unless undercut expansion anchors are used;
7. Drilled and grouted-in-place anchors for tensile load applications should use either expansive cement or expansive epoxy grout;
8. Supports should be specifically evaluated if weak-axis bending of cold-formed support steel is relied on for the seismic load path;
9. Components mounted on vibration isolation systems should have a bumper restraint or snubber in each horizontal direction. The design force should be taken as $2F_p$; and
10. Oversized washers should be used at bolted connections through the base sheet metal if the base is not reinforced with stiffeners.

Lighting fixtures resting in a suspended ceiling grid may be rehabilitated by adding wires that directly attach the fixtures to the floor above, or to the roof structure to prevent their falling.

## 11.9 ARCHITECTURAL COMPONENTS: DEFINITION, BEHAVIOR, AND ACCEPTANCE CRITERIA

### 11.9.1 Exterior Wall Components

#### 11.9.1.1 Adhered Veneer

*11.9.1.1.1 Definition and Scope* Adhered veneer shall include the following types of exterior finish materials secured to a backing material, which shall be masonry, concrete, cement plaster, or to a structural framework material by adhesives:

1. Tile, masonry, stone, terra cotta, or other similar materials;
2. Glass mosaic units;
3. Ceramic tile; and
4. Exterior plaster (stucco).

*C11.9.1.1.1 Definition and Scope* Adhered veneers are generally thinner materials, although thicker veneers (especially masonry, stone, and terra cotta) may be encountered. Although the behavior of the thicker veneers is still dominated by the behavior of the substrate, the threat to life safety due to failure may rise significantly for thicker, heavier veneers. The height of the veneer as well as the likely size of falling fragments should be considered.

Tile, masonry, stone, terra cotta, and similar materials are typically less than 1 in. thick. Glass mosaic blocks are typically 2 in. $\times$ 2 in. $\times \frac{3}{8}$ in. thick.

*11.9.1.1.2 Component Behavior and Rehabilitation Methods* Adhered veneer shall be considered deformation-sensitive.

Adhered veneer not conforming to the acceptance criteria of Section 11.9.1.1.3 shall be rehabilitated in accordance with Section 11.8.

*C11.9.1.1.2 Component Behavior and Rehabilitation Methods* Adhered veneers are predominantly deformation-sensitive. Deformation of the substrate leads to cracking or separation of the veneer from its backing. Poorly adhered veneers may be dislodged by direct acceleration.

Nonconformance requires limiting drift, special detailing to isolate the substrate from the structure to permit drift, or replacement with drift-tolerant material. Poorly adhered veneers should be replaced.

*11.9.1.1.3 Acceptance Criteria* Acceptance criteria shall be applied in accordance with Section 11.3.2.

1. **Life Safety Nonstructural Performance Level.** Backing shall be adequately anchored to resist seismic forces computed in accordance with Section 11.7.3 or 11.7.4. The drift ratio calculated in accordance with Section 11.7.5 shall be limited to 0.02.
2. **Immediate Occupancy Nonstructural Performance Level.** Backing shall be adequately attached to resist seismic design forces computed in accordance with Section 11.7.4. The drift ratio computed in accordance with Section 11.7.5 shall be limited to 0.01.

*11.9.1.1.4 Evaluation Requirements* Adhered veneer shall be evaluated by visual observation and tapping to discern looseness or cracking.

*C11.9.1.1.4 Evaluation Requirements* Tapping may indicate either defective bonding to the substrate or excessive flexibility of the supporting structure.

#### 11.9.1.2 Anchored Veneer

*11.9.1.2.1 Definition and Scope* Anchored veneer shall include the following types of masonry or stone units that are attached to the supporting structure by mechanical means:

1. Masonry units;
2. Stone units; and
3. Stone slab units.

The provisions of this section shall apply to units that are more than 48 in. above the ground or adjacent exterior area.

*C11.9.1.2.1 Definition and Scope* Masonry units are typically 5 in. or less in thickness. Stone slab units are typically 2 in. or less in thickness.

*11.9.1.2.2 Component Behavior and Rehabilitation Methods* Anchored veneer shall be considered both acceleration-sensitive and deformation-sensitive.

Anchored veneer and connections not conforming to the acceptance criteria of Section 11.9.1.2.3 shall be rehabilitated in accordance with Section 11.8.

*C11.9.1.2.2 Component Behavior and Rehabilitation Methods* Anchored veneer is both acceleration- and deformation-sensitive. Heavy units can be dislodged by direct out-of-plane acceleration, which distorts or fractures the mechanical connections. Special attention should be paid to corners and around openings, which are likely to experience large deformations. In-plane or out-of-plane deformations of the supporting structure, particularly if it is a frame, may similarly affect the connections, and the units may be displaced or dislodged by racking. Thick, anchored veneer may possess significant in-plane stiffness, which can greatly amplify the demands placed on the connections if the supporting structure racks.

Drift analysis is necessary to establish conformance with drift acceptance criteria related to performance level. The drift analysis should consider the construction and behavior of the veneer and its backing to assess the individual parts of the nonstructural component that are required to deform in order to accommodate the required drift. These parts of the nonstructural component should be checked for their capability to allow for the calculated deformation of the structure. Nonconformance requires limiting structural drift, or special detailing to isolate the substrate from the structure to permit drift. Defective connections must be replaced.

*11.9.1.2.3 Acceptance Criteria* Acceptance criteria shall be applied in accordance with Section 11.3.2.

1. **Life Safety Nonstructural Performance Level.** Backing shall be adequately anchored to resist seismic forces computed in accordance with Section 11.7.3 or 11.7.4. The drift ratio calculated in accordance with Section 11.7.5 shall be limited to 0.02.
2. **Immediate Occupancy Nonstructural Performance Level.** Backing shall be adequately attached to resist seismic design forces computed in accordance with Section 11.7.4. The drift ratio computed in accordance with Section 11.7.5 shall be limited to 0.01.

*C11.9.1.2.3 Acceptance Criteria* As an alternative to the drift limits in Section 11.9.1.2.3, the nonstructural component and its backing can be shown by approved testing or analysis to meet the intended performance level for the calculated drift.

*11.9.1.2.4 Evaluation Requirements* Stone units shall have adequate stability, joint detailing, and maintenance to prevent moisture penetration from weather that could destroy the anchors. The anchors shall be visually inspected and tested to determine capacity if any signs of deterioration are visible.

### 11.9.1.3 Glass Block Units and Other Nonstructural Masonry

*11.9.1.3.1 Definition and Scope* Glass block and other units that are self-supporting for static vertical loads,

held together by mortar and structurally detached from the surrounding structure, shall be rehabilitated in accordance with this section.

*11.9.1.3.2 Component Behavior and Rehabilitation Methods* Glass block units and other nonstructural masonry shall be considered both acceleration- and deformation-sensitive.

Rehabilitation of individual walls less than 144 sf or 15 ft in any dimension using Prescriptive Procedures based on Section 2110 of the *ICC* (2003) shall be permitted. For walls larger than 144 sf or 15 ft in any dimension, the Analytical Procedure shall be used.

Glass block units and other nonstructural masonry not conforming with the requirements of Section 11.9.1.3.3 shall be rehabilitated in accordance with Section 11.8.

*C11.9.1.3.2 Component Behavior and Rehabilitation Methods* Glass block and nonstructural masonry are both acceleration- and deformation-sensitive. Failure in-plane generally occurs by deformation in the surrounding structure that results in unit cracking and displacement along the cracks. Failure out-of-plane takes the form of dislodgment or collapse caused by direct acceleration.

Nonconformance with deformation criteria requires limiting structural drift, or special detailing to isolate the glass block wall from the surrounding structure to permit the required drift. The drift analysis should consider the construction and behavior of the veneer and its backing to assess the individual parts of the nonstructural component that are required to deform in order to accommodate the required drift. These parts of the nonstructural component should be checked for their capability to allow for the calculated deformation of the structure. Sufficient reinforcing must be provided to deal with out-of-plane forces. Large walls may need to be subdivided by additional structural supports into smaller areas that can meet the drift or force criteria.

*11.9.1.3.3 Acceptance Criteria* Acceptance criteria shall be applied in accordance with Section 11.3.2.

1. **Life Safety Nonstructural Performance Level.** Glass block and other nonstructural masonry walls and their enclosing framing, shall be capable of resisting both in-plane and out-of-plane forces computed in accordance with Section 11.7.3 or 11.7.4, or shall meet the requirements of the Prescriptive Procedure if permitted. The drift ratio

calculated in accordance with Section 11.7.5 shall be limited to 0.02.

2. **Immediate Occupancy Nonstructural Performance Level.** Glass block and other nonstructural masonry walls and their enclosing framing shall be capable of resisting both in-plane and out-of-plane forces computed in accordance with Section 11.7.4. The drift ratio calculated in accordance with Section 11.7.5 shall be limited to 0.01.

*11.9.1.3.4 Evaluation Requirements* Glass block units and other nonstructural masonry shall be evaluated based on the criteria of Section 2110 of the *ICC* (2003).

### 11.9.1.4 Prefabricated Panels

*11.9.1.4.1 Definition and Scope* The following types of prefabricated panels designed to resist wind, seismic, and other applied forces shall be rehabilitated in accordance with this section:

1. Precast concrete, and concrete panels with facing (generally stone) laminated or mechanically attached;
2. Laminated metal-faced insulated panels; and
3. Steel strong-back panels with insulated, water-resistant facing, or mechanically attached metal or stone facing.

*C11.9.1.4.1 Definition and Scope* Prefabricated panels are generally attached at discreet locations around their perimeters to the structural framing with mechanical connections.

*11.9.1.4.2 Component Behavior and Rehabilitation Methods* Prefabricated panels shall be considered both acceleration- and deformation-sensitive.

Prefabricated panels not conforming to the acceptance criteria of Section 11.9.1.4.3 shall be rehabilitated in accordance with Section 11.8.

*C11.9.1.4.2 Component Behavior and Rehabilitation Methods* Lightweight panels may be damaged by racking; heavy panels may be dislodged by direct acceleration, which distorts or fractures the mechanical connections. The imposed in-plane and out-of-plane deformations are generally accommodated by the connections and not by the prefabricated panels. These connections need to be checked for the detailing to accommodate the required drift. This is generally accomplished by a connection detailed to allow sliding

with a slotted or oversize hole. Drift can also be accommodated by deformation of the connections.

Excessive deformation of the supporting structure—most likely if it is a frame—may result in the panels imposing external racking forces on one another and distorting or fracturing their connections, with consequent displacement or dislodgment.

Drift analysis is necessary to establish conformance with drift acceptance criteria related to the Nonstructural Performance Level. The drift analysis should consider the construction and behavior of the panel and its connections to assess the individual parts of the nonstructural component that are required to deform in order to accommodate the required drift.

Nonconformance requires limiting structural drift, or special detailing to isolate panels from the structure to permit the required drift; this generally requires panel removal. Defective connections must be replaced.

*11.9.1.4.3 Acceptance Criteria* Acceptance criteria shall be applied in accordance with Section 11.3.2.

1. **Life Safety Nonstructural Performance Level.** Prefabricated panels and connections shall be capable of resisting in-plane and out-of-plane forces computed in accordance with Section 11.7.3 or 11.7.4. The drift ratio computed in accordance with Section 11.7.5 shall be limited to 0.02.
2. **Immediate Occupancy Nonstructural Performance Level.** Prefabricated panels and connections shall be capable of resisting in-plane and out-of-plane forces computed in accordance with Section 11.7.4. The drift ratio computed in accordance with Section 11.7.5 shall be limited to 0.01.

*11.9.1.4.4 Evaluation Requirements* Connections shall be visually inspected and tested to determine capacity if any signs of deterioration or displacement are visible.

### 11.9.1.5 Glazed Exterior Wall Systems

*11.9.1.5.1 Definition and Scope* Glazed exterior wall systems shall include the following types of assemblies:

1. Glazed curtain wall systems that extend beyond the edges of structural floor slabs, and are assembled from prefabricated units (e.g., "unitized" curtain wall systems) or assembled on site (e.g., "stick" curtain wall systems);
2. Glazed storefront systems that are installed between structural floor slabs and are prefabricated or assembled on site; and
3. Structural silicone glazing in which silicone sealant is used for the structural transfer of loads from the

glass to its perimeter support system and for the retention of the glass in the opening.

*C11.9.1.5.1 Definition and Scope* The following types of glass are used within each of the glazed exterior wall systems:

1. Annealed glass;
2. Heat-strengthened glass;
3. Fully tempered glass;
4. Laminated glass; and
5. Sealed insulating glass units.

The use of some of these glass types is regulated in building codes.

There are two glazing methods for installing glass in glazed curtain wall and glazed storefront systems:

1. Wet glazing, which can utilize three types of materials:
   1.1 Pre-formed tape;
   1.2 Gunable elastomeric sealants:
       1.2.1. Non-curing; and
       1.2.2. Curing;
   1.3 Putty and glazing compounds; and
2. Dry glazing, which utilizes extruded rubber gaskets as one or both of the glazing seals.

*11.9.1.5.2 Component Behavior and Rehabilitation Methods* Glazed exterior wall systems shall be considered both deformation- and acceleration-sensitive.

Glazed exterior wall systems not conforming to the acceptance criteria of Section 11.9.1.5.3 shall be rehabilitated in accordance with Section 11.8.

*C11.9.1.5.2 Component Behavior and Rehabilitation Methods* Glazed exterior wall systems are predominantly deformation-sensitive but may also become displaced or detached by large acceleration forces. Glass components within glazed exterior wall systems are deformation-sensitive. Glass performance during earthquakes, which is a function of the wall system type, glazing type, and glass type, falls into one of four categories:

1. Glass remains unbroken in its frame or anchorage;
2. Glass shatters but remains in its frame or anchorage while continuing to provide a weather barrier, and remains otherwise serviceable;
3. Glass shatters and remains in its frame or anchorage in a precarious condition, liable to fall out at any time; or
4. Glass falls out of its frame or anchorage, either in fragments, shards, or whole panels.

Drift analysis and testing or compliance with prescriptive procedures are necessary to establish conformance with drift acceptance criteria related to performance level. Nonconformance requires limiting structural drift, or special detailing to isolate the glazing system from the structure to accommodate drift, or selection of a glass type that will shatter safely or remain in the frame when shattered. This would require removal of the glass or glazed wall system and replacement with an alternative design.

*11.9.1.5.3 Acceptance Criteria* Acceptance criteria shall be applied in accordance with Section 11.3.2.

1. **Life Safety Nonstructural Performance Level.** Glazed exterior wall systems and their supporting structure shall be capable of resisting seismic design forces computed in accordance with Section 11.7.3 or 11.7.4. Glass components meeting any of the following criteria need not be rehabilitated for the Hazards Reduced or Life Safety Nonstructural Performance Level:

   1.1. Any glass component with sufficient clearance from the frame such that physical contact between the glass and the frame will not occur at the relative seismic displacement that the component must be designed to accommodate, as demonstrated by Eq. 11-10.

$$D_{clear} \geq 1.25 D_p \qquad \text{(Eq. 11-10)}$$

where

$$D_{clear} = 2c_1 \left( 1 + \frac{h_p c_2}{b_p c_1} \right)$$

   $h_p$ = height of rectangular glass;
   $b_p$ = width of rectangular glass;
   $c_1$ = clearance (gap) between vertical glass edges and the frame;
   $c_2$ = clearance (gap) between horizontal glass edges and the frame; and
   $D_p$ = relative seismic displacement that the component must be designed to accommodate. $D_p$ shall be determined by Eq. 11-9 over the height of the glass component under consideration.

   1.2. Fully tempered monolithic glass that is located no more than 10 ft above a walking surface;

   1.3. Annealed or heat-strengthened laminated glass in single thickness with interlayer no less than 0.03 in. that is captured mechanically in a wall system glazing pocket, and whose perimeter is secured to the wall system frame by a wet-glazed perimeter bead of $\frac{1}{2}$-in. minimum glass contact width, or other approved anchorage system;

1.4. Any glass component that meets the relative displacement requirement of Eq. 11-11.

$$\Delta_{fallout} \geq 1.25 D_p \quad \text{(Eq. 11-11)}$$

or 0.5 in., whichever is greater,

where

$D_p$ = relative seismic displacement that the component must be designed to accommodate; and

$D_{fallout}$ = relative seismic displacement (drift) causing glass fallout from the curtain wall, storefront, or partition, as determined in accordance with an approved engineering analysis method;

2. **Immediate Occupancy Nonstructural Performance Level.** Glazed exterior wall systems and their supporting structure shall be capable of resisting seismic design forces computed in accordance with Section 11.7.4. Glass components meeting any of the following criteria need not be rehabilitated for performance levels higher than the Life Safety Nonstructural Performance Level:

2.1. Any glass component with sufficient clearance from the frame such that physical contact between the glass and the frame will not occur at the relative seismic displacement that the component must be designed to accommodate, as demonstrated by Eq. 11-10;

2.2. Annealed or heat-strengthened laminated glass in single thickness with interlayer no less than 0.03 in. that is captured mechanically in a wall system glazing pocket, and whose perimeter is secured to the wall system frame by a wet-glazed perimeter bead of $\frac{1}{2}$-in. minimum glass contact width, or other approved anchorage system; and

2.3. Any glass component that meets the relative displacement requirement of Eq. 11-12.

$$\Delta_{fallout} \geq 1.5 \times 1.25 D_p \quad \text{(Eq. 11-12)}$$

or 0.5 in., whichever is greater.

*C11.9.1.5.3 Acceptance Criteria* One method of determining $\Delta_{fallout}$, which is used in Eq. 11-11, is to use AAMA 501.4 (AAMA 2000).

$D_{clear}$ in Eq. 11-10 is derived from a similar equation in Bouwkamp and Meehan (1960) that permits calculation of the story drift required to cause glass-to-

frame contact in a given rectangular window frame. Both equations are based on the principle that a rectangular window frame (specifically one that is anchored mechanically to adjacent stories of the primary structural system of the building) becomes a parallelogram as a result of story drift, and that glass-to-frame contact occurs when the length of the shorter diagonal of the parallelogram is equal to the diagonal of the glass panel itself.

The 1.25 factor in Eqs. 11-11 and 11-12 reflect uncertainties associated with calculated inelastic seismic displacements in building structures. Wright (1989) stated that "[P]ost-elastic deformations calculated using the structural analysis process may well underestimate the actual building deformation by up to 30%. It would therefore be reasonable to require the curtain wall glazing system to withstand 1.25 times the computed maximum story displacement to verify adequate performance." Wright's comments form the basis for using the 1.25 factor.

*11.9.1.5.4 Evaluation Requirements* To establish compliance with criteria 1.1, 1.2, 1.3, 2.1, or 2.2 in 11.9.1.5.3, glazed exterior wall systems shall be evaluated visually to determine glass type, support details, mullion configuration, sealant type, and anchors. To establish compliance with criteria 1.4 or 2.3, an approved analysis shall be used.

*C11.9.1.5.4 Evaluation Requirements* Alternatively, to establish compliance with criteria 1.4 or 2.3, glazed exterior wall systems may be tested in accordance with AAMA 501.4 (AAMA 2000).

### 11.9.2 Partitions

*11.9.2.1 Definition and Scope*

Partitions shall include vertical non-load-bearing interior components that provide space division.

Heavy partitions shall include partitions constructed of masonry materials or assemblies.

Light partitions shall include partitions constructed of metal or wood studs surfaced with lath and plaster, gypsum board, wood, or other facing materials.

*11.9.2.1.1 Evaluation Requirements* Glazed partitions that span from floor to ceiling or to the underside of floor or roof above shall be rehabilitated in accordance with Section 11.9.1.5.

*C11.9.2.1 Definition and Scope*

Heavy partitions include hollow clay tile or concrete block. Only non-load-bearing partitions are considered in this section. Structural partitions including heavy masonry partitions shall be rehabilitated in accordance with Chapter 7.

Partitions may span laterally from the floor to the underside of the floor or the roof above, with connections at the top that may or may not allow for isolation from in-plane drift. Other partitions extend only up to a hung ceiling, and may or may not have lateral bracing above that level to structural support, or may be free-standing.

Modular office furnishings that include movable partitions are considered as contents rather than partitions, and as such are not within the scope of this standard.

### 11.9.2.2 Component Behavior and Rehabilitation Methods

Partitions shall be considered both acceleration- and deformation-sensitive.

Partitions not meeting the acceptance criteria of Section 11.9.2.3 shall be rehabilitated in accordance with Section 11.8.

### C11.9.2.2 Component Behavior and Rehabilitation Methods

Partitions attached to the structural floors both above and below, and loaded in-plane, can experience shear cracking, distortion and fracture of the partition framing, and detachment of the surface finish because of structural deformations. Similar partitions loaded out-of-plane can experience flexural cracking, failure of connections to structure, and collapse. The high incidence of unsupported block partitions in low and moderate seismic levels represents a significant collapse threat.

Partitions subject to deformations from the structure can be protected by providing a continuous gap between the partition and the surrounding structure, combined with attachment that provides for in-plane movement but out-of-plane restraint. Lightweight partitions that are not part of a fire-resistive system are regarded as replaceable.

### 11.9.2.3 Acceptance Criteria

Acceptance criteria shall be applied in accordance with Section 11.3.2.

### 11.9.2.3.1 Life Safety Nonstructural Performance Level

1. **Heavy Partitions.** Nonstructural heavy partitions shall be capable of resisting out-of-plane forces computed in accordance with Section 11.7.3 or 11.7.4. The drift ratio computed in accordance with Section 11.7.5 shall be limited to 0.01.

2. **Light Partitions.** Nonstructural light partitions need not be rehabilitated for the Life Safety Nonstructural Performance Level.

### 11.9.2.3.2 Immediate Occupancy Nonstructural Performance Level

1. **Heavy Partitions.** Nonstructural heavy partitions shall be capable of resisting out-of-plane forces computed in accordance with Section 11.7.4. The drift ratio computed in accordance with Section 11.7.5 shall be limited to 0.005.

2. **Light Partitions.** Nonstructural light partitions shall be capable of resisting the out-of-plane forces computed in accordance with Section 11.7.4. The drift ratio computed in accordance with Section 11.7.5 shall be limited to 0.01.

### 11.9.2.4 Evaluation Requirements

Partitions shall be evaluated to ascertain the type of material.

### C11.9.2.4 Evaluation Requirements

For concrete block partitions, presence of reinforcing and connection conditions at edges are important. For light partitions, bracing or anchoring of the top of the partitions is important.

## 11.9.3 Interior Veneers

### 11.9.3.1 Definition and Scope

Interior veneers shall include decorative-finish materials applied to interior walls and partitions. These provisions of this section shall apply to veneers mounted 4 ft or more above the floor.

### 11.9.3.2 Component Behavior and Rehabilitation Methods

Interior veneers shall be considered deformation-sensitive.

Interior veneers not conforming to the acceptance criteria of Section 11.9.3.3 shall be rehabilitated in accordance with Section 11.8.

### C11.9.3.2 Component Behavior and Rehabilitation Methods

Interior veneers typically experience in-plane cracking and detachment, but may also be displaced or detached out-of-plane by direct acceleration. Interior partitions loaded out-of-plane and supported on flexible backup support systems can experience cracking and detachment.

Drift analysis is necessary to establish conformance with drift acceptance criteria related to the

Nonstructural Performance Level. Nonconformance requires limiting structural drift, or special detailing to isolate the veneer support system from the structure to permit drift; this generally requires disassembly of the support system and veneer replacement. Inadequately adhered veneer must be replaced.

### 11.9.3.3 Acceptance Criteria

Acceptance criteria shall be applied in accordance with Section 11.3.2.

*11.9.3.3.1 Life Safety Nonstructural Performance Level* Backing shall be adequately attached to resist seismic design forces computed in accordance with Section 11.7.3 or 11.7.4. The drift ratio computed in accordance with Section 11.7.5 shall be limited to 0.02.

*11.9.3.3.2 Immediate Occupancy Nonstructural Performance Level* Backing shall be adequately attached to resist seismic design forces computed in accordance with Section 11.7.4. The drift ratio computed in accordance with Section 11.7.5 shall be limited to 0.01.

### 11.9.3.4 Evaluation Requirements

Backup walls or other supports and the attachments to that support shall be evaluated, as well as the condition of the veneer itself.

## 11.9.4 Ceilings

### 11.9.4.1 Definition and Scope

Ceilings shall be categorized as one of the following types:

1. **Category a.** Surface-applied or furred with materials that are applied directly to wood joists, concrete slabs, or steel decking with mechanical fasteners or adhesives;
2. **Category b.** Short-dropped gypsum board sections (less than 2-ft drop) attached to wood or metal furring supported by carrier members;
3. **Category c.** Dropped gypsum board sections greater than 2 ft and suspended metal lath and plaster; or
4. **Category d.** Suspended acoustical board inserted within T-bars, together with lighting fixtures and mechanical items, to form an integrated ceiling system.

### C11.9.4.1 Definition and Scope

Furring materials include wood or metal furring acoustical tile, gypsum board, plaster, or metal panel ceiling materials.

Some older buildings have heavy decorative ceilings of molded plaster, which may be directly attached to the structure or suspended; these are typically Category a or Category c ceilings.

### 11.9.4.2 Component Behavior and Rehabilitation Methods

Ceiling systems shall be considered both acceleration- and deformation-sensitive.

Ceilings not conforming to the acceptance criteria of Section 11.9.4.3 shall be rehabilitated in accordance with Section 11.8.

Where rehabilitation is required for ceilings in Category a or b, they shall be strengthened to resist seismic design forces computed in accordance with Section 11.7.3 or 11.7.4. Where rehabilitation is required for ceilings in Category d, they shall be rehabilitated by the Prescriptive Procedure of Section 11.2.

### C11.9.4.2 Component Behavior and Rehabilitation Methods

Surface-applied or furred ceilings are primarily influenced by the performance of their supports. Rehabilitation of the ceiling takes the form of ensuring good attachment and adhesion. Metal lath and plaster ceilings depend on their attachment and bracing for large ceiling areas. Analysis is necessary to establish the acceleration forces and deformations that must be accommodated. Suspended integrated ceilings are highly susceptible to damage if not braced, causing distortion of grid and loss of panels; however, this is not regarded as a life safety threat with lightweight panels (less than 2 lbs/sf).

Rehabilitation takes the form of bracing, attachment, and edge details designed to prescriptive design standards such as *Recommendations for Direct-Hung Acoustical and Lay-in Panel Ceilings, Seismic Zones 0–2* (CISCA 1991) for seismic levels 0 through 2 and in *Recommendations for Direct-Hung Acoustical and Lay-in Panel Ceilings, Seismic Zones 3–4* (CISCA 1990) for seismic levels 3 and 4.

### 11.9.4.3 Acceptance Criteria

Acceptance criteria shall be applied in accordance with Section 11.3.2.

*11.9.4.3.1 Life Safety Nonstructural Performance Level* Ceilings in Categories a, b, or d need not be rehabilitated for the Life Safety Performance Level except as noted in the footnotes to Table 11-1. Ceilings in Category c shall be capable of accommodating the relative displacement computed in accordance with Section 11.7.3 or 11.7.4.

*11.9.4.3.2 Immediate Occupancy Nonstructural Performance Level* Ceilings in category a or b shall be capable of resisting seismic design forces computed in accordance with Section 11.7.4. Ceilings in category c shall be capable of accommodating the relative displacement computed in accordance with Section 11.7.5. Ceilings in category d shall be rehabilitated by the Prescriptive Procedure of Section 11.7.2.

### 11.9.4.4 Evaluation Requirements

The condition of the ceiling finish material, its attachment to the ceiling support system, the attachment and bracing of the ceiling support system to the structure, and the potential seismic impacts of other nonstructural systems on the ceiling system shall be evaluated.

## 11.9.5 Parapets and Appendages

### 11.9.5.1 Definition and Scope

Parapets and appendages shall include exterior nonstructural features that project above or away from the building. They shall include sculptures and ornamental features in addition to concrete, masonry, or terra cotta parapets. The following parapets and appendages shall be rehabilitated in accordance with this section:

1. Unreinforced masonry parapets with an aspect ratio greater than 1.5;
2. Reinforced masonry or reinforced concrete parapets with an aspect ratio greater than 3.0;
3. Cornices or ledges constructed of stone, terra cotta, or brick, unless supported by a steel or reinforced concrete structure; and
4. Sculptures and ornamental features constructed of stone, terra cotta, masonry, or concrete with an aspect ratio greater than 1.5.

The aspect ratio of parapets and appendages shall be defined as the height of the component above the level of anchorage (*h*) divided by the width of the component (d) as shown in Figure 11-1. For horizontal projecting appendages, the aspect ratio shall be defined as the ratio of the horizontal projection beyond the vertical support of the building to the perpendicular dimension.

### C11.9.5.1 Definition and Scope

Other appendages, such as flagpoles and signs that are similar to the above in size, weight, or potential consequence of failure may be rehabilitated in accordance with this section.

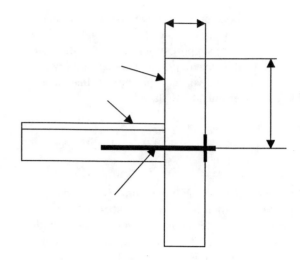

**FIGURE 11-1. Parapet Aspect Ratio.**

### 11.9.5.2 Component Behavior and Rehabilitation Methods

Parapets and appendages shall be considered acceleration-sensitive in the out-of-plane direction.

Parapets and appendages not conforming to the requirements of Section 11.9.5.3 shall be rehabilitated in accordance with Section 11.8.

### C11.9.5.2 Component Behavior and Rehabilitation Methods

Materials or components that are not properly braced may become disengaged and topple; the results are among the most seismically serious consequences of any nonstructural components.

Prescriptive design strategies for masonry parapets not exceeding 4 ft in height consist of bracing in accordance with the concepts shown in FEMA 74 (FEMA 1994) and FEMA 172 (FEMA 1992), with detailing to conform to accepted engineering practice. Braces for parapets should be spaced at a maximum of 8 ft on center and, where the parapet construction is discontinuous, a continuous backing component should be provided. Where there is no adequate connection, roof construction should be tied to parapet walls at the roof level. Other parapets and appendages should be analyzed for acceleration forces, and braced and connected according to accepted engineering principles.

### 11.9.5.3 Acceptance Criteria

Acceptance criteria shall be applied in accordance with Section 11.3.2.

*11.9.5.3.1 Life Safety Nonstructural Performance Level* Parapets and appendages exceeding the aspect ratios from Section 11.9.5.1 shall be capable of resisting seismic forces computed in accordance with Section 11.7.4.

*11.9.5.3.2 Immediate Occupancy Nonstructural Performance Level* Parapets and appendages shall be capable of resisting seismic forces computed in accordance with Section 11.7.4.

### 11.9.5.4 Evaluation Requirements

The condition of mortar and masonry, connection to supports, type and stability of the supporting structure, and horizontal continuity of the parapet coping, shall be considered in the evaluation.

## 11.9.6 Canopies and Marquees

### 11.9.6.1 Definition and Scope

Canopies shall include projections from an exterior wall that are extensions of the horizontal building structure or independent structures that are tied to the building. Marquees shall include free-standing structures. Canvas or other fabric projections need not be rehabilitated in accordance with this section.

### C11.9.6.1 Definition and Scope

Canopies and marquees are generally used to provide weather protection.

Marquees are often constructed of metal or glass.

### 11.9.6.2 Component Behavior and Rehabilitation Methods

Canopies and marquees shall be considered acceleration-sensitive.

Canopies and marquees not conforming to the acceptance criteria of Section 11.9.6.3 shall be rehabilitated in accordance with Section 11.8.

### C11.9.6.2 Component Behavior and Rehabilitation Methods

The variety of design of canopies and marquees is so great that they must be independently analyzed and evaluated for their ability to withstand seismic forces. Rehabilitation may take the form of improving attachment to the building structure, strengthening, bracing, or a combination of measures.

### 11.9.6.3 Acceptance Criteria

Acceptance criteria shall be applied in accordance with Section 11.3.2.

*11.9.6.3.1 Life Safety Nonstructural Performance Level* Canopies and marquees shall be capable of

resisting both horizontal and vertical seismic design forces computed in accordance with Section 11.7.3 or 11.7.4.

*11.9.6.3.2 Immediate Occupancy Nonstructural Performance Level* Canopies and marquees shall be capable of resisting both horizontal and vertical seismic design forces computed in accordance with Section 11.7.4.

### 11.9.6.4 Evaluation Requirements

Buckling in bracing, connection to supports, and type and stability of the supporting structure shall be considered in the evaluation.

## 11.9.7 Chimneys and Stacks

### 11.9.7.1 Definition and Scope

Chimneys and stacks that are cantilevered above building roofs shall be rehabilitated in accordance with this section. Light metal residential chimneys need not comply with the provisions of this document.

### 11.9.7.2 Component Behavior and Rehabilitation Methods

Chimneys and stacks shall be considered acceleration-sensitive.

Chimneys and stacks not conforming to the acceptance criteria of Section 11.9.7.3 shall be rehabilitated in accordance with Section 11.8.

### C11.9.7.2 Component Behavior and Rehabilitation Methods

Chimneys and stacks may fail through flexure, shear, or overturning. They may also disengage from adjoining floor or roof structures and damage them, and their collapse or overturning may also damage adjoining structures. Rehabilitation may take the form of strengthening and/or bracing and material repair. Residential chimneys may be braced in accordance with the concepts shown in FEMA 74 (FEMA 1994).

### 11.9.7.3 Acceptance Criteria

Acceptance criteria shall be applied in accordance with Section 11.3.2.

*11.9.7.3.1 Life Safety Nonstructural Performance Level* Chimneys and stacks shall be capable of resisting seismic forces computed in accordance with Section 11.7.3 or 11.7.4. Residential chimneys shall be permitted to meet the prescriptive requirements of Section 11.7.2.

*11.9.7.3.2 Immediate Occupancy Nonstructural Performance Level* Chimneys and stacks shall be capable of resisting seismic forces computed in accordance with Section 11.7.4. Residential chimneys shall be permitted to meet the prescriptive requirements of Section 11.7.2.

### 11.9.7.4 Evaluation Requirements

The condition of the mortar and masonry, connection to adjacent structure, and type and stability of foundations shall be considered in the evaluation.

Concrete shall be evaluated for spalling and exposed reinforcement. Steel shall be evaluated for corrosion.

## 11.9.8 Stairs and Stair Enclosures

### 11.9.8.1 Definition and Scope

Stairs shall include the treads, risers, and landings that make up passageways between floors, as well as the surrounding shafts, doors, windows, and fire-resistant assemblies that constitute the stair enclosure.

### 11.9.8.2 Component Behavior and Rehabilitation Methods

Each of the separate components of the stairs shall be defined as either acceleration- or deformation-sensitive depending on the predominant behavior. Components of stairs that are attached to adjacent floors or floor framing shall be considered deformation-sensitive. All other stair components shall be considered acceleration-sensitive.

Stairs not conforming to the acceptance criteria of Section 11.9.8.3 shall be rehabilitated in accordance with Section 11.8.

### C11.9.8.2 Component Behavior and Rehabilitation Methods

The stairs themselves may be independent of the structure or integral with the structure. If integral, they should form part of the overall structural evaluation and analysis, with particular attention paid to the possibility of response modification due to localized stiffness. If independent, the stairs must be evaluated for normal stair loads and their ability to withstand direct acceleration or loads transmitted from the structure through connections.

Stair enclosure materials may fall and render the stairs unusable due to debris.

Rehabilitation of integral or independent stairs may take the form of necessary structural strengthening or bracing, or the introduction of connection

details to eliminate or reduce interaction between stairs and the building structure.

Rehabilitation of enclosing walls or glazing should follow the requirements of the relevant sections of this document.

### 11.9.8.3 Acceptance Criteria

Acceptance criteria shall be applied in accordance with Section 11.3.2.

*11.9.8.3.1 Life Safety Nonstructural Performance Level* Stairs shall be capable of resisting the seismic design forces computed in accordance with Section 11.7.3 or 11.7.4 and shall be capable of accommodating the expected relative displacement computed in accordance with Section 11.7.5.

*11.9.8.3.2 Immediate Occupancy Nonstructural Performance Level* Stairs shall be capable of resisting the seismic design forces computed in accordance with Section 11.7.4 and shall be capable of accommodating the expected relative displacement computed in accordance with Section 11.7.5.

### 11.9.8.4 Evaluation Requirements

The materials and condition of stair members and their connections to supports, and the types and stability of supporting and adjacent walls, windows, and other portions of the stair shaft system shall be considered in the evaluation.

## 11.10 MECHANICAL, ELECTRICAL, AND PLUMBING COMPONENTS: DEFINITION, BEHAVIOR, AND ACCEPTANCE CRITERIA

### 11.10.1 Mechanical Equipment

### 11.10.1.1 Definition and Scope

Equipment used for the operation of the building, and that meets one or more of the following criteria shall be rehabilitated in accordance with this section:

1. All equipment weighing over 400 lbs;
2. Unanchored equipment weighing over 100 lbs that does not have a factor of safety against overturning of 1.5 or greater where design loads, calculated in accordance with Section 11.7.3 or 11.7.4, are applied;
3. Equipment weighing over 20 lbs that is attached to ceiling, wall, or other support more than 4 ft above the floor; and
4. Building operation equipment including:
   4.1. Boilers and furnaces;

4.2. Conveyors (nonpersonnel);

4.3. HVAC system equipment, vibration-isolated;

4.4. HVAC system equipment, non-vibration-isolated; and

4.5. HVAC system equipment mounted in-line with ductwork.

### C11.10.1.1 Definition and Scope

Equipment such as manufacturing or processing equipment related to the occupant's business should be evaluated separately for the effects that failure due to a seismic event could have on the operation of the building.

### 11.10.1.2 Component Behavior and Rehabilitation Methods

Mechanical equipment shall be considered acceleration-sensitive.

Mechanical equipment not conforming to the acceptance criteria of Section 11.10.1.3 shall be rehabilitated in accordance with Section 11.8.

### C11.10.1.2 Component Behavior and Rehabilitation Methods

The provisions of Section 11.10 focus on position retention, which is a primary consideration for the Life Safety Performance Level.

At the Immediate Occupancy Performance Level, position retention alone may be insufficient to assure conformance with the stated goals of the performance level. The expectation is that although some nonstructural damage is expected, the building will function following the earthquake, provided utilities are available. To achieve this level of functionality, the designer must consider the essential post-earthquake functions of the building and then identify those mechanical, electrical, and plumbing components that must operate for the building to function. Components may be identified as critical (components that must be functional) and noncritical (those components where function following an earthquake is desirable but not essential to the continued occupancy of the building). For critical components where operability is vital, the requirements of Section 2.4.5 of the 2003 *NEHRP Recommended Provisions for Seismic Regulations for New Buildings and Other Structures, 2003 Edition* [FEMA 450 (FEMA 2004)] provide methods for seismically qualifying the component.

Position retention failure of components consists of sliding, tilting, or overturning of floor- or roof-mounted equipment off its base, and possible loss of attachment (with consequent falling) for equipment attached to a vertical structure or suspended, and failure of piping or electrical wiring connected to the equipment

Construction of mechanical equipment to nationally recognized codes and standards, such as those approved by the American National Standards Institute (ANSI), provides adequate strength to accommodate all normal and upset operating loads.

For position retention, basic rehabilitation consists of securely anchoring floor-mounted equipment by bolting, with detailing appropriate to the base construction of the equipment. ASHRAE RP-812 (ASHRAE 1999) provides more information on designing and detailing seismic anchorage.

Function and operability of mechanical and electrical components is affected only indirectly by increasing design forces. However, on the basis of past earthquake experience, it may be reasonable to conclude that if structural integrity and stability are maintained, function and operability after an earthquake will be provided for many types of equipment components. For complex components, testing or experience may be the only reasonable way to improve the assurance of function and operability. Testing is a well-established alternative method of seismic qualification for small to medium-sized equipment. Several national standards have testing requirements adaptable for seismic qualification.

Seismic forces can be established by analysis using the default Eqs. 11-1 and 11-2. Equipment weighing over 400 lbs and located on the third floor or above (or on a roof of equivalent height) should be analyzed using Eqs. 11-4 and 11-5.

Existing attachments for attached or suspended equipment must be evaluated for seismic load capacity, and strengthened or braced as necessary. Attachments that provide secure anchoring eliminate or reduce the likelihood of piping or electrical distribution failure.

### 11.10.1.3 Acceptance Criteria

Acceptance criteria shall be applied in accordance with Section 11.3.2.

*11.10.1.3.1 Life Safety Nonstructural Performance Level* Equipment anchorage shall be capable of resisting seismic design forces computed in accordance with Section 11.7.3 or 11.7.4.

*11.10.1.3.2 Immediate Occupancy Nonstructural Performance Level* Equipment anchorage shall be capable of resisting seismic design forces computed in accordance with Section 11.7.4.

### *11.10.1.4 Evaluation Requirements*

Equipment shall be analyzed to establish acceleration-induced forces, and supports, hold-downs, and bracing shall be visually evaluated.

### *C11.10.1.4 Evaluation Requirements*

Existing concrete anchors may have to be tested by applying torque to the nuts to confirm that adequate strength is present.

## 11.10.2 Storage Vessels and Water Heaters

### *11.10.2.1 Definition and Scope*

Storage vessels and water heaters shall include all vessels that contain fluids used for building operation.

Vessels shall be classified into one of the following two categories:

1. **Category 1.** Vessels with structural support of contents, in which the shell is supported by legs or a skirt; or
2. **Category 2.** Flat-bottom vessels in which the weight of the contents is supported by the floor, roof, or a structural platform.

### *C11.10.2.1 Definition and Scope*

The vessel may be fabricated of materials such as steel or other metals, or fiberglass, or it may be a glass-lined tank. These requirements may also be applied, with judgment, to vessels that contain solids that act as a fluid, and vessels containing fluids not involved in the operation of the building.

### *11.10.2.2 Component Behavior and Rehabilitation Methods*

Tanks and vessels shall be considered acceleration-sensitive.

Tanks and vessels not conforming to the acceptance criteria of Section 11.10.2.3 shall be rehabilitated in accordance with Section 11.8.

### *C11.10.2.2 Component Behavior and Rehabilitation Methods*

Category 1 vessels fail by stretching of anchor bolts, buckling and disconnection of supports, and consequent tilting or overturning of the vessel. A Category 2 vessel may be displaced from its foundation, or its shell may fail by yielding near the bottom, creating a visible bulge or possible leakage. Displacement of both types of vessel may cause rupturing of connecting piping and leakage.

Category 1 residential water heaters with a capacity no greater than 100 gal may be rehabilitated by

prescriptive design methods, such as concepts described in FEMA 74 (FEMA 1994) or FEMA 172 (FEMA 1992). Category 1 vessels with a capacity less than 1,000 gal should be designed to meet the force provisions of Section 11.7.3 or 11.7.4, and bracing strengthened or added as necessary. Other Category 1 and Category 2 vessels should be evaluated against a recognized standard, such as API 650 (API 1998) for vessels containing petroleum products or other chemicals, or AWWA D100-96 (AWWA 1996) for water vessels. ASHRAE RP-812 (ASHRAE 1999) provides more information on designing and detailing seismic anchorage and bracing.

### *11.10.2.3 Acceptance Criteria*

Acceptance criteria shall be applied in accordance with Section 11.3.2.

#### *11.10.2.3.1 Life Safety Nonstructural Performance Level*

1. **Category 1 Equipment.** If the Analytical Procedure is selected based on Table 11-1, Category 1 equipment and supports shall be capable of resisting seismic forces computed in accordance with Section 11.7.3 or 11.7.4. If the Prescriptive Procedure is selected based on Table 11-1, Category 1 equipment shall meet prescriptive requirements in accordance with Section 11.7.2.
2. **Category 2 Equipment.** If the Analytical Procedure is selected based on Table 11-1, Category 2 equipment and supports shall be capable of resisting seismic forces computed in accordance with Section 11.7.3 or 11.7.4. If the Prescriptive Procedure is selected based on Table 11-1, Category 2 equipment shall meet prescriptive requirements in accordance with Section 11.7.2.

#### *11.10.2.3.2 Immediate Occupancy Nonstructural Performance Level*

1. **Category 1 Equipment.** If the Analytical Procedure is selected based on Table 11-1, Category 1 equipment and supports shall be capable of resisting seismic forces computed in accordance with Section 11.7.4. If the Prescriptive Procedure is selected based on Table 11-1, Category 1 equipment shall meet prescriptive requirements in accordance with Section 11.7.2.
2. **Category 2 Equipment.** If the Analytical Procedure is selected based on Table 11-1, Category 2 equipment and supports shall be capa-

ble of resisting seismic forces computed in accordance with Section 11.7.4. If the Prescriptive Procedure is selected based on Table 11-1, Category 2 equipment shall meet prescriptive requirements in accordance with Section 11.7.2.

### 11.10.2.4 Evaluation Requirements

All equipment shall be visually evaluated to determine the existence of hold-downs, supports, and bracing.

### C11.10.2.4 Evaluation Requirements

Existing concrete anchors may have to be tested by applying torque to the nuts to confirm that adequate strength is present.

## 11.10.3 Pressure Piping

### 11.10.3.1 Definition and Scope

The requirements of this section shall apply to all piping (except fire suppression piping) that carries fluids which, in their vapor stage, exhibit a pressure of 15 psi, gauge, or higher.

### 11.10.3.2 Component Behavior and Rehabilitation Methods

Piping shall be considered acceleration-sensitive. Piping that runs between floors or across seismic joints shall be considered both acceleration- and deformation-sensitive.

Piping not conforming to the acceptance criteria of Section 11.10.3.3 shall be rehabilitated in accordance with Section 11.8.

### C11.10.3.2 Component Behavior and Rehabilitation Methods

Appendix Chapter 6 of the 2003 *NEHRP Recommended Provisions for Seismic Regulations for New Buildings and Other Structures, 2003 Edition*[FEMA 450 (FEMA 2004)] provides preliminary criteria for the establishment of such performance criteria and their use in the assessment and design of piping systems. The performance criteria, from least restrictive to most severe, are: position retention, leak tightness, and operability. In particular, the interaction of systems and interface with the relevant piping design standards is addressed. For the Life Safety Performance level, the focus is on position retention, which is defined as the condition of a piping system characterized by the absence of collapse or fall of any part of the system.

For the Immediate Occupancy Nonstructural Performance Level, leak tightness (the condition of a piping system characterized by containment of contents or maintenance of a vacuum with no discernable leakage) is required. Operability (the condition of a piping system characterized by leak tightness as well as continued delivery, shutoff, or throttle of pipe contents flow by means of unimpaired operation of equipment and components such as pumps, compressors, and valves) is desirable but requires a significantly higher level of effort to achieve.

The most common failure of piping is joint failure, caused by inadequate support or bracing.

### 11.10.3.3 Acceptance Criteria

Acceptance criteria shall be applied in accordance with Section 11.3.2.

*11.10.3.3.1 Life Safety Nonstructural Performance Level* If the Prescriptive Procedure is selected based on Table 11-1, piping shall meet the prescriptive requirements of Section 11.7.2. If the Analytical Procedure is selected based on Table 11-1, piping shall be capable of resisting seismic forces computed in accordance with Section 11.7.3 or 11.7.4. Piping that runs between floors or across seismic joints shall be capable of accommodating relative displacements computed in accordance with Section 11.7.5.

*11.10.3.3.2 Immediate Occupancy Nonstructural Performance Level* If the Prescriptive Procedure is selected based on Table 11-1, piping shall meet the prescriptive requirements of Section 11.7.2. If the Analytical Procedure is selected based on Table 11-1, piping shall be capable of resisting seismic forces computed in accordance with Section 11.7.4. Piping that runs between floors or across seismic joints shall be capable of accommodating relative displacements computed in accordance with Section 11.7.5.

### 11.10.3.4 Evaluation Requirements

High-pressure piping shall be tested by an approved method. Lines shall be hydrostatically tested to 150% of the maximum anticipated pressure of the system.

### C11.10.3.4 Evaluation Requirements

High-pressure piping may be tested in accordance with ASME B31.9 (ASME 2000).

## 11.10.4 Fire Suppression Piping

### 11.10.4.1 Definition and Scope

Fire suppression piping shall include fire sprinkler piping consisting of main risers and laterals weighing, loaded, in the range of 30 to 100 lbs/lineal ft, with branches of decreasing size to 2 lbs/ft.

### 11.10.4.2 Component Behavior and Rehabilitation Methods

Fire suppression piping shall be considered acceleration-sensitive. Fire suppression piping that runs between floors or across seismic joints shall be considered both acceleration- and deformation-sensitive.

Fire suppression piping not conforming to the acceptance criteria of Section 11.9.4.3 shall be rehabilitated in accordance with Section 11.8.

### C11.10.4.2 Component Behavior and Rehabilitation Methods

The most common failure of fire suppression piping is joint failure, caused by inadequate support or bracing, or by sprinkler heads impacting adjoining materials.

Rehabilitation is accomplished by prescriptive design approaches to support and bracing. The prescriptive requirements of NFPA 13 (NFPA 2002) should be used.

### 11.10.4.3 Acceptance Criteria

Acceptance criteria shall be applied in accordance with Section 11.3.2.

#### 11.10.4.3.1 Life Safety Nonstructural Performance Level
If the Prescriptive Procedure is selected based on Table 11-1, fire suppression piping shall meet the prescriptive requirements of Section 11.7.2. If the Analytical Procedure is selected based on Table 11-1, fire suppression piping shall be capable of resisting seismic design forces computed in accordance with Section 11.7.3 or 11.7.4. Fire suppression piping that runs between floors or across seismic joints shall be capable of accommodating relative displacements computed in accordance with Section 11.7.5.

#### 11.10.4.3.2 Immediate Occupancy Nonstructural Performance Level
If the Prescriptive Procedure is selected based on Table 11-1, fire suppression piping shall meet the prescriptive requirements of Section 11.7.2. If the Analytical Procedure is selected based on Table 11-1, fire suppression piping shall be capable of resisting seismic design forces computed in accordance with Section 11.7.4. Fire suppression piping that runs between floors or across seismic joints shall be capable of accommodating relative displacements computed in accordance with Section 11.7.5.

### 11.10.4.4 Evaluation Requirements

The support, flexibility, protection at seismic movement joints, and freedom from impact from adjoining materials at the sprinkler heads shall be evaluated.

### C11.10.4.4 Evaluation Requirements

The support and bracing of bends of the main risers and laterals, as well as maintenance of adequate flexibility to prevent buckling, are especially important.

## 11.10.5 Fluid Piping other than Fire Suppression

### 11.10.5.1 Definition and Scope

Piping, other than pressure piping or fire suppression lines, that transfers fluids under pressure by gravity or that are open to the atmosphere—including drainage and ventilation piping, hot, cold, and chilled water piping; and piping carrying liquids, as well as fuel gas lines—shall meet the requirements of this section.

Fluid piping other than fire suppression piping shall be classified into one of the following two categories:

1. **Category 1.** Hazardous materials and flammable liquids that would pose an immediate life safety danger if exposed, because of inherent properties of the contained material; or
2. **Category 2.** Materials that, in case of line rupture, would cause property damage but pose no immediate life safety danger.

### C11.10.5.1 Definition and Scope

Hazardous materials and flammable liquids that would pose an immediate life safety danger if exposed are defined in NFPA 325-94 (NFPA 1994), 49-94 (NFPA 1994), 491M-91(NFPA 1991), and 704-90 (NFPA 2001).

### 11.10.5.2 Component Behavior and Rehabilitation Methods

Fluid piping other than fire suppression piping shall be considered acceleration-sensitive. Piping that runs between floors or across seismic joints shall be considered both acceleration- and deformation-sensitive.

Fluid piping not conforming to the acceptance criteria of Section 11.10.5.3 shall be rehabilitated in accordance with Section 11.8.

### C11.10.5.2 Component Behavior and Rehabilitation Methods

The most common failure is joint failure, caused by inadequate support or bracing.

Category 1 piping rehabilitation is accomplished by strengthening support and bracing, using the prescriptive methods of *Pipe Hangers and Supports:*

*Materials, Design and Manufacture*, SP-58 (MSS 1993). The piping systems themselves should be designed to meet the force provisions of Section 11.7.3 or 11.7.4 and relative displacement provisions of Section 11.7.5. The effects of temperature differences, dynamic fluid forces, and piping contents should be taken into account.

Category 2 piping rehabilitation is accomplished by strengthening support and bracing using the prescriptive methods of SP-58 as long as the piping falls within the size limitations of those guidelines. Piping that exceeds the limitations of those guidelines shall be designed to meet the force provisions of Section 11.7.3 or 11.7.4 and relative displacement provisions of Section 11.7.5.

More information on designing and detailing seismic bracing can be found in ASHRAE RP-812 (ASHRAE 1999).

### 11.10.5.3 Acceptance Criteria

Acceptance criteria shall be applied in accordance with Section 11.3.2.

#### 11.10.5.3.1 Life Safety Nonstructural Performance Level

1. **Category 1 piping systems.** If the Prescriptive Procedure is selected based on Table 11-1, fluid piping supports and bracing shall meet the prescriptive requirements of Section 11.7.2. If the Analytical Procedure is selected based on Table 11-1, fluid piping shall be capable of resisting seismic design forces computed in accordance with Section 11.7.3 or 11.7.4. Piping that runs between floors and across seismic joints shall be capable of accommodating relative displacements computed in accordance with Section 11.7.5.
2. **Category 2 piping systems.** If the Prescriptive Procedure is selected based on Table 11-1, fluid piping supports and bracing shall meet the prescriptive requirements of Section 11.7.2. If the Analytical Procedure is selected based on Table 11-1, fluid piping shall be capable of resisting seismic design forces computed in accordance with Section 11.7.3 or 11.7.4. Piping that runs between floors and across seismic joints shall be capable of accommodating relative displacements computed in accordance with Section 11.7.5.

#### 11.10.5.3.2 Immediate Occupancy Nonstructural Performance Level
If the Prescriptive Procedure is selected based on Table 11-1, fluid piping supports and bracing shall meet the prescriptive requirements of Section 11.7.2 for essential facilities. If the Analytical

Procedure is selected based on Table 11-1, fluid piping shall be capable of resisting seismic design forces computed in accordance with Section 11.7.4. Piping that runs between floors and across seismic joints shall be capable of accommodating relative displacements computed in accordance with Section 11.7.5.

#### 11.10.5.4 Evaluation Requirements

The support, flexibility, and protection at seismic joints of fluid piping other than fire suppression piping shall be evaluated.

Piping shall be insulated from detrimental heat effects.

#### C11.10.5.4 Evaluation Requirements

The support and bracing of bends in the main risers and laterals, as well as maintenance of adequate flexibility to prevent buckling, are especially important.

### 11.10.6 Ductwork

#### 11.10.6.1 Definition and Scope

Ductwork shall include HVAC and exhaust ductwork systems. Seismic restraints shall not be required for ductwork that is not conveying hazardous materials and that meets either of the following conditions:

1. HVAC ducts are suspended from hangers 12 in. or less in length from the top of the duct to the supporting structure. Hangers shall be installed without eccentricities that induce moments in the hangers; or
2. HVAC ducts have a cross-sectional area of less than 6 sf.

#### 11.10.6.2 Component Behavior and Rehabilitation Methods

Ducts shall be considered acceleration-sensitive. Ductwork that runs between floors or across seismic joints shall be considered both acceleration- and deformation-sensitive.

Ductwork not conforming to the acceptance criteria of Section 11.10.6.3 shall be rehabilitated in accordance with Section 11.8.

#### C11.10.6.2 Component Behavior and Rehabilitation Methods

Damage to ductwork is caused by failure of supports or lack of bracing that causes deformation or rupture of the ducts at joints, leading to leakage from the system.

Rehabilitation consists of strengthening supports and strengthening or adding bracing. Prescriptive design methods may be used in accordance with *Rectangular Industrial Duct Construction Standards* (SMACNA 1980) and *HVAC Duct Construction Standards, Metal and Flexible* (SMACNA 1985). More information on designing and detailing seismic bracing can be found in ASHRAE RP-812 (ASHRAE 1999).

### 11.10.6.3 Acceptance Criteria

Acceptance criteria shall be applied in accordance with Section 11.3.2.

### 11.10.6.3.1 Life Safety Nonstructural Performance Level
Ductwork shall meet the requirements of prescriptive standards in accordance with Section 11.7.2.

### 11.10.6.3.2 Immediate Occupancy Nonstructural Performance Level
Ductwork shall meet the requirements of prescriptive standards in accordance with Section 11.7.2.

### 11.10.6.4 Evaluation Requirements

Ductwork shall be evaluated visually to determine its length, connection type, and cross-sectional area.

## 11.10.7 Electrical and Communications Equipment

### 11.10.7.1 Definition and Scope

All electrical and communication equipment, including panel boards, battery racks, motor control centers, switch gears, and other fixed components located in electrical rooms or elsewhere in the building that meet any of the following criteria shall comply with the requirements of this section:

1. All equipment weighing over 400 lbs;
2. Unanchored equipment weighing over 100 lbs that does not have a factor of safety against overturning of 1.5 or greater where design loads computed in accordance with Section 11.7.3 or 11.7.4 are applied;
3. Equipment weighing over 20 lbs that is attached to ceiling, wall, or other support more than 4 ft above the floor; and
4. Building operation equipment.

### 11.10.7.2 Component Behavior and Rehabilitation Methods

Electrical equipment shall be considered acceleration-sensitive.

Electrical equipment not conforming to the acceptance criteria of Section 11.10.7.3 shall be rehabilitated in accordance with Section 11.8.

### C11.10.7.2 Component Behavior and Rehabilitation Methods

Failure of these components consists of sliding, tilting, or overturning of floor- or roof-mounted equipment off their bases, and possible loss of attachment (with consequent falling) for equipment attached to a vertical structure or suspended, and failure of electrical wiring connected to the equipment.

Construction of electrical equipment to nationally recognized codes and standards, such as those approved by ANSI, provides adequate strength to accommodate all normal and upset operating loads.

Basic rehabilitation consists of securely anchoring floor-mounted equipment by bolting, with detailing appropriate to the base construction of the equipment.

### 11.10.7.3 Acceptance Criteria

Acceptance criteria shall be applied in accordance with Section 11.3.2.

### 11.10.7.3.1 Life Safety Nonstructural Performance Level
If the Prescriptive Procedure is selected based on Table 11-1, electrical equipment shall meet the prescriptive requirements of Section 11.7.2. If the Analytical Procedure is selected based on Table 11-1, electrical equipment shall be capable of resisting seismic design forces computed in accordance with Section 11.7.3 or 11.7.4.

### 11.10.7.3.2 Immediate Occupancy Nonstructural Performance Level
If the Prescriptive Procedure is selected based on Table 11-1, electrical equipment shall meet the prescriptive requirements of Section 11.7.2. If the Analytical Procedure is selected based on Table 11-1, electrical equipment shall be capable of resisting seismic design forces computed in accordance with Section 11.7.4.

### 11.10.7.4 Evaluation Requirements

Equipment shall be visually evaluated to determine its category and the existence of the hold-downs, supports, and braces.

### C11.10.7.4 Evaluation Requirements

Larger equipment requiring the Analytical Procedure must be analyzed to determine forces, and be visually evaluated. Concrete anchors may have to be tested by applying torque to the nuts to confirm that adequate strength is present.

## 11.10.8 Electrical and Communications Distribution Components

### 11.10.8.1 Definition and Scope

All electrical and communications transmission lines, conduit, and cables, and their supports, shall comply with the requirements of this section.

### 11.10.8.2 Component Behavior and Rehabilitation Methods

Electrical distribution equipment shall be considered acceleration-sensitive. Wiring or conduit that runs between floors or across expansion or seismic joints shall be considered both acceleration- and deformation-sensitive.

Electrical and communications distribution components not conforming to the acceptance criteria of Section 11.10.8.3 shall be rehabilitated in accordance with Section 11.8.

### C11.10.8.2 Component Behavior and Rehabilitation Methods

Failure occurs most commonly by inadequate support or bracing, deformation of the attached structure, or impact from adjoining materials.

Rehabilitation may be accomplished by strengthening support and bracing using the prescriptive methods contained in *Rectangular Industrial Duct Construction Standards* (SMACNA 1980) and *HVAC Duct Construction Standards, Metal and Flexible* (SMACNA 1985).

### 11.10.8.3 Acceptance Criteria

Acceptance criteria shall be applied in accordance with Section 11.3.2.

#### 11.10.8.3.1 Life Safety Nonstructural Performance Level
Electrical and communications distribution components shall meet the requirements of prescriptive standards in accordance with Section 11.7.2.

#### 11.10.8.3.2 Immediate Occupancy Nonstructural Performance Level
Electrical and communications distribution components shall meet the requirements of prescriptive standards for essential facilities in accordance with Section 11.7.2.

### 11.10.8.4 Evaluation Requirements

Components shall be visually evaluated to determine the existence of supports and bracing.

## 11.10.9 Light Fixtures

### 11.10.9.1 Definition and Scope
Lighting fixtures shall be classified into one of the following categories:

1. **Category 1.** Lighting recessed in ceilings;
2. **Category 2.** Lighting surface-mounted to ceilings or walls;
3. **Category 3.** Lighting supported within a suspended ceiling system (integrated ceiling); or
4. **Category 4.** Lighting suspended from ceilings or structure by a pendant or chain.

### 11.10.9.2 Component Behavior and Rehabilitation Methods

Light fixtures not conforming to the acceptance criteria of Section 11.10.9.3 shall be rehabilitated in accordance with Section 11.8.

### C11.10.9.2 Component Behavior and Rehabilitation Methods

Failure of Category 1 and 2 components occurs through failure of attachment of the light fixture and/or failure of the supporting ceiling or wall. Failure of Category 3 components occurs through loss of support from the T-bar system, and by distortion caused by deformation of the supporting structure or deformation of the ceiling grid system, allowing the fixture to fall. Failure of Category 4 components is caused by excessive swinging that results in the pendant or chain support breaking on impact with adjacent materials, or the support being pulled out of the ceiling.

Rehabilitation of Category 1 and 2 components involves attachment repair or fixture replacement in association with necessary rehabilitation of the supporting ceiling or wall. Rehabilitation of Category 3 components involves the addition of independent support for the fixture from the structure or substructure in accordance with FEMA 74 (FEMA 1994) design concepts. Rehabilitation of Category 4 components involves strengthening of attachment and ensuring freedom to swing without impacting adjoining materials.

### 11.10.9.3 Acceptance Criteria

Acceptance criteria shall be applied in accordance with Section 11.3.2.

#### 11.10.9.3.1 Life Safety Nonstructural Performance Level

1. **Categories 1 and 2.** The connection to ceiling or wall shall be present with no visible signs of distress.

2. **Category 3.** Systems bracing and support shall meet prescriptive requirements in accordance with Section 11.7.2.
3. **Category 4.** Fixtures weighing over 20 lbs shall be adequately articulated or connections to the building shall be ductile and the fixture shall be free to swing without impacting adjoining materials.

### 11.10.9.3.2 Immediate Occupancy Nonstructural Performance Level

1. **Categories 1 and 2.** The connection to ceiling or wall shall be present with no visible signs of distress.
2. **Category 3.** Systems bracing and support shall meet prescriptive requirements for essential facilities.
3. **Category 4.** Fixtures weighing over 20 lbs shall be articulated or connections to the building shall be ductile and the fixture shall be free to swing without impacting adjoining materials.

### 11.10.9.4 Evaluation Requirements

Light fixture supports shall be visually evaluated to determine the connection type and adequacy.

## 11.11 FURNISHINGS AND INTERIOR EQUIPMENT: DEFINITION, BEHAVIOR, AND ACCEPTANCE CRITERIA

### 11.11.1 Storage Racks

#### 11.11.1.1 Definition and Scope

Storage racks shall include systems for holding materials either permanently or temporarily.

#### C11.11.1.1 Definition and Scope

Storage racks are usually constructed of metal. Storage racks are generally purchased as proprietary systems installed by a tenant and are often not under the direct control of the building owner. Thus, they are usually not part of the construction contract and often have no foundation or foundation attachment. However, they are often permanently installed and their size and loaded weight make them an important hazard to either life, property, or the surrounding structure. Storage racks in excess of 4 ft in height located in occupied locations shall be considered where the Life Safety Nonstructural Performance Level is selected.

#### 11.11.1.2 Component Behavior and Rehabilitation Methods

Storage racks shall be considered acceleration-sensitive.

Storage racks not conforming to the acceptance criteria of Section 11.11.1.3 shall be rehabilitated in accordance with Section 11.8.

#### C11.11.1.2 Component Behavior and Rehabilitation Methods

Storage racks may fail internally (through inadequate bracing or moment-resisting capacity) or externally (by overturning caused by absence or failure of foundation attachments).

Rehabilitation is usually accomplished by the addition of bracing to the rear and side panels of racks and/or by improving the connection of the rack columns to the supporting slab. In rare instances, foundation improvements may be required to remedy insufficient bearing or uplift load capacity.

Seismic forces can be established by analysis in accordance with Section 11.7.3 or 11.7.4. However, special attention should be paid to the evaluation and analysis of large, heavily loaded rack systems because of their heavy loading and lightweight structural members.

#### 11.11.1.3 Acceptance Criteria

Acceptance criteria shall be applied in accordance with Section 11.3.2.

##### 11.11.1.3.1 Life Safety Nonstructural Performance Level Storage racks shall be capable of resisting seismic design forces computed in accordance with Section 11.7.3 or 11.7.4.

##### 11.11.1.3.2 Immediate Occupancy Nonstructural Performance Level Storage racks shall be capable of resisting seismic design forces computed in accordance with Section 11.7.4.

#### 11.11.1.4 Evaluation Requirements

Buckling or racking failure of storage rack components, connection to support structures, and type and stability of supporting structure shall be considered in the evaluation.

### 11.11.2 Bookcases

#### 11.11.2.1 Definition and Scope

Bookcases constructed of wood or metal, in excess of 4 ft high, shall meet the requirements of this section.

#### 11.11.2.2 Component Behavior and Rehabilitation Methods

Bookcases shall be considered acceleration-sensitive.

Bookcases not conforming to the acceptance criteria of Section 11.11.2.3 shall be rehabilitated in accordance with Section 11.8.

### C11.11.2.2 Component Behavior and Rehabilitation Methods

Bookcases may deform or overturn due to inadequate bracing or attachment to floors or adjacent walls, columns, or other structural members. Rehabilitation is usually accomplished by the addition of metal cross-bracing to the rear of the bookcase to improve its internal resistance to racking forces, and by bracing the bookcase both in- and out-of-plane to the adjacent structure or walls to prevent overturning and racking.

### 11.11.2.3 Acceptance Criteria

Acceptance criteria shall be applied in accordance with Section 11.3.2.

#### 11.11.2.3.1 Life Safety Nonstructural Performance Level
Bookcases shall be capable of resisting seismic design forces computed in accordance with Section 11.7.3 or 11.7.4.

#### 11.11.2.3.2 Immediate Occupancy Nonstructural Performance Level
Bookcases shall be capable of resisting seismic design forces computed in accordance with Section 11.7.4.

### 11.11.2.4 Evaluation Requirements

The loading, type, and condition of bookcases, their connection to support structures, and type and stability of supporting structure shall be considered in the evaluation.

## 11.11.3 Computer Access Floors

### 11.11.3.1 Definition and Scope

Computer access floors shall include panelized, elevated floor systems designed to facilitate access to wiring, fiber optics, and other services associated with computers and other electronic components.

### C11.11.3.1 Definition and Scope

Access floors vary in height but generally are less than 3 ft above the supporting structural floor. The systems include structural legs, horizontal panel supports, and panels.

### 11.11.3.2 Component Behavior and Rehabilitation Methods

Computer access floors shall be considered both acceleration- and deformation-sensitive.

Computer access floors not conforming to the acceptance criteria of Section 11.11.3.3 shall be rehabilitated in accordance with Section 11.8.

### C11.11.3.2 Component Behavior and Rehabilitation Methods

Computer access floors may displace laterally or buckle vertically under seismic loads. Rehabilitation of access floors usually includes a combination of improved attachment of computer and communication racks through the access floor panels to the supporting steel structure or to the underlying floor system, while improving the lateral-load-carrying capacity of the steel stanchion system by installing braces or improving the connection of the stanchion base to the supporting floor, or both.

Rehabilitation should be designed in accordance with concepts described in FEMA 74 (FEMA 1994). The weight of the floor system, as well as supported equipment, should be included in the analysis.

### 11.11.3.3 Acceptance Criteria

Acceptance criteria shall be applied in accordance with Section 11.3.2.

#### 11.11.3.3.1 Life Safety Nonstructural Performance Level
Computer access floors need not be rehabilitated for the Life Safety Nonstructural Performance Level.

#### 11.11.3.3.2 Immediate Occupancy Nonstructural Performance Level
If the Prescriptive Procedure is selected based on Table 11-1, prescriptive requirements of Section 11.7.2 shall be met. If the Analytical Procedure is selected based on Table 11-1, computer access floors shall be capable of resisting seismic design forces computed in accordance with Section 11.7.4.

### 11.11.3.4 Evaluation Requirements

Buckling and racking of access floor supports, connection to the support structure, and the effects of mounted equipment shall be considered in the evaluation.

### C11.11.3.4 Evaluation Requirements

Possible future equipment should also be considered in the evaluation.

## 11.11.4 Hazardous Materials Storage

### 11.11.4.1 Definition and Scope

Hazardous materials storage shall include permanently installed containers—free-standing, on supports, or stored on countertops or shelves—that hold materials defined to be hazardous by the National Institute for Occupational Safety and Health, including the following types:

1. Propane gas tanks;
2. Compressed gas vessels; and
3. Dry or liquid chemical storage containers.

Large nonbuilding structures, such as large tanks found in heavy industry or power plants, floating-roof oil storage tanks, and large (greater than 10 ft long) propane tanks at propane manufacturing or distribution plants need not meet the requirements of this section.

### 11.11.4.2 Component Behavior and Rehabilitation Methods
Hazardous materials storage shall be considered acceleration-sensitive.

Hazardous materials storage not conforming to the acceptance criteria of Section 11.11.4.3 shall be rehabilitated in accordance with Section 11.8.

### C11.11.4.2 Component Behavior and Rehabilitation Methods
Upset of the storage container may release the hazardous material. Failure occurs because of buckling and overturning of supports and/or inadequate bracing. Rehabilitation consists of strengthening and increasing supports or adding bracing designed according to concepts described in FEMA 74 (FEMA 1994) and FEMA 172 (FEMA 1992).

### 11.11.4.3 Acceptance Criteria
Acceptance criteria shall be applied in accordance with Section 11.3.2.

*11.11.4.3.1 Life Safety Nonstructural Performance Level* Hazardous materials storage shall meet prescriptive requirements in accordance with Section 11.7.2.

*11.11.4.3.2 Immediate Occupancy Nonstructural Performance Level* Hazardous materials storage shall meet prescriptive requirements for essential facilities in accordance with Section 11.7.2.

### 11.11.4.4 Evaluation Requirements
The location and types of hazardous materials, container materials, manner of bracing, internal lateral resistance, and the effect of hazardous material spills shall be considered in the evaluation.

## 11.11.5 Computer and Communication Racks

### 11.11.5.1 Definition and Scope
Computer and communication racks shall include free-standing rack systems in excess of 4 ft in height designed to support computer and other electronic equipment. Equipment stored on computer and communication racks need not meet the requirements of this section.

### C11.11.5.1 Definition and Scope
Racks may be supported on either structural or access floors and may or may not be attached directly to these supports.

### 11.11.5.2 Component Behavior and Rehabilitation Methods
Computer and communication racks shall be considered acceleration-sensitive.

Computer communication racks not conforming to the acceptance criteria of Section 11.11.5.3 shall be rehabilitated in accordance with Section 11.8.

### C11.11.5.2 Component Behavior and Rehabilitation Methods
Computer and communication racks may fail internally (through inadequate bracing or moment-resisting capacity) or externally (by overturning caused by absence or failure of floor attachments).

Rehabilitation is usually accomplished by the addition of bracing to the rear and side panels of the racks, and/or by improving the connection of the rack to the supporting floor using concepts shown in FEMA 74 (FEMA 1994) or FEMA 172 (FEMA 1992).

### 11.11.5.3 Acceptance Criteria
Acceptance criteria shall be applied in accordance with Section 11.3.2.

*11.11.5.3.1 Life Safety Nonstructural Performance Level* Computer and communication racks need not be rehabilitated for the Life Safety Nonstructural Performance Level.

*11.11.5.3.2 Immediate Occupancy Nonstructural Performance Level* If the Prescriptive Procedure is selected based on Table 11-1, computer and communication racks shall meet the prescriptive requirements of Section 11.7.2. If the Analytical Procedure is selected based on Table 11-1, computer and communication racks shall be capable of resisting seismic design forces computed in accordance with Section 11.7.4.

### 11.11.5.4 Evaluation Requirements

Buckling or racking failure of rack components, their connection to support structures, and type and stability of the supporting structure shall be considered in the evaluation. The effect of rack failure on equipment shall also be considered.

## 11.11.6 Elevators

### 11.11.6.1 Definition and Scope

Elevators shall include cabs and shafts, as well as all equipment and equipment rooms associated with elevator operation, such as hoists, counterweights, cables, and controllers.

### 11.11.6.2 Component Behavior and Rehabilitation Methods

Components of elevators shall be considered acceleration-sensitive. Shafts and hoistway rails, which rise through multiple floors, shall be considered both acceleration- and deformation-sensitive.

Elevator components not conforming to the acceptance criteria of Section 11.11.6.2 shall be rehabilitated in accordance with Section 11.8.

### C11.11.6.2 Component Behavior and Rehabilitation Methods

Components of elevators may become dislodged or derailed. Shaft walls and the construction of machinery room walls are often not engineered and must be considered in a way similar to that for other partitions. Shaft walls that are of unreinforced masonry or hollow tile must be considered with special care, since failure of these components violates Life Safety Nonstructural Performance Level criteria.

Elevator machinery may be subject to the same damage as other heavy floor-mounted equipment. Electrical power loss renders elevators inoperable.

Rehabilitation measures include a variety of techniques taken from specific component sections for partitions, controllers, and machinery. Rehabilitation specific to elevator operation can include seismic shutoffs, cable restrainers, and counterweight retainers; such measures should be in accordance with ASME A17.1 (ASME 2000).

### 11.11.6.3 Acceptance Criteria

Acceptance criteria shall be applied in accordance with Section 11.3.2.

#### 11.11.6.3.1 Life Safety Nonstructural Performance Level If the Prescriptive Procedure is selected based on Table 11-1, elevator components shall meet the pre-

scriptive requirements of Section 11.7.2. If the Analytical Procedure is selected based on Table 11-1, elevator components shall be capable of resisting seismic design forces computed in accordance with Section 11.7.3 or 11.7.4.

#### 11.11.6.3.2 Immediate Occupancy Nonstructural Performance Level If the Prescriptive Procedure is selected based on Table 11-1, elevator components shall meet the prescriptive requirements of Section 11.7.2. If the Analytical Procedure is selected based on Table 11-1, elevator components shall be capable of resisting seismic design forces computed in accordance with Section 11.7.4.

### 11.11.6.4 Evaluation Requirements

The construction of elevator shafts shall be considered in the evaluation.

### C11.11.6.4 Evaluation Requirements

The possibility of displacement or derailment of hoistway counterweights and cables should be considered, as should the anchorage of elevator machinery.

## 11.11.7 Conveyors

### 11.11.7.1 Definition and Scope

Conveyors shall include material conveyors, including all machinery and controllers necessary to operation.

### 11.11.7.2 Component Behavior and Rehabilitation Methods

Conveyors shall be considered both acceleration- and deformation-sensitive.

Conveyors not conforming to the acceptance criteria of Section 11.11.7.3 shall be rehabilitated in accordance with Section 11.8.

### C11.11.7.2 Component Behavior and Rehabilitation Methods

Conveyor machinery may be subject to the same damage as other heavy floor-mounted equipment. In addition, deformation of adjoining building materials may render the conveyor inoperable. Electrical power loss renders the conveyor inoperable.

Rehabilitation of the conveyor involves prescriptive procedures using special skills provided by the conveyor manufacturer.

### 11.11.7.3 Acceptance Criteria

Acceptance criteria shall be applied in accordance with Section 11.3.2.

*11.11.7.3.1 Life Safety Nonstructural Performance Level* Conveyors need not be rehabilitated for the Life Safety Nonstructural Performance Level.

*11.11.7.3.2 Immediate Occupancy Nonstructural Performance Level* If the Analytical Procedure is selected based on Table 11-1, conveyors shall be capable of resisting seismic design forces computed in accordance with Section 11.7.4. If the Prescriptive Procedure is selected based on Table 11-1, conveyors shall meet prescriptive standards in accordance with Section 11.7.2.

### 11.11.7.4 Evaluation Requirements

The stability of machinery shall be considered in the evaluation.

## A. USE OF THIS STANDARD FOR LOCAL OR DIRECTED RISK MITIGATION PROGRAMS

### A.1 GENERAL

This ASCE Standard for Seismic Rehabilitation of Buildings is written in mandatory language suitable for adoption and enforcement by code officials in local risk mitigation programs, by organizations or governmental agencies in directed mitigation programs covering many buildings, or for reference by building owners voluntarily undertaking rehabilitation of buildings. This appendix provides guidance on the use of this standard for local or directed risk mitigation programs.

Local or directed risk mitigation programs may target certain building types for rehabilitation or require complete rehabilitation coupled with other renovation work. The incorporation of variable Rehabilitation Objectives and the use of Model Building Types in this standard allows creation of subsets of rehabilitation requirements to suit local conditions of seismicity, building inventory, social and economic considerations, and other factors. Provisions appropriate for local situations can be extracted, put into regulatory language, and adopted into appropriate codes, standards, or local ordinances.

### A.2 INITIAL CONSIDERATIONS FOR MITIGATION PROGRAMS

Local or directed programs can either target high-risk building types or set overall priorities. These decisions should be made with full consideration of physical,

social, historic, and economic characteristics of the building inventory. Although financial incentives can induce voluntary risk mitigation, carefully planned mandatory or directed programs, developed in cooperation with those whose interests are affected, are generally more effective. Potential benefits of such programs include reduction of direct earthquake losses—casualties, costs to repair damage, and loss of use of buildings—as well as more rapid overall recovery. Rehabilitated buildings may also increase in value and be assigned lower insurance rates. Additional issues that should be considered for positive or negative effects include the interaction of rehabilitation with overall planning goals, historic preservation, and the local economy. These issues are discussed in FEMA 275 (FEMA 1998).

### A.2.1 Potential Costs of Local or Directed Programs

The primary costs of seismic rehabilitation—the construction work itself, including design, inspection, and administration—are normally paid by the owner. Additional costs that should be weighed when creating seismic risk reduction programs are those associated with developing and administering the program, such as the costs of identifying high-risk buildings, environmental or socioeconomic impact reports, training programs, plan checking, and construction inspection.

The construction costs include not only the cost of the pure structural rehabilitation, but also the costs associated with new or replaced finishes that may be required. In some cases, seismic rehabilitation work will trigger other local jurisdictional requirements, such as hazardous material removal or partial or full compliance with the Americans with Disabilities Act. The costs of seismic or functional improvements to nonstructural systems should also be considered. There may also be costs to the owner associated with temporary disruption or loss of use of the building during construction. To offset these costs, there may be low-interest earthquake rehabilitation loans available from state or local government, or building tax credits.

If seismic rehabilitation is the primary purpose of construction, the costs of various nonseismic work that may be required should be included as direct consequences. On the other hand, if the seismic work is an added feature of a major renovation, the nonseismic improvements probably would have been required anyway and therefore should not be attributed to seismic rehabilitation.

A discussion of these issues, as well as guidance on the range of costs of seismic rehabilitation, are included in FEMA 156 (FEMA 1994) and 157 (FEMA

1995), and in FEMA 276 (FEMA 1999). Since the data for these documents were developed prior to this standard, the information is not based on buildings rehabilitated specifically in accordance with the current document. However, performance levels defined in this standard are not intended to be significantly different than parallel levels used previously, and costs should still be reasonably representative.

## A.2.2 Timetables and Effectiveness

Presuming that new buildings are being constructed with adequate seismic protection and that older buildings are occasionally demolished or replaced, the inventory of seismically hazardous buildings in any community will be gradually reduced. This attrition rate is normally small, since the structures of many buildings have useful lives of 100 years or more and very few buildings are actually demolished. If buildings or districts become historically significant, they may not be subject to attrition at all. Thus, in many cases, doing nothing (or waiting for an outside influence to force action) may present a large cumulative risk to the inventory.

It has often been pointed out that exposure time is a significant element of risk. The time aspect of risk reduction is so compelling that it often appears as part of book and workshop titles; for example, *Between Two Earthquakes: Cultural Property in Seismic Zones* (Feilden 1987); *Competing Against Time* (California Governor's Board of Inquiry 1990); and *In Wait for the Next One* (EERI 1995). Therefore, an important consideration in the development of programs is the time allotted to reach a certain risk reduction goal. It is generally assumed that longer programs create less hardship than short ones by allowing more flexibility in planning for the cost and possible disruption of rehabilitation, as well as by allowing natural or accelerated attrition to reduce undesirable impacts. On the other hand, the net reduction of risk is smaller due to the increased exposure time of the seismically deficient building stock.

Given a high perceived danger and certain advantageous characteristics of ownership, size, and occupancy of the target buildings, mandatory programs have been completed in as little as five to ten years. More extensive programs—involving complex buildings such as hospitals, or with significant funding limitations—may have completion goals of 30 to 50 years. Deadlines for individual buildings are also often determined by the risk presented by building type, occupancy, location, soil type, funding availability, or other factors.

## A.2.3 Historic Preservation

Seismic rehabilitation of buildings can affect historic preservation in two ways. First, the introduction of new components that will be associated with the rehabilitation may in some way impact the historic fabric of the building. Second, the seismic rehabilitation work can serve to better protect the building from possibly unrepairable future earthquake damage. The effects of any seismic risk reduction program on historic buildings or preservation districts should be carefully considered during program development, and subsequent work should be carefully monitored to assure compliance with national preservation guidelines discussed in Section A.6.

## A.3 USE IN PASSIVE PROGRAMS

Programs that only require seismic rehabilitation in association with other activity on the building are often classified as "passive." "Active" programs, on the other hand, are those that mandate seismic rehabilitation for targeted buildings in a certain time frame, regardless of other activity associated with the building (see Section 1.6.3). Activities in a building that may passively generate a requirement to seismically rehabilitate—such as an increase in occupancy, structural modification, or a major renovation that would significantly extend the life of the building—are called "triggers." The concept of certain activities triggering compliance with current standards is well established in building codes. However, the details of the requirements have varied widely. These issues have been documented with respect to seismic rehabilitation in California (Hoover 1992). Passive programs reduce risk more slowly than do active programs.

## A.3.1 Selection of Seismic Rehabilitation Triggers

This standard does not cover triggers for seismic rehabilitation. The extent and detail of seismic triggers will greatly affect the speed, effectiveness, and impacts of seismic risk reduction, and the selection of triggers is a policy decision expected to be made locally by the person, agency, or jurisdiction responsible for the inventory. Triggers that have been used or considered in the past include revision of specified proportions of the structure; renovation of specified percentages of the building area; work on the building that costs more than a specified percentage of the building value; change in use that increases the occupancy or importance of the building; and changes of ownership.

### A.3.2 Selection of Passive Seismic Rehabilitation Standards

This standard purposely affords a wide variety of options that can be adopted for seismic rehabilitation to facilitate risk reduction. Standards can be selected with varying degrees of risk reduction and varying costs by designating different Rehabilitation Objectives. As described previously, a Rehabilitation Objective is created by specifying a desired target Building Performance Level for specified earthquake ground motion criteria. A jurisdiction can thus specify appropriate standards by extracting applicable requirements and incorporating them into its own code or standard, or by reference.

A single Rehabilitation Objective could be selected under all triggering situations [the Basic Safety Objective (BSO), for example], or more stringent objectives can be used for important changes to the building and less stringent objectives for minor changes. For example, it is sometimes necessary for design professionals, owners, and building officials to negotiate the extent of seismic improvements done in association with building alterations. Complete rehabilitation is often required by local regulation for complete renovation or major structural alterations. It is the intent of this standard to provide a common framework for all of these various uses.

### A.4 USE IN ACTIVE OR MANDATED PROGRAMS

Active programs are most often targeted at high-risk building types or occupancies. Active seismic risk reduction programs are those that require owners to rehabilitate their buildings to specified Rehabilitation Objectives in a certain time frame or, in the case of government agencies or other owners of large inventories, to set self-imposed deadlines for completion.

### A.4.1 Selection of Buildings to be Included

Programs would logically target only the highest-risk buildings or at least create priorities based on risk. Risk can be based on the likelihood of building failure, the occupancy or importance of buildings, soil types, or other factors. This standard is primarily written to be used in the process of rehabilitation and does not directly address the comparative risk level of various building types or other risk factors. Certain building types, such as unreinforced masonry bearing wall buildings and older, improperly detailed reinforced concrete frame buildings, have historically presented a high risk, depending on local seismicity and building

practice. Therefore, these building types have sometimes been targeted in active programs.

A more pragmatic consideration is the ease of locating targeted buildings. If certain building types cannot be easily identified, either by the local jurisdiction or by the owners and their engineers, enforcement could become difficult and costly. In the extreme, every building designed prior to a given acceptable code cycle would require a seismic evaluation to determine whether targeted characteristics or other risk factors are present, the cost of which may be significant. An alternate procedure might be to select easily identifiable building characteristics to set timelines, even if more accurate building-by-building priorities are somewhat compromised.

### A.4.2 Selection of Active Seismic Rehabilitation Standards

As discussed for passive programs in Section A.3.2, this standard is written to facilitate a wide variation in risk reduction. Factors used to determine an appropriate Rehabilitation Objective include local seismicity, the costs of rehabilitation, and local socioeconomic conditions.

It may be desirable to use Simplified Rehabilitation Methods for active or mandated programs. Only Limited Performance Objectives are included in this standard for this method. However, if a program has identified a local building type with few variations in material and configuration, a study of a sample of typical buildings using Systematic Methods may establish that compliance with the requirements of Simplified Rehabilitation meets the BSO, or better, for this building type in this location. Such risk and performance decisions can only be made at the local level.

### A.5 SOCIAL, ECONOMIC, AND POLITICAL CONSIDERATIONS

The scope of this standard is limited to the engineering basis for seismically rehabilitating a building, but the user should also be aware of significant nonengineering issues and social and economic impacts presented in this section. These problems and opportunities, which vary with each situation, are discussed in FEMA 275 (FEMA 1998).

### A.5.1 Construction Cost

If seismic rehabilitation were always inexpensive, the social and political costs and controversies would largely disappear. Unfortunately, seismic rehabilitation often requires removal of architectural materials to access the vulnerable portions of the structure. Nonseismic upgrading (e.g., electrical, handicapped access, historic restoration) is frequently triggered by a building code's renovation permit requirements and is desirable to undertake at the same time.

### A.5.2 Housing

Although seismic rehabilitation ultimately improves the housing stock, units can be temporarily lost during the construction phase, which may be very lengthy, and can require relocation of tenants.

### A.5.3 Impacts on Lower-Income Groups

Lower-income residents and commercial tenants can be displaced by seismic rehabilitation and nonseismic upgrading, which can raise rents and real estate prices because of the need to recover the costs of the work. Possible hardships on these groups need to be given heavy consideration because they may affect the very societal fabric of a community.

### A.5.4 Regulations

As with efforts to impose safety regulations in other fields, mandating seismic rehabilitation is often controversial. This standard is written as mandatory code provisions for possible application and adaptation for that use. In such cases, political controversy should be expected and nonengineering issues of all kinds should be carefully considered.

### A.5.5 Architecture

Even if a building is not historic, there are often significant architectural impacts. The exterior and interior appearance may change, and the division of spaces and arrangement of circulation routes may be altered.

### A.5.6 Community Revitalization

Seismic rehabilitation not only poses issues and implies costs, but also confers benefits. In addition to enhanced public safety and economic protection from earthquake loss, seismic rehabilitation can play a leading role in the revitalization of older commercial and industrial areas as well as residential neighborhoods. Potential synergies between these two programs in a community should be carefully explored by local planners, officials, and design professionals.

## A.6 CONSIDERATIONS FOR HISTORIC BUILDINGS

It must be determined early in the process whether a building is "historic." A building is historic if it is at least 50 years old, and is listed in or is potentially eligible for the National Register of Historic Places and/or a state or local register as an individual structure, or as a contributing structure in a district. Structures less than 50 years old may also be historic if they possess exceptional significance. For historic buildings, users should develop and evaluate alternative solutions with regard to their effect on the loss of historic character and fabric. This section provides guidance for developing such alternative solutions.

### A.6.1 Secretary of the Interior's Standards

For historic buildings, users should develop alternative solutions using the *Standards for the Treatment of Historic Properties with Guidelines for Preserving, Rehabilitating, Restoring, and Reconstructing Historic Buildings* (Secretary of the Interior 1995).

In addition to rehabilitation, the Secretary of the Interior also has standards for preservation, restoration, and reconstruction [*Standards for the Treatment of Historic Properties* (Secretary of the Interior 1992)]. A seismic rehabilitation project may include work that falls under the Rehabilitation Standards, the Treatment Standards, or both. This standard is intended for use as part of rehabilitation, preservation, and restoration work done on historic buildings.

For historic buildings as well as for other structures of architectural interest, it is important to note that the Secretary of the Interior's Standards define rehabilitation as "the process of returning a property to a state of utility, through repair or alteration, which makes possible an efficient contemporary use while preserving those portions and features of the property which are significant to its historic, architectural and cultural values." Further guidance on the treatment of historic properties is contained in publications listed in the *Technical Preservation Services for Historic Buildings Sales Publication Catalog* available online at http://www.cr.nps.gov/hps/tps/tpscat.htm.

### A.6.2 Application of Building Codes and Standards

It should be noted that many codes covering historic buildings allow some amount of flexibility in required performance, depending on the effect of rehabilitation on important historic features.

If a building contains items of unusual architectural interest, consideration should be given to the value of these items. It may be desirable to rehabilitate the building to the Damage Control Structural Performance Range as specified in this standard to ensure that the architectural fabric survives certain earthquakes.

### A.6.3 Rehabilitation Strategies

In development of initial risk mitigation strategies, consideration must be given to the architectural and historic value of the building and its fabric. Development of a Historic Structure Report identifying the primary historic fabric may be essential in the preliminary planning stages for certain buildings. Some structurally adequate solutions may nevertheless be unacceptable because they involve destruction of historic fabric or character. Alternate rehabilitation methods that lessen the impact on the historic fabric should be developed for consideration. Partial demolition may be inappropriate for historic structures. Components that create irregularities may be essential to the historic character of the structure. The advice of historic preservation experts may be necessary early in the rehabilitation process.

Structural rehabilitation of historic buildings may be accomplished by hiding the new structural members or by exposing them as admittedly new components in the building's history. Often, the exposure of new structural members is preferred because alterations of this kind are reversible (i.e., they could conceivably be undone at a future time with no loss of historic fabric to the building). The decision to hide or expose structural members is a complex one and is best made by a preservation professional.

### A.6.4 Rehabilitation Objectives

If seismic rehabilitation is required by the governing building jurisdiction, the minimum seismic requirements should be matched with a Rehabilitation Objective defined in this standard.

## SYMBOLS

$A$    Cross-sectional area of a pile, Eq. 4-9
Cross-sectional area of shear wall boundary members or diaphragm chords, in.$^2$, Eqs. 8-2, 8-4, 8-5

$A_b$    Gross area of bolt or rivet, Eqs. 5-18, 5-22, 5-24
Sum of net mortared area of bed joints above and below the test unit, Eq. 7-2

$A_c$    Area of column, Eq. 5-8

$A_e$    Effective net area of the horizontal leg, Eq. 5-20

$A_f$    Area of foundation footprint if the foundation components are interconnected laterally, Eq. 4-14

$A_g$    Gross area of the horizontal leg, Eq. 5-19
Gross area of cast iron column, Eq. 5-36
Gross area of column, in.$^2$, Eq. 6-4

$A_j$    Effective cross-sectional area of a beam–column joint, in.$^2$, in a plane parallel to the plane of reinforcement generating shear in the joint, calculated as specified in Section 6.5.2.3.1, Eq. 6-5

$A_n$    Area of net mortared/grouted section, Eqs. 7-1, 7-2, 7-4, 7-6

$A_{ni}$    Area of net mortared/grouted section of masonry infill, Eq. 7-15

$A_s$    Area of non-prestressed tension reinforcement, in.$^2$, Tables 6-18, 6-20
Area of reinforcement, Eq. 7-13

$A_s'$    Area of compression reinforcement, in.$^2$, Tables 6-18, 6-20

$A_v$    Area of shear reinforcement, Eq. 6-4

$A_w$    Nominal area of the web, Eq. 5-7
Area of link stiffener web, Eq. 5-28, 5-31
Area of the web cross section, $= b_w d$, Chapter 6

$A_x$    Accidental torsion amplification factor, Eq. 3-1

$B$    Width of footing, Eqs. 4-6, 4-7, 4-8

$B_1$    Damping coefficient used to adjust one-second period spectral response for the effect of viscous damping, Eqs. 1-10, 1-11

$B_{D1}$    Numerical damping coefficient taken equal to the value of $B_1$, as determined in Section 1.6.1.5.1, at $\beta$
Effective damping $\beta$ equal to the value of $\beta_D$, Eq. 9-2

$B_{M1}$    Numerical damping coefficient taken equal to the value of $B_1$, as determined in Section 1.6.1.5.1, at $\beta$
Effective damping $\beta$ equal to the value of $\beta_M$, Equation 9-4

$B_S$    Coefficient used to adjust short-period spectral response for the effect of viscous damping, Eqs. 1-8, 1-9, 1-11

$C$ (or $C_j$)    Damping coefficient for viscoelastic device (or device $j$), Eqs. 9-22, 9-24, 9-29, 9-30, 9-35, 9-37

$C_0$    Modification factor to relate spectral displacement of an equivalent single-degree-of-freedom (SDOF) system to the roof displacement of the building multi-degree-of-freedom (MDOF) system, Eq. 3-14
Damping coefficient for fluid-viscous device, Eq. 9-25

$C_1$    Modification factor to relate expected maximum inelastic displacements to displacements calculated for linear elastic response, Eqs. 3-4, 3-5, 3-9, 3-14, 3-19

$C_2$ — Modification factor to represent the effects of pinched hysteresis shape, cyclic stiffness degradation and strength deterioration on the maximum displacement response, Eqs. 3-4, 3-5, 3-9, 3-14, 3-19

$C_b$ — Coefficient to account for effect of nonuniform moment given in *Load and Resistance Factor Design Specification for Structural Steel Buildings (LRFD)* (AISC 1999), Eq. 5-9

$CF_i$ — Stage combination factors for use with velocity-dependent energy dissipation devices as calculated by Eq. 9-31 or 9-32

$C_m$ — Effective mass factor from Table 3-1, Eqs. 3-9, 3-15

$C_t$ — Numerical value for adjustment of period $T$, Eq. 3-6

$C_{vx}$ — Vertical distribution factor for the pseudo-lateral force, Eqs. 3-10, 3-11

$D$ — Generalized deformation, unitless Relative displacement between two ends of an energy dissipation unit, Eqs. 9-1, 9-20, 9-22

$D^-$ — Maximum negative displacement of an energy dissipation unit, Eqs. 9-21, 9-23

$D^+$ — Maximum positive displacement of an energy dissipation unit, Eqs. 9-21, 9-23

$\dot{D}$ — Relative velocity between two ends of an energy dissipation unit, Eqs. 9-22, 9-25

$D_{ave}$ — Average displacement of an energy dissipation unit, equal to $(|D^+| + |D^-|)/2$, Eq. 9-24

$D_{clear}$ — Required clearance between a glass component and the frame, Eq. 11-9

$DCR$ — Demand-capacity ratio, computed in accordance with Eq. 2-1 or required in Eq. 2-2

$\overline{DCR}$ — Average demand-capacity ratio for a story, computed in accordance with Eq. 2-2

$D_D$ — Design displacement, in. (mm), at the center of rigidity of the isolation system in the direction under consideration, Eqs. 9-2, 9-6, 9-8, 9-10, 9-14, 9-15, 9-18, 9-22

$D_D'$ — Design earthquake target displacement, in. (mm), at a control node located at the center of mass of the first floor above the isolation system in the direction under consideration, as prescribed by Eq. 9-10

$D_M$ — Maximum displacement, in. (mm), at the center of rigidity of the isolation system in the direction under consideration, Eqs. 9-4, 9-7, 9-11, 9-16, 9-17, 9-19

$D_M'$ — BSE-2 target displacement, in. (mm), at a control node located at the center of mass of the first floor above the isolation system in the direction under consideration, as prescribed by Eq. 9-11

$D_p$ — Relative seismic displacement that the component must be designed to accommodate, Eqs. 11-8, 11-9, 11-10, 11-11

$D_r$ — Drift ratio for nonstructural components, Eq. 11-7

$D_{TD}$ — Total design displacement, in. (mm), of a component of the isolation system, including both translational displacement at the center of rigidity and the component of torsional displacement in the direction under consideration, as specified by Eq. 9-6

$D_{TM}$ — Total maximum displacement, in. (mm), of a component of the isolation system, including both translational displacement at the center of rigidity and the component of torsional displacement in the direction under consideration, as specified by Eq. 9-7

$E$ — Young's modulus of elasticity, Eqs. 4-9, 5-1, 5-2, 5-17, 8-2, 8-4, 8-5

$E_c$ — Modulus of elasticity of concrete, psi, Eq. 6-6

$E_{fe}$ — Expected elastic modulus of frame material, ksi, Eq. 7-7

$E_{Loop}$ — Energy dissipated, kip-in. (kN-mm), in an isolator unit during a full cycle of reversible load over a test displacement range from $\Delta^+$ to $\Delta^-$, as measured by the area enclosed by the loop of the force–deflection curve, Eq. 9-13

$E_m$ — Masonry elastic modulus

$E_{me}$ — Expected elastic modulus of masonry in compression as determined per Section 7.2.2.4, Eq. 7-7

$E_s$ — Modulus of elasticity of reinforcement, psi, Chapter 6

$E_{se}$ — Expected elastic modulus of reinforcing steel per Section 7.2.2.8

$F$ — Force in an energy dissipation unit, Eqs. 9-1, 9-20, 9-22, 9-25

$F^-$ — Negative force, k, in an isolator or energy dissipation unit during a single cycle of prototype testing at a displacement amplitude of $\Delta^-$, Eqs. 9-12, 9-21, 9-23, 9-38

$F^+$ — Positive force, k, in an isolator or energy dissipation unit during a single cycle of prototype testing at a displacement amplitude of $\Delta^+$, Eqs. 9-12, 9-21, 9-23, 9-38

$F_a$ — Factor to adjust spectral acceleration in the short-period range for site class, Eq. 1-7

$F_{cr}$ — Allowable axial buckling stress, Eq. 5-36

$F_i$ — Inertia force at floor level $i$, Eq. 9-27

| | | | |
|---|---|---|---|
| $F_{mi}$ | Lateral load applied at floor level $i$, Eqs. 3-12, C3-2 | $H_{rw}$ | Height of the retaining wall, Eq. 4-16 |
| | $m$-th mode horizontal inertia force at floor level $i$, Eq. 9-34 | $I$ | Moment of inertia, Eq. 6-6 |
| $F_p$ | Horizontal seismic force for design of a structural or nonstructural component and its connection to the structure, Eqs. 2-3, 2-4, 2-5, 2-6, 2-7 | $I_b$ | Moment of inertia of a beam, Eqs. 5-1, 5-17 |
| | | $I_c$ | Moment of inertia of a column, Eq. 5-2 |
| | | $I_{col}$ | Moment of inertia of column section, Eq. 7-7 |
| | Component seismic design force applied horizontally at the center of gravity of the component or distributed according to the mass distribution of the component, Eqs. 11-1, 11-2, 11-3, 11-4 | $I_f$ | Moment of inertia of most flexible frame member confining infill panel, Chapter 7 |
| $F_{pv}$ | Component seismic design force applied vertically at the center of gravity of the component or distributed according to the mass distribution of the component, Eqs. 11-2, 11-5, 11-6 | $I_g$ | Moment of inertia of gross concrete section about centroidal axis, neglecting reinforcement, Chapter 6 |
| | | $I_p$ | Component performance factor; 1.0 shall be used for the Life Safety Nonstructural Performance Level and 1.5 shall be used for the Immediate Occupancy Nonstructural Performance Level, Eqs. 11-1, 11-3, 11-4, 11-5, 11-6 |
| $F_{px}$ | Diaphragm lateral force at floor level $x$, Eq. 3-12 | $J$ | A coefficient used in linear procedures to estimate the actual forces delivered to force-controlled components by other (yielding) components, Eqs. 3-4, 3-19 |
| $F_{te}$ | Expected tensile strength, Eqs. 5-20, 5-22, 5-24 | | |
| $F_v$ | Factor to adjust spectral acceleration at 1 sec for site class, Eq. 1-8 | $K$ | Length factor for brace; defined in *Load and Resistance Factor Design Specification for Structural Steel Buildings (LRFD)* (AISC 1999), Chapter 5 |
| | Design shear strength of bolts or rivets, Chapter 5 | | |
| $F_{ve}$ | Unfactored nominal shear strength of bolts or rivets given in *Load and Resistance Factor Design Specification for Structural Steel Buildings (LRFD)* (AISC 1999), Eq. 5-18 | $K'$ | Storage stiffness as prescribed by Eq. 9-23 |
| | | $K''$ | Loss stiffness as prescribed by Eq. 9-24 |
| | | $K_\theta$ | Rotational stiffness of a partially restrained connection, Eqs. 5-15, 5-16, 5-17 |
| $F_x$ | Lateral load applied at floor level $x$, Eq. 3-11, Fig. C7-2 | | Effective rotational stiffness of the foundation, Eq. 4-14 |
| $F_y$ | Specified minimum yield stress for the type of steel being used, Eq. 5-7 | $K^*_{Fixed}$ | $= M^*\left(\dfrac{2\pi}{T}\right)^2$, Eq. 4-14 |
| $F_{yb}$ | $F_y$ of a beam, Chapter 5 | $K_b$ | Flexural stiffness, Eqs. 5-27, 5-29 |
| $F_{yc}$ | $F_y$ of a column, Chapter 5 | $K_{Dmax}$ | Maximum effective stiffness, k/in., of the isolation system at the design displacement in the horizontal direction under consideration, as prescribed by Eq. 9-14 |
| $F_{ye}$ | Expected yield strength, Eqs. 5-1 through 5-8, 5-19, 5-23, 5-25, 5-31, 5-34 | | |
| $F_{yf}$ | $F_y$ of a flange, Chapter 5 | $K_{Dmin}$ | Minimum effective stiffness, k/in. (kN/mm), of the isolation system at the design displacement in the horizontal direction under consideration, as prescribed by Eq. 9-15 |
| $F_{yLB}$ | Lower-bound yield strength, Chapter 5 | | |
| $G$ | Soil shear modulus, Eqs. 4-6, 4-12, 4-14, Shear modulus of steel, Eqs. 5-28, 5-33 Modulus of rigidity of wood structural panels, psi, Eqs. 8-2, 8-4, 8-5 | | |
| | | $K_E$ | Format conversion factor for calculating LRFD reference resistance based on allowable stress factor, Section C8.3.2.5 |
| $G_d$ | Shear stiffness of shear wall or diaphragm assembly, Eqs. 8-1, 8-3 | | |
| $G_m$ | Masonry shear modulus | $K_e$ | Effective stiffness of the building in the direction under consideration, for use with the NSP, Eq. 3-14 |
| $G_{me}$ | Shear modulus of masonry as determined per Section 7.2.2.7 | | |
| $G_o$ | Initial or maximum shear modulus, Eqs. 4-4, 4-5, 4-12 | | Elastic stiffness of a link beam, Eqs. 5-27, 5-30 |
| $H$ | Horizontal load on footing, Chapter 4 | | |

| | | | |
|---|---|---|---|
| $K_i$ | Elastic stiffness of the building in the direction under consideration, for use with the NSP, Eq. 3-13 | $M_c$ | Ultimate moment capacity of footing, Eq. 4-8 |
| $K_{Mmax}$ | Maximum effective stiffness, k/in., of the isolation system at the maximum displacement in the horizontal direction under consideration, as prescribed by Eq. 9-16 | $M_{CE}$ | Expected flexural strength of a member or joint, Eqs. 5-3, 5-4, 5-6, 5-15, 5-16, 5-18, 5-22, 5-24, 5-25, 5-26, 5-32 |
| | | $M_{CEx}$ | Expected bending strength of a member about the x-axis, Eqs. 5-10, 5-11, 5-13, 6-1 |
| | | $M_{CEy}$ | Expected bending strength of a member about y-axis, Eqs. 5-10, 5-11, 5-13, 6-1 |
| $K_{Mmin}$ | Minimum effective stiffness, k/in., of the isolation system at the maximum displacement in the horizontal direction under consideration, as prescribed by Eq. 9-17 | $M_{CLx}$ | Lower-bound flexural strength of the member about the x-axis, Eq. 5-12 |
| | | $M_{CLy}$ | Lower-bound flexural strength of the member about the y-axis, Eq. 5-12 |
| $K_s$ | Shear stiffness, Eqs. 5-27, 5-28 | $M_{gCS}$ | Moment acting on the slab column strip, Chapter 6 |
| $K_{sh}$ | Horizontal spring stiffness, Chapter 4 | $M_n$ | Nominal moment strength at section, Chapter 6 |
| $K_W$ | Global stiffness of steel plate shear wall, Chapter 5 | | |
| $K_x$ | Effective translational stiffness of the foundation, Eq. 4-14 | $M_{nCS}$ | Nominal moment strength of the slab column strip, Chapter 6 |
| $L$ | Length of footing in plan dimension, Eqs. 4-7, 4-8 | $M_{OT}$ | Total overturning moment induced on the element by seismic forces applied at and above the level under consideration, Eqs. 3-4, 3-5 |
| | Length of pile in vertical dimension, Eq. 4-9 | | |
| | Length of member along which deformations are assumed to occur, Chapter 6 | $M_{pCE}$ | Expected plastic moment capacity, Eq. 5-6 |
| | Length of wall or wall pier, Eqs. 7-3, 7-5 | $M_{ST}$ | Stabilizing moment produced by dead loads acting on the element, Eqs. 3-4, 3-5 |
| | Diaphragm span, distance between shear walls or collectors, Eqs. 8-3, 8-4, 8-5 | $M_{UD}$ | Design moment, Chapter 6 |
| $L_b$ | Length or span of beam, Eqs. 5-6, 5-17 | $M_{UDx}$ | Design bending moment about the x-axis for axial load $P_{UF}$, kip-in., Eq. 6-1 |
| | Distance between points braced against lateral displacement of the compression flange or between points braced to prevent twist of the cross sections; given in *Load and Resistance Factor Design Specification for Structural Steel Buildings (LRFD)* (AISC 1999), Eq. 5-9 | $M_{UDy}$ | Design bending moment about the y-axis for axial load $P_{UF}$, kip-in., Eq. 6-1 |
| | | $M_{UFx}$ | Bending moment in the member about the x-axis, calculated in accordance with Section 3.4.2.1.2, Eq. 5-12 |
| | | $M_{UFy}$ | Bending moment in the member about the y-axis, calculated in accordance with Section 3.4.2.1.2, Eq. 5-12 |
| $L_c$ | Length of beam, clear span between columns, Chapter 5 | $M_x$ | Bending moment in a member for the x-axis, Eqs. 5-10, 5-11, 5-13 |
| $L_{inf}$ | Length of infill panel, Eqs. 7-10, 7-12 | | |
| $L_p$ | The limiting unbraced length between points of lateral restraint for the full plastic moment capacity to be effective; given in *Load and Resistance Factor Design Specification for Structural Steel Buildings (LRFD)* (AISC 1999), Eqs. 5-6, 5-9 | $M_y$ | Bending moment in a member for the y-axis, Eqs. 5-10, 5-11, 5-13 |
| | | | Yield moment strength at section, Eq. 6-6 |
| | | $N$ | Number of piles in a pile group, Eq. 4-9 |
| | | $\overline{N}$ | Average Standard Penetration Test (SPT) blow count in soil within the upper 100 ft of soil, calculated in accordance with Eq. 2-8 |
| $L_r$ | The limiting unbraced length between points of lateral support beyond which elastic lateral torsional buckling of the beam is the failure mode; given in *Load and Resistance Factor Design Specification for Structural Steel Buildings (LRFD)* (AISC 1999), Eq. 5-9 | | |
| | | $(N_1)_{60}$ | SPT blow count normalized for an effective stress of 1 ton psf and corrected to an equivalent hammer energy efficiency of 60%, Eq. 4-5 |
| $M$ | Design moment at a section, Eq. 6-4 | $N_b$ | Number of bolts or rivets, Eqs. 5-18, 5-22, 5-24 |
| $M^*$ | Effective mass for the first mode, Eq. 4-14 | | |

$N_u$ — Factored axial load normal to cross section occurring simultaneously with $V_u$. To be taken as positive for compression, negative for tension, and to include effects of tension due to creep and shrinkage, Eq. 6-4

$P$ — Vertical load on footing, Eq. 4-8
Axial force in a member, Eqs. 5-2, 5-4

$P_c$ — Lower-bound of vertical compressive strength for wall or wall pier, Eqs. 7-7, 7-13

$P_{CE}$ — Expected axial strength of a member or joint, Eqs. 5-19, 5-20, 5-21, 5-26

$P_{CL}$ — Lower-bound axial strength of column, Eqs. 5-10, 5-11, 5-12, 5-36
Lower-bound axial compressive force due to gravity loads specified in Eq. 3-4

$P_D$ — Superimposed dead load at the top of the wall or wall pier under consideration, Eqs. 7-1, 7-3

$P_{EY}$ — Probability of exceedance in $Y$ years, expressed as a decimal, Eq. 1-2

$P_I$ — Plasticity index for soil, determined as the difference in water content of soil at the liquid limit and plastic limit, Section 1.6.1.4.1

$P_0$ — Nominal axial load strength at zero eccentricity, Chapter 6

$P_R$ — Mean return period, Eq. 1-2

$P_{UF}$ — Design axial force in a member, Eqs. 5-10, 5-11, 5-12

$P_{ye}$ — Expected yield axial strength of a member, Eqs. 5-2, 5-4

$Q$ — Generalized force in a component, Figs. 2-3, 2-5, 5-1, 6-1, 7-1, 8-1

$Q_{allow}$ — Allowable bearing load specified for the design of deep foundations for gravity loads (dead plus live loads) in the available design documents, Eq. 4-2

$Q_c$ — Expected bearing capacity of deep or shallow foundation, Eqs. 4-2, 4-3, 4-7

$Q_{CE}$ — Expected strength of a component at the deformation level under consideration, Eqs. 2-1, 3-20, 5-3 through 5-8, 5-18, 5-22, 5-24, 5-25, 5-26, 5-30, 5-31, 5-32, 5-34, 5-35, 7-3, 7-15

$Q_{CEb}$ — Expected bending strength of the beam, Eq. 5-14

$Q_{CL}$ — Lower-bound estimate of the strength of a component at the deformation level under consideration, Eqs. 3-21, 5-36, 6-5, 7-4, 7-5, 7-6, 7-14

$Q_{CLc}$ — Lower-bound strength of the connection, Eq. 5-14

$Q_D$ — Design action due to dead load, Eqs. 3-2, 3-3

$Q_E$ — Design action due to design earthquake loads, Eqs. 3-18, 3-19

$Q_G$ — Design action due to gravity loads, Eqs. 3-2, 3-3, 3-18, 3-19, 7-5

$Q_L$ — Design action due to live load, Eqs. 3-2, 3-3

$Q_S$ — Design action due to snow load, Eqs. 3-2, 3-3

$Q_{UD}$ — Deformation-controlled design action due to gravity and earthquake loads, Eqs. 2-1, 3-18, 3-20

$Q_{UF}$ — Force-controlled design action due to gravity and earthquake loads, Eqs. 3-19, 3-21

$Q_y$ — Yield strength of a component, Figs. 2-3, 2-5

$Q'_y$ — Substitute yield strength, Fig. 2-5

$R$ — Ratio of the elastic-strength demand to the yield-strength coefficient, Eqs. 3-14, 3-15

$R_{max}$ — Maximum strength ratio, Eq. 3-16

$R_{OT}$ — Response modification factor for overturning moment $M_{OT}$, Eq. 3-5

$R_p$ — Component response modification factor from Table 11-2, Eq. 11-3

$RRS_{bsa}$ — Ratio of response spectra factor for base slab averaging, Eq. 4-11, Fig. 4-7

$RRS_e$ — Ratio of response spectra factor for embedment, Eq. 4-12

$S_1$ — Spectral response acceleration parameter at a one-second period, obtained from response acceleration maps, Eqs. 1-1, 1-3, 1-5

$S_a$ — Spectral response acceleration, Eqs. 1-8, 1-9, 1-10, 3-10, 3-15, 3-16

$S_n$ — Distance between $n$-th pile and axis of rotation of a pile group, Eq. 4-10

$S_S$ — Spectral response acceleration parameter at short periods, obtained from response acceleration maps, Eqs. 1-1, 1-3, 1-7

$S_{X1}$ — Spectral response acceleration parameter at a 1-sec period for any Earthquake Hazard Level and any damping, adjusted for site class, Eqs. 1-5, 1-10, 1-11, 1-13, 1-14, 1-15, 1-16

$S_{XS}$ — Spectral response acceleration parameter at short periods for the selected Earthquake Hazard Level and damping, adjusted for site class, and determined in accordance with Section 1.6.1.4 or 1.6.2.1, Eqs. 1-4, 1-8, 1-9, 1-11, 1-13, 1-14, 1-15, 1-16, 4-16, 11-1, 11-3, 11-4, 11-5, 11-6

$T$ — Fundamental period of the building in the direction under consideration, seconds,

Eqs. 1-8, 1-10, 3-6, 3-7, 3-8, 3-9, 3-10,
3-11, 4-10, 4-11, 4.12, 9-29
Tensile load in column, Eq. 5-13
Fundamental period of the building using a
model with a fixed base, seconds,
Eqs. 4-14, 4-15

$\tilde{T}$ — Fundamental period of the building using a
model with a flexible base, seconds,
Eqs. 4-14, 4-15

$\tilde{T}/T_{eff}$ — Effective period lengthening ratio,
Eqs. 4-13, 4-14, 4-15, Fig. 4-8

$T_0$ — Period at which the constant acceleration
region of the design response spectrum
begins at a value $= 0.2T_S$, Eqs. 1-8, 1-12

$T_{CE}$ — Expected tensile strength of column computed in accordance with Eq. 5-8

$T_D$ — Effective period, in seconds, of the seismic-
isolated structure at the design
displacement in the direction under consideration, as prescribed by Eq. 9-3

$T_e$ — Effective fundamental period of the building in the direction under consideration, for
use with the NSP, Eqs. 3-13, 3-14
Effective fundamental period, in seconds,
of the building structure above the
isolation interface on a fixed base in the
direction under consideration,
Eqs. 9-10, 9-11

$T_i$ — Elastic fundamental period of the building
in the direction under consideration, for use
with the NSP, Eq. 3-13

$T_M$ — Effective period, in seconds, of the seismic-
isolated structure at the maximum displacement in the direction under consideration,
as prescribed by Eq. 9-5

$T_m$ — m-th mode period of the rehabilitated building including the stiffness of the velocity-
dependent devices, Eq. 9-35

$T_S$ — Period at which the constant acceleration
region of the design response spectrum
transitions to the constant velocity region,
Eqs. 1-8, 1-9, 1-10, 1-11, 1-12, 1-13, 3-9,
3-14

$T_{ss}$ — Secant fundamental period of a rehabilitated building calculated using Eq. 3-14 but
replacing the effective stiffness ($K_e$) with
the secant stiffness ($K_s$) at the target displacement, Eq. 9-37

$V$ — Pseudo-lateral load, Eqs. 3-9, 3-10
Design shear force at section, Eq. 6-4

$V*$ — Modified equivalent base shear, Chapter 9

$V_b$ — The total lateral seismic design force or
shear on elements of the isolation system or

elements below the isolation system, as
prescribed by Eq. 9-8

$V_c$ — Nominal shear strength provided by concrete, Eq. 6-4

$V_{CE}$ — Expected shear strength of a member,
Eqs. 5-11, 5-31, 5-32, 5-34

$V_{CL}$ — Lower-bound shear strength, Eq. 7-4

$V_d$ — Base shear at $\Delta_d$, Fig. 3-1, Chapter 3

$V_{dt}$ — Lower-bound shear strength based on diagonal tension stress for wall or wall pier,
Chapter 7

$V_{fre}$ — Expected story shear strength of the bare
steel frame taken as the shear capacity of
the column, Chapter 7

$V_g$ — Shear acting on slab critical section due to
gravity loads, Chapter 6

$V_i$ — The total calculated lateral shear force in
the direction under consideration in an element or at story i due to earthquake
response to the selected ground shaking
level, as indicated by the selected linear
analysis procedure, Eqs. 2-2

$V_{ine}$ — Expected shear strength of infill panel,
Eq. 7-8

$V_n$ — Nominal shear strength at section, Eq. 6-4,
6-5

$V_o$ — Shear strength of slab at critical section,
Chapter 6

$V_{pz}$ — Panel zone shear, Chapter 5

$V_r$ — Expected shear strength of wall or wall pier
based on rocking shear, Eq. 7-3

$V_s$ — Nominal shear strength provided by shear
reinforcement, Chapter 6
The total lateral seismic design force or
shear on elements above the isolation system, as prescribed by Section 9.2.4.4.2,
Eq. 9-9

$V_t$ — Base shear in the building at the target displacement, Chapter 3

$V_{tc}$ — Lower-bound shear strength based on toe
compressive stress for wall or wall pier,
Chapter 7

$V_{test}$ — Test load at first movement of a masonry
unit, Eq. 7-2

$V_u$ — Factored shear force at section, Chapter 6

$V_y$ — Yield strength of the building in the direction under consideration, for use with the
NSP, Eq. 3-15

$V_{ya}$ — Nominal shear strength of a member
modified by the axial load magnitude,
Chapter 5

$W$ — Weight of a component, calculated as specified in this standard, Chapter 2

Effective seismic weight of a building, including total dead load and applicable portions of other gravity loads listed in Section 3.3.1.3.1, Eqs. 3-9, 3-15, 4-14

The total seismic dead load, kips (kN). For design of the isolation system, $W$ is the total seismic dead-load weight of the structure above the isolation interface, Eqs. 9-3, 9-5

$W_D$    Energy dissipated in a building or element thereof or energy dissipation device during a full cycle of displacement, Eqs. 9-24, 9-39

$W_j$    Work done by an energy dissipating device, $j$, in one complete cycle corresponding to floor displacement, Eqs. 9-26, 9-28, 9-29, 9-36, 9-37

$W_k$    Maximum strain energy in a frame as calculated by Eq. 9-27

$W_{mj}$    Work done by device $j$ in one complete cycle corresponding to modal floor displacements $\delta_{mi}$, Eq. 9-33

$W_{mk}$    Maximum strain energy in the frame in the $m$-th mode determined using Eq. 9-34

$W_p$    Component operating weight, Eqs. 11-1, 11-3, 11-4, 11-5, 11-6

$X$    Height of upper support attachment at level $x$ as measured from grade, Eq. 11-7

$Y$    Time period in years corresponding to a mean return period and probability of exceedance, Eq. 1-2

Height of lower support attachment at level $y$ as measured from grade, Eq. 11-7

$Z$    Plastic section modulus, Eqs. 5-1, 5-2, 5-3, 5-4, 5-6

$Z'$    Adjusted resistance for mechanical fastener, Chapter 8

$a$    Parameter used to measure deformation capacity in component load–deformation curves, Figures 2-3, 5-1, 6-1

Site class factor, Eq. 3-14

Clear width of wall between vertical boundary elements, Eqs. 5-33, 5-34

Equivalent width of infill strut, Eqs. 7-7, 7-9, 7-10, 7-11, 7-12

Longitudinal dimension of full footprint of building foundation, Eq. 4-11

$a'$    Parameter used to measure deformation capacity in component load–deformation curve, Fig. 2-5

$a_1$    $= c_e \exp(4.7 - 1.6h/r_\theta)$, Eq. 4-14

$a_2$    $= c_e[25 \ln(h/r_\theta) - 16]$, Eq. 4-14

$a_p$    Component amplification factor from Table 11-2, Eq. 11-3

$b$    Parameter used to measure deformation capacity in component load–deformation curves, Figs. 2-3, 5-1, 6-1

Shear wall length or width, Eqs. 8-1, 8-2

Diaphragm width, Eqs. 8-4, 8-5

The shortest plan dimension of the rehabilitated building, ft (mm), measured perpendicular to $d$, Eqs. 9-6, 9-7

Transverse dimension of full footprint of building foundation, Eq. 4-11

$b_a$    Connection dimension, Eqs. 5-22, 5-23

$b_{bf}$    Beam flange width in equations for beam–column connections in Sections 5.5.2.4.2 and 5.5.2.4.3

$b_{cf}$    Column flange width in equations for beam–column connections in Sections 5.5.2.4.2 and 5.5.2.4.3

$b_e$    Effective foundation size, ft, Eq. 4-11, Fig. 4-7

$b_f$    Flange width, Tables 5-5, 5-6, 5-7

$b_p$    Width of rectangular glass, Eq. 11-9

$b_t$    Connection dimension, Eqs. 5-24, 5-25

$b_w$    Web width, in., Eq. 6-4

$c$    Parameter used to measure residual strength, Figs. 2-3, 5-1, 6-1, 7-1, 8-1, Chapter 4

$c_1$    Size of rectangular or equivalent rectangular column, capital, or bracket measured in the direction of the span for which moments are being determined, in., Section 6.5.4.3

Clearance (gap) between vertical glass edges and the frame, Eq. 11-9

$c_2$    Clearance (gap) between horizontal glass edges and the frame, Eq. 11-9

$c_e$    $= 1.5(e/r_x) + 1$, Eq. 4-14

$d$    Depth of soil sample for calculation of effective vertical stress, Eq. 4-5

Parameter used to measure deformation capacity, Figs. 2-3, 5-1, 6-1, 7-1, 8-1

Distance from extreme compression fiber to centroid of tension reinforcement, in., Eq. 6-4

The longest plan dimension of the rehabilitated building, ft (mm), Eqs. 9-6, 9-7

$d_a$    Elongation of anchorage at end of wall determined by anchorage details and load magnitude, Eq. 8-1

Deflection at yield of tie-down anchorage or deflection at load level to anchorage at end of wall determined by anchorage details and dead load, in., Eq. 8-2

$d_b$    Overall beam depth, Eqs. 5-7, 5-8, 5-21, 5-22, 5-23, 5-24, 5-25, 5-26, 5-29
Nominal diameter of bar, in., Eq. 6-3

$d_{bg}$    Depth of the bolt group, Table 5-5

$d_c$    Column depth, Eq. 5-5

$d_i$    Depth, ft, of a layer of soils having similar properties, and located within 100ft of the surface, Eqs. 1-6, 1-7

$d_w$    Depth to groundwater level, Eq. 4-5

$d_z$    Overall panel zone depth between continuity plates, Chapter 5

$e$    Length of eccentric braced frame (EBF) link beam, Eqs. 5-28, 5-29, 5-30, 5-32
Parameter used to measure deformation capacity, Figs. 2-3, 5-1, 6-1, 7-1, 8-1
Actual eccentricity, ft (mm), measured in plan between the center of mass of the structure above the isolation interface and the center of rigidity of the isolation system, plus accidental eccentricity, ft (mm), taken as 5% of the maximum building dimension perpendicular to the direction of force under consideration, Eqs. 9-6, 9-7
Foundation embedment depth, ft, Eqs. 4-12, 4-14, Fig. 4-8

$e_n$    Nail deformation at yield load per nail for wood structural panel sheathing, Eqs. 8-2, 8-4, 8-5

$f_l$    Fundamental frequency of the building, Eq. 9-24

$f_a$    Axial compressive stress due to gravity loads specified in Eqs. 3-2, 7-5

$f_{ae}$    Expected vertical compressive stress, Chapter 7

$f_c$    Compressive strength of concrete, psi, Eqs. 6-4, 6-5

$f'_c$    Compressive strength of concrete, psi, Table 6-5

$f'_M$    Lower-bound masonry compressive strength, Eqs. 7-5, 7-6, 7-14

$f_{me}$    Expected compressive strength of masonry as determined in Section 7.2.2.3

$f_{pc}$    Average compressive stress in concrete due to effective prestress force only (after allowance for all prestress losses), Chapter 6

$f_s$    Stress in reinforcement, psi, Eqs. 6-2, 6-3

$f'_t$    Lower-bound masonry tensile strength, Chapter 7

$f_{te}$    Expected masonry flexural tensile strength as determined in Section 7.2.2.5

$f_{vie}$    Expected shear strength of masonry infill, Eq. 7-8

$f_y$    Yield strength of tension reinforcement, Eqs. 6-2, 6-3
Yield strength of shear reinforcement, Eq. 6-4

$f_{ye}$    Expected yield strength of reinforcing steel as determined in Section 7.2.2.8

$g$    Acceleration of gravity 386.1 in./sec² (or 9,807 mm/sec² for SI units), Eqs. 3-14, 4-14, 9-2, 9-3, 9-4, 9-5, 9-30

$h$    Period effect factor = $1 + 0.15 \cdot \ln T_e$, Eq. 3-16
Average story height above and below a beam–column joint, Eq. 5-17
Clear height of wall between beams, Eqs. 5-33, 5-35
Distance from inside of compression flange to inside of tension flange, Eq. 5-7
Height of member along which deformations are measured, Chapter 6
Overall thickness of member, in., Eq. 6-4
Height of a column, pilaster, or wall, Chapter 7
Shear wall height, Eqs. 8-1, 8-2
Average roof elevation of structure, relative to grade elevation, Eq. 11-3
Effective structure height, Eq. 4-14, Fig. 4-8

$h_c$    Assumed web depth for stability, Chapter 5
Gross cross-sectional dimension of column core measured in the direction of joint shear, in., Chapter 6

$h_{col}$    Height of column between beam centerlines, Eq. 7-7

$h_{eff}$    Effective height of wall or wall pier components under consideration, Eqs. 7-3, 7-5

$h_i$    Height from the base of a building to floor level $i$, Eqs. 3-12, 4-14, 9-9

$h_{inf}$    Height of infill panel, Eqs. 7-7, 7-10, 7-12, 7-13, 7-14

$h_n$    Height to roof level, ft, Eq. 3-6

$h_p$    Height of rectangular glass, Eq. 11-9

$h_w$    Height of wall or segment of wall considered in the direction of shear force, Chapter 6

$h_x$    Height from base to floor level $x$, ft, Eqs. 3-11, 9-9

$k$    Exponent used for determining the vertical distribution of lateral forces, Equation 3-11
Coefficient used for calculation of column shear strength, Chapter 6

$k_l$    Distance from the center of the split tee stem to the edge of the split tee flange fillet, Eq. 5-25

$k_{eff}$    Effective stiffness of an isolator unit, as prescribed by Eq. 9-12, or an energy dissipation unit, as prescribed by Eq. 9-23 or 9-38

$k_h$    Horizontal seismic coefficient in soil acting on retaining wall, Eq. 4-16

$k_{sr}$    Winkler spring stiffness in overturning (rotation) for pile group, expressed as moment/unit rotation, Eq. 4-10

$k_{sv}$    Winkler spring stiffness in vertical direction, expressed as force/unit displacement/unit area, Eq. 4-6

Pile group axial spring stiffness expressed as force/unit displacement, Eq. 4-9

$k_v$    Shear buckling coefficient, Chapter 5

$k_{vn}$    Axial stiffness of $n$-th pile in a pile group, Eq. 4-10

$l_b$    Length of beam, Eq. 5-1

Provided length of straight development, lap splice, or standard hook, in., Eq. 6-2

$l_{beff}$    Assumed distance to infill strut reaction point for beams, Eq. 7-11

$l_c$    Length of column, Eqs. 5-2, 5-36

$l_{ceff}$    Assumed distance to infill strut reaction point for columns, Eq. 7-9

$l_d$    Required length of development for a straight bar, in., Eq. 6-2

$l_e$    Length of embedment of reinforcement, in., Eq. 6-3

$l_p$    Length of plastic hinge used for calculation of inelastic deformation capacity, in., Eq. 6-6

$l_w$    Length of entire wall or a segment of wall considered in the direction of shear force, in., Chapter 6

$m$    Component demand modification factor to account for expected ductility associated with this action at the selected Structural Performance Level. $m$-factors are specified in Chapters 4 through 8, Eqs. 3-20, 5-9

$m_e$    Effective $m$-factor, Eq. 5-9

$m_t$    Value of $m$-factor for the column in tension, Eq. 5-13

$m_x$    Value of $m$ for bending about the $x$-axis of a member, Eqs. 5-10, 5-11, 5-13, 6-1

$m_y$    Value of $m$ for bending about the $y$-axis of a member, Eqs. 5-10, 5-11, 5-13, 6-1

$n$    Total number of stories in the vertical seismic framing, Eq. C3-2

Shear wave velocity reduction factor, Eq. 4-12

$p_{D+L}$    Gravity compressive stress at the test location considering actual dead plus live loads in place at time of testing, Eq. 7-2

$q$    Vertical bearing pressure, Eq. 4-8

$q_{allow}$    Allowable bearing pressure specified in the available design documents for the design of shallow foundations for gravity loads (dead plus live loads), Eq. 4-1

$q_c$    Expected bearing capacity of shallow foundation expressed in load per unit area, Eqs. 4-1, 4-3, 4-7, 4-8

$q_{in}$    Expected transverse strength of an infill panel, Eq. 7-14

$r$    Governing radius of gyration, Eq. 5-36

$r_\theta$    Equivalent foundation radius for rotation, Eq. 4-14, Fig. 4-8

$r_{inf}$    Diagonal length of infill panel, Eq. 7-7

$r_x$    Equivalent foundation radius for translation, Eq. 4-14, Fig. 4-8

$s$    Spacing of shear reinforcement, Eq. 6-4

$s_i$    Minimum separation distance between adjacent buildings at level $i$, Eq. 2-8

$\overline{s_u}$    Undrained shear strength of soil, lbs/ft$^2$, Chapter 1

$s_u$    Average value of the undrained soil shear strength in the upper 100 ft of soil, calculated in accordance with Eq. 1-6, lbs/ft$^2$

$t$    Thickness of continuity plate, Chapter 5

Effective thickness of wood structural panel or plywood for shear, in., Eqs. 8-2, 8-4, 8-5

$t_a$    Thickness of angle, Eqs. 5-21, 5-23

$t_{bf}$    Thickness of beam flange, Chapter 5

$t_{bw}$    Thickness of beam web, Chapter 5

$t_{cf}$    Thickness of column flange, Chapter 5

$t_{cw}$    Thickness of column web, Chapter 5

$t_f$    Thickness of flange, Eqs. 5-25, 5-29

$t_{inf}$    Thickness of infill panel, Eqs. 7-7, 7-13, 7-14

$t_p$    Thickness of panel zone including doubler plates, Eq. 5-5

Thickness of flange plate, Eq. 5-26

$t_s$    Thickness of split tee stem, Eqs. 5-24, 5-25

$t_w$    Thickness of web, Eqs. 5-7, 5-29

Thickness of plate wall, Eq. 5-33

Thickness of wall web, in., Chapter 6

$t_z$    Thickness of panel zone (doubler plates not necessarily included), Chapter 5

$v$    Maximum shear in the direction under consideration, Eq. 8-5

$v_{mL}$    Lower-bound masonry shear strength, Eqs. 7-1, 7-4

$v_{tL}$    Lower-bound bed-joint shear strength defined as lower 20th percentile of $v_{to}$, Eq. 7-1

$v_{to}$    Bed-joint shear stress from single test, Eq. 7-2

| | |
|---|---|
| $v_y$ | Shear at yield in the direction under consideration in lb/ft, Eqs. 8-1, 8-2, 8-3, 8-4, 8-5 |
| $w$ | Water content of soil, calculated as the ratio of the weight of water in a unit volume of soil to the weight of soil in the unit volume, expressed as a percentage, Section 1.6.1.4.1 <br> Length of connection member, Eqs. 5-23, 5-25 |
| $w_i$ | Portion of the effective seismic weight located on or assigned to floor level $i$, Eqs. 3-11, 3-12, C3-2, 4-14, 9-9 |
| $w_x$ | Portion of the effective seismic weight located on or assigned to floor level $x$, Eqs. 3-11, 3-12, C3-2, 9-9 |
| $w_z$ | Width of panel zone between column flanges, Chapter 5 |
| $x$ | Elevation in structure of component relative to grade elevation, Eq. 11-3 |
| $y$ | The distance, ft (mm), between the center of rigidity of the isolation system rigidity and the element of interest, measured perpendicular to the direction of seismic loading under consideration, Eqs. 9-6, 9-7 |
| $\Delta$ | Generalized deformation, Figs. 2-3, 2-5, 5-1, 6-1, 8-1 <br> Total elastic and plastic displacement, Chapter 5 <br> Calculated deflection of diaphragm, wall, or bracing element, in., Chapter 6 |
| $\Delta^-$ | Negative displacement amplitude, in. (mm), of an isolator or energy dissipation unit during a cycle of prototype testing, Eqs. 9-12, 9-13, 9-38 |
| $\Delta^+$ | Positive displacement amplitude, in. (mm), of an isolator or energy dissipation unit during a cycle of prototype testing, Eqs. 9-12, 9-13, 9-38 |
| $\Delta_{ave}$ | Average displacement of an energy dissipation unit during a cycle of prototype testing, equal to $(|\Delta^+| + |\Delta^-|)/2$, Eq. 9-39 |
| $\Delta_c$ | Axial deformation at expected buckling load, Section 5.6.2 |
| $\Delta_d$ | Diaphragm deformation, Eqs. 3-7, 3-8 <br> Lesser of target displacement or displacement at maximum base shear, Fig. 3-1, Eq. 3-16 |
| $\Delta_{eff}$ | Differentiated displacement between the top and bottom of the wall or wall pier components under consideration over a height, $h_{eff}$, Fig. 7-1 |
| $\Delta_{fallout}$ | Relative seismic displacement (drift) causing glass fallout from the curtain wall, |

| | |
|---|---|
| | storefront, or partition, as determined in accordance with an approved engineering analysis method, Eqs. 11-10, 11-11 |
| $\Delta_i$ | Story displacement (drift) of story $i$ divided by the story height, Chapter 5 |
| $\Delta_{i1}$ | Estimated lateral deflection of Building 1 relative to the ground at level $i$, Eq. 2-8 |
| $\Delta_{i2}$ | Estimated lateral deflection of Building 2 relative to the ground at level $i$, Eq. 2-8 |
| $\Delta_{inf}$ | Deflection of infill panel at mid-length when subjected to transverse loads, Eq. 7-13 |
| $\Delta_p$ | Additional earth pressure on retaining wall due to earthquake shaking, Eq. 4-16 |
| $\Delta_t$ | Axial deformation at expected tensile yield load, Section 5.6.2 |
| $\Delta_w$ | Average in-plane wall displacement, Eq. 3-8 |
| $\Delta_y$ | Displacement at effective yield strength, Fig. 3-1, Eq. 3-16 <br> Generalized yield deformation, unitless, Fig. 5-1 <br> Calculated deflection of diaphragm, shear wall, or bracing element at yield, Eqs. 8-1, 8-2, 8-3, 8-4, 8-5 |
| $\Sigma(\Delta_c X)$ | Sum of individual chord-splice slip values on both sides of the diaphragm, each multiplied by its distance to the nearest support, Eqs. 8-4, 8-5 |
| $\Sigma E_D$ | Total energy dissipated, in.-kips, in the isolation system during a full cycle of response at the design displacement, $D_D$, Eq. 9-18 |
| $\Sigma E_M$ | Total energy dissipated, in.-kips, in the isolation system during a full cycle of response at the maximum displacement, $D_M$, Eq. 9-19 |
| $\Sigma|F^+_D|_{max}$ | Sum, for all isolator units, of the maximum absolute value of force, kips (kN), at a positive displacement equal to $D_D$, Eq. 9-14 |
| $\Sigma|F^+_D|_{min}$ | Sum, for all isolator units, of the minimum absolute value of force, kips (kN), at a positive displacement equal to $D_D$, Eq. 9-15 |
| $\Sigma|F^+_M|_{max}$ | Sum, for all isolator units, of the maximum absolute value of force, kips (kN), at a positive displacement equal to $D_M$, Eq. 9-16 |
| $\Sigma|F^+_M|_{min}$ | Sum, for all isolator units, of the minimum absolute value of force, kips (kN), at a positive displacement equal to $D_M$, Eq. 9-17 |
| $\Sigma|F^-_D|_{max}$ | Sum, for all isolator units, of the maximum absolute value of force, kips (kN), at a negative displacement equal to $D_D$, Eq. 9-14 |

$\Sigma|F^-{}_D|_{min}$    Sum, for all isolator units, of the minimum absolute value of force, kips (kN), at a negative displacement equal to $D_D$, Eq. 9-15

$\Sigma|F^-{}_M|_{max}$    Sum, for all isolator units, of the maximum absolute value of force, kips (kN), at a negative displacement equal to $D_M$, Eq. 9-16

$\Sigma|F^-{}_M|_{min}$    Sum, for all isolator units, of the minimum absolute value of force, kips (kN), at a negative displacement equal to $D_M$, Eq. 9-17

$\alpha$    Factor equal to 0.5 for fixed-free cantilevered shear wall, or 1.0 for fixed-fixed wall pier, Eqs. 7-3, 7-5

Velocity exponent for a fluid-viscous device, Eq. 9-25

$\alpha_1$    Positive post-yield slope ratio equal to the positive post-yield stiffness divided by the effective stiffness, Fig. 3-1, Chapter 3

$\alpha_2$    Negative post-yield slope ratio equal to the negative post-yield stiffness divided by the effective stiffness, Fig. 3-1, Eq. 3-17

$\alpha_e$    Effective negative post-yield slope ratio equal to the effective post-yield negative stiffness divided by the effective stiffness, Eqs. 3-16, 3-17

$\alpha_{P-\Delta}$    Negative slope ratio caused by P-$\Delta$ effects, Fig. 3-1, Eq. 3-17

$\beta$    Modal damping ratio, Chapter 1, Eqs. 1-13, 4-13

Factor to adjust fundamental period of the building, Eq. 3-6

Ratio of expected frame strength, $v_{fre}$, to expected infill strength, $v_{ine}$, Chapter 7

Damping inherent in the building frame (typically equal to 0.05), Eqs. 9-26, 9-28, 9-30

$\beta_0$    Effective damping ratio of the structure–foundation system, Eq. 4-13

$\beta_D$    Effective damping of the isolation system at the design displacement, as prescribed by Eq. 9-18

$\beta_{eff}$    Effective damping of isolator unit, as prescribed by Eq. 9-13, or an energy dissipation unit, as prescribed by Eq. 9-39; also used for the effective damping of the building, as prescribed by Eqs. 9-26, 9-28, 9-30, 9-31, 9-32, 9-36

$\beta_{eff-m}$    Effective damping in $m$-th mode prescribed by Eq. 9-33

$\beta_f$    Foundation–soil interaction damping ratio, Eq. 4-13, 4-14, Fig. 4-8

$\beta_M$    Effective damping of the isolation system at the maximum displacement, as prescribed by Eq. 9-19

$\beta_m$    $m$-th mode damping in the building frame, Eq. 9-33

$\gamma$    Unit weight, weight/unit volume (lbs/ft$^3$ or N/m$^3$), Eq. 4-4

Coefficient for calculation of joint shear strength, Eq. 6-5

$\gamma_f$    Fraction of unbalanced moment transferred by flexure at slab–column connections, Chapter 6

$\gamma_t$    Total unit weight of soil, Eqs. 4-5, 4-16

$\gamma_w$    Unit weight of water, Eq. 4-5

$\delta_i$    Displacement at floor $i$, Eqs. 9-26, 9-27.

Displacement at floor $i$ due to lateral load $F_i$, Eqs. C3-2

$\delta_{mi}$    $m$-th mode horizontal displacement at floor $i$, Eq. 9-34

$\delta_{mrj}$    $m$-th relative displacement between the ends of device $j$ along its axis, Eq. 9-35

$\delta_{rj}$    Relative displacement between the ends of energy dissipating device $j$ along the axis of the device, Eqs. 9-29, 9-37

$\delta_t$    Target displacement, Fig. 3-1

$\delta_{xA}$    Deflection at level $x$ of Building A, determined by an elastic analysis as defined in Chapter 3, Eqs. 11-7, 11-8

$\delta_{xB}$    Deflection at level $x$ of Building B, determined by an elastic analysis as defined in Chapter 3, Eq. 11-8

$\delta_{yA}$    Deflection at level $y$ of Building A, determined by an elastic analysis as defined in Chapter 3, Eq. 11-7

$\eta$    Displacement multiplier, greater than 1.0, to account for the effects of torsion, Eq. 3-1

$\theta$    Generalized deformation, radians, Figs. 5-1, 6-1

Angle between infill diagonal and horizontal axis, $\tan\theta = h_{inf}/L_{inf}$, radians, Eq. 7-7

$\theta_b$    Angle between lower edge of compressive strut and beam, radians, Eqs. 7-11, 7-12

$\theta_c$    Angle between lower edge of compressive strut and column, radians, Eqs. 7-9, 7-10

$\theta_i$    Story drift ratio, radians, Chapter 5

$\theta_j$    Angle of inclination of energy dissipation device, Eq. 9-30

$\theta_y$    Generalized yield deformation, radians, Fig. 5-1

Yield rotation, radians, Eqs. 5-1, 5-2, 5-30, 5-35, 6-6

| | |
|---|---|
| $\kappa$ | A knowledge factor used to reduce component strength based on the level of knowledge obtained for individual components during data collection, Eqs. 3-20, 3-21, 6-1 |
| $\lambda$ | Near field effect factor, Eq. 3-17<br>Correction factor related to unit weight of concrete, Eq. 6-5 |
| $\lambda_1$ | Coefficient used to determine equivalent width of infill strut, Eq. 7-7 |
| $\lambda_2$ | Infill slenderness factor, Eq. 7-14 |
| $\mu$ | Coefficient of shear friction, Chapter 6<br>Expected ductility demand, Eq. 4-15 |
| $\nu$ | Poisson's ratio, Eq. 4-6, 4-14 |
| $\nu_s$ | Shear wave velocity in soil, ft/sec, Section 1.6.1.4.1<br>Shear wave velocity at low strains, Eq. 4-4<br>Shear wave velocity for site soil conditions taken as average value of velocity to a depth of $b_e$ below foundation, ft/sec, Eq. 4-12 |
| $\overline{\nu_s}$ | Average value of the soil shear wave velocity in the upper 100 ft of soil, calculated in accordance with Eq. 1-6, ft/sec |
| $\rho$ | Ratio of non-prestressed tension reinforcement, Chapter 6 |
| $\rho_{bal}$ | Reinforcement ratio producing balanced strain conditions, Chapter 6 |
| $\rho_g$ | Total of vertical reinforcement ratio plus horizontal reinforcement ratio in a wall or wall pier, Chapter 7 |
| $\rho_{lp}$ | Yield deformation of a link beam, Chapter 5 |
| $\rho_n$ | Ratio of distributed shear reinforcement in a plane perpendicular to the direction of the applied shear, Chapter 6 |
| $r'$ | Ratio of non-prestressed compression reinforcement, Chapter 6 |
| $r''$ | Reinforcement ratio for transverse joint reinforcement, Chapter 6 |
| $\sigma$ | Standard deviation of the variation of the material strengths, Chapter 2 |
| $\sigma_o'$ | Effective vertical stress, Eq. 4-5 |
| $\phi$ | Strength reduction factor<br>Angle of shearing resistance for soil, Chapter 4 |
| $\phi_i$ | Modal displacement of floor $i$, Eq. 9-30 |
| $\phi_{il}$ | First mode displacement at level $i$, Eq. 4-14 |
| $\phi_{rj}$ | Relative modal displacement in horizontal direction of energy dissipation device $j$, Eq. 9-30 |
| $\chi$ | A factor to calculate horizontal seismic force, $F_p$, Eqs. 2-6, 2-7 |
| $\omega_1$ | Fundamental angular frequency equal to $2\pi f_1$, Eq. 9-24 |

## ACRONYMS

| | |
|---|---|
| **AAMA** | American Architectural Manufacturers Association |
| **ABK** | Agbabian, Barnes and Kariotis Joint Venture |
| **ACI** | American Concrete Institute |
| **AISC** | American Institute of Steel Construction |
| **AISI** | American Iron and Steel Institute |
| **ANSI** | American National Standards Institute |
| **APA** | American Plywood Association |
| **API** | American Petroleum Institute |
| **ASCE** | American Society of Civil Engineers |
| **ASHRAE** | American Society of Heating, Refrigeration and Air Conditioning Engineers |
| **ASME** | American Society of Mechanical Engineers |
| **ASTM** | American Society for Testing and Materials |
| **ATC** | Applied Technology Council |
| **AWS** | American Welding Society |
| **AWWA** | American Water Works Association |
| **BNC** | Ball and Cone System, a rolling seismic isolator |
| **BOCA** | Building Officials and Code Administrators |
| **BRANZ** | Building Research Association of New Zealand |
| **BSE-1** | Basic Safety Earthquake-1 |
| **BSE-2** | Basic Safety Earthquake-2 |
| **BSO** | Basic Safety Objective |
| **BSSC** | Building Seismic Safety Council |
| **CBF** | Concentric Braced Frame |
| **CISCA** | Ceilings and Interior Systems Construction Association |
| **CP** | Collapse Prevention |
| **CQC** | Complete Quadratic Combination |
| **CRSI** | Concrete Reinforcing Steel Institute |
| **CUREE** | California Universities for Research in Earthquake Engineering |
| **CUSEC** | Central United States Earthquake Consortium |
| **DCR** | Demand Capacity Ratio |
| **DDS** | Double Diagonal Sheathing |
| **EBF** | Eccentric Braced Frame |
| **EDD** | Energy Dissipation Device |
| **EDS** | Energy Dissipation System |
| **EERC** | Earthquake Engineering Research Center |
| **EERI** | Earthquake Engineering Research Institute |
| **FEMA** | Federal Emergency Management Agency |

| | |
|---|---|
| **FPS** | Friction-Pendulum System, a sliding seismic isolator |
| **FR** | Fully Restrained |
| **GFRC** | Glass Fiber Reinforced Concrete |
| **HDR** | High-Damping Rubber Bearings, an elastomeric seismic isolator |
| **HR** | Hazards Reduced |
| **HVAC** | Heating, Ventilation, and Air Conditioning |
| **IBC** | International Building Code |
| **ICBO** | International Conference of Building Officials |
| **IO** | Immediate Occupancy |
| **LDP** | Linear Dynamic Procedure |
| **LRB** | Low-Damping Rubber Bearings with a Lead Core, an elastomeric seismic isolator |
| **LRFD** | Load and Resistance Factor Design |
| **LS** | Life Safety |
| **LSP** | Linear Static Procedure |
| **MCE** | Maximum Considered Earthquake |
| **MDOF** | Multi-Degree of Freedom System |
| **MEP** | Mechanical, Electrical, and Plumbing |
| **MSJC** | Masonry Standards Joint Committee |
| **NAVFAC** | Publication of the U.S. Department of the Navy |
| **NBS** | National Bureau of Standards |
| **NCEER** | National Center for Earthquake Engineering Research |
| **NDE** | Nondestructive Examination |
| **NDP** | Nonlinear Dynamic Procedure |
| **NDS** | National Design Specification |
| **NEHRP** | National Earthquake Hazard Reduction Program |
| **NFPA** | National Fire Protection Association |
| **NIST** | National Institute of Standards and Technology (formerly NBS) |
| **NSP** | Nonlinear Static Procedure |
| **OMF** | Ordinary Moment Frame |
| **PCI** | Precast Concrete Institute |
| **PR** | Partially Restrained |
| **PS-#** | Product Standard-# |
| **QAP** | Quality Assurance Plan |
| **RB** | Low-Damping Rubber Bearings, an elastomeric seismic isolator |
| **RM** | Reinforced Masonry |
| **SAC** | SEAOC, ATC, and CUREe Joint Venture |
| **SBC** | Southern Building Code |
| **SBCCI** | Southern Building Code Congress International |
| **SDI** | Steel Deck Institute |
| **SDOF** | Single-Degree of Freedom |
| **SEAOC** | Structural Engineers Association of California |

| | |
|---|---|
| **SJI** | Steel Joist Institute |
| **SMACNA** | Sheet Metal and Air Conditioning Contractors National Association |
| **SMF** | Special Moment Frame |
| **SPT** | Standard Penetration Test |
| **SSI** | Soil–Structure Interaction |
| **TMS** | The Masonry Society |
| **UBC** | Uniform Building Code |
| **UCBC** | Uniform Code for Building Conservation |
| **URM** | Unreinforced Masonry |
| **USGS** | United States Geological Survey |
| **WWPA** | Western Wood Products Association |

## DEFINITIONS

**Acceleration-Sensitive Nonstructural Component:** A nonstructural component that is sensitive to, and subject to, damage from inertial loading.

**Acceptance Criteria:** Limiting values of properties such as drift, strength demand, and inelastic deformation used to determine the acceptability of a component at a given performance level.

**Action:** An internal moment, shear, torque, axial load, deformation, displacement, or rotation corresponding to a displacement due to a structural degree of freedom; designated as force- or deformation-controlled.

**Active Fault:** A fault for which there is an average historic slip rate of 1 mm per year or more, and evidence of seismic activity within Holocene times (past 11,000 years).

**Adjusted Resistance:** The reference resistance adjusted to include the effects of applicable adjustment factors resulting from end use and other modifying factors excluding time-effect adjustments, which are considered separately and are not included.

**Aspect Ratio:** Ratio of height-to-width for shear walls and span-to-width for horizontal diaphragms.

**Assembly:** Two or more interconnected components.

**Authority Having Jurisdiction:** The organization, political subdivision, office, or individual legally charged with responsibility for administering and enforcing the provisions of this standard.

**Balloon Framing:** Continuous stud framing from sill to roof, with intervening floor joists nailed to studs and supported by a let-in ribbon.

**Base:** The level at which earthquake effects are imparted to the building.

**Beam:** A structural member whose primary function is to carry loads transverse to its longitudinal axis.

**Bearing Wall:** A wall that supports gravity loads of at least 200 lbs/lineal ft from floors and/or roofs.

**Bed Joint:** The horizontal layer of mortar on which a masonry unit is laid.

**Boundary Component:** A structural component at the boundary of a shear wall or a diaphragm or at an edge of an opening in a shear wall or a diaphragm that possesses tensile and/or compressive strength to transfer lateral forces to the lateral-force-resisting system.

**Braced Frame:** A vertical lateral-force-resisting element consisting of vertical, horizontal, and diagonal components joined by concentric or eccentric connections.

**BSE-1:** Basic Safety Earthquake-1, taken as the lesser of the ground shaking for a 10%/50-year earthquake or two-thirds of the BSE-2 at a site.

**BSE-2:** Basic Safety Earthquake-2, taken as the ground shaking based on the Maximum Considered Earthquake (MCE) at a site.

**BSO:** Basic Safety Objective is a Rehabilitation Objective that achieves the dual rehabilitation goals of the Life Safety Building Performance Level for the BSE-1 Earthquake Hazard Level and the Collapse Prevention Building Performance Level for the BSE-2 Earthquake Hazard Level.

**Building Occupancy:** The purpose for which a building, or part thereof, is used or intended to be used, designated in accordance with the applicable building code.

**Building Performance Level:** A limiting damage state for a building, considering structural and non-structural components, used in the definition of Rehabilitation Objectives.

**Cast Iron:** A hard, brittle, nonmalleable iron-carbon alloy containing 2.0% to 4.5% carbon. Shapes are obtained by reducing iron ore in a blast furnace, forming it into bars (or pigs), and remelting and casting it into its final form.

**Cavity Wall:** A masonry wall with an air space between wythes.

**Chord:** See **Diaphragm Chord**.

**Clay Tile Masonry:** Masonry constructed with hollow units made of clay tile.

**Clay-Unit Masonry:** Masonry constructed with solid, cored, or hollow units made of clay; can be ungrouted or grouted.

**Closed Stirrups or Ties:** Transverse reinforcement defined in Chapter 7 of ACI 318 (ACI 2002) consisting of standard stirrups or ties with 90-degree hooks and lap splices in a pattern that encloses longitudinal reinforcement.

**Code Official:** The individual representing a local jurisdiction who is legally charged with responsibility for administering and enforcing the provisions of a legally adopted building code.

**Coefficient of Variation:** For a sample of data, the ratio of the standard deviation for the sample to the mean value for the sample.

**Collar Joint:** Vertical longitudinal joint between wythes of masonry or between masonry wythe and back-up construction; can be filled with mortar or grout.

**Collector:** See **Drag Strut**.

**Column (or Beam) Jacketing:** A rehabilitation method in which a concrete column or beam is encased in a steel or concrete "jacket" to strengthen and/or repair the member by confining the concrete.

**Component:** A part of an architectural, mechanical, electrical, or structural system of a building.

**Component, Flexible:** A component, including its attachments, having a fundamental period greater than 0.06 sec.

**Component, Nonstructural:** An architectural, mechanical, or electrical component of a building that is permanently installed in, or is an integral part of, a building system.

**Component, Primary:** A structural component that is required to resist seismic forces in order for the structure to achieve the selected performance level.

**Component, Rigid:** A component, including its attachments, having a fundamental period less than or equal to 0.06 sec.

**Component, Secondary:** A structural component that is not required to resist seismic forces in order for the structure to achieve the selected performance level.

**Component, Structural:** A component of a building that provides gravity- or lateral-load resistance as part of a continuous load path to the foundation, including beams, columns, slabs, braces, walls, wall piers, coupling beams, and connections; designated as primary or secondary.

**Composite Masonry Wall:** Multi-wythe masonry wall acting with composite action.

**Composite Panel:** A structural panel composed of thin wood strands or wafers bonded together with exterior adhesive.

**Concentric Braced Frame:** Braced frame element in which component worklines intersect at a single point or at multiple points such that the distance between intersecting worklines (or eccentricity) is less than or equal to the width of the smallest component connected at the joint.

**Concrete Masonry:** Masonry constructed with solid or hollow units made of concrete; can be ungrouted or grouted.

**Condition of Service:** The environment to which the structure will be subjected.

**Connection:** A link that transmits actions from one component or element to another component or element, categorized by type of action (moment, shear, or axial).

**Connection Hardware:** Proprietary or custom-fabricated body of a component that is used to link wood components.

**Connectors:** Nails, screws, lags, bolts, split rings, shear plates, headed studs, and welds used to link components to other components.

**Contents:** Movable items within the building introduced by the owner or occupants, weighing 400 lbs or more.

**Continuity Plates:** Column stiffeners at the top and bottom of a panel zone.

**Control Node:** A node located at the center of mass at the roof of a building used in the Nonlinear Static Procedure (NSP) to measure the effects of earthquake shaking on a building.

**Corrective Measure:** Any modification of a component or element, or the structure as a whole, implemented to improve building performance.

**Coupling Beam:** A component that ties or couples adjacent shear walls acting in the same plane.

**Cripple Studs:** Short studs between a header and top plate at openings in wall framing, or studs between the base and sill of an opening.

**Cripple Wall:** Short wall between the foundation and first floor framing.

**Critical Action:** The component action that reaches its elastic limit at the lowest level of lateral deflection or loading of the structure.

**Cross Tie:** A component that spans the width of the diaphragm and delivers out-of-plane wall forces over the full depth of the diaphragm.

**Decay:** Decomposition of wood caused by action of wood-destroying fungi. The term "dry rot" is used interchangeably with decay.

**Decking:** Solid sawn lumber or glue-laminated decking, nominally 2 to 4 in. thick and 4 or more in. wide. Decking shall be tongue-and-groove or connected at longitudinal joints with nails or metal clips.

**Deep Foundation:** Driven piles made of steel, concrete, or wood, or cast-in-place concrete piers or drilled shafts of concrete.

**Deformability:** The ratio of the ultimate deformation to the limit deformation.

**Deformation-Sensitive Nonstructural Component:** A nonstructural component that is sensitive to deformation imposed by the drift or deformation of the structure, including deflection or deformation of diaphragms.

**Demand:** The amount of force or deformation imposed on an element or component.

**Design Displacement:** The design earthquake displacement of an isolation or energy dissipation system, or elements thereof, excluding additional displacement due to actual and accidental torsion.

**Design Earthquake:** A user-specified earthquake for the design of a building having ground shaking criteria described in Chapter 1.

**Design Resistance (Force or Moment, as appropriate):** Resistance provided by member or connection; the product of adjusted resistance, the resistance factor, and time-effect factor.

**Diagonal Bracing:** Inclined components designed to carry axial load, enabling a structural frame to act as a truss to resist lateral forces.

**Diaphragm:** A horizontal (or nearly horizontal) structural element used to transfer inertial lateral forces to vertical elements of the lateral-force-resisting system.

**Diaphragm Chord:** A boundary component perpendicular to the applied load that is provided to resist tension or compression due to the diaphragm moment.

**Diaphragm Collector:** A component parallel to the applied load that is provided to transfer lateral forces in the diaphragm to vertical elements of the lateral-force-resisting system.

**Diaphragm Ratio:** See **Aspect Ratio.**

**Diaphragm Strut:** See **Diaphragm Tie.**

**Diaphragm Tie:** A component parallel to the applied load that is provided to transfer wall anchorage or diaphragm inertial forces within or across the diaphragm. Also called diaphragm strut.

**Differential Compaction:** An earthquake-induced process in which soils become more compact and settle in a nonuniform manner across a site.

**Dimensioned Lumber:** Lumber from nominal 2 through 4 inches thick and nominal 2 or more inches wide.

**Displacement-Dependent Energy Dissipation Devices:** Devices having mechanical properties such that the force in the device is related to the relative displacement in the device.

**Displacement Restraint System:** Collection of structural components and elements that limit lateral displacement of seismically-isolated buildings during the BSE-2.

**Dowel-Bearing Strength:** The maximum compression strength of wood or wood-based products when subjected to bearing by a steel dowel or bolt of specific diameter.

**Dowel-Type Fasteners:** Bolts, lag screws, wood screws, nails, and spikes.

**Dressed Size:** The dimensions of lumber after surfacing with a planing machine.

**Dry Rot:** See **Decay**.

**Dry Service:** Structures wherein the maximum equilibrium moisture content does not exceed 19%.

**Earthquake Hazard Level:** Ground shaking demands of specified severity, developed on either a probabilistic or deterministic basis.

**Eccentric Braced Frame:** Braced frame element in which component worklines do not intersect at a single point and the distance between the intersecting worklines (or eccentricity) exceeds the width of the smallest component connecting at the joint.

**Edge Distance:** The distance from the edge of the member to the center of the nearest fastener.

**Effective Damping:** The value of equivalent viscous damping corresponding to the energy dissipated by the building, or element thereof, during a cycle of response.

**Effective Stiffness:** The value of the lateral force in the building, or an element thereof, divided by the corresponding lateral displacement.

**Effective Void Ratio:** Ratio of collar joint area without mortar to the total area of the collar joint.

**Element:** An assembly of structural components that act together in resisting forces, including gravity frames, moment-resisting frames, braced frames, shear walls, and diaphragms.

**Energy Dissipation Device:** Non-gravity-load-supporting element designed to dissipate energy in a stable manner during repeated cycles of earthquake demand.

**Energy Dissipation System:** Complete collection of all energy dissipation devices, their supporting framing, and connections.

**Expected Strength:** The mean value of resistance of a component at the deformation level anticipated for a population of similar components, including consideration of the variability in material strength as wells as strain-hardening and plastic section development

**Fair Condition:** Masonry found during condition assessment to have mortar and units intact but with minor cracking.

**Fault:** Plane or zone along which earth materials on opposite sides have moved differentially in response to tectonic forces.

**Flexible Connection:** A link between components that permits rotational and/or translational movement without degradation of performance, including universal joints, bellows expansion joints, and flexible metal hose.

**Flexible Diaphragm:** A diaphragm with horizontal deformation along its length more than twice the average story drift.

**Foundation System:** An assembly of structural components, located at the soil–structure interface, that transfer loads from the superstructure into the supporting soil.

**Fundamental Period:** The longest natural period of the building in the direction under consideration.

**Gauge or Row Spacing:** The center-to-center distance between fastener rows or gauge lines.

**Glulam Beam:** Shortened term for glue-laminated beam, which is a wood-based component made up of layers of wood bonded with adhesive.

**Good Condition:** Masonry found during condition assessment to have mortar and units intact and no visible cracking.

**Grade:** The classification of lumber with regard to strength and utility, in accordance with the grading rules of an approved agency.

**Grading Rules:** Systematic and standardized criteria for rating the quality of wood products.

**Gypsum Wallboard or Drywall:** An interior wall surface sheathing material; can sometimes be considered for resisting lateral forces.

**Head Joint:** Vertical mortar joint placed between masonry units in the same wythe.

**High-Deformability Component**: A component whose deformability is not less than 3.5 when subjected to four fully reversed cycles at the limit deformation.

**Hollow Masonry Unit:** A masonry unit with net cross-sectional area in every plane parallel to the bearing surface less than 75% of the gross cross-sectional area in the same plane.

**Hoops:** Transverse reinforcement defined in Chapter 21 of ACI 318 (ACI 2002) consisting of closed ties with 135-degree hooks embedded into the core and no lap splices.

**Infill:** A panel of masonry placed within a steel or concrete frame. Panels separated from the surrounding frame by a gap are termed "isolated infills." Panels that are in full contact with a frame around its full perimeter are termed "shear infills."

**In-Plane Wall:** See **Shear Wall**.

**Isolation Interface:** The boundary between the upper portion of the structure (superstructure), which is isolated, and the lower portion of the structure, which is assumed to move rigidly with the ground.

**Isolation System:** The collection of structural components that includes all individual isolator units, all structural components that transfer force between components of the isolation system, and all connections to other structural components. The isolation system also includes the wind-restraint system, if such a system is used to meet the design requirements of this section.

**Isolator Unit:** A horizontally flexible and vertically stiff structural component of the isolation system that permits large lateral deformations under seismic load. An isolator unit shall be used either as part of or in addition to the weight-supporting system of the building.

**Joint:** An area where ends, surfaces, or edges of two or more components are attached; categorized by type of fastener or weld used and method of force transfer.

**King Stud:** Full-height studs adjacent to openings that provide out-of-plane stability to cripple studs at openings.

**Knee Joint:** A joint that in the direction of framing has one column and one beam.

**Landslide:** A down-slope mass movement of earth resulting from any cause.

**Lateral-Force-Resisting System:** Those elements of the structure that provide its basic lateral strength and stiffness.

**Light Framing:** Repetitive framing with small, uniformly spaced members.

**Lightweight Concrete:** Structural concrete that has an air-dry unit weight not exceeding 115 pcf.

**Limit Deformation:** Two times the initial deformation that occurs at a load equal to 40% of the maximum strength.

**Limited-Deformability Component:** A component that is neither a low-deformability nor a high-deformability component.

**Link Beam:** A component between points of eccentrically connected members in an eccentric braced frame element.

**Link Intermediate Web Stiffeners:** Vertical web stiffeners placed within a link.

**Link Rotation Angle:** Angle of plastic rotation between the link and the beam outside of the link, derived using the specified base shear, $V$.

**Liquefaction:** An earthquake-induced process in which saturated, loose, granular soils lose shear strength and liquefy as a result of increase in pore-water pressure during earthquake shaking.

**Load and Resistance Factor Design:** A method of proportioning structural components (members, connectors, connections, and assemblages) using load factors and strength reduction factors such that no applicable limit state is exceeded when the structure is subjected to all design load combinations.

**Load Duration:** The period of continuous application of a given load, or the cumulative period of intermittent applications of load. See **Time-Effect Factor.**

**Load Path:** A path through which seismic forces are delivered from the point at which inertial forces are generated in the structure to the foundation and, ultimately, the supporting soil.

**Load Sharing:** The load redistribution mechanism among parallel components constrained to deflect together.

**Load/Slip Constant:** The ratio of the applied load to a connection and the resulting lateral deformation of the connection in the direction of the applied load.

**Low-Deformability Component**: A component whose deformability is 1.5 or less.

**Lower-Bound Strength**: The mean minus one standard deviation of the yield strengths, $Q_y$, for a population of similar components.

**Lumber:** The product of the sawmill and planing mill, usually not further manufactured other than by sawing, resawing, passing lengthwise through a standard planing machine, crosscutting to length, and matching.

**Masonry:** The assemblage of masonry units, mortar, and possibly grout and/or reinforcement; classified with respect to the type of masonry unit, including clay-unit masonry, concrete masonry, or hollow-clay tile masonry.

**Mat-Formed Panel:** A structural panel manufactured in a mat-formed process including oriented strand board and waferboard.

**Maximum Considered Earthquake (MCE):** An extreme earthquake hazard level defined by MCE maps which are based on a combination of mean 2%/50-year probabilistic spectra and 150% of median deterministic spectra at a given site.

**Maximum Displacement:** The maximum earthquake displacement of an isolation or energy dissipation system, or elements thereof, excluding additional displacement due to actual or accidental torsion.

**Mean Return Period:** The average period of time, in years, between the expected occurrences of an earthquake of specified severity.

**Model Building Type:** One of the common building types listed and described in Table 10-2.

**Moisture Content:** The weight of the water in wood expressed as a percentage of the weight of the oven-dried wood.

**Moment Frame:** A building frame system in which seismic shear forces are resisted by shear and flexure in members and joints of the frame.

**Narrow Wood Shear Wall:** Wood shear walls with an aspect ratio (height-to-width) greater than 2:1.

**Nominal Size:** The approximate rough-sawn commercial size by which lumber products are known and sold in the market. Actual rough-sawn sizes vary from nominal. Reference to standards or grade rules is required to determine nominal to actual finished size relationships, which have changed over time.

**Nominal Strength:** The capacity of a structure or component to resist the effects of loads, as determined by (1) computations using specified material strengths and dimensions, and formulas derived from accepted principles of structural mechanics; or (2) field tests or laboratory tests of scaled models, allowing for modeling effects and differences between laboratory and field conditions.

**Nonbearing Wall:** A wall that supports gravity loads less than 200 lbs/lineal ft.

**Noncompact Member:** A steel section that has width-to-thickness ratios exceeding the limiting values for compactness specified in *Load and Resistance Factor Design Specification for Structural Steel Buildings (LRFD)*(AISC 1999).

**Noncomposite Masonry Wall:** Multi-wythe masonry wall acting without composite action.

**Nonstructural Component:** See **Component, Nonstructural**.

**Nonstructural Performance Level:** A limiting damage state for nonstructural building components used to define Rehabilitation Objectives.

**Ordinary Moment Frame:** A moment frame system that meets the requirements for Ordinary Moment Frames as defined in seismic provisions for new construction in AISC 341, *Seismic Provisions*, Chapter 5 (AISC 2002) .

**Oriented Strand Board:** A structural panel composed of thin, elongated wood strands with surface layers arranged in the long panel direction and core layers arranged in the cross-panel direction.

**Out-of-Plane Wall:** A wall that resists lateral forces applied normal to its plane.

**Overturning:** Behavior that results when the moment produced at the base of vertical lateral-force-resisting elements is larger than the resistance provided by the building weight and the foundation resistance to uplift.

**Panel:** A sheet-type wood product.

**Panel Rigidity or Stiffness:** The in-plane shear rigidity of a panel; the product of panel thickness and modulus of rigidity.

**Panel Shear:** Shear stress acting through the panel thickness.

**Panel Zone:** Area of a column at a beam-to-column connection delineated by beam and column flanges.

**Parapet:** Portions of a wall extending above the roof diaphragm.

**Partially Grouted Masonry Wall:** A masonry wall containing grout in some of the cells.

**Particleboard:** A panel manufactured from small pieces of wood, hemp, and flax, bonded with synthetic or organic binders, and pressed into flat sheets.

**Perforated Wall or Infill Panel:** A wall or panel not meeting the requirements for a solid wall or infill panel.

**Pitch or Spacing:** The longitudinal center-to-center distance between any two consecutive holes or fasteners in a row.

**Platform Framing:** Construction method in which stud walls are constructed one floor at a time, with a floor or roof joist bearing on top of the wall framing at each level.

**Ply:** A single sheet of veneer, or several strips laid with adjoining edges that form one veneer lamina in a glued plywood panel.

**Plywood:** A structural panel composed of plies of wood veneer arranged in cross-aligned layers bonded with adhesive cured upon application of heat and pressure.

**Pole:** A round timber of any size or length, usually used with the larger end in the ground.

**Pole Structure:** A structure framed with generally round, continuous poles that provide the primary vertical frame and lateral-load-resisting system.

**Poor Condition:** Masonry found during condition assessment to have degraded mortar, degraded masonry units, or significant cracking.

**Pounding:** The action of two adjacent buildings coming into contact with each other during earthquake excitation as a result of their close proximity and differences in dynamic response characteristics.

**Preservative:** A chemical that, when suitably applied to wood, makes the wood resistant to attack by fungi, insects, marine borers, or weather conditions.

**Pressure-Preservative-Treated Wood:** Wood products pressure-treated by an approved process and preservative.

**Primary Component:** See **Component, Primary**.

**Primary (Strong) Panel Axis:** The direction that coincides with the length of the panel.

**Probability of Exceedance:** The chance, expressed as a percentage (%), that a more severe event will occur within a specified period, expressed in number of years.

**Punched Metal Plate:** A light steel plate fastener with punched teeth of various shapes and configurations that are pressed into wood members to effect force transfer.

**P-Δ Effect:** The secondary effect of vertical loads and lateral deflection on the shears and moments in various components of a structure.

**Redundancy:** The quality of having alternative load paths in a structure by which lateral forces can be transferred, allowing the structure to remain stable following the failure of any single element.

**Re-Entrant Corner:** Plan irregularity in a diaphragm, such as an extending wing, plan inset, or E-, T-, X-, or L-shaped configuration, where large tensile and compressive forces can develop.

**Rehabilitation Measures:** Modifications to existing components, or installation of new components, that correct deficiencies identified in a seismic evaluation as part of a scheme to rehabilitate a building to achieve a selected Rehabilitation Objective.

**Rehabilitation Method:** One or more procedures and strategies for improving the seismic performance of existing buildings.

**Rehabilitation Objective:** One or more rehabilitation goals, each goal consisting of the selection of a target Building Performance Level and an Earthquake Hazard Level.

**Rehabilitation Strategy:** A technical approach for developing rehabilitation measures for a building to improve seismic performance.

**Reinforced Masonry Wall:** A masonry wall with the following minimum amounts of vertical and horizontal reinforcement: vertical reinforcement of at least 0.20 in.$^2$ in cross section at each corner or end, at each side of each opening, and at a maximum spacing of 4 ft throughout. Horizontal reinforcement of at least 0.20 in.$^2$ in cross section at the top of the wall, at the top and bottom of wall openings, at structurally connected roof and floor openings, and at a maximum spacing of 10 ft.

**Repointing:** A method of repairing cracked or deteriorating mortar joints in which the damaged or deteriorated mortar is removed and the joints are refilled with new mortar.

**Required Member Resistance (or Required Strength):** Action on a component or connection, determined by structural analysis, resulting from the factored loads and the critical load combinations.

**Resistance:** The capacity of a structure, component, or connection to resist the effects of loads.

**Resistance Factor:** A reduction factor applied to member resistance that accounts for unavoidable deviations of the actual strength from the nominal value and for the manner and consequences of failure.

**Rigid Diaphragm:** A diaphragm with horizontal deformation along its length less than half the average story drift as specified in Section 3.2.4.

**Rough Lumber:** Lumber as it comes from the saw prior to any dressing operation.

**Row of Fasteners:** Two or more fasteners aligned with the direction of load.

**Running Bond:** A pattern of masonry where the head joints are staggered between adjacent courses by at least one-quarter of the length of a masonry unit.

**Scragging:** The process of subjecting an elastomeric bearing to one or more cycles of large-amplitude displacement.

**Seasoned Lumber:** Lumber that has been dried either by open-air drying within the limits of moisture contents attainable by this method, or by controlled air drying.

**Secondary Component:** See **Component, Secondary**.

**Seismic Evaluation:** An approved process or methodology of evaluating deficiencies in a building which prevent the building from achieving a selected Rehabilitation Objective.

**Shallow Foundation:** Isolated or continuous-spread footings or mats.

**Shear Wall:** A wall that resists lateral forces applied parallel with its plane. Also known as an in-plane wall.

**Sheathing:** Lumber or panel products that are attached to parallel framing members, typically forming wall, floor, ceiling, or roof surfaces.

**Short Captive Column:** A column with a height-to-depth ratio less than 75% of the nominal height-to-depth ratios of the typical columns at that level.

**Shrinkage:** Reduction in the dimensions of wood due to a decrease of moisture content.

**Simplified NSP Analysis:** A nonlinear static analysis in which only primary lateral-force-resisting elements are modeled, and component degradation is not explicitly modeled.

**Simplified Rehabilitation Method:** An approach applicable to certain types of buildings and Rehabilitation Objectives in which an analysis of the response of the entire building to earthquake hazards is not required.

**Slip-Critical Joint:** A bolted joint in which slip resistance of the connection is required.

**Solid Masonry Unit:** A masonry unit with net cross-sectional area in every plane parallel to the bearing surface equal to 75% or more of the gross cross-sectional area in the same plane.

**Solid Wall or Solid Infill Panel:** A wall or infill panel with openings not exceeding 5% of the wall surface area. The maximum length or height of an opening in a solid wall must not exceed 10% of the wall width or story height. Openings in a solid wall or infill panel must be located within the middle 50% of a wall length and story height, and must not be contiguous with adjacent openings.

**Special Moment Frame (SMF):** A moment frame system that meets the special requirements for frames as defined in seismic provisions for new construction.

**Stack Bond:** A placement of masonry units such that the head joints in successive courses are aligned vertically.

**Stiff Diaphragm:** A diaphragm that is neither flexible nor rigid.

**Storage Racks:** Industrial pallet racks, movable shelf racks, and stacker racks made of cold-formed or hot-rolled structural members. Does not include other types of racks such as drive-in and drive-through racks, cantilever wall-hung racks, portable racks, or racks made of materials other than steel.

**Story:** The portion of a structure between the tops of two successive finished floor surfaces and, for the top-most story, from the top of the floor finish to the top of the roof structural element.

**Strength:** The maximum axial force, shear force, or moment that can be resisted by a component.

**Stress Resultant:** The net axial force, shear, or bending moment imposed on a cross section of a structural component.

**Strong-Back System:** A secondary system, such as a frame, commonly used to provide out-of-plane support for an unreinforced or under-reinforced masonry wall.

**Strong Column–Weak Beam:** A connection where the capacity of the column in any moment frame joint is greater than that of the beams, ensuring inelastic action in the beams.

**Structural Component:** See **Component, Structural**.

**Structural Performance Level:** A limiting structural damage state; used in the definition of Rehabilitation Objectives.

**Structural Performance Range:** A range of structural damage states; used in the definition of Rehabilitation Objectives.

**Structural System:** An assemblage of structural components that are joined together to provide regular interaction or interdependence.

**Stud:** Vertical framing member in interior or exterior walls of a building.

**Subassembly:** A portion of an assembly.

**Sub-Diaphragm:** A portion of a larger diaphragm used to distribute loads between members.

**Systematic Rehabilitation Method:** An approach to rehabilitation in which complete analysis of the response of the building to earthquake hazards is performed.

**Target Displacement:** An estimate of the maximum expected displacement of the roof of a building calculated for the design earthquake.

**Tie:** See **Drag Strut**.

**Tie-Down:** A device used to resist uplift of the chords of shear walls.

**Tie-Down System:** For seismically isolated structures, the collection of structural connections, components, and elements that provide restraint against uplift of the structure above the isolation system.

**Timber:** Lumber of nominal cross-section dimensions of 5 in. or more.

**Time-Effect Factor:** A factor applied to adjusted resistance to account for effects of duration of load. (See **Load Duration**.)

**Total Design Displacement:** The design earthquake displacement of an isolation or energy dissipation system, or components thereof, including additional displacement due to actual and accidental torsion.

**Total Maximum Displacement:** The maximum earthquake displacement of an isolation or energy dissipation system, or components thereof, including additional displacement due to actual and accidental torsion.

**Transverse Wall:** A wall that is oriented transverse to in-plane shear walls, and resists lateral forces applied normal to its plane. Also known as an out-of-plane wall.

**Unreinforced Masonry (URM) Wall:** A masonry wall containing less than the minimum amounts of reinforcement as defined for reinforced masonry walls; assumed to resist gravity and lateral loads solely through resistance of the masonry materials.

**V-Braced Frame:** A concentric braced frame (CBF) in which a pair of diagonal braces located either above or below a beam is connected to a single point within the clear beam span.

**Velocity-Dependent Energy Dissipation Devices:** Devices having mechanical characteristics such that the force in the device is dependent on the relative velocity in the device.

**Veneer:** A masonry wythe that provides the exterior finish of a wall system and transfers out-of-plane load directly to a backing, but is not considered to add load-resisting capacity to the wall system.

**Vertical Irregularity:** A discontinuity of strength, stiffness, geometry, or mass in one story with respect to adjacent stories.

**Waferboard:** A non-veneered structural panel manufactured from 2- to 3-in. flakes or wafers bonded together with a phenolic resin and pressed into sheet panels.

**Wall Pier:** Vertical portion of a wall between two horizontally adjacent openings.

**Wind-Restraint System:** The collection of structural components that provides restraint of the seismic-isolated structure for wind loads; may be either an integral part of isolator units or a separate device.

**Wood Structural Panel:** A wood-based panel product bonded with an exterior adhesive, meeting the requirements of PS 1-95 (NIST 1995) or PS 2-92 (NIST 1992), including plywood, oriented strand board, waferboard, and composite panels.

**Wrought Iron:** An easily welded or forged iron containing little or no carbon. Initially malleable, it hardens quickly when rapidly cooled.

**Wythe:** A continuous vertical section of a wall, one masonry unit in thickness.

**X-Braced Frame:** A concentric braced frame (CBF) in which a pair of diagonal braces crosses near the mid-length of the braces.

**Y-Braced Frame:** An eccentric braced frame (EBF) in which the stem of the Y is the link of the EBF system.

## REFERENCES

American Concrete Institute (ACI). (2001). "Acceptance criteria for moment frames based on structural testing." *ACI T1.1-01*, Detroit.

American Concrete Institute (ACI). (2002). "Building code requirements for reinforced concrete." *ACI 318*, Detroit.

American Concrete Institute (ACI). (2002). Masonry Standards Joint Committee (MSJC). "Building code requirements for masonry structures." *ACI 530/ASCE 5/TMS 402*, Detroit; American Society of Civil Engineers, Reston, Va; and The Masonry Society, Boulder, Colo.

American Concrete Institute (ACI). (2002). Masonry Standards Joint Committee (MSJC), "Specification for masonry structures." *ACI 530.1/ASCE 6/TMS 602*, Detroit; American Society of Civil Engineers, Reston, Va; and The Masonry Society, Boulder, Colo.

American Institute of Steel Construction (AISC). (1983). *Iron and steel beams 1873–1952*, Chicago.

American Institute of Steel Construction (AISC). (1999). *Load and resistance factor design specification for structural steel buildings (LRFD)*, Chicago.

American Institute of Steel Construction (AISC). (2002). "Seismic provisions for structural steel buildings." *AISC 341*, Chicago.

American Society of Civil Engineers (ASCE). (1996). "Standard for load and resistance factor design (LRFD) for engineered wood construction." *ASCE 16, AF&PA/ASCE Standard No. 16-95,* New York.

American Society of Civil Engineers (ASCE). (2002). "Seismic evaluation of existing buildings." *ASCE 31*, Reston, Va.

American Society of Civil Engineers (ASCE). (2005). "Minimum design loads for buildings and other structures." *ASCE 7*, Reston, Va.

American Society of Mechanical Engineers (ASME). (2000). "Safety code for elevators and escalators." *ASME A17.1*, New York.

ASTM. (1996). "Standard test method for splitting tensile strength of cylindrical concrete specimens." *ASTM C496-96*, West Conshohocken, Pa.

ASTM. (1998). "Standard specification for computing the reference resistance of wood-based materials and structural connections for load and resistance factor design." *ASTM D5457-93*, West Conshohocken, Pa.

ASTM. (1999). "Standard test method for penetration test and split-barrel sampling of soils." *ASTM D1586-99*, West Conshohocken, Pa.

ASTM. (2000). "Standard specification for fusion-bonded epoxy-coated pipe piles." *ASTM A972/A972M-00*, West Conshohocken, Pa.

ASTM. (2000). "Standard test method for measurement of masonry flexural bond strength." *ASTM C1072-00A*, West Conshohocken, Pa.

ASTM. (2000). "Standard test methods for conducting creep, creep-rupture, and stress-rupture tests of metallic materials." *ASTM E139-00*, West Conshohocken, Pa.

ASTM. (2001). "Standard specification for high-strength low-alloy steel shapes of structural quality, produced by quenching and self-tempering process (QST)." *ASTM A913/A913M-01*, West Conshohocken, Pa.

ASTM. (2001). "Standard test method for compressive strength of cylindrical concrete specimens." *ASTM C39/C39M-01*, West Conshohocken, Pa.

ASTM. (2002). "Standard specification for carbon steel bolts and studs, 60,000 psi tensile strength." *ASTM A307-02*, West Conshohocken, Pa.

ASTM. (2002). "Standard specification for steel strand, uncoated seven-wire for prestressed concrete." *ASTM A416/A416M-02*, West Conshohocken, Pa.

ASTM. (2002). "Standard specification for uncoated stress-relieved steel wire for prestressed concrete." *ASTM A421/A421M-02*, West Conshohocken, Pa.

ASTM. (2002). "Standard methods for establishing structural grades and related allowable properties for visually graded lumber." *ASTM D245-00*, West Conshohocken, Pa.

ASTM. (2002). "Standard test measures for flexural bond strength of masonry." *ASTM E518-02*, West Conshohocken, Pa.

ASTM. (2003). "Standard specification for high-strength low-alloy structural steel." *ASTM A242/A242M-03,* West Conshohocken, Pa.

ASTM. (2003). "Standard test methods and definitions for mechanical testing of steel products." *ASTM A370-03,* West Conshohocken, Pa.

ASTM. (2003). "Standard specification for deformed and plain billet-steel bars for concrete reinforcement." *ASTM A615/A615M-03A,* West Conshohocken, Pa.

ASTM. (2003). "Standard specification for uncoated high-strength steel bar for prestressing concrete." *ASTM A722/A722M-98,* West Conshohocken, Pa.

ASTM. (2003). "Standard test method for obtaining and testing drilled cores and sawed beams of concrete." *ASTM C42/C42M-03,* West Conshohocken, Pa.

ASTM. (2003). "Standard test method for in situ compressive stress within solid unit masonry estimated using flatjack measurements." *ASTM C1196-03,* West Conshohocken, Pa.

ASTM. (2003). "Standard test method for in situ measurement of masonry deformability properties using flatjack method." *ASTM C1197-03,* West Conshohocken, Pa.

ASTM. (2003). "Standard test methods for strength of anchors in concrete and masonry elements." *ASTM E488-96,* West Conshohocken, Pa.

ASTM. (2004). "Standard specification for carbon structural steel." *ASTM A36/A36M-04,* West Conshohocken, Pa.

ASTM. (2004). "Standard specification for high-strength low-alloy columbium-vanadium structural steel." *ASTM A572/A572M-04* (formerly *A441/A441M*), West Conshohocken, Pa.

ASTM. (2004). "Standard specification for structural steel shapes." *ASTM A992/A992M-04,* West Conshohocken, Pa.

American Welding Society (AWS). (2002). "Structural welding code-steel." *AWS D1.1,* Miami, Fla.

American Welding Society (AWS). (1998). "Structural welding code-sheet steel." *AWS D1.3,* Miami, Fla.

Federal Emergency Management Agency (FEMA). (1992). "NEHRP handbook of techniques for the seismic rehabilitation of existing buildings." Prepared by the Building Seismic Safety Council for the Federal Emergency Management Agency. *FEMA 172,* Washington, D.C.

Federal Emergency Management Agency (FEMA). (2001). "NEHRP recommended provisions for seismic regulations for new buildings and other structures, 2000 Ed., Part 1: Provisions and Part 2:
Commentary." Prepared by the Building Seismic Safety Council for the Federal Emergency Management Agency. *FEMA 368,* Washington, D.C.

Federal Emergency Management Agency (FEMA). (2004). "NEHRP recommended provisions for seismic regulations for new buildings and other structures, 2003 Ed., Part 1: Provisions and Part 2: Commentary." Prepared by the Building Seismic Safety Council for the Federal Emergency Management Agency. *FEMA 450,* Washington, D.C.

International Conference of Building Officials (ICBO). (1997). *Uniform Building Code,* Whittier, Calif.

International Code Council (ICC). (2003). *International Building Code,* Falls Church, Va.

## COMMENTARY REFERENCES

ABK Joint Venture (ABK). (1984). "Methodology for mitigation of seismic hazards in existing unreinforced masonry buildings." (*Topical Report 08*). Agbabian & Associates, SB Barnes & Associates, and Kariotis & Associates, El Segundo, Calif.

Al-Hussaini, T., Zayas, V., and Constantinou, M. C. (1994). "Seismic isolation of multi-story frame structures using spherical sliding isolation systems." *Rep No. NCEER-94-0007,* National Center for Earthquake Engineering Research, State University of New York at Buffalo.

American Architectural Manufacturers Association (AAMA). (2000). "Recommended static test method for evaluating curtain wall and storefront systems subjected to seismic and wind induced interstory drift." *AAMA 501.4,* Schaumberg, Ill.

American Concrete Institute (ACI). (2001). "Acceptance criteria for moment frames based on structural testing." *ACI T1.1-01,* Detroit.

American Concrete Institute (ACI). (2002). "Building code requirements for reinforced concrete." *ACI 318,* Detroit.

American Concrete Institute (ACI). (2002). Masonry Standards Joint Committee (MSJC), "Specification for masonry structures." *ACI 530.1/ASCE 6/TMS 602,* Detroit; American Society of Civil Engineers, Reston, Va.; and The Masonry Society, Boulder, Colo.

American Forest & Paper Association (AF&PA). (1996). "LRFD manual for engineered wood construction," including supplements and guidelines. *AF&PA LRFD,* Washington, D.C.

American Forest & Paper Association (AF&PA). 1997, "National design specification for wood construction." *ANSI/AF&PA NDS-1997,* Washington, D.C.

American Forest & Paper Association (AF&PA). (2001). "ASD manual for engineered wood construction," including supplements and guidelines. *AF&PA ASD*, Washington, D.C.

American Petroleum Institute (API). (1998). "Welded steel tanks for oil storage," 10th Ed. *API 650*, Washington, D.C.

American Plywood Association (APA). (1995). "Design capacities of APA performance rated structural-use panels." *Tech Note N375B*, Tacoma, Wash.

American Plywood Association (APA). (1997). *Plywood design specification*, Tacoma, Wash.

American Plywood Association (APA). (2001). *Diaphragms and shear walls design/construction guide* (APA Form L350), Tacoma, Wash.

American Society of Civil Engineers (ASCE). (1996). "Standard for load and resistance factor design (LRFD) for engineered wood construction." *AF&PA/ASCE Standard No. 16-95, ASCE 16*, New York.

American Society of Civil Engineers (ASCE). (1998). "Seismic analysis of safety-related nuclear structures." *ASCE 4*, Reston, Va.

American Society of Civil Engineers (ASCE). (1999). "Standard guideline for structural condition assessment of existing buildings." *ASCE 11*, New York.

American Society of Civil Engineers (ASCE). (2002). "Seismic evaluation of existing buildings." *ASCE 31*, Reston, Va.

American Society of Heating, Refrigeration, and Air Conditioning Engineers (ASHRAE). (1999). "A practical guide to seismic restraint." *ASHRAE RP-812*, Atlanta.

American Society of Mechanical Engineers (ASME). (2000). "Safety code for elevators and escalators." *ASME A17.1*, New York.

American Society of Mechanical Engineers (ASME). (2000). "Code for pressure piping." *ASME B31*, New York.

American Society of Mechanical Engineers (ASME). (2001). "Power piping." *ASME B31.1*, New York.

American Society of Mechanical Engineers (ASME). (2002). "Process piping." *ASME B31.3*, New York.

American Society of Mechanical Engineers (ASME). (2002). "Liquid transportation systems for hydrocarbons, liquid petroleum gas, anhydrous ammonia, and alcohols." *ASME B31.4*, New York.

American Society of Mechanical Engineers (ASME). (2001). "Refrigeration plant." *ASME B31.5*, New York.

American Society of Mechanical Engineers (ASME). (2000). "Gas transmission and distribution piping systems." *ASME B31.8*, New York.

American Society of Mechanical Engineers (ASME). (2000). "Building services piping." *ASME B31.9*, New York.

American Society of Mechanical Engineers (ASME). (2002). "Slurry transportation systems." *ASME B31.11*, New York.

American Water Works Association (AWWA). (1996). "Welded steel tanks for water storage." *AWWA D100*, Denver, Colorado.

American Welding Society (AWS). (1998). "Structural welding code—sheet steel." *AWS D1*, Miami, Fla.

Applied Technology Council (ATC). (1981). "Guidelines for the design of horizontal wood diaphragms." *ATC-7*, Redwood City, Calif.

Applied Technology Council (ATC). (1992). "Guidelines for seismic testing of components of steel structures." *Report No. ATC-24*, Redwood City, Calif.

Applied Technology Council (ATC). (1993). "Proceedings of the workshop to resolve seismic rehabilitation subissues—July 29 and 30, 1993; development of guidelines for seismic rehabilitation of buildings, phase I: Issues identification and resolution." *ATC-28-2*, Redwood City, Calif.

Applied Technology Council (ATC). (1996). "Seismic evaluation and retrofit of concrete buildings." *ATC-40*, prepared by the Applied Technology Council, Redwood City, Calif. for the California Seismic Safety Commission (*Report No. SSC 96-01*).

ASTM (1955) "Recommended Practice for Conducting Long-Time High-Temperature Tension Test of Metallic Materials (Withdrawn 1959)", *ASTM E22-41*, West Conshohocken, Pa.

ASTM. (1998). "Standard specification for computing the reference resistance of wood-based materials and structural connections for load and resistance factor design." *ASTM D5457-93*, West Conshohocken, Pa.

ASTM. (2002). "Standard test method for pulse velocity through concrete." *ASTM C597-02*, West Conshohocken, Pa.

ASTM. (2002). "Standard test method for diagonal tension (shear) in masonry assemblages." *ASTM E519-02*, West Conshohocken, Pa.

Ayres, J. M., and Sun, T. Y. (1973a). "Nonstructural damage to buildings, the great Alaska earthquake of 1964." *Engineering*, National Academy of Sciences, Washington, D.C.

Ayres, J. M., and Sun, T. Y. (1973b). "Nonstructural damage." *The San Fernando,*

*California Earthquake of February 9. (1971),* Vol. 1B, National Oceanic and Atmospheric Administration, Washington, D.C.

Bartlett, S. F., and Youd, T. L. (1992). "Empirical prediction of lateral spread displacement." *Proc. 4th Japan–U.S. Workshop on Earthquake-Resistant Design of Lifeline Facilities and Countermeasures for Soil Liquefaction*, M. Hamada and T. D. O'Rourke, eds. Report No. NCEER-92-0019, Vol. I, National Center for Earthquake Engineering Research, Buffalo, N.Y., 351–365.

Baziar, M. H., Dobry, R., and Elgamal, A-W. M. (1992). "Engineering evaluation of permanent ground deformations due to seismically-induced liquefaction." *Report No. NCEER-92-0007*, National Center for Earthquake Engineering Research, Buffalo, N.Y.

Berry, M., and Eberhard, M. (2005). "Practical performance model for bar buckling." *J. Struct. Engrg. (ASCE)*, 131(7), 1060–1070.

Biskinis D. E., Roupakias G. K., and Fardis, M. N. (2004). "Degradation of shear strength of reinforced concrete members with inelastic cyclic displacements." *ACI Struct. J.,* 101(6).

Bouwkamp, J. G., and Meehan, J. F. (1960). "Drift limitations imposed by glass." *Proc. 2nd World Conf. on Earthquake Engineering*, Tokyo, 1763–1778.

Bowles, J.E. (1988). *Foundation Analysis and Design*, 4th Ed., McGraw-Hill, New York.

Bozorgnia, Y., Mansour, N., and Cambell, K. W. (1996). "Relationship between vertical and horizontal response spectra for the Northridge earthquake." *Proc. 11th World Cong. on Earthquake Engineering*, Acapulco, Mexico.

Breyer, D. E., Fridley, K. J., and Cobeen, K. E. (1999). *Design of wood structures,* 4th Ed., McGraw-Hill, New York.

Building Officials and Code Administrators International (BOCAI). (1999), *National building code*, Country Club Hills, Ill.

Building Seismic Safety Council (BSSC). (2000). *NEHRP recommended provisions for seismic regulations for new buildings and other structures*, 2000 Ed. Prepared by the Building Seismic Safety Council for the Federal Emergency Management Agency, Washington, D.C.

California Governor's Board of Inquiry on the 1989 Loma Prieta Earthquake. (1990). *Competing against time.* Report to Governor George Deukmejian, State of California, Office of Planning and Research, Sacramento, Calif.

Calvi, G. M. (1988). "Correlation between ultrasonic and load tests on old masonry specimens." *Proc.*

*8th Intl. Brick/Block Masonry Conf.,* Elsevier Applied Science, Essex, U.K., 1665–1672.

Cho, J. -Y., and Pincheira, J. A. (2006). "Inelastic analysis of reinforced concrete columns with short lap splices subjected to reversed cyclic loads." *ACI Struct. J.,* 103(2).

Cho, D. M., and Retamal, E. (1993). "The Los Angeles County Emergency Operations Center on high-damping rubber bearings to withstand an earthquake bigger than the big one." *Proc. Seminar on Seismic Isolation, Passive Energy Dissipation, and Active Control*, Applied Technology Council Report No. ATC-17-1, Redwood City, Calif., 209–220.

Ceilings and Interior Systems Construction Association (CISCA). (1990). *Recommendations for direct-hung acoustical and lay-in panel ceilings, seismic zones 3–4*, Deerfield, Ill.

Ceilings and Interior Systems Construction Association (CISCA). (1991). *Recommendations for direct-hung acoustical and lay-in panel ceilings, seismic zones 0–2*, Deerfield, Ill.

Computers and Structures, Inc. (CSI). (2003). *ETABS (Version 8.0): Linear and nonlinear, static and dynamic analysis and design of building systems*, Berkeley, Calif.

Concrete Reinforcing Steel Institute (CRSI). (1981). *Evaluation of reinforcing steel systems in old reinforced concrete structures*, Chicago.

Constantinou, M. C., Tsopelas, P. C., Kim, Y.-S., and Okamoto, S. (1993). "NCEER–TAISEI Corporation research program on sliding seismic isolation systems for bridges: Experimental and analytical study of friction pendulum system (FPS)." *Report No. NCEER-93-0020*, National Center for Earthquake Engineering, State University of New York at Buffalo.

Construction Engineering Research Laboratory (CERL). (1997). "The CERL equipment fragility and protection, experimental definition of equipment vulnerability to transient support motions." *Tech Rep 97/58*, ADA No. 342277, U.S. Army Corps of Engineers, Champaign, Ill.

Dept. of the Army, Navy, and Air Force. (1986). "Seismic design of buildings." *Publication Nos. Air Force AFM 88-3, Army TM5-809-10.1, Navy NAVFAC P-355.1*, Chapter 13.1, Dept. of the Army, Navy, and Air Force, Washington, D.C.

Drake, R. M., and Bachman, R. E. (1995). "Interpretation of instrumental building seismic data and implications for building codes." *Proc. SEAOC Annual Conv.*, Structural Engineering Association of California, Sacramento, Calif.

Earthquake Engineering Research Institute (EERI). (1995). "In wait for the next one." *Proc. 4th*

*Japan–U.S. Workshop on Urban Earthquake Hazard Reduction*, Earthquake Engineering Research Institute and Japan Institute of Social Safety Science, sponsors, Osaka.

Elwood, K. J., and Moehle, J. P. (2004). "Evaluation of existing reinforced concrete columns." *Proc. 13th World Conf. on Earthquake Engineering*, Vancouver, B.C.

Elwood, K. J., and Moehle, J. P. (2005a). "Drift capacity of reinforced concrete columns with light transverse reinforcement." *Earthquake Spectra* (Earthquake Engineering Research Institute, Oakland, Calif.), 21(1), 71–89.

Elwood, K. J., and Moehle, J. P. (2005b). "Axial capacity model for shear-damaged columns." *ACI Struct. J.*, 102(4), 578–587.

Elwood, K. J., and Eberhard, M. O. (2006). "Effective stiffness of reinforced concrete columns." *PEER Research Digest*, Pacific Earthquake Engineering Research Center, University of California, Berkeley.

Epperson, G. S., and Abrams, D. P. (1989). "Nondestructive evaluation of masonry buildings, Advanced Construction Technology Center." *Rep No. 89-26-03*, College of Engineering, University of Illinois at Urbana.

Fardis, M. N., and D. E. Biskinis. (2003). "Deformation capacity of RC members, as controlled by flexure or shear." *Otani Symposium 2003*, Tokyo, 511–530.

Feilden, B. M. (1987). *Between two earthquakes: Cultural property in seismic zones*, Getty Conservation Institute, Marina del Rey, Calif.

Federal Emergency Management Agency (FEMA). (1988). "Rapid visual screening of buildings for potential seismic hazards: A handbook." *FEMA 154*, prepared by the Applied Technology Council (*Rep No. ATC-21*) for the Federal Emergency Management Agency, Washington, D.C.

Federal Emergency Management Agency (FEMA). (1992). "NEHRP handbook of techniques for the seismic rehabilitation of existing buildings." *FEMA 172*, prepared by the Building Seismic Safety Council for the Federal Emergency Management Agency, Washington, D.C.

Federal Emergency Management Agency (FEMA). (1992). "NEHRP handbook for the seismic evaluation of existing buildings." *FEMA 178*, prepared by the Building Seismic Safety Council for the Federal Emergency Management Agency, Washington, D.C.

Federal Emergency Management Agency (FEMA). (1992). "Development of guidelines for seismic rehabilitation of buildings, phase I: Issues identification and resolution." *FEMA 237*, prepared by the Applied Technology Council for the Federal Emergency Management Agency, Washington, D.C.

Federal Emergency Management Agency (FEMA). (1994). "Reducing the risks of nonstructural earthquake damage, a practical guide." *FEMA 74*, Washington, D.C.

Federal Emergency Management Agency (FEMA). (1995). "Typical costs for seismic rehabilitation of buildings, 2nd Ed., Vol. II: Supporting documentation." *FEMA 157*, prepared by the Hart Consultant Group for the Federal Emergency Management Agency, Washington, D.C.

Federal Emergency Management Agency (FEMA). (1995). "Typical costs for seismic rehabilitation of buildings, 2nd Ed., Vol. 1: Summary." *FEMA 156*, prepared by the Hart Consultant Group for the Federal Emergency Management Agency, Washington, D.C.

Federal Emergency Management Agency (FEMA). (1995). NEHRP recommended provisions for seismic regulations for new buildings, 1994 Ed., Part 1: Provisions and Part 2: Commentary." *FEMA 222A* and *FEMA 223A*, prepared by the Building Seismic Safety Council for the Federal Emergency Management Agency, Washington, D.C.

Federal Emergency Management Agency (FEMA). (1997). "NEHRP guidelines for the seismic rehabilitation of buildings." *FEMA 273*, prepared by the Building Seismic Safety Council for the Federal Emergency Management Agency, Washington, D.C.

Federal Emergency Management Agency (FEMA). (1997). "NEHRP commentary on the guidelines for seismic rehabilitation of buildings." *FEMA 274*, prepared by the Building Seismic Safety Council for the Federal Emergency Management Agency, Washington, D.C.

Federal Emergency Management Agency (FEMA). (1997). "NEHRP recommended provisions for seismic regulations for new buildings and other structures, 1997 Ed., Part 1: Provisions, and Part 2: Commentary." *FEMA 302* and *FEMA 303*, prepared by the Building Seismic Safety Council for the Federal Emergency Management Agency, Washington, D.C.

Federal Emergency Management Agency (FEMA). (1998). "Planning for seismic rehabilitation: Societal issues." *FEMA 275*, prepared by VSP Associates for the Building Seismic Safety Council and Federal Emergency Management Agency, Washington, D.C.

Federal Emergency Management Agency (FEMA). (1998). "Evaluation of earthquake-damaged concrete and masonry wall buildings—Basic procedures manual." *FEMA 306,* prepared by the Applied Technology Council (ATC-43 Project) for the Federal Emergency Management Agency, Washington, D.C.

Federal Emergency Management Agency (FEMA). (1998). "Evaluation of earthquake-damaged concrete and masonry wall buildings—Technical resources." *FEMA 307,* prepared by the Applied Technology Council (ATC-43 Project) for the Federal Emergency Management Agency, Washington, D.C.

Federal Emergency Management Agency (FEMA). (1998). "Repair of earthquake-damaged concrete and masonry wall buildings." *FEMA 308,* prepared by the Applied Technology Council (ATC-43 Project) for the Federal Emergency Management Agency, Washington, D.C.

Federal Emergency Management Agency (FEMA). (1999). "Case studies: An assessment of the NEHRP Guidelines for the Seismic Rehabilitation of Buildings" *FEMA 343,* Federal Emergency Management Agency, Washington, D.C.

Federal Emergency Management Agency (FEMA). (1999). "Guidelines for the seismic rehabilitation of buildings: Example applications." *FEMA 276,* prepared by the Applied Technology Council for the Building Seismic Safety Council and the Federal Emergency Management Agency, Washington, D.C.

Federal Emergency Management Agency (FEMA). (2000). "Recommended seismic design criteria for moment-resisting steel frame structures." *FEMA 350,* prepared by the SEAOC, ATC, and CUREe Joint Venture for the Federal Emergency Management Agency, Washington, D.C.

Federal Emergency Management Agency (FEMA). (2000). "Recommended seismic evaluation and upgrade criteria for existing welded moment resisting steel structures." *FEMA 351,* prepared by the SEAOC, ATC, and CUREe Joint Venture for the Federal Emergency Management Agency, Washington, D.C.

Federal Emergency Management Agency (FEMA). (2000). "Recommended specifications and quality assurance guidelines for steel moment-frame construction for seismic applications." *FEMA 353,* prepared by the SEAOC, ATC, and CUREe Joint Venture for the Federal Emergency Management Agency, Washington, D.C.

Federal Emergency Management Agency (FEMA). (2000). "State of art report on connection performance." *FEMA 355D,* prepared by the SEAOC, ATC, and CUREe Joint Venture for the Federal Emergency Management Agency, Washington, D.C.

Federal Emergency Management Agency (FEMA). (2000). "State of art report on performance prediction and evaluation." *FEMA 355F,* prepared by the SEAOC, ATC, and CUREe Joint Venture for the Federal Emergency Management Agency, Washington, D.C.

Federal Emergency Management Agency (FEMA). (2000). "Prestandard and commentary for the seismic rehabilitation of buildings." *FEMA 356,* prepared by the American Society of Civil Engineers for the Federal Emergency Management Agency, Washington, D.C.

Federal Emergency Management Agency (FEMA). (2000). "Global topics report on the prestandard and commentary for the seismic rehabilitation of buildings." *FEMA 357,* prepared by the American Society of Civil Engineers for the Federal Emergency Management Agency, Washington, D.C.

Federal Emergency Management Agency (FEMA). (2001). "NEHRP recommended provisions for seismic regulations for new buildings and other structures, 2000 Ed., Part 1: Provisions and Part 2: Commentary." *FEMA 368* and *FEMA 369,* prepared by the Building Seismic Safety Council for the Federal Emergency Management Agency, Washington, D.C.

Federal Emergency Management Agency (FEMA). (2004). "NEHRP recommended provisions for seismic regulations for new buildings and other structures, 2003 Ed., Part 1: Provisions and Part 2: Commentary." *FEMA 450* and *FEMA 451,* prepared by the Building Seismic Safety Council for the Federal Emergency Management Agency, Washington, D.C.

Federal Emergency Management Agency (FEMA). (2005). "Improvement of nonlinear static seismic analysis procedures." *FEMA 440,* prepared by the Applied Technology Council for the Federal Emergency Management Agency, Washington, D.C.

Franklin, A. G., and Chang, F. K. (1977). "Earthquake resistance of earth and rock-fill dams: Permanent displacements of earth embankments by Newmark sliding block analysis." *Misc. Paper S-71-17, Rep 5,* U.S. Army Corps of Engineers, Waterways Experiment Station, Vicksburg, Miss.

Gazetas, G. (1991). "Foundation vibrations." *Foundation engineering handbook,* H. Y. Fang, ed., Van Nostrand Reinhold, New York, 553–593.

Hamada, M., Yasuda, S., Isoyama, R., and Emoto, K. (1986). *Study on liquefaction induced permanent ground displacements.* Report for the Association for the Development of Earthquake Prediction, Tokyo.

Hamburger, R. O. (1993). "Methodology for seismic capacity evaluation of steel-frame buildings with infill unreinforced masonry." *Proc. 1993 Natl. Earthquake Conf.*, Central U.S. Earthquake Consortium, Memphis, Tenn., Vol. II, 173-191.

Hanna, A. M. (1981). "Foundations on strong sand overlying weak sand." *J. Geotech. Engrg. Div.* (ASCE), 107(GT7), 915–927.

Hanna, A. M., and Meyerhof, G. G. (1980). "Design charts for ultimate bearing capacity of foundations on sand overlying soft clay." *Can. Geotech. J.*, 17, 300–303.

Hoover, C. A. (1992). *Seismic retrofit policies: An evaluation of local practices in zone 4 and their application to zone 3*, Earthquake Engineering Research Institute, Oakland, Calif.

Housner, G. W. (1963). "The behavior of inverted pendulum structures during earthquakes." *Bull. Seismological Soc. of Am.*, 53(2), 403–417.

International Conference of Building Officials (ICBO). (1994). *Uniform Building Code*, Whittier, Calif.

International Conference of Building Officials (ICBO). (1997a). "In-place masonry shear tests." *UBC 21-6*, Whittier, Calif.

International Conference of Building Officials (ICBO). (1997b). *Uniform Code for Building Conservation*, Whittier, Calif.

International Conference of Building Officials (ICBO). (2000). "Acceptance criteria for seismic qualification testing of nonstructural components." *AC-156*, International Conference of Building Officials Evaluation Services, Whittier, Calif.

International Code Council (ICC). (2003). *International Building Code*, Falls Church, Va.

Institute of Electrical and Electronics Engineers, Inc. (IEEE). (1997). "IEEE recommended practice for seismic design of substations." *IEEE 693*, New York.

Ishihara, K., and Yoshimine, M. (1992). "Evaluation of settlements in sand deposits following liquefaction during earthquakes." *Soils and Foundations* (Japanese Society of Soil Mechanics and Foundation Engineering), 32(1), 173–188.

Johnson, J. J., Conoscente, J. P., and Hamburger, R .O. (1992). "Dynamic analysis of impacting structural systems." *Proc. 10th World Conf. on Earthquake Engineering*, Madrid, Spain.

Kasai, K., Maison, B. F., and Patel, D. J. (1990). "An earthquake analysis for buildings subjected to a type of pounding." *Proc. 4th U.S. National Conf. of Earthquake Engineering*, Earthquake Engineering Research Institute, Oakland, Calif.

Kehoe, B. E., and Attalla, M. R. (2000). "Considerations of vertical acceleration on structural response." *Proc. 12th World Cong. on Earthquake Engineering*, Auckland, New Zealand.

Kelly, J. M. (1988). "Base isolation in Japan." *Rep. No. UCB/EERC-88/20*, Earthquake Engineering Research Center, University of California, Berkeley.

Kelly, J. M. (1993). *Earthquake-resistant design with rubber*, Springer-Verlag, London.

Kingsley, G. R., Noland, J. L., and Atkinson, R. H. (1987). "Nondestructive evaluation of masonry structures using sonic and ultrasonic pulse velocity techniques." *Proc. 4th North American Masonry Conf.*, The Masonry Society, Boulder, Colo.

Lagorio, H. J. (1990). *Earthquakes: An architect's guide to nonstructural seismic hazards*, John Wiley & Sons, Inc., New York.

Lynn, A. C. (2001). "Seismic evaluation of existing reinforced concrete building columns." PhD dissertation, Dept. of Civil and Environmental Engineering, University of California, Berkeley.

Lynn, A. C., Moehle, J. P., Mahin, S. A., and Holmes, W. T. (1996). "Seismic evaluation of existing reinforced concrete columns." *Earthquake Spectra* (Earthquake Engineering Research Institute, Oakland, Calif.), 12(4), 715–739.

Makdisi, F. I., and Seed, H. B. (1978). "Simplified procedure for estimating dam and embankment earthquake-induced deformations." *J. Geotech. Engrg. Div.* (ASCE), 104(GT7), 849–867.

Makris, N., and Konstantinidis, D. (2001). *The rocking spectrum and the shortcomings of design guidelines*, Pacific Earthquake Engineering Research Center, Berkeley, California.

Makris, N. and Roussos, Y. (1998). *Rocking response and overturning of equipment under horizontal, pulse-type motions*, Pacific Earthquake Engineering Research Center, Berkeley, Calif.

Manufacturers Standardization Society (MSS). (1993). "Pipe hangers and supports: Materials, design and manufacture." *SP-58*, Manufacturers Standardization Society of the Valve and Fitting Industry, Vienna, Va.

Melek, M. and Wallace, J. W. (2004). "Cyclic behavior of columns with short lap splices." *ACI Struct. J.*, 101(6).

Meyerhof, G. G. (1974). "Ultimate bearing capacity of footings on sand layer overlying clay." *Can. Geotech. J.*, 11(2), 223–229.

Murota, N., Goda, K., Suzusi, S., Sudo, C., and Suizu, Y. (1994). "Recovery characteristics of dynamic properties of high-damping rubber bearings." *Proc. 3rd U.S.–Japan Workshop on Earthquake Protective*

*Systems for Bridges, Berkeley, California (1994).* Report No. NCEER 94-0009, National Center for Earthquake Engineering Research, State University of New York at Buffalo, 1–63 to 2–76.

Nagarajaiah, S., Reinhorn, A., and Constantinou, M. C. (1991). "3D-BASIS: Nonlinear dynamic analysis of three dimensional base isolated structures." *Rep. No. NCEER-91-0005*, National Center for Earthquake Engineering Research, State University of New York at Buffalo.

National Research Council (NRC). (1985). Committee on Earthquake Engineering, Commission on Engineering and Technical Systems. *Liquefaction of soils during earthquakes*, National Academy Press, Washington, D.C.

National Bureau of Standards (NBS), now National Institute for Science and Technology. (1977). "Evaluation of structural properties of masonry in existing buildings." *Building Science Series 62*, U.S. Dept. of Commerce, Washington, D.C.

Newmark, N. M. (1965). "Effect of earthquake on dams and embankments." *Geotechnique*, 15, 139–160.

National Fire Protection Association (NFPA). (1991). "Manual of hazardous chemical reactions." *NFPA, 49IM-91*, Quincy, Mass.

National Fire Protection Association (NFPA). (1994). "Hazardous chemicals data." *NFPA, 49–94*, Quincy, Mass.

National Fire Protection Association (NFPA). (1994). "Guide to fire hazard properties of flammable liquids, gases, and volatile solids." *NFPA, 325–94*, Quincy, Mass.

National Fire Protection Association (NFPA). (2001). "Standard system for the identification of the fire hazards of materials." *NFPA, 704*, Quincy, Mass.

National Fire Protection Association (NFPA). (2002). "Standard for the installation of sprinkler systems." *NFPA 13*, Quincy, Mass.

National Institute of Standards and Technology (NIST). (1992). "Performance standard for wood-based structural use panels." *U.S. Product Standard PS 2-92*, Washington, D.C.

National Institute of Standards and Technology (NIST). (1995). "Construction and industrial plywood with typical APA trademarks." *U.S. Product Standard PS 1-95*, Washington, D.C.

Nims, D. F., Richter, P. J., and Bachman, R. E. (1993). "The use of the energy dissipation restraint for seismic hazard mitigation." *Earthquake Spectra* (Earthquake Engineering Research Institute, Oakland, Calif.), 9(3), 467–498.

Noland, J. L., Atkinson, R. H., and Kingsley, G. R. (1987). "Nondestructive methods for evaluating masonry structures." *Proc. Intl. Conf. on Structural Faults and Repair*, London.

Pais, A., and Kausel, E. (1988). "Approximate formulas for dynamic stiffnesses of rigid foundations." *Soil Dynamics and Earthquake Engrg.*, 7(4), 213–227.

Panagiotakos, T. B., and Fardis, M. N. (2001). "Deformation of reinforced concrete members at yielding and ultimate." *ACI Struct. J.*, 98(2).

Pekcan, G., Mander, J., and Chen, S. (1995). "The seismic response of a 1:3 scale model RC structure with elastomeric spring dampers." *Earthquake Spectra* (Earthquake Engineering Research Institute, Oakland, Calif.), 11(2), 249–267.

Precast/Prestressed Concrete Institute (PCI) (1999). *PCI design handbook: Precast and prestressed concrete*, 5th ed, Precast/Prestressed Concrete Institute, Chicago.

Priestley, M. J. N., Evison, R. J., and Carr, A. J. (1978). "Seismic response of structures free to rock on their foundations." *Bull. New Zealand Natl. Soc. for Earthquake Engrg.*, 11(3), 141–150.

Reinhorn, A. M., Nagarajaiah, S., Constantinou, M. C., Tsopelas, P., and Li, R. (1994). "3D-BASIS-TABS (Version 2.0): Computer program for nonlinear dynamic analysis of three-dimensional base isolated structures." *Rep. No. NCEER-94-0018*, National Center for Earthquake Engineering Research, State University of New York at Buffalo.

Sansalone, M., and Carino, N. (1988). "Impact-echo method: Detecting honeycombing, the depth of surface opening cracks, and ungrouted tendons." *Concrete Int.* (ACI), 10(4), 38–46

Secretary of the Interior. (1992). *Standards for the treatment of historic properties*, National Park Service, Washington, D.C.

Secretary of the Interior. (1995). *Standards for the treatment of historic properties with guidelines for preserving, rehabilitating, restoring, and reconstructing historic buildings*, National Park Service, Washington, D.C.

Seed, R. B., and Harder, L. F. (1990). "SPT-based analysis of cyclic pore pressure generation and undrained residual strength." *Proc. H. B. Seed Memorial Symposium*, St. Thomas, U.S. Virgin Islands, Vol. 2, 351–376.

Seed, H. B., and Idriss, I. M. (1971). "Simplified procedure for evaluating soil liquefaction potential." *J. Soil Mech. and Found. Div.* (ASCE) 97(SM9), 1249–1273.

Seed, H. B., and Idriss, I. M. (1982). "Ground motions and soil liquefaction during earthquakes."

*Monograph Series*, Earthquake Engineering Research Institute, Oakland, Calif.

Seed, H. B., Tokimatsu, K., Harder, L. F., and Chung, R. M. (1985). "Influence of SPT procedures in soil liquefaction resistance evaluations." *J. Geotech. Engrg. Div.* (ASCE), 111(12), 1425–1445.

Sezen, H. (2002). "Seismic response and modeling of lightly reinforced concrete building columns," PhD dissertation, Dept. of Civil and Environmental Engineering, University of California, Berkeley.

Sheet Glass Association of Japan. (1982). *Earthquake safety design of windows*, Japan.

Sheet Metal and Air Conditioning Contractors National Association (SMACNA). (1980). *Rectangular industrial duct construction standards*, Chantilly, Va.

Sheet Metal and Air Conditioning Contractors National Association (SMACNA). (1985). *HVAC duct construction standards, metal and flexible*, Chantilly, Va.

Sheet Metal and Air Conditioning Contractors National Association (SMACNA). (1991). *Seismic restraint manual guidelines for mechanical equipment, and Appendix E-1993 addendum*, Chantilly, Va.

Soong, T. T., and Constantinou, M. C. (1994). *Passive and active structural vibration control in civil engineering,* Springer-Verlag, Wien, New York.

Southern Building Code Congress International (SBCCI). (1999). *Standard building code*, Birmingham, Ala.

Stark, T. D., and Mesri, G. (1992). "Undrained shear strength of liquefied sands for stability analysis," *J. Geotech. Engrg. Div.* (ASCE), 118(11), 1727–1747.

Steel Deck Institute (SDI). (1981). *Diaphragm Design Manual*, 1st Ed., Fox River Grove, Ill.

Timeshenko, S., and Woinowsky-Krieger, S. (1959). *Theory of plates and shells*, 2nd Ed., McGraw-Hill, New York.

Timler, P. A. (2000). "Design evolution and state-of-the-art development of steel plate shear wall construction in North America." *Proc. 69th Ann. SEAOC Conv.*, Vancouver, B.C., 197–208.

Tissell, J. R. (1993). "Wood structural panel shear walls." *Research Rep 154*, American Plywood Association, Tacoma, Wash.

Tissell, J. R., and Elliott, J. R. (1997). "Plywood diaphragms." *Research Rep 138*, American Plywood Association, Tacoma, Wash.

Tokimatsu, A. M., and Seed, H. B. (1987). "Evaluation of settlements in sands due to earthquake shaking." *J. Geotech. Engrg. Div.* (ASCE), 113(8), 681–878.

TSM (1992). "Seismic design for buildings." *Tri-Services Manual TM5-809-10 NAV FAC P 355*, AFM 88-3, Chapter 13, Washington, D.C.

Tsopelas, P., and Constantinou, M. C. (1994). "Experimental and analytical study of systems consisting of sliding bearings and fluid restoring force-damping devices." *Report No. NCEER 94-0010*, National Center for Earthquake Engineering Research, State University of New York at Buffalo.

Tsopelas, P. C., Constantinou, M. C., and Reinhorn, A. M. (1994) "3D-BASIS-ME computer program for nonlinear dynamic analysis of seismically isolated single and multiple structures and liquid storage tanks." *Report No. NCEER-94-0014*, National Center for Earthquake Engineering Research, State University of New York at Buffalo.

U.S. Dept. of Energy. (1997). "Seismic evaluation procedure for equipment in U.S. Department of Energy facilities." *DOE/EH-545*, Washington, D.C.

U.S. Dept. of the Navy (NAVFAC). (1986a). "Soil mechanics: Naval facilities engineering command design manual." *NAVFAC DM-7.01*, Alexandria, Va.

U.S. Dept. of the Navy (NAVFAC). (1986b). "Foundation and earth structures: Naval facilities engineering command design manual." *NAVFAC DM-7.02*, Alexandria, Va.

U.S. Nuclear Regulatory Commission (USNRC). (1976). "Combining modal responses and spatial components in seismic response analysis." *Regulatory Guide 1.92*, Washington, D.C.

Wallace, J. W. (1994). "New methodology for seismic design of RC shear walls." *J. Struct. Engrg. Div.* (ASCE), 120(3), 863–884.

Wallace, J. W. and Thomsen IV, J. H. (1995). "Seismic Design of RC Shear Walls (Parts I and II)," *J. Struct Engrg.* (ASCE), 75–101.

Western Wood Products Association (WWPA). (1996). *Western woods use book*, Portland, Ore.

Wood, S. L. (1990). "Shear strength of low-rise reinforced concrete walls." *ACI Struct. J.,* 87(1), 99–107.

Wright, P. D., 1989. "The development of a procedure and rig for testing the racking resistance of curtain wall glazing." *Study Rep No. 17*, Building Research Association of New Zealand (BRANZ), 18.

Yegian, M. K., Marciano, E. A., and Gharaman, V. G. (1991). "Earthquake-induced permanent deformation: Probabilistic approach." *J. Struct. Engrg. Div.* (ASCE), 117(1), 35–50.

Yim, S. C., and Chopra, A. K. (1985). "Simplified earthquake analysis of multistory structures with foundation uplift." *J. Struct. Engrg. Div.* (ASCE), 111(12), 2708-2731.

Youd, T. L., and Perkins, D. M. (1978). "Mapping liquefaction induced ground failure potential." *J. Struct. Engrg. Div.* (ASCE), 104(GT4), 433-446.

Youd, T.L., Idriss, I. M., et al. (2001). "Liquefaction resistance of soils: Summary report from the 1996 NCEER and 1998 NCEER/NSF work-shops on evaluation of liquefaction resistance of soils." *J. Geotech. and Geoenvir. Engrg.* (ASCE), 127(4), 297–313, October.

Zayas V., Low, S. S., and Mahin, S. A. (1987). *The FPS earthquake resisting system, experimental report*, Report No. UBC/EERC-87/01. Earthquake Engineering Research Center, University of California, Berkeley, CA.

# INDEX